Handbook of Immunochemistry

Handbook of Immunochemistry

Miroslav Ferenčík

Institute of Immunology
Faculty of Medicine
Comenius University
Bratislava
Czechoslovakia

SPRINGER-SCIENCE+BUSINESS MEDIA, B.V.

First edition 1993

Originally published by Chapman & Hall in 1993

Typeset by Alfa Verlag

ISBN 978-94-010-4678-7 ISBN 978-94-011-1552-0 (eBook)
DOI 10.1007/ 978-94-011-1552-0

A catalogue record for this book is available from the British Library

Library of Congress Cataloging-in-Publication data available

To my family

Contents

Preface

The book appeared in two previous Slovak editions for university students in Czechoslovakia. This edition presents a completely new version updated according to recent advances not only in immunochemistry and essential immunology but also in molecular biology, biochemistry and molecular genetics. The scope of the book is considerable since the goal was to cover the field of immunochemistry from the widest point of view including both the topic and methods of contemporary immunochemistry. Each chapter provides basic information on a specific subtopic, clearly and understandably, and presents principles of individual immunochemical methods. I am confident that the book will fill the gap between the books on essential immunology and highly specialised books on individual areas of immunochemistry (*e.g.* on antibodies, antigens, numerous immunochemical techniques, *etc.*). It may also prove useful for beginning investigators from different biological and medical fields as it supplies basic information needed for solving their scientific problems by immunochemical approaches. I do hope that readers will find the text stimulatory and pleasury to read.

I wish to thank all colleagues and friends for supplying their own results, suggestions and for their encouraging comments. My thanks go also to the editors and publishers for their valuable contribution to the preparation of the book.

1 Introduction

The term immunochemistry was coined by the Swedish chemist ARRHENIUS who used it for the first time in his lectures in 1907. ARRHENIUS characterized immunochemistry as a science dealing with "chemical reactions of substances that occur in the blood of animals after injection of foreign substances, *i.e.* after immunization. The products of immunization react with these foreign substances similarly as proteins or enzymes according to their chemical properties." This, 85-year-old definition, can be expressed using contemporary terminology as follows: "Immunochemistry studies not only the chemistry of antigens and antibodies and the mechanism of their interaction, but also many other molecules participating in immune reactions." Thus, the original definition of ARRHENIUS embodies the main aspects of contemporary modern immunochemistry.

Immunochemistry appears to be a typical interdisciplinary science. It forms a boundary between immunology and chemistry but is closely associated with other disciplines, *e.g.* molecular biology, microbiology, biochemistry, biophysics and many other medical sciences. Despite the fact that a number of outstanding chemists and biochemists have considerably contributed to contemporary, "state of the art" immunochemistry, it has nevertheless developed as a branch of immunology.

Experimental and theoretical knowledge, which had been obtained in many laboratories and clinics all over the world, particularly during the past two decades, contributed to the development of immunology which thus lost its empirical character. It ceased to be a science dealing only with the prevention and therapy of infectious diseases. At present, modern immunology occupies one of the leading positions among basic biomedical sciences. One may presume that immunology is likely to solve a number of basic problems which have permanently fascinated mankind, and have been at the centre of scientific interest. Of greatest interest are, for example, various aspects of embryology, cancer, ageing and replacement of damaged organs. Thus, the programmed differentiation of lymphocytes from the resting cell into a highly specialized cell may serve as a general model of morphogenesis. The explanation of the mechanism of interactions among various lymphocyte subpopulations may be a basis for understanding the relationship between embryonic inducers and differentiation of tissues. As regards tumours, the immune system is now considered to be the main defence mechanism, which is capable, under normal conditions, of successfully destroying malignant transformed cells that begin to multiply without control. Furthermore, reactivity of the immune system is directly related to the process of ageing. The success of tissue and organ transplantation among individuals belonging to the same animal species is dependent not only on the surgeon's skill, but primarily on the degree of similarity or identity of antigen composition (histocompatibility markers) of the donor and recipient, or on the possibility of suppressing the immune system of the recipient, or inducing tolerance to foreign tissue.

It seems that the resistance of the organism to a disease is genetically determined and is somehow associated with the phenotype expression of histocompatibility markers (these markers are responsible for the tissue compatibility and they express the genetic uniqueness of every individual). Therefore, the possibility of controlling the purposeful regulation of the immune response may become a very effective tool, not only against infectious diseases, but also against various diseases in general.

Immunologists have devised methods which allow antibodies possessing almost absolute specificity and purity *in vitro* (*i.e.* outside the organism) to be obtained. These monoclonal antibodies not only became effective diagnostic and therapeutic tools, but the method of their preparation has facilitated the isolation of previously unobtainable biologically important substances, *e.g.* certain hormones, enzymes, *etc.* The discovery of monoclonal antibodies, using the hybridoma technique, was a strong stimulus for the contemporary intensive development of biotechnology.

These examples suggest that immunology occupies one of the leading positions in the contemporary development of biomedical and biotechnological sciences. Immunological knowledge contributes not only to the maintenance of human health and welfare, but also to increased economic efficiency.

Contemporary immunology deals firstly with research into defence reactions; those control and regulatory mechanisms responsible for the reaction of the organism to the foreign substance (antigen) entering the body, or to its own altered antigen (with altered structure and properties). Such an immune response is a complex of mutual reactions between the molecules and cells of the immune system. It does not operate separately, but is an integral component of defence and adaptation mechanisms maintaining the overall homeostasis of the organism, thus ensuring survival of an individual and of biological species. In addition, it is also regulated by other physiological systems, particularly by the neurohumoral system. Reactions at the molecular level and properties of regulatory molecules are the subject of immunochemistry.

The methodological basis of contemporary immunochemistry exceeds the needs of immunology and is used by various life sciences, including biochemistry, molecular biology and genetics. It includes, for example, the use of immunological markers and immunochemical methods for structural studies of biological membranes, mechanism of transport of various substances, biological functions of subcellular structures, mechanism of protein synthesis *etc*. Immunochemistry has an indirect impact on other scientific disciplines employing and modifying some of its results and methods. Immunoradiometric and immunoenzyme methods used in endocrinology (hormone determination), pharmacology, forensic medicine and toxicology (determination of drugs and toxins), clinical biochemistry and analytical chemistry may serve as examples.

Thus, modern immunochemistry may be considered to be an integral part of immunology; conversely, however, its results and techniques form an autonomous unit employed by other scientific disciplines. Scientists active in related disciplines need not be familiar with the whole subject of immunology, although a basic immunological knowledge is required. It is the aim of this book to provide the reader with this basic knowledge.

When required for a better understanding of the text, individual problems are discussed in a broader immunological scope. Thus, the reader can obtain basic information not only about antigens and antibodies ("molecules of immunity"), their properties, formation and interactions but also about immunochemical methods and, finally, about the main mechanisms of immunity, including various effector and regulatory molecules which are an integral part of these mechanisms.

This book is dedicated to university students in natural sciences, chemistry and the food industry, as well as to advanced college students with a particular emphasis on natural or medical sciences. In addition, it may serve as a handbook for teachers of such students and for researchers in biochemical and clinical laboratories who are going to use immunochemical or immunological methods. Because of limited space, only the basic principles of individual methods are given, detailed description being omitted. Methodological details can be found in other specialized handbooks and monographs. To help with the specialized technical terminology, a glossary of basic terms can be found at the end of the book.

2 Importance of the immune system to life

According to present concepts, the development of three basic mechanisms is required for the origin of life.

1. Free energy supply and its transformation to a form capable of driving biochemical reactions. Biocatalysis plays a key role.
2. Transcription and transfer of genetic information (gene replication and transcription).
3. Information from the external and internal environment and its further processing for coordination and regulation of important life processes in a given system.

During evolution, these three components have continually changed and improved until they reached the contemporary form. Only the most advantageous mechanisms — with regard to conditions of external environment — have been preserved. The first two mechanisms are the subject of biochemistry, biophysics and molecular genetics. As well as biochemistry, physiology, neurology, endocrinology and immunology have contributed considerably to our knowledge of information systems.

The amount and quality of information have played an important role during evolution. Even now the obtaining and processing of information are an inevitable condition of existence of any living system, cell or individual, because they ensure its adaptation to changes of external and internal environment.

	↗	chemical (humoral)
Information systems of a living organism	→	electrochemical (nervous)
	↘	immunochemical (immune)

The humoral system appears to be the oldest and most primitive information system. During phylogenetic development, the humoral system did not disappear but occurred at a qualitatively higher level (enzymes — substrates — inhibitors, inducers — repressors, phytohormones, hormones, mediators, *etc.*). In worms and arthropods, a second information system — nervous — began to occur. Its basis is the nerve cell — the neuron. The human brain has a mean mass of 1 400 g and contains approximately 10^{10} interconnected neurons. The connection is mediated by electric impulses and special chemical substances; neurotransmitters and neuroendocrine peptides.

From the evolutionary point of view, the immune system is very old. Its basic property — the recognition of "non-self" and "self" — can be found in the simplest multicellular organisms, *e.g.* porifera and coelenterata. Such recognition results in the rejection (non-fusion) of tissues or cells of genetically unrelated individuals. This is caused by the presence of individual markers on the cell surface. These markers — tissue antigens — are products of genes belonging to the major histocompatibility complex and are typical for every individual.

The immune system is a diffuse organ which weighs about 1 000 g in adults. It is composed of approximately 10^{12} lymphocytes (a specific type of leukocyte) and accessory cells (macrophages, polymorphonuclear leukocytes), 10^{20} antibody (immunoglobulin) molecules and millions of molecules of various effector and regulatory substances, *e.g.* immunohormones (regulating proliferation, maturation and differentiation of cells of the immune system), complement components, microbicidal and cytotoxic substances *etc.* The lymphocytes circulate freely in the blood and lymph, or are taken up by primary and secondary lymphoid organs.

In any individual, the complex network of immune mechanisms, formed during phylogenetic development, maintains the integrity of the internal environment against changes originating during embryogenesis at certain developmental stages, or against changes, caused by environmental factors, *e.g.* viruses, microorganisms, chemical substances, unfavourable physical or psychological effects. During the development of an individual (ontogeny), the immune mechanisms develop both on the genetic basis and as an adaptation process under the influence of antigenic stimuli.

The reaction of the immune system (immune response) can be useful for the organism (immunity), or harmful (immunopathological reaction). The immune system fulfils two main functions, *i.e.* the immunological surveillance and defence against parasites and pathogenic microorganisms (bacteria, viruses, yeasts and fungi).

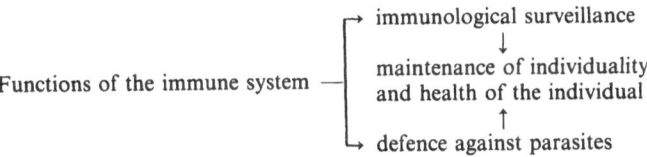

Functions of the immune system —
→ immunological surveillance
↓
maintenance of individuality and health of the individual
↑
→ defence against parasites

The term immunological surveillance refers to the ability of the immune system to detect damaged, worn-out or altered autologous cells and structures and to ensure their elimination. Similarly, the immune system is able to recognize tumour cells or cells from a genetically different individual of the same species. Thus, immunological surveillance ensures the maintenance of a perfect, error-proof and permanent composition of cells, tissues and the whole organism according to a given genome. In fact this system is responsible for the maintenance of the identity and integrity of the individual, *i.e.* its biological and biochemical individuality. This function is performed by its components, *i.e.* lymphocytes and antibodies, systematically penetrating all tissues of the organism *via* the blood and lymphatic circulation where they contact and recognize each other and altered or non-self cells or molecules. The key function of the immune system is therefore the discrimination of "self" from "non-self" (foreign).

It is of interest that this ability is already partially expressed in single-cell organisms, *e.g.* bacteria. The primitive ability of bacteria to recognize "foreign" (non-self) is manifested by significantly faster degradation of foreign or altered autologous proteins (*e.g.* after incorporation of amino acid analogues) as compared with that of normal autologous proteins. The basis of this mechanism is the ability of macromolecule-degrading enzymes (particularly proteinases) to distinguish between the normal and altered structure of their substrates and react to these changes by altered (usually enhanced) activity.

Evidence has been obtained that unicellular organisms, as well as individual cells of a multicellular organism, may mutually recognize each other through specific sensitive sites on their surface (receptors and effectors).

Even lymphocytes and antibodies mutually recognize each other. Thus according to JERNE (1974) the whole immune system forms a network which takes up anything "foreign" or "altered" and which permits easy bidirectional information exchange. The immune system controls both "self" and "non-self" cells and molecules. The "self" structures are tolerated under normal conditions. On the other hand, all foreign or altered autologous structures are recognized as antigens which results in their marking, uptake, destruction and elimination from the organism. Antibodies and lymphocytes do not recognize whole foreign cells or molecules, but only certain sites on their surface, called determinant groups or determinants. JERNE called them epitopes. Molecules containing determinants (epitopes) are antigens. Similarly, antibodies do not bind to the determinant group by the whole surface of their molecule (which is impossible for spatial reasons), but only through a specific binding site (paratope). Antibodies are a heterogeneous population of glycoprotein molecules, belonging, according to physicochemical properties, to the globulins. They occur in the blood serum, in other body fluids and even in intracellular spaces, as well as in the cytoplasmic membrane of *B*-lymphocytes producing antibodies. During electrophoretic fractionation of serum, antibodies migrate primarily in the gamma-globulin fraction. Because of their biological properties, this particular type of globulin was called immunoglobulins.

Both the phylogenetically most advanced information systems, *i.e.* the nervous and immune systems, possess certain common properties. They are designed not only to accept information entry (external and internal stimuli), but also to process and store the acquired information in their memory. Another common property is the capability to potentiate (amplify) or weaken the information signals according to the acquired experience, ensuring the optimum adaptation of the organism to different life conditions.

The humoral and nervous systems have "learned" to cooperate closely and, therefore, they are often referred to jointly as the neurohumoral (neuroendocrine) system.

The relationship between the immune and nervous systems has not yet been unambiguously established. Originally it was presumed that the immune system is self-regulated to a considerable degree. However, it has been shown that there is a bidirectional interconnection between both systems, and that they can mutually influence their activity (BESEDOVSKY and SORKIN, 1977; BLALOCK, 1984, 1989). Such an interconnection is achieved by at least two mechanisms. The first mechanism is mediated by nerves present in lymphoid organs, *e.g.* thymus, bone marrow, lymphatic glands, Peyer's patches, appendix and spleen. The second mechanism is of a humoral nature and is mediated by hormones, derived from the hypothalamus, pituitary gland, adrenal cortex and peptides of the opiomelanocortin family (*e.g.* adrenocorticotropin-releasing hormone, somatostatin, growth hormone, prolactin, adrenocorticotropin, glucocorticoids). In general, these hormones are released during stress. It has recently been found, however, that the release of adrenocorticotropic hormone (ACTH) from the pituitary gland may also be elicited by particular "immunohormones", produced by lymphocytes (lymphokines) or thymus (thymosin alpha-1), and that the cells of the immune system themselves produce ACTH and other neuropeptides (endorphins, enkephalins) that can influence the activity of both immune cells and neurons (BESEDOVSKY *et al.*, 1985; ADER *et al.*, 1990). Therefore, it is not surprising that, during the immune response, the level of circulating glucocorticoids increases together with the increasing antibody concentration, similarly to the situation induced by stress. In contrast to stress, however, these glucocorticoids are released from the adrenal cortext by ACTH which has been generated not in the nervous system, but from the immune system.

These findings became the basis of two extreme theories of immune function. Thus, BLALOCK and SMITH (1985) suggest that the immune system serves as a sensory organ for stimuli such as bacteria, viruses and tumours, *i.e.* stimuli that cannot be recognized by the central and peripheral nervous systems. According to an alternative proposal, a particular population of the immune cells are "free swimming nervous cells". A set of these cells represents a kind of "mobile brain". The latter idea is supported by the fact that the cells of the immune system possess several receptors with the same specificity as the cells of the nervous system, permitting their regulation by identical neurotransmitters, neuroendocrine peptides and immunohormones.

It should be stressed that the information exchange between the immune and nervous systems is bidirectional and obviously advantageous for both systems. This means that not only can the nervous system influence the function of the immune system, but also that the immune system can influence the activity of the nervous system. Therefore, not just neurohumoral, but also neuroimmune regulation is now recognized and a new interdisciplinary science — *psychoneuroimmunology* — has recently emerged, which deals with this highest type of homeostatic regulation of the organism (RABIN *et al.*, 1989). Under physiological conditions, the mutual interactions between the immune and nervous systems are fine and well-balanced. The intensity of mutual information exchange increases under pathological conditions, *e.g.*

during infection or malignant processes. Incorrect or insufficient interaction results in damage to the nervous system caused by the immune system. This occurs in various nervous diseases with an immunopathological mechanism, *e.g.* demyelinization diseases, experimental allergic encephalomyelitis, multiple sclerosis, *etc.* Stress-related neuroendocrine mediators (ACTH, corticosteroids, catecholamines) suppress a wide variety of immunological responses, including lymphokine and antibody production and macrophage-mediated tumouricidal activity. This neuroendocrine-induced modulation of macrophage function may play a significant role in the stress-induced enhancement of neoplastic disease.

The importance of the immune system for life lies in the fact that it belongs to three basic information systems (humoral, nervous, immune), ensuring the coordination of life processes during the ontogenetic development and later adaptation to various changes during the lifetime. These three information systems are discussed separately for purely didactic reasons. Despite the fact that each of them fulfils relatively independent functions and operates autonomously to a certain degree, they are in fact mutually interconnected, and in higher animals, they form a superinformation and superregulation system.

2.1 Why did the immune system originate?

Man and other higher animals are morphologically and chemically similar. However, no two individuals with identical qualitative and quantitative compositions exist. Such polymorphism is the basis of the chemical uniqueness of each individual; individual composition being controlled by the individual genetic equipment — the genome. The genetic variability is not purposeless, since it increases the probability of survival of a given species. The best adapted individuals to environmental changes are more common among a large number of genetically different individuals.

Data obtained by molecular genetics have shown that every cell nucleus contains material that controls all phenotypic markers of the individual. Thus, higher organisms strongly resist fusion with other, albeit related individuals and it is therefore impossible to transplant tissues or organs from one man to another. It is obvious that in an opposite case, properties of an individual or species would be determined randomly, rather than by precise genetic laws. In order to maintain a stable composition of any individual, and thus of the whole species, mechanisms, able to recognize "self" from "non-self" had to develop.

In ancient times, simple organisms did not need such a system. On the contrary, many of them had lived in symbiosis in order to maintain conditions essential for life. There is a lot of evidence to support the hypothesis that plants and animals have previously been the hosts of unicellular organisms which had found "shelter" within their cells and, in turn, provided the organism with their products. Chloroplasts in plants (remnants of blue-green algae) and mitochondria within animal and plant cells (descendants of primeval microbes) may serve as examples. According to one theory, even the nucleus of eukaryotic cells is a remnant of primitive unicellular organisms.

Natural evolution and origin of more complex life forms induced changes in this originally "peaceful coexistence". It has gradually changed to a chemical isolationism, which is now regularly found, particularly in higher organisms. The immune system appears to be an information mechanism maintaining this isolationism.

2.2 What is the essence of individual uniqueness which is mediated by the immune system?

The uniqueness of an individual is not only based on superficial differences, *e.g.* shape of nose, colour of eyes, body height *etc.*; differences at the molecular level are most important. The mammalian genome is formed by 3×10^9 pairs of purine and pyrimidine nucleotides, bound to form a DNA molecule. This number of nucleotides is sufficient to encode several thousands of proteins. Each of these proteins possesses its own shape (three-dimensional conformation), which is determined by the primary amino acid sequence. As the primary structure of all known proteins is microheterogeneous, it is obvious that certain proteins may have several — and antibodies even more — alternative conformations that are compatible with their normal biological function. Such alternative conformations differ only slightly. This wide protein repertoire results in the uniqueness of the individual at the molecular level, as each individual has only one or several conformations for each protein out of all possibilities. A similar property is possessed not only by proteins but also by polysaccharides, thus also contributing to individual uniqueness.

The immune system is responsible for discrimination of protein and polysaccharide conformation structures that are not encoded in the genome of a given individual. It is obvious that there must be a lower limit to the size of molecules recognized as foreign. One amino acid or dipeptide does not yield sufficient information to allow the immune recognition system to determine its origin. Such a dipeptide may originate, for example, from the human liver cell, but also from the influenza virus or mouse leukocyte. The shape of its molecule is identical in all these cases. However, if the peptide is larger, *e.g.* containing 20–30 amino acids, their sequence forms a sufficiently large complex to permit the immune system to recognize whether a certain conformation is self or non-self. An exception are small molecules of artificially synthesized substances, *e.g.* dinitrophenol, or substances that do not occur in the animal organism. These small molecules are foreign and can induce the immune response, although they must first be bound to the surface of larger molecules. Under such conditions, the immune response can also be induced by more complicated molecules, *e.g.* cholesterol, phosphatidylcholines, *etc.*

References

Ader, R., Felten, D. and Cohen, N. (1990) Interactions between the brain and the immune system. *Annu. Rev. Pharmacol. Toxicol.*, **30**, 561–602.

Arrhenius, S. (1907) *Immunochemistry*. New York, Macmillan, 309 pp.

Besedovsky, H. O., del Rey, A. E. and Sorkin, E. (1985) Immune-neuroendocrine interactions. *J. Immunol.*, **135**, 750s–54s.

Besedovsky, H. and Sorkin, E. (1977) Network of immune-neuroendocrine interactions. *Clin. Exp. Immunol.*, **27**, 1–12.

Blalock, J. E. (1984) The immune system as a sensory organ. *J. Immunol.*, **132**, 1067–70.

Blalock, J. E. (1989) A molecular basis for bidirectional communication between the immune and neuroendocrine system. *Physiol. Rev.*, **69**, 1–32.

Blalock, J. E. and Smith, E. M. (1985) The immune system: our mobile brain? *Immunol. Today*, **6**, 115–17.

Jerne, N. K. (1974) Towards a network theory of the immune response. *Ann. Immunol.* (Inst. Pasteur), **1250**, 373–89.

Rabin, B. S., Cohen, S., Ganguli, R., Lysle, D. T. and Cunnick, J. E. (1989) Bidirectional interaction between the central nervous system and the immune system. *CRC Crit. Rev. Immunol.*, **9**, 279–312.

3 Immunology and immunochemistry

The subject of immunology belongs to the biological and medical sciences. The relationship between immunology and other areas of science is schematically depicted in *Fig. 3.1*. This scheme shows that contemporary immunology is a true interdisciplinary science. Modern immunology studies cellular and molecular reactions of the human and animal organism, which occur after foreign antigens have been recognized by the immune system. A complex of these reactions represents an immune response which can, as for as the

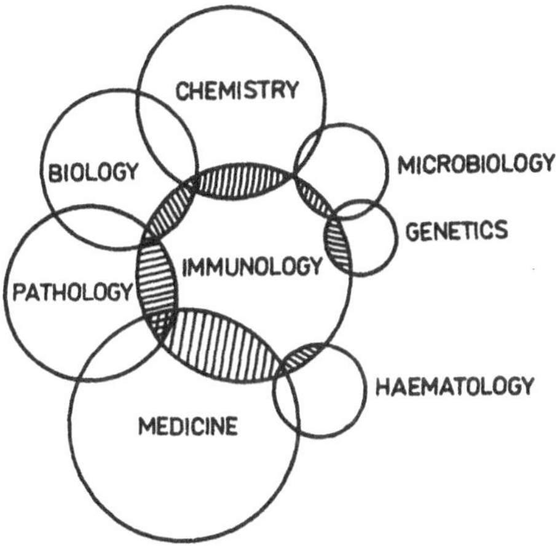

Fig. 3.1. Relationship of immunology to other scientific disciplines.

organism is concerned, be useful, indifferent, or harmful. The specific defence (immunity) against infectious diseases is an example of an advantageous immune response, whereas the harmful type of immune response is represented, for example, by various allergic reactions (hypersensitivity). The immune response can be induced not only by foreign antigens, but also by self, antigenically altered cells and molecules that are subsequently recognized as foreign. Even in this latter case the reaction may by useful (immunity against tumours) or harmful (autoimmune diseases).

The immune response is initiated by interaction of antigen with antigen-sensitive cells. Antigen-sensitive, or immunocompetent cells form a distinct lymphocyte population, bearing antigen receptors on their surface. Lympho-

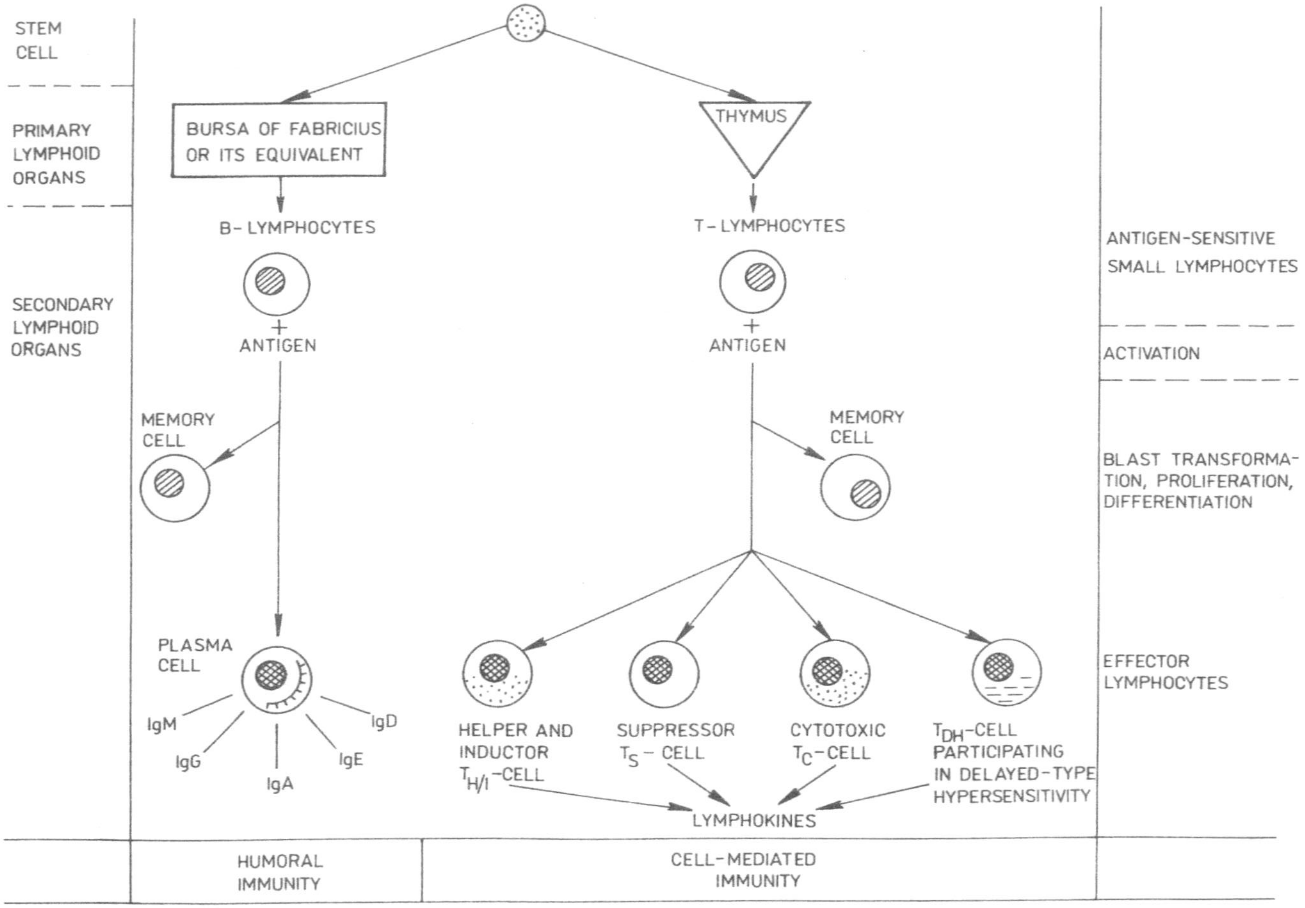

Fig. 3.2. Origin of *B*- and *T*-lymphocytes and their role in the immune mechanism.
It has not yet been firmly established whether T_{DH} cells form an independent subpopulation.

cytes gain immunocompetence (the ability to recognize and specifically respond to antigen) during their development in primary lymphoid organs. Nouza (1981) compared primary lymphoid organs with universities. Every lymphocyte is submitted not only to formal learning but also to strict examinations. In contrast to true universities, lymphocytes are allowed to leave the primary lymphoid organs only after learning their profession and function perfectly.

Vertebrates have two types of primary lymphoid organs (*Fig. 3.2*). The first is the thymus and lymphocytes, called *T*-lymphocytes (*T*-cells), that have been "educated" in the thymus. The second primary lymphoid organ is the bursa of Fabricius in birds, or its equivalent in mammals, and lymphocytes, called *B*-lymphocytes (*B*-cells), that have passed their university studies in these organs. Both lymphocyte types, after gaining immunocompetence, settle in secondary peripheral lymphoid organs (lymph nodes, spleen, certain other aggregates of lymphoid tissue), or circulate freely in the blood or lymphatic vessels.

The antigen receptors on the surface of *B*- and *T*-lymphocytes are glycoprotein molecules, anchored by one end to the phospholipid bilayer of the cytoplasmic membrane. The second end protrudes from the cell surface like TV antennae on a roof. The part of the receptor molecule protruding from the lymphocyte surface contains the combining site possessing a structure complementary to a certain antigenic determinant. Each key fits a single lock, and similarly the antigen determinant can only bind to a single antigen receptor. This is the basis of the specificity of their binding. This means that there exist millions of different binding (combining) sites — as many as there are antigen determinants. However, one lymphocyte bears only one type of receptor, *i.e.* with an identical combining site. Therefore, of the total pool of circulating lymphocytes in an organism, only a small proportion (several hundred) possess receptors with an identical combining site. Other lymphocytes have other types of combining sites.

More precisely, one lymphocyte with receptors of a given specificity not only binds antigens with exactly complementary determinants, but also structurally similar determinants. However, the binding strength with these slightly different antigens is lower.

Antigen receptors on the surface of *B*-cells are in fact antibody (immunoglobulin) molecules, which, after being synthesized within the cell, did not pass across the cytoplasmic membrane, but were incorporated into it as receptors. The antigen receptors on *T*-lymphocytes also possess similar types of combining sites, similar to receptors on *B*-lymphocytes. However, other parts of their molecule are different; they do not belong to immunoglobulins, but to genetically different types of glycoproteins. Antigen receptors on *T*-cells do not recognize the complementary antigen determinants directly, but in association with histocompatibility markers (antigens), typical for cells of each individual, together with accessory cells (particularly certain subpopulations of macrophages).

After the antigen determinant has been bound to the appropriate receptor, the consequent chain of biochemical reactions results in proliferation of lymphocytes possessing this receptor type, and in differentiation of their descendants into effector cells. The activated lymphocytes give rise to a cell line (clone) of identical cells, whereas the majority of other lymphocytes remain in a relative resting state and expect their "own" antigen. As the antigen determinant reacts specifically with the combining site of the lymphocyte receptor this type of immunity is designated as the specific immune response. The specific immune response is mediated by the specific immune mechanisms. They are encoded in the genome of every individual, although to be expressed phenotypically, they must come into contact with the antigen.

3.1 Specific immune mechanisms

Specific immune mechanisms consist of two basic types (*Fig. 3.2*) — mechanisms mediated by antibodies (*humoral immunity*) and cell-mediated immunity (*cellular immunity*). When the antigen activates certain *B*-lymphocytes, the latter differentiate into plasma cells, that synthesize and secrete antibodies with a binding site identical to that carried by the original *B*-cells that recognized the antigen. Thus, antibodies are responsible for the humoral type of specific immune reaction. When the antigen stimulates the immunocompetent *T*-cells, specific antibodies are not formed, but various types of regulatory, cytotoxic or delayed-type hypersensitivity reaction occur. These are mediated particularly by the inducer or helper, suppressor and cytotoxic *T*-lymphocytes and their products — the lymphokines and interleukins.

The former two subpopulations, *i.e.* the helper and suppressor *T*-lymphocytes, play the regulatory role in the induction of specific immune response. The *helper T-cells* assist the *B*-cells to recognize the antigen and are essential for initiation of antibody formation and origin of cytotoxic *T*-lymphocytes. They are also therefore designated as inducer *T*-cells (T_H — or $T_{H/I}$ cells). The role of *suppressor T-cells* (T_S) is to suppress the initial phase of the immune response. Thus, one T_S-cell subpopulation blocks antibody formation, whereas other subpopulations suppress the generation of effector cells of specific cellular immunity, particularly of cytotoxic *T*-lymphocytes. Recent data demonstrate that the function of T_S-cells is actually performed by the cytotoxic T_S-cells in collaboration with the helper *T*-cells. $T_{H/I}$-lymphocythes can be divided into at least two functionally different subpopulations. One of these helps the development of specific humoral or cellular immunity, whereas the other activates the cytotoxic *T*-lymphocytes to suppress the immune response.

Cytotoxic T-cells (T_C-cells or CTL — cytotoxic *T*-lymphocytes) possess a cytotoxic activity, *i.e.* they damage and kill various target cells, which bear on their surface the antigen which originally induced their formation. The destruction is specific and, therefore, other cells lacking the respective antigen, are not damaged by the T_C-cells. The target cells for CTL include, for example, tumour cells, cells transplanted from a genetically non-identical individual (with different histocompatibility antigens), or autologous cells with a virus-modified surface.

The effector and regulatory *T*-lymphocytes are involved in the mechanisms of specific immunity either directly of *via* various chemical substances secreted extracellularly. These substances are called lymphokines and have the character of immunohormones. They specifically influence the activity and function not only of individual cell populations of the immune system, but also of other cell types. The main producers of lymphokines appear to be $T_{H/I}$-cells.

A particular lymphocyte population — *the memory cells* — "remember" the antigen after the first encounter. After repeated contact with the same antigen, the memory cells are responsible for more rapid and intensive immune response. Since the lymphocytes remember the antigen they have already met, this specific type of immunity is also called *acquired immunity*.

Cellular immunity is developmentally older than the ability to form antibodies. Therefore, the latter only occurs in phylogenetically more advanced animals.

The contact of antigen with immunocompetent cells need not only result in antibody formation or generation of effector lymphocytes, but it may also induce *immunological tolerance, i.e.* a state of non-reactivity of the immune system to repeated antigen administration, or, finally, various immunopathological phenomena:

$$\text{Antigen} + \text{immunocompetent cell} \begin{array}{l} \nearrow \text{immunity} \\ \rightarrow \text{tolerance} \\ \searrow \text{immunopathology} \end{array}$$

Recently, the mechanism of immunological tolerance has been intensively studied. The success of tissue and organ transplantation, even with maximal antigen similarity (compatibility) between donor and recipient, depends on the suppression of the immune response which attempts to reject the foreign graft. Therefore, the immune response has to be suppressed after transplantation or immunopathological conditions by immunosuppressive therapy. Induction of immunological tolerance against the graft, however, appears to be more physiological, and, as far as the organism is concerned, it would be a much better way of suppressing the transplantation reaction.

The immune response can be suppressed by physical (irradiation), chemical (inhibitors of protein synthesis, RNA and DNA synthesis, corticosteroids, cytotoxic drugs *etc.*), or biological (thymectomy, antilymphocyte serum *etc.*) methods. Immunosuppression results in a short-term paralysis (inhibition) of the immune system.

3.2 Mechanisms of natural (non-specific) immunity

The term natural immunity designates the non-specific resistance of the organism against various microorganisms (viruses, rickettsiae, mycoplasmas, bacteria, fungi, yeasts, protozoa), foreign and tumour cells. Even this type of resistance is controlled by the genome of any individual, but compared with acquired immunity, the mechanisms of natural immunity operate spontaneously, regardless of previous contact with a particular antigen. Similarly to the non-specific mechanism, natural immunity acts not only against a single, but against many different antigens.

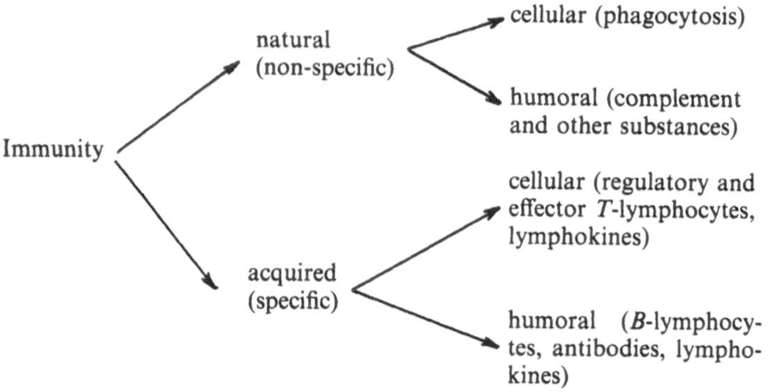

Mechanisms of natural immunity are developmentally older. They may be divided again, similarly to specific immunity, into the cellular and humoral type. The basic mechanism of non-specific cellular immunity appears to be *phagocytosis*, whereas the humoral branch of non-specific immunity is represented by the *complement system*, interferons, chemotactic factors, opsonins, mediators of the inflammatory reaction, multiple enzymes and their inhibitors and some other factors. The main role of phagocytosis is to take up, ingest, kill or inactivate and degrade foreign particles, including invading pathogenic microorganisms. Phagocytosis is performed by particular types of leukocytes, called *phagocytes*.

It should be stressed, however, that no immune mechanism can perform its function separately. All mechanisms cooperate mutually and complete each other. The complex of substances, cells and tissues participating in the natural immunity, potentiates the efficiency of specific immunity and, therefore, they are referred to as *immunological amplification systems*. On the other hand, products of specific immune reactions, particularly antibodies and some lymphokines, facilitate and enhance the course of many reactions of non-specific immunity.

3.3 Branches of immunology

In order to fulfil the requirements of social usefulness and scientific benefit, the results of theoretical studies should ultimately have practical importance and should contribute to the progress of knowledge and development of human civilization. Thus, in any branch of science, one can separate basic (theoretical) from applied (practical) research, although the difference is not always clear-cut. Immunology can also be divided into two main parts — theoretical (experimental) and practical (applied) immunology. Until recently, practical immunology has dealt almost exclusively with the application of experimental immunology in clinical medicine and, therefore, it was called clinical or medical immunology. Nowadays, the results of immunology are also applied in other fields, *e.g.* certain biotechnologies and animal husbandry.

Immunology can also be divided according to other criteria. According to the type of immune reactions studied, the IUIS (International Union of Immunological Societies) originally recommended dividing immunology into six areas: immunochemistry, microbial immunology, cell immunology, allergology, transplantation immunology and tumour immunology. Nowadays, however, these areas of immunology overlap and do not correspond exactly to the subject of research. Furthermore, the accumulation of new data gave rise to new branches of immunology, such as molecular immunology, immunology of ageing, immunology of pregnancy, the foetus and the newborn, ecological immunology *etc.* In addition, at the boundary with other scientific disciplines, many hybrid branches, such as immunobiology, immunogenetics, immunopharmacology, immunotoxicology, immunopathology, immunohistology, psychoneuroimmunology *etc.* developed.

Microbial immunology studies bacterial, viral and protozoal antigens, their relationship with the origin of infectious disease, host defence against infectious agents (anti-infectious immunity), mechanisms of resistance of microbes to defence mechanisms of the host, development of vaccines and immune sera.

Cellular immunology deals with studies on cells, tissues and organs participating in the immune response, and particularly with the development and functional activity of antibody-forming cells and effector cells of cell-mediated immunity.

Allergology studies changes induced by hypersensitivity, *i.e.* the excessive reaction of the immune system to certain antigens, called *allergens*. These reactions include, for example, hay fever, rhinitis, atopic eczema, atopic bronchial asthma and others. The term *atopy* refers to a genetically encoded hypersensitivity to various antigens, which permits excessive formation of reagins (antibodies of the IgE class). Allergic reactions of this type are generated very quickly (within a few minutes) after antigen invasion into the body and are thus called *immediate hypersensitivity reactions*. By analogy the *delayed-type hypersensitivity* is generated within several hours or days; instead of reagins, this reaction is mediated by T_{DH}-lymphocytes. Allergology is a part of **clinical immunology**, together with diagnosis and treatment of immunodeficiencies, autoimmunity and other immunopathological diseases. Clinical immunology has two main components — laboratory (primarily for diagnostic work) and clinical (performed by clinical immunologists — physicians — in cooperation with laboratory workers). This part of immunology partially overlaps *immunopathology*, which, contrary to clinical immunology, is more oriented towards recognition of causes and mechanisms of abnormal function of the immune system.

Transplantation immunology has only developed recently and has substantially influenced transplantation of organs (particularly of kidneys) and cells (bone marrow). It also includes research on typing of transplantation (histocompatibility) antigens, immunological tolerance, suppression of the immune response, and even immunological aspects of blood transfusion and immunological relationships between the mother and foetus. Advances in research into histocompatibility antigens have contributed to the present rapid development of **immunogenetics**, studying the laws of genetic transfer of various markers present on immune cell surfaces and antibody molecules, as well as the role of these markers in the susceptibility to certain diseases.

Tumour immunology studies tumour (cancer cell)-associated antigens, mechanisms of antitumour immunity, possibilities of immunological diagnosis and immunotherapy of malignant tumours.

The **immunology of ageing** studies changes in the reactivity and capacity of immune mechanisms during ontogenic development of the individual.

Immunobiology studies particularly the phylogenetic development of the immune system and differences in the immunological apparatus among various biological species.

Immunopharmacology is the study of suppressive (immunosuppressive) and stimulatory (immunostimulatory) effects of natural and synthetic substances (drugs) on the immune system with the aim of therapeutic modulation of immune activity.

Immunotoxicology studies the toxic effects of various industrial toxins, xenobiotics and toxic agrochemicals in the environment, as well as the side (undesirable) effects of drugs on the immune system. It is closely connected with **ecoimmunology** (ecological immunology) which studies the adverse effects of human economic activities on the immune system of both individuals and population groups in affected areas.

Psycho(neuro)immunology investigates the relationship between psychosocial processes and function of the immune system, and its influence on changes in susceptibility to certain infectious diseases and malignant tumours.

3.4 The object of immunochemistry

Immunochemistry has developed as a part of immunology and largely includes the study of immune processes at the molecular level, particularly the structure, properties and interaction of molecules participating in immune reactions. In order to fulfil its function, the immune system must regulate the activity of these molecules in space and time. Molecular immunology studies all events required for mutual interaction and organization of immune mechanisms. Immunochemistry and molecular immunology are closely related and have identical interests in most cases.

The objects of contemporary immunochemistry include:

1. antigen and antibody structure, and the relationship between their structure and biological activity;
2. biosynthesis of antibodies at the molecular level, their biological turnover and mechanisms of secretion out of the cell;
3. preparation and use of monoclonal antibodies;
4. kinetics and mechanism of antigen–antibody reactions;
5. structure and properties of differentiation markers (antigens), as well as cellular receptors of the immune system (antigen receptors on *B*- and *T*-cells, receptors for Fc-domains of immunoglobulins, receptors for complement components *etc.*);
6. structure and properties of regulatory substances (immunohormones) such as lymphokines, monokines, interleukins, thymic hormones, various growth and differentiation factors of cells of the immune system *etc.*;
7. enzymatic and non-enzymatic systems participating directly in defence reactions (complement system, microbicidal, cytotoxic and degradation mechanisms of phagocytes and certain lymphocytes);
8. effect of antibodies on biologically active antigens such as enzymes, inhibitors, hormones, toxins, cell receptors *etc.*;
9. development of immunological and immunochemical analytical methods and their use in solving chemical and biological problems.

Immunochemistry employs many biochemical and biophysical techniques, but since these methods are not applicable to all immunochemical problems, various specific methods have been developed, which, in turn, have been used in other scientific fields. These specific methods provide immunochemistry with a certain degree of independence. Nevertheless, immunochemistry belongs unequivocally to the complex of modern immunology.

Reference

Nouza, K. (1981) *The hidden power of immunity*. Prague, Mladá fronta, 272 pp. (*in Czech*).

4 Antigens

Antigens (from the Greek "anti" = against and "gen" = to form) are macromolecular substances of natural or artificial (synthetic) origin, recognized by the immune system as foreign (non-self). After administration to a suitable (competent) organism, the antigens stimulate formation of antibodies, lymphokines, regulatory and effector *T*-lymphocytes, *i.e.* they induce an immune response.

Antigen ↗ complete (functional) — immunogen
Antigen ↘ incomplete — hapten

For the functional sense, an antigen may be complete (functional), or incomplete, which is usually termed a hapten. The complete antigen is also referred to as an immunogen. The term "antigen" has often been used incorrectly in the sense of a complete antigen. This term is more general at present since it designates a substance which may possess properties of both immunogens and haptens.

Immunogen ↗ immunogenicity
Immunogen ↘ specificity

Hapten → specificity

An **immunogen** (complete antigen) has two basic properties: immunogenicity — the ability to induce an immune response (origin of antibodies, regulatory and effector lymphocytes); and specificity — the ability to react with these lymphocytes and antibodies. The antigen reacts only with those lymphocytes and antibodies whose formation it had induced. It does not usually react with other lymphocytes and antibodies. This is the basis of the specificity of this reaction.

A **hapten** is unable to induce the immune response; it can only specifically react with cells and antibodies which originated after the interaction of the immunogen with the immunocompetent cell. From these definitions of antigen and hapten, basically formulated by the Austrian immunologist and pathologist Karl Landsteiner, a Nobel prize winner in 1930, it follows that a hapten is a part of the molecule of the complete antigen.

Each immunogen is composed of the macromolecular *carrier* and low-molecular determinant groups (*Fig. 4.1*). The *determinants* are either a natural component of the immunogen, or are artificially bound to it; in the latter case, they are the haptens. Obviously, even natural determinants may become

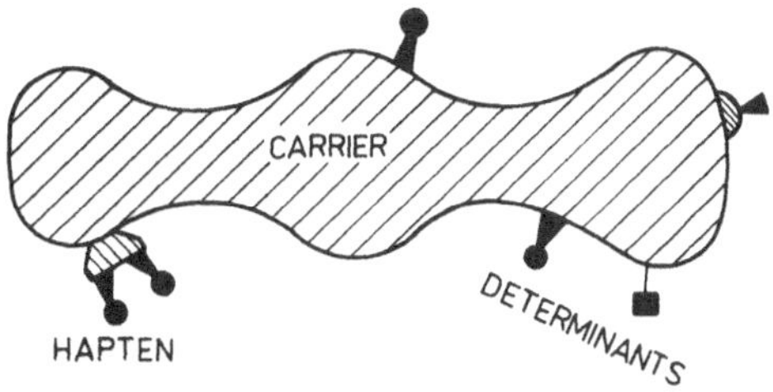

Fig. 4.1. Structure of a complete antigen (immunogen).

haptens provided they are chemically or enzymatically cleaved from the macromolecular carrier. Only the whole complex (carrier-determinants) is immunogenic.

$$\text{Hapten} \begin{array}{l} \nearrow \text{complete} \\ \searrow \text{simple (inhibiting, semi-hapten)} \end{array}$$

There are two types of hapten. The reaction between a *complete hapten* and specific antibody can be directly detected *in vitro* using common serological or other immunochemical techniques. Simple haptens or *semi-haptens* cannot be detected directly. They can be detected indirectly only on the basis of inhibition of the reaction between antibodies and antigens (or complete haptens). Therefore, semi-haptens are also called *inhibiting haptens*. This property of semi-haptens can be expressed schematically:

antibody + antigen = positive reaction
antibody + semi-hapten + antigen = no reaction (inhibition)

When the antibody and antigen or complete hapten mixture is present, precipitation of the complex may occur. However, when the semi-hapten is added to this mixture, no reaction takes place, *i.e.* there is inhibition. This phenomenon is caused by the fact that the semi-hapten binds preferentially to the antibody combining site, which, in turn, becomes inaccessible for the antigen or complete hapten. The antibody–semi-hapten complex does not form a precipitate and, therefore, cannot be detected by common serological and immunochemical methods. The precipitation reaction occurs with complexes of antibodies with antigen or complete haptens only.

4.1 What kind of substances can be immunogenic?

It follows from the discussion above that an immunogen cannot be an arbitrary substance. The basis of immunogenicity has not yet been completely clarified. However, it has been established that a substance may become *immunogenic* when it possesses certain physical, chemical and biological properties. The degree of immunogenicity also depends on genetically determined responsiveness and maturity of animals and on the route of administration.

Antigen properties
↗ physical (relative molecular weight, solubility, electric charge)
→ chemical (structure, degradability)
↘ biological (species remoteness)

As far as physical properties are concerned, the relative molecular weight is very important. Immunogens are macromolecular substances and usually have molecular weights above 10 000.

Immunogens
↗ cellular (insoluble)
↘ colloidal (soluble)

According to solubility, immunogens are divided into cellular (microorganisms, cells and subcellular structures) and colloidal (soluble macromolecules) types. By decreasing the solubility of colloidal immunogens, their immunogenicity is usually increased. Therefore, the colloid immunogens are often precipitated with bentonite or potassium aluminium sulphate before immunization.

The presence of an electric charge on the immunogen molecule is not esential for its immunogenicity. If an immunogen has a positive charge, the antibody will have a negative charge and *vice versa,* as shown by Sela and Mozes (1969).

The degree of immunogenicity is also influenced by the shape of the immunogen molecule (conformation) and availability of determinant groups.

Immunogens
↗ proteins, polypeptides
→ polysaccharides
↘ nucleoproteins and other complexes

From the chemical standpoint, immunogens may be various biopolymers, particularly proteins, polypeptides and polysaccharides (*Table 4.1*). Among nucleic acids, some RNAs and single-stranded (denatured) DNA, but particularly their complexes with proteins (nucleoproteins) can be immunogenic. Pure lipids are usually only haptens, but in complexes with proteins or polysaccharides they become relatively good immunogens. Sometimes phospholipids can also become immunogenic.

Table 4.1 Chemical composition of immunogens

Type of immunogen	*Example*
Polypeptides	proteohormones, synthetic antigens
Proteins	plasma, microbial, enzymes
Polysaccharides	bacterial capsules, synthetic (dextrans)
Glycoproteins	immunoglobulins, blood group antigens, histocompatibility antigens
Peptidoglycans	in bacterial cell walls
Nucleoproteins	chromatin, ribosomes
Lipoproteins	plasma, cell membranes
Lipopolysaccharides	cell walls of Gram-negative bacteria (endotoxins)

In principle, immunogenic activity may be exhibited by all biopolymers present in living systems, as well as by synthetic and conjugated (semi-synthetic) antigens. The ability to induce an immune response, however, is not equal. Proteins are the most potent immunogens, followed by polysaccharides and some polysaccharide complexes. However, this does not necessarily mean that all proteins or polysaccharides are good immunogens. Thus, gelatin, for example, is a weak immunogen, whereas serum albumin has strong immunogenicity. What is the basis of this difference? The presence of determinant groups on the molecule surface is particularly important. Albumin contains enough determinants in suitable conformation and availability to bind to receptors of immunocompetent cells; gelatin on the other hand has relatively few determinant groups. If low-molecular-weight haptens (*e.g.* dinitrophenol) are bound to gelatin, they assume the function of determinant groups. The modified gelatin then becomes a good immunogen.

The degree of immunogenicity is also influenced by the *degradability* of the immunogen molecule. During degradation, destruction of determinant groups may result in decreased immunogenicity and specificity changes. On the other hand, partial degradation, *i.e.* the kind of "processing", to which antigens are subjected in macrophages, results in the expression of new determinants. As long as these new determinants are hidden in the native molecule and therefore inaccessible for receptors of immunocompetent cells, such antigen-processing may enhance their immunogenicity. The mode of antigen treatment in the organism prior to its contact with antigen-sensitive cells is genetically determined and represents a part of the *immunological amplification system.*

Among important biological properties of the immunogen, genetic diversity (*e.g. species remoteness*) should be mentioned first. The greater the phylogenetic difference between the organism of origin and the organism penetrated by the immunogen, the greater its immunogenicity. However, not all foreign substances are immunogens. For example, carbon in the form of soot is foreign for the animal organism but does not induce an immune response. It is instead removed from the body by mechanisms of natural immunity (phagocytosis). Thus, one property is not enough for the substance to become immunogenic. For this a suitable combination of physical, chemical and biological properties is required.

Immunogens occur in nature in a very complex form as microbes, animal cells, viruses or their components. Each microbe or cell is composed of numerous antigens with various carriers and determinants.

The ability to react by immune response to certain immunogens is inherited as an autosomal dominant marker, in both qualitative and quantitative senses. Thus, individuals may respond, depending on their genetic apparatus, better, worse or even not at all (tolerance of antigen). The quality and direction of the immune response is controlled, in addition to the genome of the immunized individual, by properties of the antigen and other factors, particularly the site of antigen contact (route of immunization) and antigen dose.

4.2 What are antigenic determinants?

It follows from the previous chapter that antigenic determinants are certain groups of atoms, present on the surface of the antigen molecule, which are capable of a specific reaction with receptors on immunocompetent lymphocytes, as well as with antibodies or effector lymphocytes induced by these determinants.

What is the chemical structure of the determinant group? Its structure depends on the origin and type of antigen. For protein antigens, the determinants are formed by amino acid residues (usually several amino acids at the ends of polypeptide chains). In polysaccharides, they are formed by monosaccharide units, and in nucleic acids by several nucleotides or purine and pyrimidine bases.

Determinants ↗ sequential
↘ conformational

SELA (1971) differentiates sequential and conformational determinants. The *sequential determinant* is determined by the sequence of basic subunits in the biopolymer (amino acids, monosaccharides, nucleotides). These subunits may exist in various random conformations.

In contrast, the *conformational determinant* is formed by only one of all possible conformations of a certain part of the antigen molecule (of the polypeptide or polysaccharide chain). The conformational determinants are typical for native antigens. Denaturation of protein antigens changes their spatial structure together with the specificity of conformational determinants. On the other hand, the specificity of sequential determinants is not changed by denaturation. Antibodies directed against conformational determinants usually do not react with small peptides derived from the antigen molecule region containing conformational determinants. These peptides possess an identical primary structure, but they lack the conformational structure (spatial architecture), which is characteristic for the whole peptide chain or its long section only.

Conformational structures or **conformations** are molecular forms of the same substance that are not energetically equivalent and their individual atom groups lack identical spatial orientation. Multiplicity of conformational structures is based on the fact that atom groups rotate freely around certain bonds in the molecule. As long as atoms, located at the bond around which they rotate, are of different size, electric charge *etc.*, they may occupy mutually more or less suitable positions. In certain positions, the atomic groups would, for example, hinder each other and thus more energy will be required for maintaining such a position than the position without steric hindrance. These latter positions are energetically more convenient and, therefore, they will be occupied by most molecules at lower energetic level. The conformational structures should be understood in the statistical sense, *i.e.* that at a certain moment, the majority of molecules will occupy the most energetically convenient position, but other positions (conformations) will also be present in the system in lower numbers. Their mutual ratio will depend on the total energy present in the system.

Complex molecules, *e.g.* antigens, have a high number of bonds, around which the terminal atomic groups can freely rotate. For this reason also their energetically most advantageous formations result from the very complex effect of various forces and steric factors.

The sequential determinants may be cleared from the carrier and become typical haptens, whereas conformational determinants are mostly an integral part of the carrier.

The present knowledge concerning the determinant groups has been obtained primarily by using chemically altered antigens, *i.e.* antigens to which chemically defined low-molecular-weight haptens were bound. In the past few years, this model function was assumed almost exclusively by synthetic polypeptides with precisely defined structure.

For example, a tripeptide, composed of three amino acids — tyrosine, alanine and glutamic acid (Tyr-Ala-Glu, abbreviated to TAG), can bind to polyalanine side chains of the polypeptide, with a basic skeleton formed by polylysine, or directly polymerize into high-molecular-weight periodic polymers. The first polypeptide is abbreviated to TAG-A-L; the second to $(TAG)_n$. SELA (1971) established that the TAG tripeptide in the former polymer behaves as the sequential determinant, but in the latter as the conformational determinant (*Fig. 4.2*). Under physiological conditions, the polypeptide $(TAG)_n$ occurs in the form of an α-helix. In this case the structure is typical for the conformational determinant. Antibodies with completely different specificity (no cross-precipitin reactions can be observed) against both determinant types forming TAG tripeptide can be prepared. Peptides TAG and $(TAG)_2$ are efficient inhibitors of precipitation in the antibody–

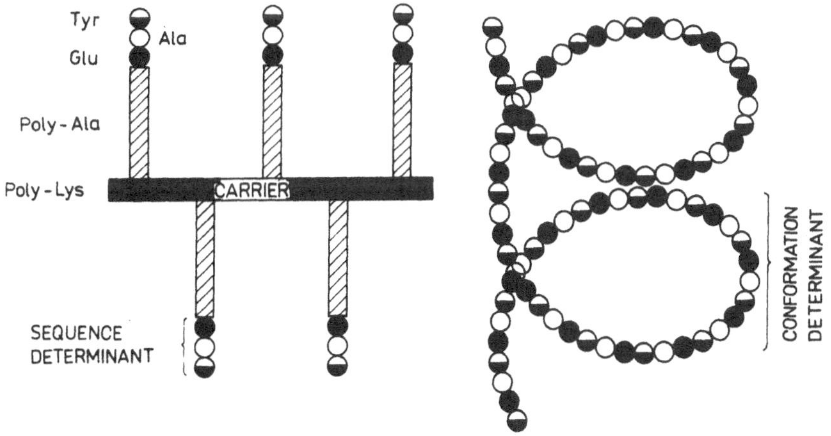

Fig. 4.2. Schematic representation of a sequential and conformational antigenic determinant, formed by a tripeptide Tyr-Ala-Glu (TAG).
Left, polypeptide TAG-A-L, right, polypeptide $(TAG)_n$

–TAG-A-L system, but they do not influence the course of precipitation in the antibody–$(TAG)_n$ system at all. The precipitation in the latter system is effectively inhibited by the peptides $(TAG)_7$ and $(TAG)_9$ that can create a structure indicative of the α-helix. This hepta- and nonapeptide, however, does not inhibit the precipitation in the antibody–TAG-A-L system.

Synthetic antigens are more suitable for studies of immunogenicity and specificity than natural antigens. In addition to a defined structure, they make it possible to change the structure of the antigen molecule by different chemical treatments. Such interventions make it possible to localize more precisely groups of atoms or molecules that are responsible for the biological activity of the antigen. Homopolymers of amino acids (containing many units of identical amino acids) are very weak immunogens. Their immunogenicity increases proportionally with the increasing spectrum of amino acid composition and steric structure.

What is the size of an antigen determinant? Exact measurements are still lacking. From experiments using particular sections of polypeptide and polysaccharide chains (oligopeptides, oligosaccharides) as inhibitory haptens, one can assume that sequentional determinants are of the size of trimers up to octamers. Thus, in the protein antigens, the determinants are formed by between three and eight amino acid residues. The size of the conformational determinants is even less known. It is presumed that in protein antigens the size does not exceed 20 amino acid units. Most fibrillar and all denatured globular proteins have sequentional determinants only, whereas native globular proteins have conformational determinants. This can be illustrated by the following example: it is known that there is a high degree of homology (similarity in amino acid composition) between lysozyme, isolated from egg white, and lactalbumin. Antibodies, prepared against both proteins (lysozyme and lactalbumin) after denaturation, react not only with the protein (antigen) which induced their formation, but also cross-react with the other protein. It thus follows that after denaturation, both lysozyme and lactalbumin determinants become similar and are of sequential type. However, when native proteins are used for antibody preparation, there are no cross-reactions, indicating the presence of various determinants of conformational type.

$$\text{Determinants} \begin{array}{l} \nearrow \text{ immunopotent (free)} \\ \searrow \text{ immunologically mute (hidden)} \end{array}$$

From the functional point of view, SELA (1969) distinguished two determinant types: *free determinants*, that in a complex with a suitable carrier induce the production of large amounts of antibodies, and *hidden determinants* that do not normally induce antibody formation. After chemical treatment or enzymatic fragmentation hidden determinants may become immunopotent.

One of the first synthetic antigens, prepared by SELA *et al.* in 1962: poly-(Tyr,Glu)-poly-Ala-poly-Lys (abbreviated (T,G)-A-L, serves as an example of such hidden determinants. The antigenic determinants in this polypeptide are formed by a tripeptide which consists of tyrosine units and glutamic acid. As soon as the determinant binds to the terminal ends of the polyalanine side chains forming a branch of the basic polylysine chain, free immunocomponent determinants are generated. When the determinant is localized between polyalanine and polylysine chains, it becomes hidden and cannot be immunologically manifested — it is immunologically mute. By chemical modification of the carrier forming the poly-Lys-poly-Ala chain instead of poly-Lys, these determinants re-acquire immunopotency (*Fig. 4.3*).

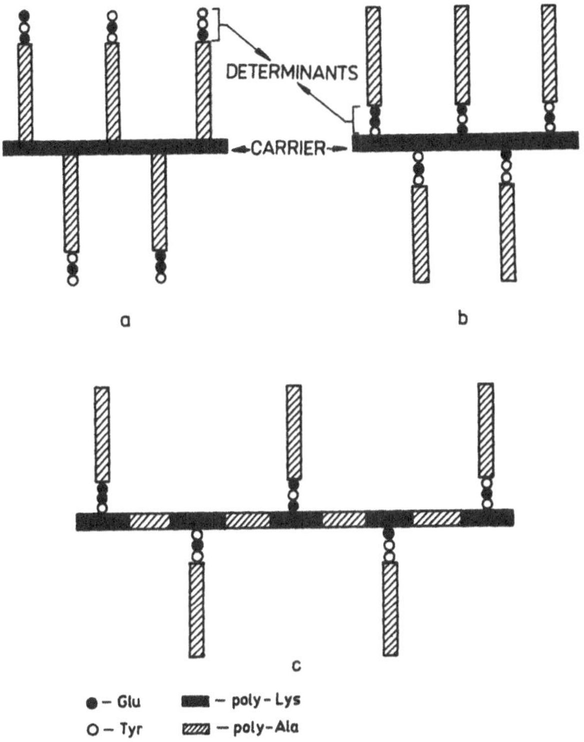

Fig. 4.3. Schematic view of free (immunopotent) and hidden antigenic determinants (according to Sela, 1969).
(a) Free (immunopotent) determinants; specific antibodies are formed after immunization of rabbits; (b) hidden determinants (immunologically mute) — no immune response; (c) side poly-Ala chains bind to lysine residues only. If poly-Ala segments are inserted into the poly-Lys carrier, the space between side chains becomes wider and the originally hidden determinants become immunopotent and may induce antibody formation.

It therefore follows that the term "*immunopotency*" reflects the ability of certain regions of the antigen molecule to serve as an antigenic determinant and induce formation of specific antibodies or other immune products. Immunopotency is thus a quantitative expression of the strength of the antigenic determinant. Every determinant is composed of several subunits although these do not participate equally in binding with the antibody. The extent to which each unit of antigen determinant contributes to this bond is the measure of its *immunodominance*.

4.3 Classification of antigens

Antigens can be classified according to various criteria: their chemical and physical properties and origin, their relationship to the host organism, and their association with diseases *etc.* In section 4.1 antigens were classified according to solubility (cellular and soluble antigens) and chemical differences (protein, polysaccharide, nucleoprotein and other antigens).

With respect to their origin, antigens may be divided into:

1. *natural* — practically all biopolymers and their complexes occurring in nature;
2. *synthetic* — polypeptides, polysaccharides and certain artificially synthetized polynucleotides;
3. *artificial* — natural antigens altered by a physical or chemical treatment. In particular, this group includes denatured antigens and conjugates of natural antigens with precisely defined haptens (*conjugated antigens*).

According to their relationship to the organism, antigens can be classified as:

1. *exogenous* (of external origin), such as foreign cells, cell structures and molecules. Exogenous antigens for man include antigens of bacteria, yeasts, viruses, plant pollen and antigens of transplanted tissues;
2. *endogenous* (components of own cells and tissues): these are also known as autologous antigens (*autoantigens*).

With respect to the relationship between two organisms, between which antigen transfer is performed (tissue transplantation), antigens can be divided into five groups:

1. *xenogeneic* (heterologous) — originating from an organism of another species;
2. *allogeneic* (homologous) — derived from a genetically non-identical individual of the same species. In haematology, these antigens are often termed *isoantigens;*
3. *isogeneic* — derived from genetically identical individuals of the same species (homozygotic twins);
4. *syngeneic* — derived from a genetically identical individual within an inbred strain. The inbred strain arises, for example, by repeated crossbreeding of siblings for many generations;
5. *autochtonous* — antigens of own organism (autoantigens).

Heterophilic antigens are found in various phylogenetically unrelated species and in some cases even in cells and tissues of certain animals and microbes. Antibodies induced by these antigens cross-react with individual heterophilic antigens. This antigen type may be involved in the pathogenesis of certain disease, *e.g.* infections mononucleosis, rheumatic fever, glomerulonephritis *etc.* For example, there is a relationship between antigens of group A beta-haemolytic streptococci and antigens of the human myocardium, *i.e.* they are heterophilic. Antibodies, induced by streptococcal antigens, can cross-react with human myocardial antigens. If such a reaction takes place *in vivo* (after streptococcal infection when anti-streptococcal antibodies are present in the organism), it may damage the heart tissue.

The *Forssman antigen,* found in tissues of guinea pigs, sheep erythrocytes, tissues and erythrocytes of dogs, cats, fowl and human type A and AB red blood cells, is another example of a heterophilic antigen. It is also present in numerous bacteria.

Alloantigens (homologous antigens) have genetically controlled determinants, differing in any individual of a certain species. This is the case in humans with the exception of homozygotic twins. Alloantigens are not immunogenic for the organism of their origin, but are immunogenic for other individuals of the same species. These antigens are *polymorphic, i.e.* they occur in a certain population in multiple (different) forms.

Polymorphism is the existence of two or more genetically different forms of a definite marker (in this case alloantigen) within one population. These different forms are determined by the presence of two or more alleles on one chromosome locus. However, each locus of any individual can contain only one allele of a certain set occurring in the whole population. A population is a group of individuals having a common genetic apparatus which is determined by random mutual pair-crossing.

Blood group antigens or histocompatibility (transplantation) antigens are examples of alloantigens. The presence of incompatible blood group antigens may cause haemolytic disease of the newborn or the post-transfusion reaction. The success of transplantation of foreign tissues and organs is determined by the histocompatibility antigens. Alloantigens induce formation of alloantibodies and a particularly strong cellular immune response.

Autologous antigens (autoantigens) are generally tolerated by the organism and become immunogenic only under certain conditions. Thus they may induce an immunological reaction in the organism of their origin. Autoantibodies and autoreactive cells are formed against autoantigens and may cause autoimmune or autoaggressive diseases. Some autoantigens are organ-specific. A part of them is generated during the later phase of ontogenic development representing a form separated from the immune system (cornea, spermatozoa, thyroglobulin *etc.*). When these antigens penetrate, *e.g.* after trauma, into the immune system, induction of antibody formation may occur. Autoantigens are also generated under the influence of physical, chemical and biological factors on cells and their structures (burns, ionizing radiation, certain drugs, toxins and microbial enzymes *etc.*).

4.4 Synthetic and conjugated antigens

Synthetic antigens are biopolymers with a known structure of the carrier and determinant groups such as polypeptides with linear or branched chain. Two examples of synthetic polypeptides with branched multiple chains are shown in *Figs. 4.2* and *4.3*. Besides antigens composed of amino acids only, antigens containing saccharides, nucleosides and other substances in their molecule, may also be synthesized.

Synthesized naturally occurring antigens, their fragments or various derivatives may be of great biological importance. Such synthetic antigens induce formation of antibodies and other immune products with a specificity identical to that of corresponding natural antigens. Thus they can replace unavailable natural antigens for immunization. Alternately, antibodies formed against synthetic antigens can cross-react with natural antigens. Synthetic antigens inducing formation of antibodies which cross-react with bacterial cell wall or other microbial antigens are an example. This antigen type indicates the way that *synthetic vaccines* are prepared. Antibodies induced by natural antigen-derived derivatives have a relatively altered specificity which can be exploited in various ways. Alternately, the immune response against the natural antigen can be modified by such a derivative. In the case of a harmful immune response, such a modification can be therapeutically important. Thus, for example, the administration of a synthetic fragment of the basic myelin protein (the insulating coat of nerve fibres) prevents the induction of experimental allergic encephalomyelitis, which occurs after immunization of experimental animals with this protein or with brain tissue homogenate.

Conjugated antigens are represented by complexes, in which chemically defined determinant groups (haptens) are bound to a natural or synthetic carrier. The carrier is usually a complete natural antigen which, after hapten binding, actually becomes an artificial antigen. Basic immunological studies on haptens conjugated to protein carriers were performed by LANDSTEINER (1936) and PAULING (1940).

In general, any low-molecular-weight substance is a hapten and can become immunogenic after binding to a suitable carrier, *i.e.* it can induce an immune response. Conjugation of the hapten with the carrier occurs both *in vitro* and *in vivo*. Conjugation of low-molecular-weight substances such as some drugs (penicillin, sulphonamides, *etc.*) with cells or macromolecules in the host organism initiates allergic (immunopathological, hypersensitivity) reactions to these substances.

Conjugated antigens, prepared *in vitro*, can be used for immunological research purposes as they possess precisely defined determinants. In addition, antibodies can be prepared against them, which are then employed for the accurate and specific immunochemical detection of appropriate haptens.

Proteins such as serum albumins, globulins, ovalbumin and fibrinogen belong to the carriers used most frequently for the preparation of conjugated antigens. The hapten is not bound to any site on the carrier, but only to certain groups of atoms, possessing a desirable reactivity. A classical protein carrier, bovine serum albumin (BSA) contains, according to ERLANGER (1973), 104 such reactive groups: ε-amino groups of lysine residues (59), α-amino group (1), hydroxyphenyl groups of tyrosine residues (21), sulphydryl groups of cysteine (6) and imidazole groups of histidine residues (17).

Even the hapten possesses reactive groups involved in the binding with the carrier. The spectrum of these groups is wider than on the protein carrier and, therefore, there are numerous possible reactions between the hapten and the carrier.

When the hapten possesses free carboxyl groups, these can bind to amino groups of the carrier in many ways. The reaction with mixed anhydride or carbodiimides is most frequent:

Reaction with mixed anhydride

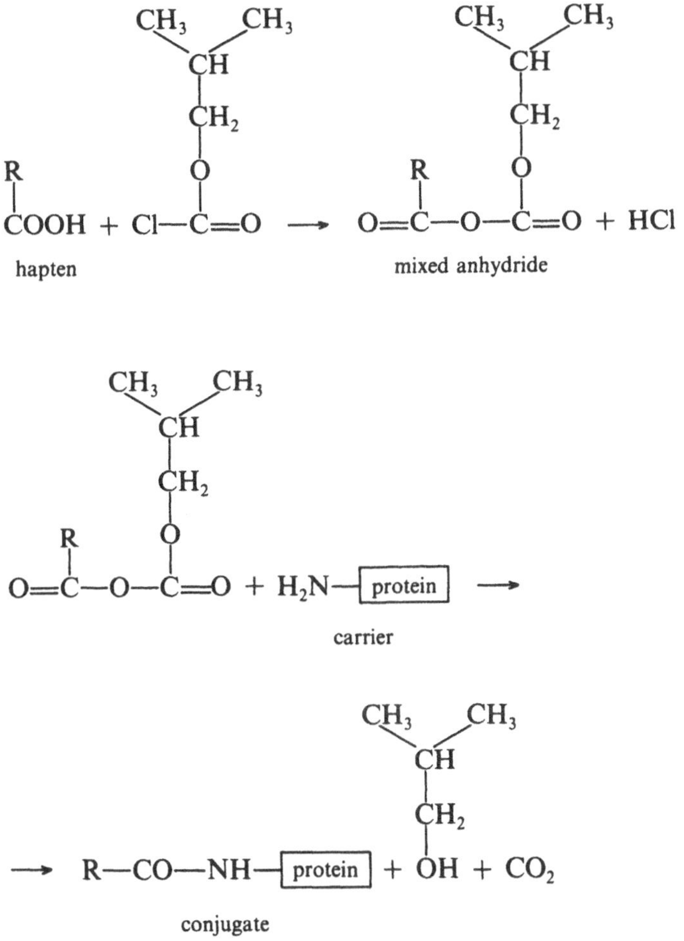

When serum albumin is used as a carrier, 15–30 hapten groups are bound.

Using the reaction with carbodiimides, protein is first bound to the carbodiimide molecule, and this intermediate product reacts with the hapten, possessing a free carboxyl group, and *O*-acylurea is generated. *O*-acylurea can be converted intramolecularly into *N*-acylurea. The latter compound reacts with the amino groups of the protein carrier, generating a conjugate:

$$R'-N=C=N-R' \xrightarrow{H^+} R'-\overset{+}{N}H=C=N-R'$$

carbodiimide

$$R'-\overset{+}{N}H=C=N-R' + R-CO-OH \longrightarrow$$

hapten

$$\longrightarrow R'-NH-\underset{\underset{N-R'}{\|}}{C}-O-CO-R + \overset{+}{H}$$

O-acylurea

$$R'-NH-\underset{\underset{N-R'}{\|}}{C}-O-CO-R$$

$$\downarrow$$

$$R'-NH-\underset{R-CO-}{C}=O \atop \underset{R-CO-N-R'}{} + H_2N-\boxed{\text{protein}} \longrightarrow R-CO-NH-\boxed{\text{protein}} + \overset{\text{NH}-R'}{\underset{\text{NH}-R'}{\overset{\text{CO}}{|}}}$$

N-acylurea   carrier   conjugate   substituted urea

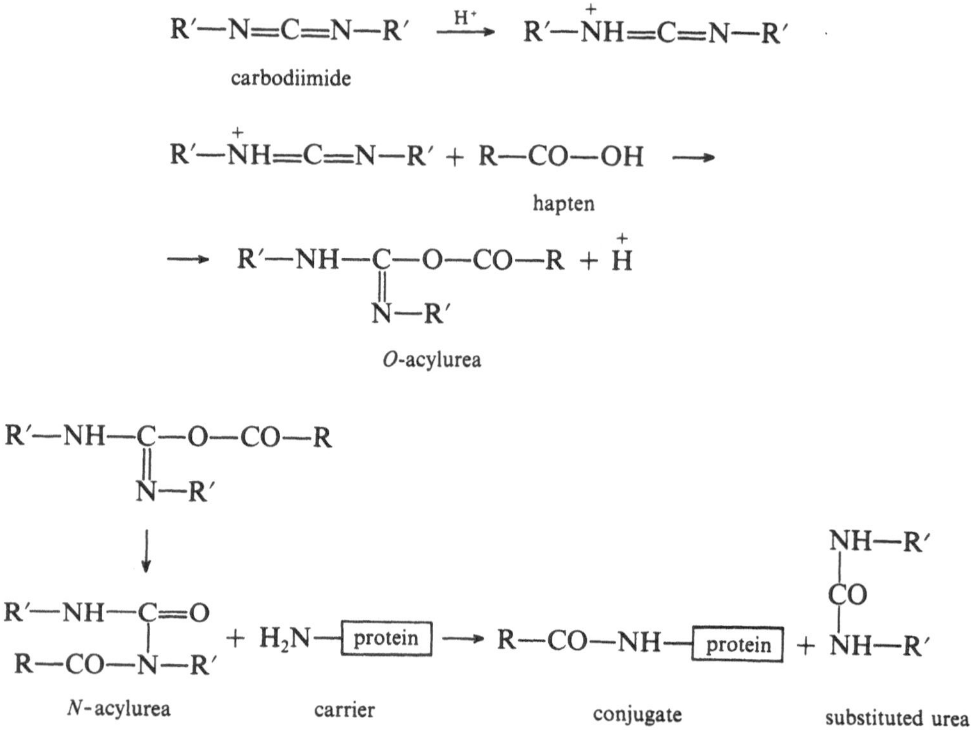

O-acylurea, however, may also react with the carboxyl group of a further hapten which results in generation of acid anhydride which, in turn, then reacts with the amino acid of the carrier:

$$R'-NH-\underset{\underset{N-R'}{\|}}{C}-O-CO-R + R-COOH \longrightarrow$$

$$\longrightarrow R-CO-O-CO-R + R'-NH-CO-NH-R'$$

acid anhydride

$$R-CO-O-CO-R + H_2N-\boxed{\text{protein}} \longrightarrow$$

$$\longrightarrow R-CO-NH-\boxed{\text{protein}} + R-COOH$$

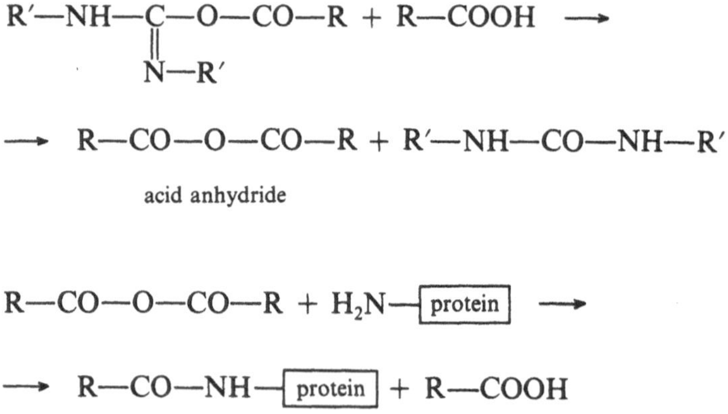

The reaction takes place in water provided that water-soluble carbodiimides, *e.g.* 1-ethyl-3-(3-dimethylaminopropyl)-carbodiimide hydrogenchloride (EDC),

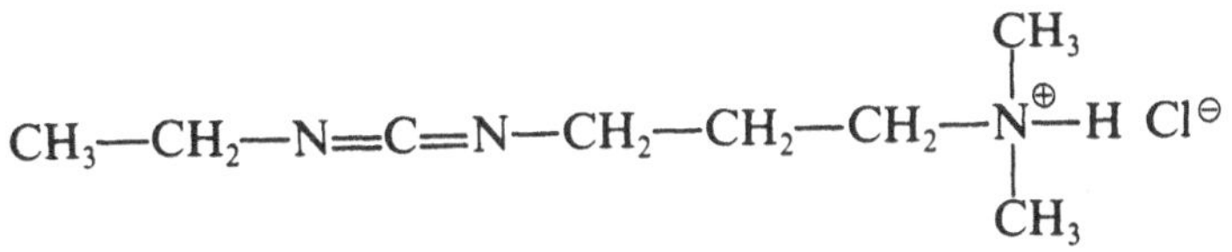

or 1-cyclohexyl-3-(2-morpholinoethyl)-carbodiimide-methyl-*p*-toluenesulphate (CMC)

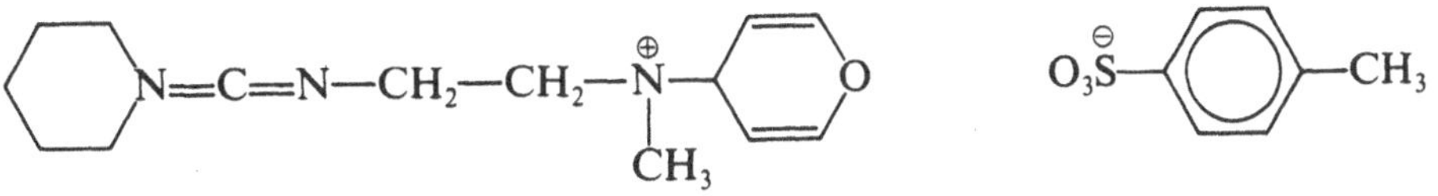

are used.

In addition to binding various low-molecular-weight substances, such as adenosine-3′,5′-monophosphate, morphine, lysergic acid diethylamide, serotonin, prostaglandins, barbiturates *etc.*, this technique permits the binding of numerous peptides (angiotensin, bradykinin, ACTH) to the protein carrier. It can even be used for the preparation of protein–protein conjugates. In such a case, a protein reacting similarly, *i.e.* through carboxyl groups, is used instead of the hapten.

Protein–protein conjugates may also be prepared with the aid of diisocyanates:

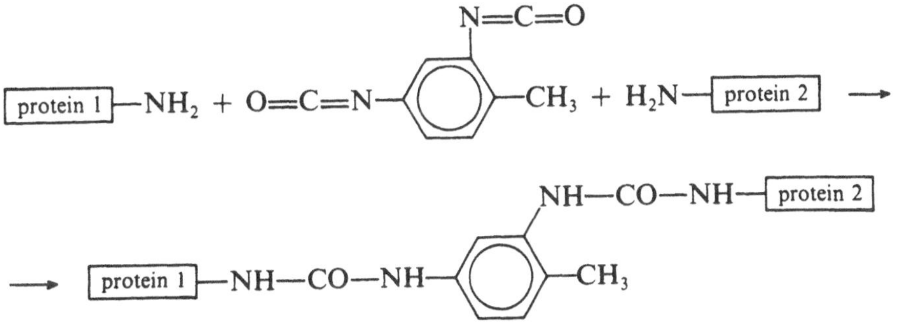

Protein–protein conjugates are covalently bound multimolecular complexes. Small proteins and polypeptides, *e.g.* insulin and cytochrome *c*, are conjugated with larger proteins in order to enhance their immunogenicity.

The second reactive group, through which haptens may bind to the carrier, is the amino group. Haptens having an amino group may be divided into two groups — aromatic and aliphatic amines.

Haptens with aromatic amino groups are conjugated with the protein carrier using the classical diazo-technique introduced by LANDSTEINER (1936). The principle of this method is based on the reaction of an aromatic amine with nitrous acid yielding the diazonium salt which then couples at weakly alkaline pH with tyrosine, histidine or lysine residues of the protein carrier molecule. In some cases, even the arginine and tryptophan residues may react. Azoproteins are generated by diazotation and coupling, *e.g.*:

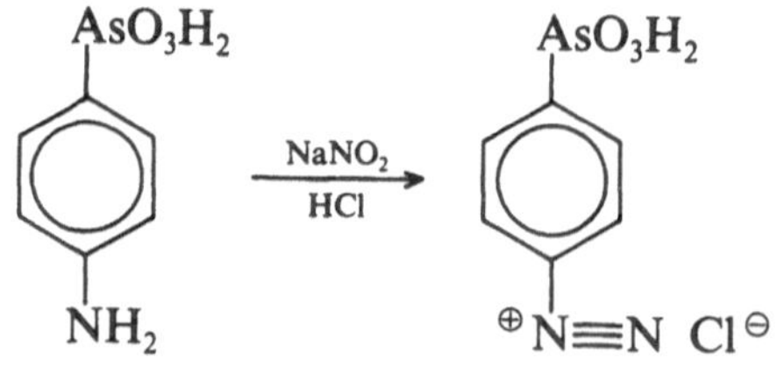

Arsanylic acid   Diazonium salt of arsanylic acid

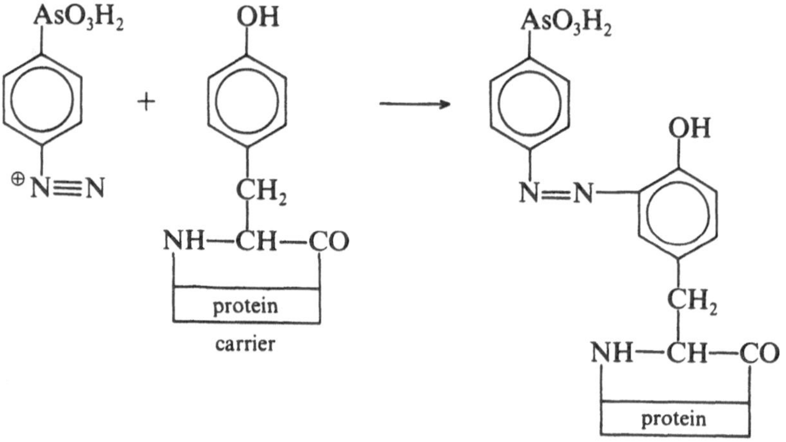

conjugate (azoprotein)

Aliphatic amines can be bound to the macromolecular protein carrier using several methods. First, there is the above-mentioned reaction with carbodiimides; furthermore, amines can be transformed by *p*-nitrobenzoyl-chloride to *p*-nitrobenzoylamides that are reduced to *p*-aminobenzoyl derivates and these are coupled to the protein carrier using the diazotization technique; this bond can be easily accomplished with glutaraldehyde:

$$\underset{\text{hapten}}{R{-}NH_2} + \underset{\text{glutaraldehyde}}{O{=}CH{-}(CH_2)_3{-}CH{=}O} \longrightarrow$$

$$\longrightarrow R{-}N{=}CH{-}(CH_2)_3{-}CH{=}O + H_2O$$

$$R{-}N{=}CH{-}(CH_2)_3{-}CH{=}O + H_2N{-}\boxed{\text{protein}} \xrightarrow{-H_2O}$$

carrier

$$\longrightarrow R{-}N{=}CH{-}(CH_2)_3{-}CH{=}N{-}\boxed{\text{protein}}$$

$$R{-}N{=}CH{-}(CH_2)_3{-}CH{=}N{-}\boxed{\text{protein}} \xrightarrow{NaBH_4}$$

$$\longrightarrow R{-}NH{-}CH_2{-}(CH_2)_3{-}CH_2{-}NH{-}\boxed{\text{protein}}$$

conjugate

A disadvantage of the reaction with glutaraldehyde is the production of undesirable by-products, *e.g.* homoconjugates and various polymers. Therefore, heterobifunctional substances of the maleinimide-*N*-hydroxysuccinimide ester type, forming bridges between the hapten and carrier, have recently been used to prepare quality conjugates suitable for enzyme immunoanalysis. Such a procedure is also suitable for the conjugation of two proteins. During this reaction, *N*-hydroxysuccinimide ester reacts first with the amino group of one protein or hapten, followed by reaction of the maleinimide component with a free sulphydryl group of another protein or hapten:

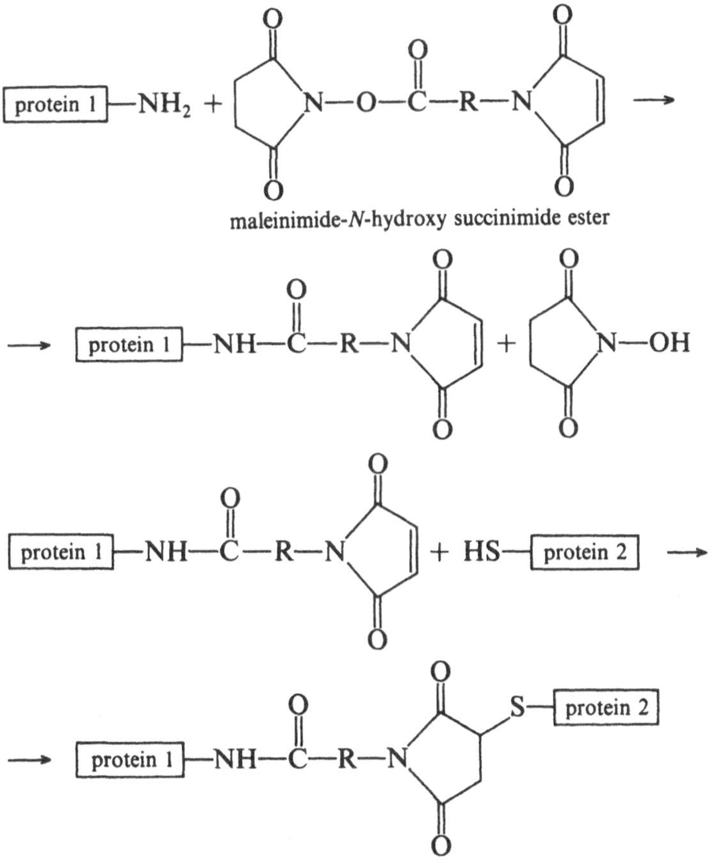

It is advantageous to use maleinimide-*N*-butyryloxysuccinimide ester (in the given formulas R = $-CH_2-CH_2-CH_2-$), a crystalline substance, which is soluble in aqueous buffer solutions.

In many cases, the protein to be conjugated does not contain a free sulphydryl group. However, this group can be introduced to the molecules of various proteins using *N*-succinimidyl-*S*-acetyl thioacetate (SATA) that reacts with protein amino groups and thereby yields acetylthio derivatives. Sulphydryl groups are then released following the reaction with hydroxylamine:

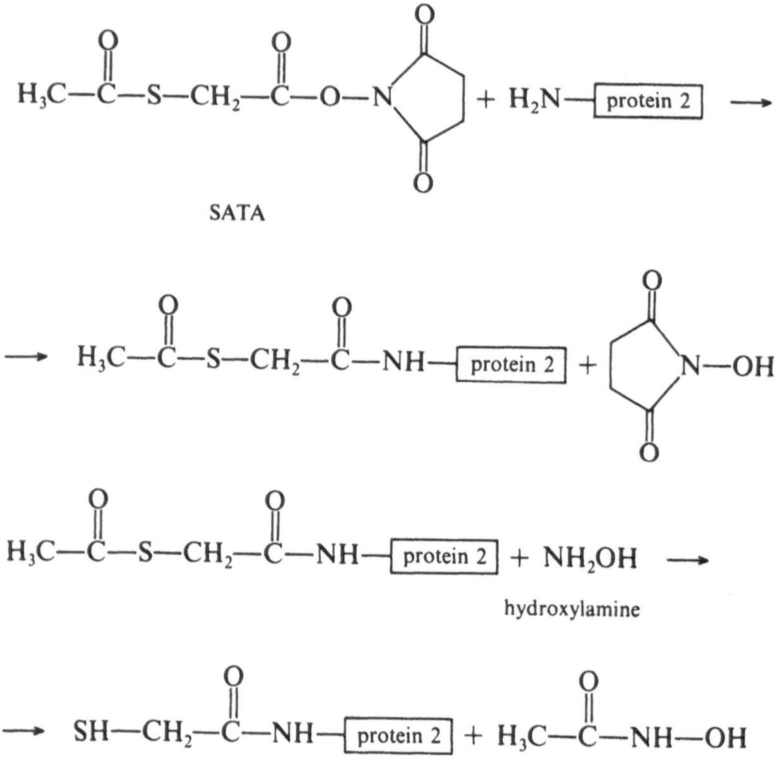

The hydroxyl group of haptens (alcohols, phenols, saccharides, nucleosides *etc.*) is relatively low-reactive, and must therefore be activated. Phenols may react with diazotized *p*-aminobenzoic acid and thus a carboxyl group, readily reacting with the carrier, is introduced to the molecule. Hydroxyl groups of saccharides and polysaccharides may be activated by the classical LANDSTEINER (1936) method using nitrophenyl derivatives, when *p*-nitrophenyl glycosides are produced; these are transformed to *p*-aminophenyl glycosides by hydrogenation and the latter compounds are bound to the carrier *via* coupling and diazotization. An alternative method employs cyanobromide; during this reaction dicarboxyimide reacting with the primary amino group of the carrier is produced:

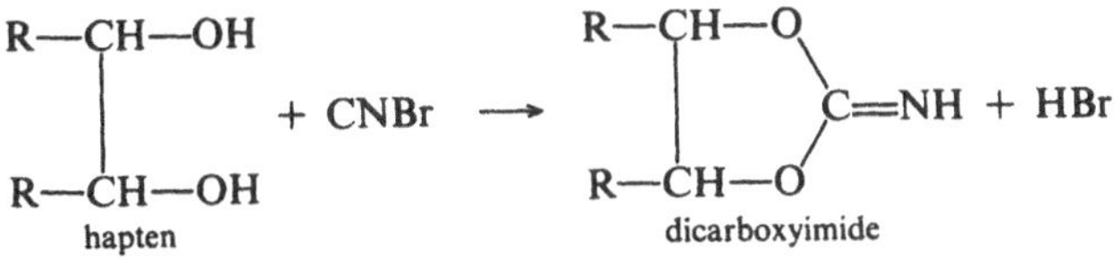

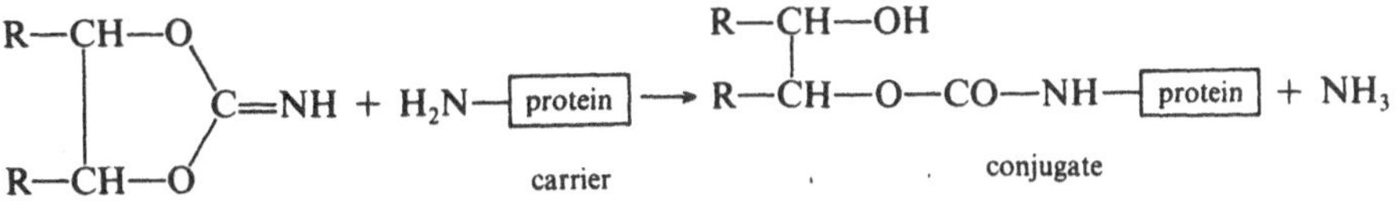

The cyanobromide method may also be used for binding a hapten with a primary amino group to the polysaccharide carrier. In this case, hydroxyl groups of the carrier are first activated by cyanobromide.

Nitrophenyl derivatives can be readily bound to natural or synthetic protein antigens. Thus, proteins with covalently bound 2,4-dinitrophenyl (DNP) or 2,4,6-trinitrophenyl (TNP) groups form effective immunogens in numerous animal species. The DNP and TNP groups bind *via* nucleophilic substitution to $—NH_2$, —OH and —SH groups of lysine and cysteine residues in protein molecules. The nucleophilic substitution is facilitated by the presence of an electron donor on C_1. Therefore, nitrobenzenes, substituted by a halogen on C_1, are more reactive than non-substitued nitrobenzenes:

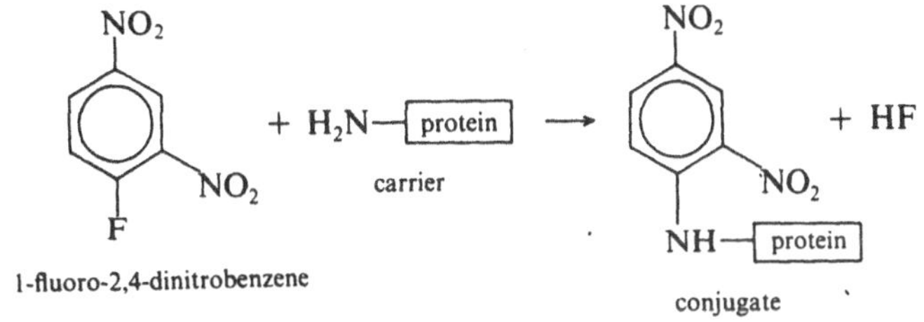

Nitrophenyl substituents may be bound to proteins even through the diazo reaction, *e.g.* DNP by means of diazo-2,4-dinitroaniline.

A rather simple method for the preparation of nucleoside- or nucleotide-protein conjugates is based on the reaction of two adjoining hydroxyl groups of a pentose (ribose or deoxyribose) with periodate. Thus, dialdehydes are generated which react with the carrier amino group; the originating aldimines are stabilized by reduction with sodium borohydride:

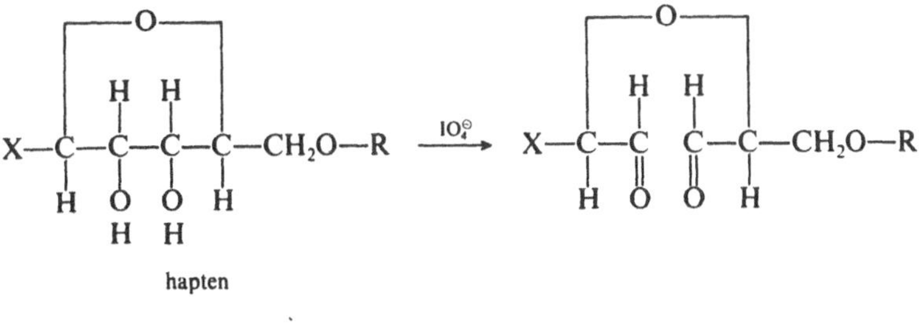

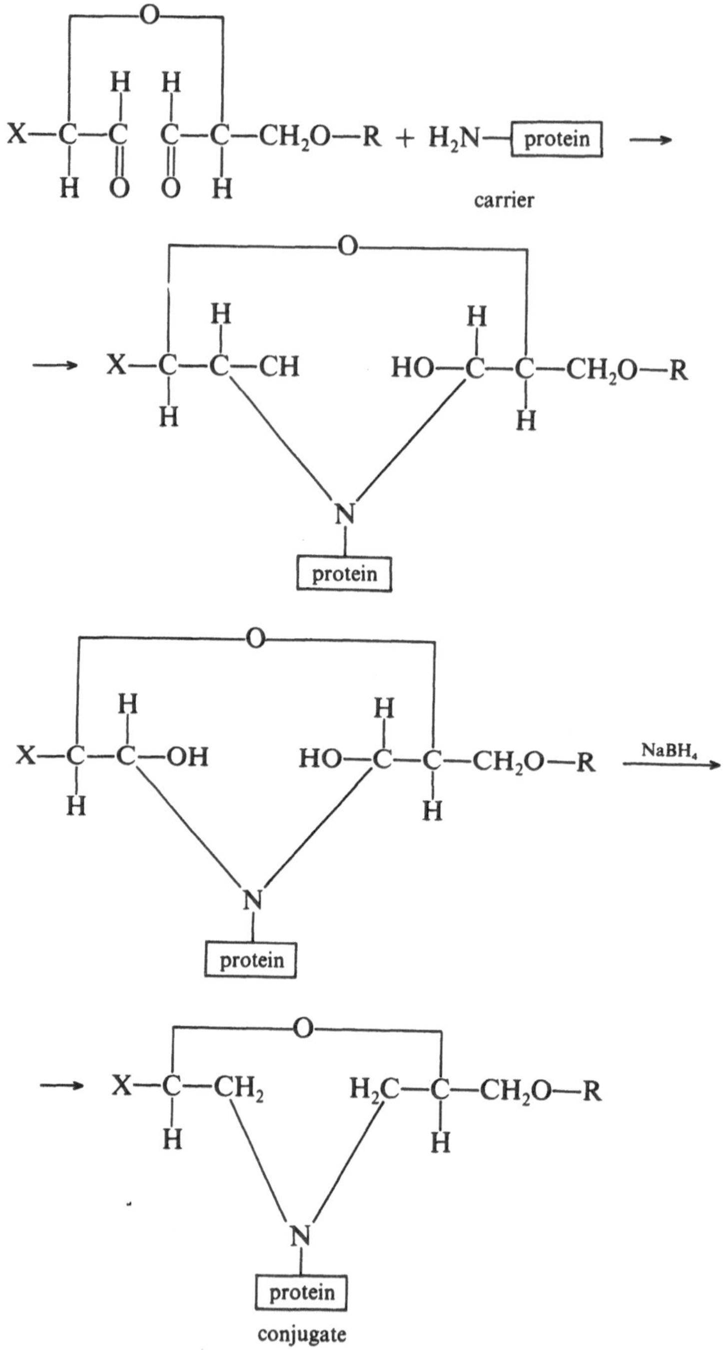

where X is a purine or pyrimidine base,

$R = -H$ or $-PO_3H_2$.

Using this method, 30 nucleotide or nucleoside determinant groups may, on average, be bound to one albumin molecule.

Immunochemical studies sometimes require a hapten reacting during conjugation with only one amino acid of the carrier rather than with numerous residues, as in most cases. An example of a suitable reagent is *S*-acetylmercaptosuccinate anhydride (SAMSA) which reacts only with free ε-amino groups of lysine residues. Using SAMSA, a thiol derivative of the protein can be prepared that may react with any hapten possessing the ability to react with sulphydryl groups.

During the reaction between proteins and fluorescein isocyanate, conjugates which fluoresce after exposure to ultraviolet light are produced. These can be used for the immunofluorescence determination of antigens and antibodies.

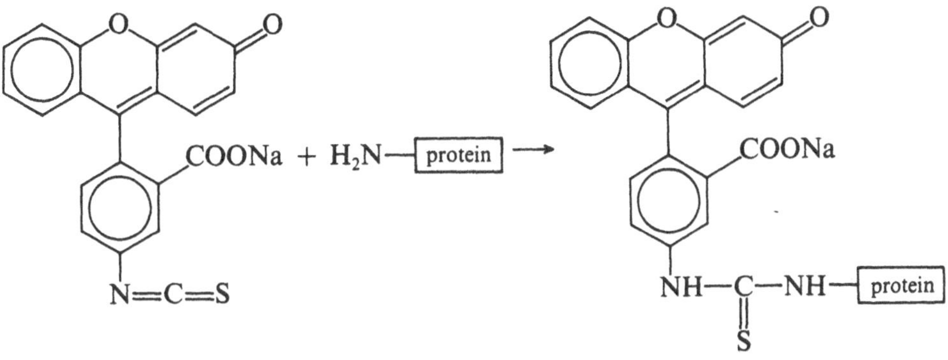

fluorescein isothiocyanate conjugate

4.5 Occurrence of the most important antigens and the principles of their isolation

The determination of many antigens is currently an important diagnostic tool for physicians, immunologists, microbiologists and virologists. The importance of antigens lies in the detection of bacterial, viral and other microbial antigens for the diagnosis of infectious diseases, prevention of certain diseases (vacccine production), determining the suitability of blood donors (AB0, Rh blood groups) or tissues for transplantation (histocompatibility antigens) and the investigation of autoimmune diseases *etc.* Antigens with a precisely known structure enable studies on their biological properties, mechanism of induction of the immune response, and antigen–antibody interaction. It is, therefore, not surprising that the antigen structure has recently been studied intensively.

In order to determine their structure, antigens must first be isølated in a pure form. The tissues, cells, various body fluids and secretions, which represent the basic material for antigen isolation, contain a wide spectrum of various antigens and secondary substances. Therefore, isolation of a single antigen from such mixtures is a tedious biochemical procedure.

Important antigens are present in bacterial cells and animal tissues. The bacterial cell contains several antigen types such as extracellular protein antigens, exopolysaccharides, flagellar antigens, antigens of bacterial cell walls, membranes and intracellular structures. These antigens are important not only for the physiology of the bacterial cell, but many of them participate both in host defence reactions as well as reactions that are harmful for the host.

The basic scheme of preparation of pure antigens depends on their localization in the bacterial cell. Bacteria must be cultured under conditions that permit the maximal production of the desired antigen. Extracellular protein antigens diffuse into the culture medium as products of bacterial cells. Exotoxins and extracellular enzymes belong to the most important bacterial products and can be separated from bacteria by centrifuging the liquid cultures. However, the bacterial cell synthesizes several extracellular products, and intracellular components may also be released into the culture medium. Therefore the extracellular material must be further fractionated (ion exchange or gel chromatography, electrophoresis on various media, isoelectric focusing, affinity chromatography) after separation by centrifugation. *Exopolysaccharides* are polysaccharides secreted by the bacterial cell across the cell membrane into the external environment. They are different from other polysaccharides, *e.g.* the lipopolysaccharides of Gram-negative bacteria, that are an integral component of cell walls. The method of preparation of pure exopolysaccharides depends on their physicochemical properties. They may appear in the form of free mucus or capsules demarcating the bacterial cells. Exopolysaccharides of pneumococci, *Klebsiella* and *Escherichia coli* are the best defined.

The main problem in the preparation of *flagellar antigens*, important for the serological determination of some species (*e.g. Salmonella*), is the separation of flagella from bacterial cells without contaminating the flagellar antigens with other components of the bacterial cell.

Cell wall antigens are important not only from the immunochemical point of view, but also for preventive vaccination, because some of them (*e.g.* from *Brucella abortus, Bordetella pertussis*) can protect experimental animals and man against infection. Isolation of antigens from the walls of Gram-positive bacteria and separation of individual cell wall components, is relatively easy, whereas in Gram-negative bacteria, it is very difficult. Separation can usually be accomplished only after selective degradation of some components using enzymatic or relatively drastic chemical methods. The cell walls are isolated from homogenates obtained by mechanical disruption of bacterial cells. The disruption can be accomplished in various ways: in glass homogenizers with teflon pistons, ultrasonic disintegration, shaking with small glass beads (ballotini), repeated freezing and thawing, osmotic lysis, *etc.* Bacterial cells are usually separated from other material present in the homogenate by simple differential centrifugation in a density gradient or in a two-phase liquid system.

Before preparing antigens from cytoplasmic membranes, the membranes must first be purified. Cytoplasmic membranes are easily isolated in lysozyme-sensitive bacterial species. This enzyme cleaves cell wall peptidoglycans into glycosaminepeptides (mucopeptides) and thus and osmotically fragile protoplast (in Gram-positive bacteria) or spheroplast (in Gram-negative bacteria) is generated. Lysis of bacteria (bacteriolysis) can be achieved by treatment with lysozyme or other proteolytic enzymes in an isotonic medium, which is then replaced with distilled water; the membrane fractions can then be isolated by centrifugation. In Gram-negative bacteria, however, the enzymatic degradation of the cell wall is incomplete and the membrane fractions are contaminated by cell wall components.

Intracellular antigens are obtained from cell residues after separation of exoantigens, cell walls and cytoplasmic membranes. The residues are separated by ultracentrifugation into a fraction containing the cell organelles and a soluble fraction. Individual intracellular antigens are obtained from these fractions by further purification.

Among higher plant antigens, *pollen antigens* causing various allergic diseases are particularly important. Even certain insect components, *e.g.* mites (*Acarina*) in house dust, may act as allergens.

Antigens of animal origin are isolated from the tissues and organs. Some antigens are present in all tissues of certain animal species or an individual, whereas others are organ or tissue-specific. Thus, they occur only in certain organs or tissues. A number of antigens can be obtained from body fluids, such as blood serum, cerebrospinal fluid, lymph, urine *etc.* Before purifying tissue antigens, the tissue must be disintegrated. Some organs, *e.g.* the liver, dissociate easily into cells after relatively mild homogenization; rather more vigorous methods must be used for some other organs. Thus the isolation of cells from organs with a high concentration of connective tissue requires treatment with proteolytic enzymes (*e.g.* trypsin). It should be noted, however, that such enzymes may damage the isolated cells. Fractionation of animal cells is performed by similar methods to bacterial cells. The first step is disintegration. The cell homogenate is then centrifuged at various centrifugal forces (*g*), which allows the separation of individual fractions of cell organelles: nuclear and membrane fraction, the mitochondrial and lysosomal fraction, ribosomal and soluble fraction; the latter being the cytoplasmic and soluble components. These basic fractions can be further separated using density gradient centrifugation, or by other methods. Antigens are then prepared from a particular fraction, either by direct extraction, or after disintegration of isolated organelles (nuclei, mitochondria, lysosomes). After the fractionation, many antigens remain insoluble because they bind to membrane residues. In such cases the antigens can be released into solution using various detergents (*e.g.* Triton X-100), or by limited proteolysis. Each cell fraction, however, contains a highly heterogeneous mixture of antigens and other substances. A pure antigen is isolated from this mixture using various biochemical techniques for separation of macromolecules, particularly proteins. Recently, various types of affinity chromatography, including immunoaffinity chromatography (immunoadsorption), have been used for this purpose.

4.6 Endotoxins of Gram-negative bacteria

Macromolecular substances present in cell membranes belong to the most important bacterial antigens. These antigens not only allow serological typing of individual species and strains of bacteria, but some of them participate in various biological activities, including their pathogenic effects. According to the composition of the cell membrane and its reaction to Gram staining, two main groups of bacteria — Gram-positive and Gram-negative — can be discriminated (*Fig. 4.4*).

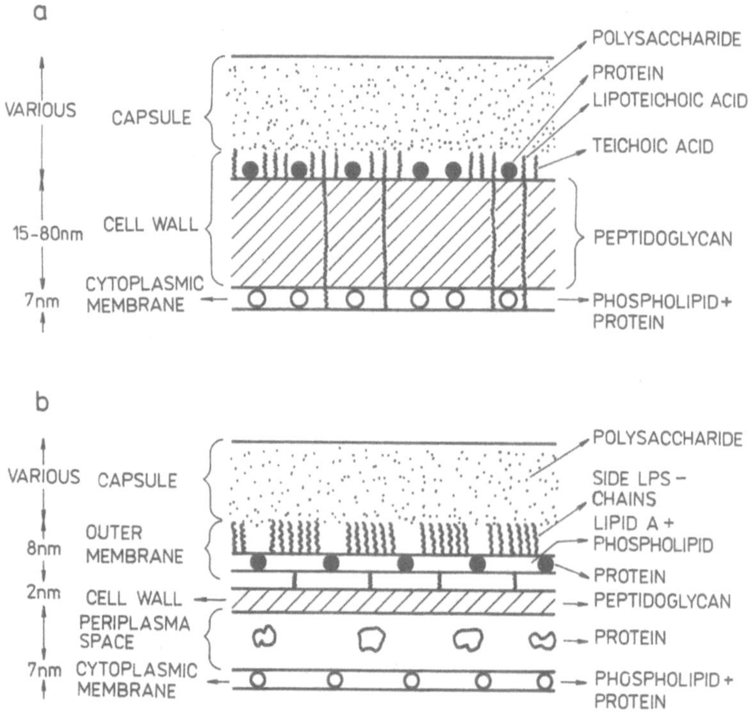

Fig. 4.4. The cell envelope of (a) Gram-positive and (b) Gram-negative bacteria.
The capsule need not be always present (adapted from Taussig, 1984).

Gram-negative bacteria have a more complex cell membrane composed of four layers (from the inside): the cytoplasmic (inner) membrane, the periplasmic space, a very thin peptidoglycan cell wall and an outer membrane, usually closely associated with the cell wall. Behind the outer membrane, the capsule may be localized. The outer membrane contains lipopolysaccharide (LPS), phospholipid and membrane proteins, acting as receptors, and molecular pores (porins) for the non-selective transport of small molecules from the extracellular solution. The outer membrane and peptidoglycan of the cell wall are closely connected through divalent cations (Mg^{2+}) and specific lipoproteins (*Fig. 4.5*). The periplasmic space contains several hydrolytic enzymes and specific transport proteins participating in the chemotactic movement of the bacterial cell. In contrast to other components of the cell envelope, the structure of the periplasmic space is not visible on electron microscopy.

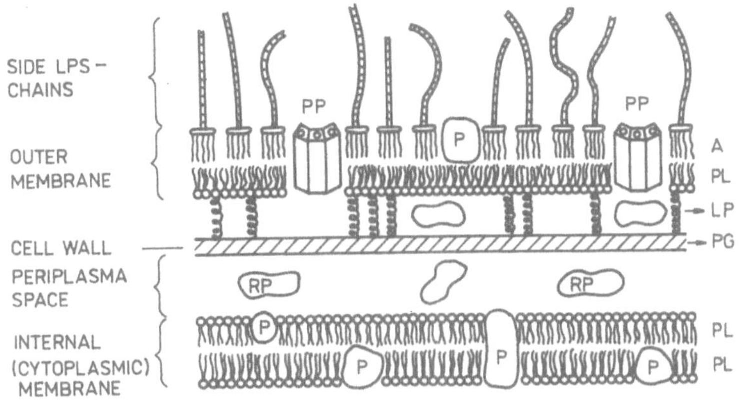

Fig. 4.5. Detailed structure of the Gram-negative bacterial coat (according to Taussig, 1984). PP, porin proteins (trimer); P, membrane protein; SP, soluble protein; A, lipid A; PL, phospholipid; LP, lipoprotein; PG, peptidoglycan.

The *lipopolysaccharide* is a permeable barrier for the transport of substances from the cell into the external environment and *vice versa*. With respect to its biological activity, it is called Gram-negative bacterial endotoxin. Thus, the terms LPS and endotoxin are synonymous, although the former expresses the chemical character and the latter the functional activity of this substance. It should be stressed, however, that endotoxin, despite its name, is not an intracellular bacterial toxin, but is actually a complex firmly bound to its outer layer.

The term *endotoxin* is generally used to refer to the thermostable polysaccharide toxin, firmly bound to the bacterial cell, in contrast to the thermolabile protein *exotoxin*, secreted into the external environment.

Endotoxin is responsible for many pathophysiological symptoms observed during Gram-negative bacterial infections. These reactions are mostly elicited by free endotoxin, liberated after bacterial lysis by host defence mechanisms, or during their artificial disintegration. Free endotoxin is also released from growing cultures of bacteria into the culture medium and may exhibit chemical and biological properties different from that of the membrane-bound endotoxin. Endotoxins with different compositions can be produced by various bacteria, particularly *Enterobacteriaceae* (*Salmonella, Shigella, Escherichia, Proteus, Pseudomonas, Klebsiella, Pasteurella ect.*).

LPS is the main component of the somatic *O-antigen* of Gram-negative bacteria. It contains various determinants, governing the specificity of the whole *O*-antigen, which contains besides LPS, phospholipid (*lipid B*) and protein. LPS is a typical amphipathic molecule, having both hydrophobic and hydrophilic portions. The hydrophilic portion is formed by a polysaccharide chain, whereas the hydrophobic end of the molecule is represented by *lipid A* (*Fig. 4.6*). The polysaccharide can be divided into a polysaccharide chain with *O*-specificity and core oligosaccharide (*R*-specific core).

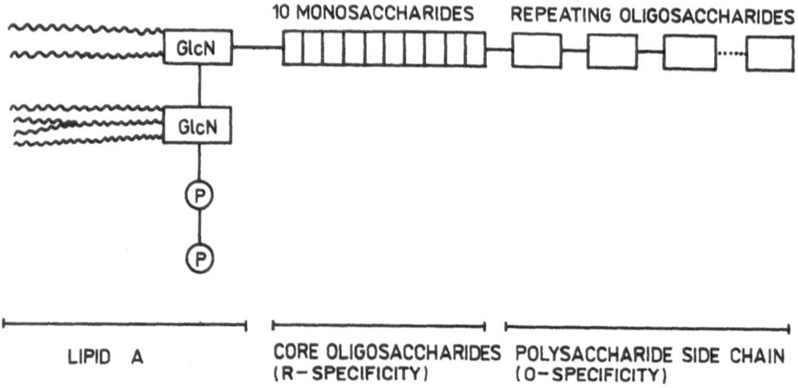

Fig. 4.6. Schematic structure of the LPS molecule.

After administration into an organism, LPS elicits various biological reactions with a significant participation of its lipid component. In addition, LPS induces an immune response in which the antigenic determinants, localized in its polysaccharide portion, play an important role. *Lipid A*, however, also has antigenic determinants.

Of the biological activities of endotoxin, the following should be mentioned: pyrogenicity (the ability to cause an increase in body temperature), changes in the number of circulating leukocytes (leukocytopenia, leukocytosis), complement activation *via* the alternative pathway, activation of macrophages, aggregation of platelets, mitogenic effects, adjuvant effects, induction of interferon formation, release of interleukin 1, stimulation of non-specific anti-infectious resistance, tumour necrosis (by direct effect on tumour cell or mediated by tumouricidal factors released from macrophages) and others. Administration of a higher dose of endotoxin may produce lethal shock.

From the immunological point of view, endotoxins belong to a widely distributed group of antigens that come into contact with the warm-blooded animals immediately after birth — during bacterial colonization of their intestinal tract. Indeed, a prerequisite for survival of these animals and man appears to be the formation of effective defence and detoxification mechanisms against endotoxin. On the other hand, endotoxins stimulate formation of antibodies with a strong opsonizing effect and, together with complement and lysozyme, they can kill bacteria containing endotoxin in their cell walls.

The physical properties and chemical composition of endotoxins depend on the species and strain of bacteria employed for their preparation. Thanks to the works of Westphal (1975), Westphal *et al.* (1983), Lüderitz *et al.* (1978, 1982), Galanos and Lüderitz (1984) and many others, the structure of endotoxins prepared from *Salmonella* and some other enterobacteria has been well defined.

The side polysaccharide chain, located at the outer end of LPS, is composed of 40 repeating oligosaccharides and each of them has between three and five monosaccharide units. The length of side chains may be different, even in the same bacterial cell. The composition of oligosaccharide units is characteristic for individual genera and species of Gram-negative bacteria.

Antibodies, induced by immunization with endotoxin, have specific combining sites for these oligosaccharide units indicating that they contain the antigen determinants. The existence of these *O*-specific determinants is the basis for serological classification of many enterobacteria. For example, the well-known Kauffmann–White scheme allows the classification of several serologically defined groups (serotypes) of *Salmonella*.

Common monosaccharides, *e.g.* glucose, galactose, rhamnose, glucosamine, galactosamine, are the constituents of the LPS polysaccharide component; in addition, the presence of various 3,6-didesoxyhexoses (abequose, colitose, tyvelose, 2-O-acetyl-abequose) and uronic acids, can be demonstrated. In each oligosaccharide unit, one monosaccharide can be considered to be immunodominant. Such immunodominant monosaccharides contribute substantially to the total binding affinity between the oligosaccharide determinant and binding site of the specific antibody.

In *Salmonella*, the oligosaccharide which is responsible for the *O*-specificity most frequently contains the trisaccharide mannose-rhamnose-galactose (Man-Rha-Gal). The repeating oligosaccharide unit of *Salmonella*, belonging to group B with *O*-antigen factors 4 and 12 according to the Kauffmann–White scheme, has the following composition:

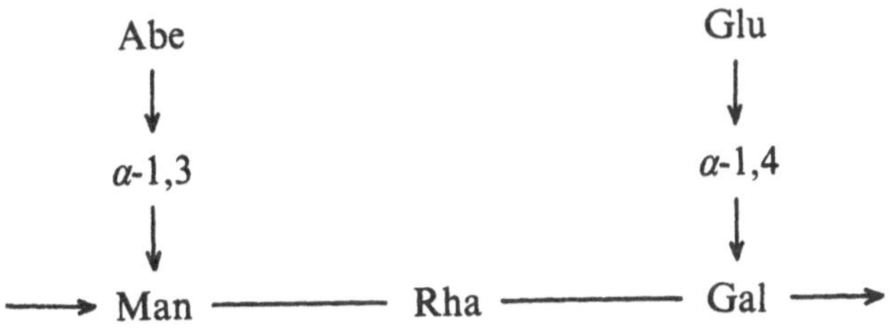

In this oligosaccharide, abequose (Abe) binds by 1,3-α-glycosidic bond to mannose, which represents an immunodominant monosaccharide for factor 4. Several tens of such specific factors can be demonstrated in *Salmonella*.

Oligosaccharide units with *O*-specificity are synthesized only in cell walls of bacteria, growing in *S-phase*. However, they are not present in LPS isolated from *R*-strains. Endotoxins of strains growing in *R-phase* contain only certain basic saccharides, and their chemical structure is simpler and their antigenic structure similar.

Colonies of bacteria growing in *S-phase* have a smooth, glossy shape, whereas colonies in *R-phase* have a rough surface with irregular margins. They are often radially wrinkled with elevated centres. These two colony types differ not only morphologically, but they also exhibit a different virulence. The *R*-mutants of original *S*-strains are generally non-pathogenic.

Antisera agaist LPS, prepared from *R*-strains, do not possess an interspecies specificity. In the whole genus *Salmonella*, for example, only two serologically different antigen structures are found in *R*-strains.

It therefore follows that the outer moiety of the LPS chain is responsible for the *O*-specificity, whereas the inner part represents the *R*-specific core. This *R*-specific oligosaccharide in endotoxins from *Salmonella* is composed of 10 monosaccharides; in addition to glucose, galactose and *N*-acetylglucosamine they contain a heptose and a C_8 monosaccharide 2-keto-3-deoxyoctonic acid (KDO) which mediates the binding of the core oligosaccharide to *lipid A*. The core oligosaccharide is acidic, since its monosaccharides are phosphorylated.

The lipid component of bacterial lipopolysaccharides is formed by *lipid A* with a basic polysaccharide chain containing a number of fatty acid branches. The building unit of the *lipid A* polysaccharide chain is a disaccharide — 4-phosphoglucoseaminyl-β-1,6-glucoseamine-1-phosphate. In each subunit, the amino groups are substituted by 3-hydroxymyristic acid and the hydroxyl groups are esterified by lauric, myristic, palmitic or another molecule of 3-hydroxymyristic acid. One structural subunit of *lipid A* is composed of two moles of glucosamine, two moles of phosphoric acid and six moles of saturated fatty acids. In addition, it contains phosphorylethanolamine and 4-amino-4-deoxy-L-arabinose.

4.7 Erythrocyte antigens

Every animal cell has numerous antigen markers on its surface. Some of them mark their relationship with a given animal species, others are organ- or tissue-specific, some can be found in few individuals of a given species only and some are characteristic for a single individual only. From the chemical point of view, these antigens are glycoproteins, proteins or glycolipids and are integral parts of the cytoplasmic membrane. They belong to various systems; the membrane antigens on the erythrocyte (red blood cell) surface being the first recognized. These erythrocyte antigens have been traditionally known as *blood groups* despite the fact that some of them occur also in other, even plant, cells.

The first such system of three markers was first described in 1901 by LANDSTEINER, and designated AB0. The whole AB0 system was independently described in 1907 by the Czech psychiatrist JAN JANSKÝ. The *AB0 system* is composed of four groups — 0 ("zero"), A, B and AB which are characterized by antigen determinants present on the surface of erythrocytes. Antibodies can be formed against these determinants. Erythrocytes of a particular individual always possess one of these groups only. As a rule, the serum of an individual does not contain antibodies against erythrocytes of their own blood group. However, antibodies against red blood cells of other blood groups (*i.e.* alloantigens) are present. When erythrocytes are mixed with serum containing the corresponding anti-erythrocyte antibodies, agglutination occurs. The possible combinations of reactions are shown in *Table 4.2*. This scheme demonstrates that group 0 erythrocytes are not agglutinated by serum of any other blood group. Conversely serum from group 0 individuals contains both anti-A and anti-B antibodies and therefore agglutinates erythrocytes of all other blood groups. This means that an individual with blood group 0 can be used as a "universal donor" (his blood can be transfused to recipients of any of the four basic AB0 groups). However, such an individual may only receive blood from a donor of group 0. In addition to group 0 serum, group A red blood cells are also agglutinated by serum from group B (containing anti-A antibodies) and group B erythrocytes are similarly agglutinated by serum from group A subjects. Group AB erythrocytes are agglutinated by all antisera with the exception of that from their own serum. There are several subgroups of blood group A, the most important being A_1 and A_2.

Table 4.2 Reaction of erythrocytes with antisera against the AB0 blood groups

Erythrocytes	*Antiserum from groups*			
	0	*A*	*B*	*AB*
0	–	–	–	–
A	+	–	+	–
B	+	+	–	–
AB	+	+	+	–

It was originally presumed that group 0 red blood cells do not contain an agglutinogen (giving rise to agglutination after reaction with the corresponding antibody) and this is why they were designated "zero". It was found later, however, that they do, in fact, contain a weak agglutinogen, now designated H.

Antigenic markers of the AB0 system are determined in humans by two genes. These genes have three alleles, A, B and 0; alleles A and B are inherited codominantly, whereas the allele 0 is recessive. Because of this fact, there are six basic genotypes, but only four basic phenotypes in the AB0 system (*Table 4.3*).

Table 4.3 Basic phenotypes and genotypes of the AB0 blood group system

Phenotypes	*Genotypes*	
	Homozygous	*Heterozygous*
0	00	—
A	AA	A0
B	BB	B0
AB	—	AB

Antigens bearing markers of the AB0 system are glycoproteins and their determinants are formed by oligosaccharide chains. Antigen H has the basic function, which is present, with one exception (so-called "Bombay blood"), in all red blood cells, *i.e.* even in erythrocytes of the 0 phenotype. Antigen H is generated from a precursor or pre-antigen H under the control of the H gene. Gene H is amorphous, because it does not actively influence the structure and modification of the H antigen. This ability, however, is exhibited by the A and B genes, which regulate complementation of oligosaccharide chains of the H antigen, so that this antigen acquires the specificity of A or B. The oligosaccharides are composed of glucose, galactose, fucose, *N*-acetylglucosamine and *N*-acetylgalactosamine. The immunodominant monosaccharide of the H antigen is L-fucose, of the A antigen it is D-*N*-acetylgalactosamine and of the B antigen it is D-galactose.

The blood serum of any individual contains antibodies against those antigen markers of the AB0 system, that are absent on his own red blood cells. These are alloantibodies (they react against alloantigens) and are also referred to as *isohaemagglutinins* in immunohaematology. Originally, they were considered to be natural antibodies; however, it is now known that they are formed as a response to immunization with the A and B antigens in the intestine, because these antigens are present on the cell surface of other animals and plants contained in human food, as well as on microorganisms, colonizing the intestinal tract. These regularly occurring antibodies belong to the IgM class. In addition, sera of some individuals may also contain immune isohaemagglutinins, belonging to the IgG class, which have originated in response to transfusion of AB0-incompatible blood.

After transfusion of incompatible blood, haemolysis of donor red blood cells occurs in the recipient, which may cause serious clinical complications. Thus, compatibility within the AB0 system is more important than other blood group systems, not only for blood transfusion, but also for successful organ transplantation.

Besides the AB0 system, the saccharide structure also contains antigen determinants of other systems, *e.g.* the P, Lewis and Lutheran systems. In contrast, determinants of the rhesus (Rh) and MNSs systems are probably composed of amino acid residues only.

Of these systems, the *Rh system* is the most important. By immunization of rabbits with erythrocytes of the monkey *Macaca mulatta* (originally *Macaca rhesus*), antiserum is obtained that agglutinates most human red cells. They were therefore divided into Rh-positive and Rh-negative (Rh from "rhesus"). In our population, approximately 85% of individuals have Rh-positive erythrocytes, the rest are Rh-negative. A more detailed analysis revealed that the Rh system is not simple. There exist several antigenic determinants controlling its specificity (C, D, E, c and e). By their combination, over 30 phenotypes are generated. For common transfusion practice, only Rh-positivity (Rh+) and Rh-negativity (Rh−) are determined. Rh+ is also designated according to the immunodominant determinant termed D. The Rh antigens are only present in erythrocytes.

In contrast to the AB0 system, no natural alloantibodies are formed against determinants of the Rh system. Immune anti-Rh antibodies appear only in sera of Rh-negative individuals immunized with Rh+ red blood cells, *e.g.* in women, immunized during pregnancy by erythrocytes of the Rh+ foetus. Anti-Rh antibodies are the main cause of the haemolytic disease of the newborn and a frequent cause of adverse post-transfusion reactions.

References

Erlanger, B. F. (1973) Principles and methods for the preparation of drug protein conjugates for immunological studies. *Pharmacol. Rev.*, **25**, 271–80.

Galanos, C. and Lüderitz, O. (1984) Lipopolysaccharide: properties of an amphipathic molecule. In: Rietschel, E. J. (ed.), *Handbook of Endotoxin. Vol. 1. Chemistry of Endotoxin*, Elsevier Sci. Publ., pp. 46–58.

Janský, J. (1907) Haematologic studies in psychiatric patients. *Klinický zborník* (Prague), **8**, 85–139. (*in Czech*).

Landsteiner, K. (1901) Ueber Agglutinatioserscheinungen normalen menschlicher Blutes. *Wiener Klin. Wchschr.*, **14**, 1132–4.

Landsteiner, K. (1936) *Specificity of Serological Reactions*. Springfield, IL, Charles C. Thomas, 178 pp.

Lüderitz, O., Freudenberg, M. A., Galanos, C., Lehmann, V., Rietschel, E. T. and Shaw, D. H. (1982) Lipopolysaccharides of gram-negative bacteria. In: Razim, S. and Rottem, S. (eds.), *Current Topics in Membranes and Transport, Vol. 17: Membrane Lipids of Prokaryotes*, New York, Acad. Press, pp. 75–151.

Lüderitz, O., Galanos, C., Lehmann, V. and Mayer, H. (1978) Chemical structure and biological activities of lipid A's from various bacterial families. *Naturwissensch.*, **65**, 578–85.

Pauling, L. (1940) A theory of the structure and process of formation of antibodies. *J. Am. Chem. Soc.*, **62**, 2643–57.

Sela, M. (1969) Antigenicity: some molecular aspects. *Science*, **166**, 1365–8.

Sela, M. (1971) Effect of antigenic structure on antibody biosynthesis. *Ann. N. Y. Acad. Sci.*, **190**, 181–202.

Sela, M., Fuchs, S. and Arnon, R. (1962) Studies on the chemical basis of the antigenicity of proteins. 5. Synthesis, characterization and immunogenicity of some multichain and linear polypeptides containing tyrosine. *Biochem. J.*, **85**, 223−31.

Sela, M. and Mozes, E. (1969) Dependence of the chemical nature of antibodies on the net electrical charge of antigens. *Proc. Natl. Acad. Sci. USA*, **55**, 445–51.

Westphal, O. (1975) Bacterial endotoxins. *Int. Archs. Allergy Appl. Immunol.*, **49**, 1–43.

Westphal, O., Jann, K. and Himmelspach, K. (1983) Chemistry and immunochemistry of bacterial lipopolysaccharides as cell wall antigens and endotoxins. *Progr. Allergy*, **33**, 9–39.

5 The major histocompatibility complex

5.1 What are the histocompatibility (transplantation) antigens?

It has long been known that the fate of transplanted foreign cells, tissues and organs depends on genetic factors. Basic experiments concerning this problem were performed in laboratory animals, particularly in mice. The genetic basis of tissue incompatibility (histoincompatibility) was demonstrated in *inbred animal strains*. These are homogeneous animal lines, generated by repeated crossing of brothers and sisters or parents and their progeny over many generations. When the transfer of tissue is performed between individuals of various inbred strains, the graft is generally rejected. Similarly, rejection occurs after tissue transplantation between individuals of normal non-inbred (outbred) populations, *e.g.* in the human population. On the other hand, grafts transferred between individuals of the inbred line or between monozygotic twins in man, survive permanently. It has been shown that particular antigens are responsible for tissue compatibility (histocompatibility) or incompatibility (histoincompatibility) and determine rejection or survival of the graft. These antigens are products of histocompatibility genes and are therefore called **histocompatibility antigens**.

The attempts to replace (transplant) skin were an important stimulus which led to the discovery of histocompatibility antigens. P. B. MEDAWAR, a Nobel prize winner in 1960, studying skin-transplanted soldiers who had suffered various wounds and burns during World War II, showed that autologous skin grafts (taken from their own body) were accepted and survived, whereas grafts obtained from other individuals — donors — were regularly rejected within about four weeks. After the war, BILLINGHAM *et al.* (1954) studied this phenomenon experimentally. They employed two strains of mice — black (strain C_{57}) and white (strain A). They established that the skin graft, taken from black mice and transplanted to white mice, was rejected in approximately 10 days. However, when the white mice had been immunized with a skin homogenate or homogenate of spleen or lymph nodes obtained from black mice prior to skin transplantation, the skin grafts were rejected faster (within 3–7 days). This experiment demonstrated that the rate of rejection is associated with specific immune processes.

In 1958, DAUSSET, VAN ROOD *et al.*, and PAYNE and ROLFS showed that sera of women with multiple pregnancies contained antibodies against human leukocytes despite compatibility in the AB0 and Rh systems. The formation of these antibodies in mothers is a result of alloimmunization during pregnancy.

Antigens of paternal origin, present in foetal leukocytes, stimulate formation of specific antibodies in the maternal organism. These antibodies are rarely of pathogenic importance, but their discovery permitted serotyping of some histocompatibility antigens, present on the surface of leukocytes or other cells.

The term "histocompatibility antigens" was coined by SNELL in 1948. J. DAUSSET and G. D. SNELL were awarded the Nobel prize in 1980 for original discoveries in the field of histocompatibility antigens, together with B. BENACERRAF.

The histocompatibility antigens (designated H) are typical alloantigens. Individuals in the population can be differentiated according to the composition of H antigens on their cells, which is the basis for rejection of grafts after allotransplantation of tissues and organs. **Allotransplantation** is the transfer of a graft between two genetically different individuals of the same biological species. The graft is in this case called an *allotransplant* or *allograft* (*Table 5.1*). In most tissues and organs, the transplantation can only be successful if the donor has identical, or at least highly related, histocompatibility antigens to the recipient. Thus, histocompatibility antigens are responsible for the transplantation reactions and were, therefore, originally called *transplantation antigens*. It is known, however, that H antigens have a much wider and more important function than just their participation in transplantation reactions.

Table 5.1 Terminology of transplantation immunology

Graft designation	*Type of tissue*	*Receipient–donor genetic relation*
Autograft	Autogenous Autochtonous (autologous)	The same individual (donor = recipient)
Isograft	Syngeneic Isogeneic (isologous)	Genetically identical individuals of the same species (homozygous twins, animals of the inbred strain)
Allograft	Allogeneic (homologous)	Genetically different individuals of the same species
Xenograft	Xenogeneic (heterologous)	Individuals of two different biological species

Obsolete terms are given in parentheses.

The **transplantation reaction** is a complex response of an immunologically mature organism to foreign histocompatibility antigens present in the graft of a genetically different individual. This reaction is manifested in two ways:
1. graft rejection (host-versus-graft reaction);
2. graft-versus-host reaction (GVHR).

In most cases, the reaction of the recipient against the graft predominates. However, if the recipient is incapable of an immunological reaction (embryos, newborns, or individuals influenced by immunosuppressive treatment), the reaction may proceed in the opposite direction in which immunocompetent cells, present in the graft, react against the host (recipient). In both reactions, the immune response is mediated by specific antibodies against transplantation antigens and effector lymphocytes that destroy the graft directly by cytotoxicity or by lymphokines. The memory cells, maintaining the state of transplantation immunity, are generated simultaneously.

The histocompatibility antigens are products of histocompatibility genes, found in multiple systems. Their presence has been detected in all animal species examined so far. It is estimated that vertebrates possess over 30 such systems. One of them, however, occupies a dominant position in all species and is termed the **major histocompatibility system (MHS)** or **major histocompatibility complex (MHC)**. In man, this complex is called HLA (Human Leukocyte Antigens), in mice the H-2 complex and in chickens, the B-complex *etc.* (*Table 5.2*).

Table 5.2 Major histocompatibility complexes in man and in some animals

Species	*Designation*	*Gene regions*	*Total number of known alleles*
Man	HLA	DP-DQ-DR — B — C — A	over 100
Chimpanzee	ChLA	D — B — A	about 15
Cattle	BoLA	D_1, D_2 — A	about 15
Pig	SLA	D/DR — B — C — A	about 40
Dog	DLA	D — B — C — A	about 25
Rabbit	RLA	D — B — A	about 20
Guinea-pig	GPLA	I — B	about 10
Rat	RT1	B — A	about 25
Mouse	H-2	K — I — D — L -Qa-Tla	about 150
Chicken	B	B-F — B-G	over 10

During studies on the function of H antigens, Snell (1958, 1976) devised a method for obtaining strains of mice that are different in a single gene H only — *the congenic strains*. It is thanks to these strains that the mouse became the most widely used experimental model in immunogenetics and immunobiology. Snell observed that, in skin transplantation between mice of two congenic strains, the different H genes are not equally important for the speed of graft rejection. Indeed, the gene coding for antigen II is more important in the rate of graft rejection, and in 1956, the whole gene complex was termed H-2. The differences in H-2 antigens between the donor and recipient caused graft rejection within two weeks after transplantation, whereas if other H genes were different, graft rejection was delayed or the graft survived. Therefore, the system of major H genes is called the "major histocompatibility complex". Because of its obvious function in transplantation, the MHC is sometimes designated as "strong". Other systems belong to "weak" or non-MHC groups.

5.2 The H-2 complex

The complex of H-2 genes is localized on chromosome 17 and represents a DNA segment, approximately 4 000 kbp long (paired kilobases, or 4 000 000 nucleotide pairs). The whole complex can be divided into six regions — K, I, S, D, Qa and Tla (*Table 5.3*). The classical H-2 complex contains genes between the regions H-2K (closer to the centromere) and H-2D. It codes for the antigens belonging to the histocompatibility antigens of classes I and II and products of the S region. Class I antigens are typical transplantation antigens. They induce formation of specific antibodies and can therefore be detected serologically (*SD antigens*, serologically detectable). They are similar to products of loci Qa-2, Tla and Qa-1, which are present downstream of region D. Antigens, coded by K and D regions, are present on the surface of almost all somatic cells, whereas Qa antigens are primarily on *B*- and *T*-lymphocytes and Tla antigens on thymocytes and some leukaemic cells. The transplantation antigen-encoding genes are very polymorphous. The K and D loci contain more than 50 alleles. In contrast, Qa and Tla antigens are significantly less polymorphous.

Table 5.3 Simplified genetic map of the H-2 complex

Regions	K	I	S	D Qa Tla
Loci	K	$A_{\beta1}$ $A_{\beta2}$ $A_{\beta3}$ E_β E_α	C4 Slp B	D L
Antigens class I				
Antigens class II				

The somatic cells of mice are diploid and therefore possess two antigens, encoded by alleles of the H-2K and H-2D loci. The heredity of histocompatibility antigens is controlled according to MENDEL'S laws. The offspring receives a haploid number of chromosomes from both the mother and father. This means that it receives two H-2 antigens from each of them (one from the H-2K locus and one from the H-2D locus). These are the products of one of two allele combinations that are present in the homologous chromosome pair of the father and mother.

The locus is a site on the chromosome where a certain gene is present. *The gene* designates a DNA segment containing specifically encoded information for a single genetic marker. The existence of a gene can be demonstrated only when it occurs in at least two alternate forms, called *alleles*. An individual, however, can possess only one of the possible alleles on one locus. All somatic cells, with the exception of gonads, always have a pair of identical (homologous) chromosomes — they are therefore diploid. This is why any individual always has two genes for any property (prospective potency). The genes coding for a particular property may be identical and such an individual is then called a homozygote. Conversely, the individual may have two different alleles of the same gene, known as a heterozygotic combination. The gonad has a half-haploid number of chromosomes.

The class II histocompatibility antigens are also integral membrane glycoproteins. However, they are not transplantation antigens and fulfil other functions. They are encoded by loci, present in the I region ("i") and are therefore called *Ia antigens* ("I-region associated"). The H-2 complex contains two Ia-molecule types encoded by I-A and I-E regions (*Table 5.3*). Ia antigens are primarily present on *B*-cells, some *T*-cells and macrophages; these antigens are also highly serologically polymorphous. The genes encoding for Ia antigens are called *Ir genes* (Ir = immune response) and determine the ability of the individual to respond highly, poorly or not at all to various antigens. Presumably, the I region also contains *Is genes* which suppress the specific immune response, *i.e.* which determine the extent of the suppressive regulation. However, the products of Is genes have not been demonstrated experimentally to date. Regulation of the immune response by Ir genes is achieved at the level of antigen presentation and interaction between macrophages and *T*- and *B*-cells.

In addition to class I and II antigens, proteins representing certain complement components (C4 and factor B), or similar molecules (Slp, sex-limited protein), are encoded by the H-2S region. They are present in the blood serum and can bind passively to erythrocytes. The products of the H-2S region were originally designated as class III antigens. Nowadays, however, they are not considered to be components of the H-2 complex and this term has therefore been abandoned.

5.3 The HLA system

The HLA system represents the major histocompatibility complex (MHC) of man and consists of a series of gene-defining molecules providing the context for the recognition of antigens by *T*-lymphocytes (HÁNA *et al.*, 1982; BUC, 1989; MAYR, 1990). It encompasses approximately one thousandth of the human genome (3.5 centimorgans = 3 500 kilobases, kb) and is localized on the short arm of chromosome 6 (band 6p21) (*Fig. 5.1*). A simplified scheme of the HLA system is shown in *Table 5.4*.

At present, four regions of the HLA gene complex are relatively well characterized. Starting at the centromere, there is the HLA-D region, which is followed by HLA-B, HLA-C and HLA-A. Between the HLA-A and HLA-C regions there is a fifth region E and beyond the HLA-A there are two additional regions, F and G, containing genes for antigens that are similar to the Qa and Tla antigens of the mouse H-2 complex. Their detailed characteristics are unknown at present. Between the HLA-D and HLA-B regions there is a DNA segment containing genes for certain complement components (C2, C4 and factor B), tumour necrosis factors, heat shock proteins and 21-hydroxylases. Originally their products were designated class III HLA antigens although this term is no longer used.

All HLA antigens can be divided, according to their structure, into two groups — class I and class II antigens. *Class I antigens* are present in all human cells except mature erythrocytes, whereas *class II antigens* show a restricted tissue distribution. They are primarily found on the surface of immunocompetent cells, particularly of *B*-lymphocytes, monocytes and macrophages, some endothelial and epithelial cells, dendritic cells and activated *T*-lymphocytes.

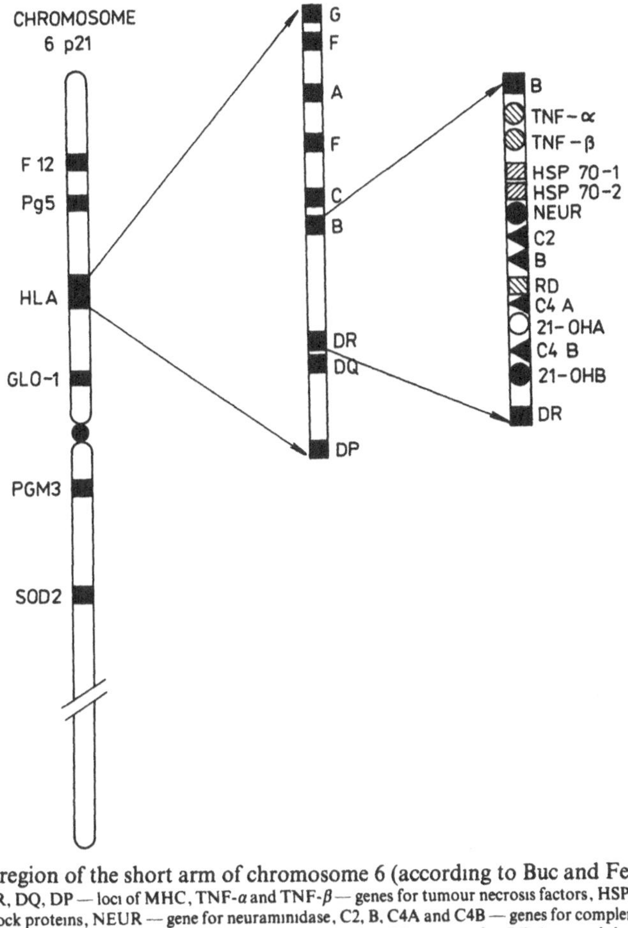

Fig. 5.1. HLA region of the short arm of chromosome 6 (according to Buc and Ferenčík, 1993). A, B, C, E, F, G, DR, DQ, DP — loci of MHC, TNF-*α* and TNF-*β* — genes for tumour necrosis factors, HSP70-1 and HSP70-2 — genes for heat shock proteins, NEUR — gene for neuraminidase, C2, B, C4A and C4B — genes for complement components, 21-OHA and 21-OHB — genes for 21-hydroxylases of the P450 system, RD — gene for RD (repeated dimers) protein, F12 — gene for factor 12 of the blood coagulation system, Pg5 — gene for pepsinogen, GLO-1 — gene for glyoxalase, PGM3 — gene for phosphoglucomutase, SOD2 — gene for superoxide dismutase

Until now, three subregions — HLA-DP, HLA-DQ and HLA-DR have been well identified in the HLA-D region. Their products represent class II antigens and they are typical *Ia antigens; i.e.* antigens of the immune response. HLA-DQ antigens are structurally similar to the products of I-A locus from murine complex H-2, and HLA-DR antigens are similar to antigens coded by the I-E locus.

Table 5.4 Simplified genetic map of the HLA system (Buc, 1989; Bodmer *et al.*, 1990)

Regions	*D*																*Complement components*[a]				*B*	*C*	*E*	*A*	*F*	*G*
Subregions	DP				DN	DO	DQ				DR															
Loci[b]	**B2**	**A2**	B1	A1	**A**	**B**	**B2**	**A2**	B1	A1	A	B1	**B2**	B3	B4	B5	C4B	C4A	B	C2						
Number of alleles identified until 1989	—	—	19	4	—	—	—	—	13	8	1	34	—	4	1	4	22	13	11	4	32	11	—	25	—	—
Antigen class I																					///	///	///	///	///	///
Antigen class II	///	///	///	///	///	///	///	///	///	///	///	///	///	///	///	///										

[a] As well as the functional alleles, at least one non-functional (zero) allele has been found for every C4B, C4A, B and/or C2 locus.

[b] The loci in bold type contain pseudogenes.

Within these subregions several loci are present. The HLA-DR alloantigens are situated on molecules consisting of the gene products of HLA-DRA and DRB1 (DR1–DRw18), DRB3 (DRw52), DRB4 (DRw53) or DRB5 (DRw15, DRw16). The product of DRA shows no variability, DRB2 is a pseudogene.

In the HLA-DQ subregion, there are four loci: DQA1, DQA2, DQB1 and DQB2. Both the DQA1 and DQB1 chains are polymorphic. DQA2 and DQB2 represent probably pseudogenes or intact genes but the products of these two loci are not yet known.

The HLA-DP subregion contains four loci: DPA1, DPB1, DPA2 and DPB2. The DP molecules are built up from the DPA1α and the DPB1β chain. DPA2 and DPB2 are pseudogenes.

Three other loci, DNA, DOB and DVB, are also situated in the HLA-D region, however, their products are not yet known.

Table 5.5 Designation of HLA-A, -B and -C alleles and class I HLA antigens (Bodmer *et al.*, 1990)

HLA alleles	*HLA specificity*	*HLA alleles*	*HLA specificity*	*HLA alleles*	*HLA specificity*
A*0101	A1	B*0701	B7	Cw*0101	Cw1
A*0201	A2	B*0702	B7	Cw*0201	Cw2
A*0202	A2	B*0801	B8	Cw*0202	Cw2
A*0203	A2	B*1301	B13	Cw*0301	Cw3
A*0204	A2	B*1302	B13	Cw*0501	Cw5
A*0205	A2	B*1401	B14	Cw*0601	Cw6
A*0206	A2	B*1402	Bw65(14)	Cw*0701	Cw7
A*0207	A2	B*1501	Bw62(15)	Cw*1101	Cw11
A*0208	A2	B*1801	B18	Cw*1201	—
A*0209	A2	B*2701	B27	Cw*1301	—
A*0210	A2	B*2702	B27	Cw*1401	—
A*0301	A3	B*2703	B27		
A*0302	A3	B*2704	B27		
A*1101	A11	B*2705	B27		
A*2401	A24(9)	B*2706	B27		
A*2501	A25(10)	B*3501	B35		
A*2601	A26(10)	B*3701	B37		
A*2901	A29(w19)	B*3801	B38(16)		
A*3001	A30(w19)	B*3901	B39(16)		
A*3101	A31(w19)	B*4001	Bw60(40)		
A*3201	A32(w19)	B*4002	B40		
A*3301	Aw33(w19)	B*4101	Bw41		
A*6801	Aw68(28)	B*4201	Bw42		
A*6802	Aw68(28)	B*4401	B44(12)		
A*6901	Aw69(28)	B*4402	B44(12)		
		B*4601	Bw46		
		B*4701	Bw47		
		B*4901	B49(21)		
		B*5101	B51(5)		
		B*5201	Bw52(5)		
		B*5701	Bw57(17)		
		B*5801	Bw58(17)		

Nomenclature for factors of the HLA system includes designation of both HLA alleles and specificities. The naming of alleles is defined by nucleotide or inferred amino acid sequences. The specificities of individual histocompatibility antigens are designated by similar symbols as individual MHC loci. For example, the product of the HLA-A*0101 allele will have a similar designation with a number index, characterizing an appropriate phenotype, *i.e.* HLA-A1. Several alleles may have the same phenotypic specificity, *e.g.* A*0201 – A*0210 alleles express only one serological specificity HLA-A2 (*Table 5.5*). When designating some phenotypes, the number index is preceded by a letter "w", *e.g.* HLA-Bw42. This letter means that the given antigen has been defined preliminarily in a meeting of an expert group (workshop).

Class I HLA antigens can be determined by serological techniques using specific antibodies. Products of the HLA-D region had originally been determined by the mixed lymphocyte culture (MLC) test using homozygotic typing cells (HTC). Therefore, they are referred to as LD antigens (lymphocyte-detectable).

LD antigens were discovered following the observation that mixing lymphocytes from two donors results in blastogenesis, despite the fact that both donors have identical HLA-A, HLA-B and HLA-C antigens. *Blastogenesis* is a developmental stage of the cell, characterized by increased metabolism and enhanced biosynthesis of proteins, RNA and DNA. Morphologically, it is manifested by increased cell size (lymphoblasts) and by cell division. Blastogenesis in mixed lymphocyte cultures is caused by different LD antigens, present on the surface of lymphocytes of different individuals.

Until 1987, a total of 26 different LD antigens of the HLA-D locus (HLA-Dw1 up to HLA-Dw26) had been determined. It was found, however, that serologically detectable antigens (*SD antigens*) are also coded in loci HLA-DQ and HLA-DR. These are typical Ia antigens. Until now, 19 SD products of the HLA-DR locus consisting of 43 known alleles (*Table 5.6*) and seven SD products of the HLA-DQ locus (21 alleles) have been detected (*Table 5.7*). Each SD antigen, coded by these loci, can have one or several LD subtypes. HLA-DQA*1 alleles code only HLA-*D*-associated specificities. It therefore follows that using serological and lymphocyte techniques, no identical antigen determinants (epitopes), can be detected. It should be emphasized that epitopes detected by homozygote typization cells (HTC), are present in molecules of all class II antigens and, to a lesser degree, even in class I antigens. The Dw specificities then correlate with LD subtypes. It means, for example, that HLA-Dw1 antigenic specificity is encoded by DRB1*0101, DQA1*0101, or DQB1*0501 alleles (*Tables 5.6* and *5.7*). Thus, an independent HLA-D locus does not exist. The Dw specificities are represented by certain epitopes only that are an integral part of HLA-DR or HLA-DQ antigens. However, their limited number are also present in HLA-A, HLA-B and HLA-C antigens. HLA-DP antigens are only detectable by means of HTC and, therefore, do not form LS subtypes, but LD types. So far 23

HLA-DP alleles and seven LD types (HLA-DPw1 up to HLA-DPw6 and HLA-DP"Cp63") have been identified (*Table 5.7*).

Table 5.6 Designations of HLA-DR alleles and specificities (Bodmer *et al.*, 1990)

HLA alleles	*HLA-DR specificities*	*HLA*-D-*associated (T-cell-defined) specificities*
DRB1*0101	DR1	Dw1
DRB1*0102	DR1	Dw20
DRB1*0103	DR'BR	Dw'BON'
DRB1*1501	DRw15(2)	Dw2
DRB1*1502	DRw15(2)	Dw12
DRB1*1601	DRw16(2)	Dw21
DRB1*1602	DRw16(2)	Dw22
DRB1*0301	DRw17(3)	Dw3
DRB1*0302	DRw18(3)	Dw'RSH'
DRB1*0401	DR4	Dw4
DRB1*0402	DR4	Dw10
DRB1*0403	DR4	Dw13
DRB1*0404	DR4	Dw14
DRB1*0405	DR4	Dw15
DRB1*0406	DR4	Dw'KT2'
DRB1*0407	DR4	Dw13
DRB1*0408	DR4	Dw14
DRB1*1101	DRw11(5)	Dw5
DRB1*1102	DRw11(5)	Dw'JVM'
DRB1*1102	DRw11(5)	Dw'JVM'
DRB1*1103	DRw11(5)	—
DRB1*1104	DRw11(5)	Dw'FS'
DRB1*1201	DRw12(5)	Dw'DB6'
DRB1*1301	DRw13(w6)	Dw18
DRB1*1302	DRw13(w6)	Dw19
DRB1*1303	DRw13(w6)	Dw'HAG'
DRB1*1401	DRw14(w6)	Dw9
DRB1*1402	DRw14(w6)	Dw16
DRB1*0701	DR7	Dw17
DRB1*0702	DR7	Dw'DB1'
DRB1*0801	DRw8	Dw8.1
DRB1*0802	DRw8	Dw8.2
DRB1*0803	DRw8	Dw8.3
DRB1*0901	DR9	Dw23
DRB1*1001	DRw10	—
DRB3*0101	DRw52a	Dw24
DRB3*0201	DRw52b	Dw25
DRB3*0202	DRw52b	Dw25
DRB3*0301	DRw52c	Dw26
DRB4*0101	DRw53	Dw4, Dw10, Dw13, Dw14, Dw15, Dw17, Dw23
DRB5*0101	DRw15(2)	Dw2
DRB5*0102	DRw15(2)	Dw12
DRB5*0201	DRw16(2)	Dw21
DRB5*0202	DRw16(2)	Dw22

Table 5.7 Designation of HLA-DQ and HLA-DP alleles and specificities (Bodmer *et al.*, 1990)

HLA alleles	*HLA-DQ specificities*	*HLA-D-associated (T-cell-defined) specificities*	*HLA alleles*	*HLA-DP specificities (T-cell-defined)*
DQA1*0101	—	Dw1,w9	DPA1*0101	—
DQA1*0102	—	Dw2,w21,w19	DPA1*0102	—
DQA1*0103	—	Dw18,w12,w8	DPA1*0103	—
		Dw'FS'	DPA1*0201	—
DQA1*0201	—	Dw7,w11		
DQA1*0301	—	Dw4,w10,w13,	DPB1*0101	DPw1
		w14,w15,w23	DPB1*0201	DPw2
DQA1*0401	—	Dw8,Dw'RSH'	DPB1*0202	DPw2
DQA1*0501	—	Dw3,w5,w22	DPB1*0301	DPw3
DQA1*0601	—	Dw8	DPB1*0401	DPw4
			DPB1*0402	DPw4
DQB1*0501	DQw5(w1)	Dw1	DPB1*0501	DPw5
DQB1*0502	DQw5(w1)	Dw21	DPB1*0601	DPw6
DQB1*0503	DQw5(w1)	Dw9	DPB1*0801	—
DQB1*0601	DQw6(w1)	Dw12,w8	DPB1*0901	DP'Cp63'
DQB1*0602	DQw6(w1)	Dw2	DPB1*1001	—
DQB1*0603	DQw6(w1)	Dw18,Dw'FS'	DPB1*1101	—
DQB1*0604	DWw6(w1)	Dw19	DPB1*1301	—
DQB1*0201	DQw2	Dw3,w7	DPB1*1401	—
DQB1*0301	DQw7(w3)	Dw4,w5,w8,w13	DPB1*1501	—
DQB1*0302	DQw8(w3)	Dw4,w10,w13,w14	DPB1*1601	—
DQB1*0303	DQw9(w3)	Dw23,w11	DPB1*1701	—
DQB1*0401	DQw4	Dw15	DPB1*1801	—
DQB1*0402	DQw4	Dw8,Dw'RSH'	DPB1*1901	—

The genes, localized on one chromosome pair, form a unit, called a *haplotype*. The haplotype is always transferred to the offspring as one unit. Therefore, somatic cells of any individual contain two haplotypes — one from the mother and the second from the father. Since in a family there are just two maternal and two paternal haplotypes, the children can only inherit four possible combinations. If one designates maternal haplotype "ab" and paternal haplotype "cd", the following combinations may be inherited by their children: ac, ad, bc or bd. The haplotype may be represented, for example, by a complex: HLA-A1, HLA-B13, HLA-Cw6, HLA-DPw4, HLA-DQw2, HLA-DRw8. According to Mendel's laws, two siblings have a 25% probability of identity in HLA antigens, 50% probability of identity in one haplotype and 25% probability of difference in both haplotypes.

5.4 The structure of HLA and H-2 antigens and their gene organization

Class I and II histocompatibility antigens have a similar structure (*Fig. 5.2*). Both are glycoproteins composed of two polypeptide chain types and have four outer domains (segments of the polypeptide chain, possessing a characteristic conformation).

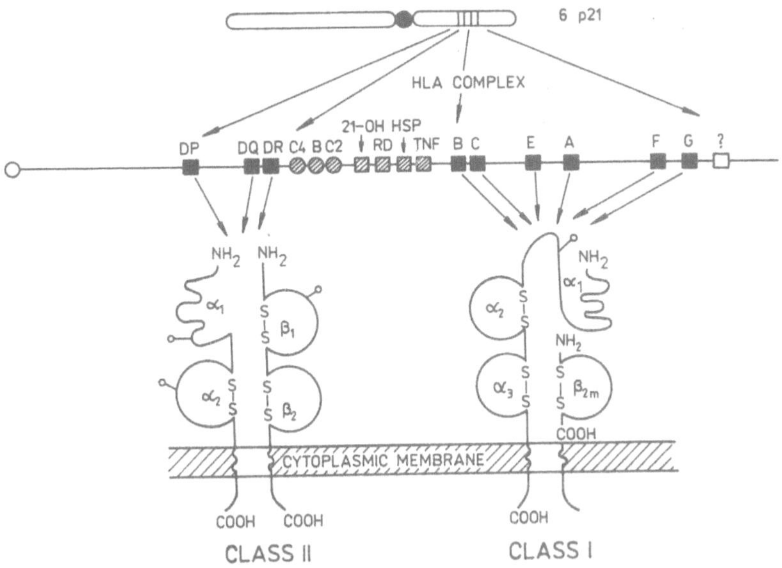

Fig. 5.2. HLA complex and structure of class I and class II histocompatibility antigens. α_1, α_2, α_3, β_1, β_2-domains, β_2m — β_2-microglobulin

Class I antigens are heterodimer molecules containing polypeptide chain α, which has three outer domains (α_1, α_2, α_3) and β-microglobulin. Only α_2 and α_3 domains have intrachain disulphide bonds. Each domain consists of approximately 90 amino acid residues and the α chain contains a total of 338 residues for HLA antigens or 348 for H-2 antigens. The α chain has, in addition to outer domains, a transmembrane segment (passes across the cytoplasmic membranes and consists of 28 hydrophobic amino acids) and a cytoplasmic domain (32 amino acids including the COOH-terminal end of the molecule) with a structure different from that of the outer domains. The alloantigenic determinants are mainly carried by α_1 and α_2 domains. The relative molecular weight of the chain is 44 000, 10% being saccharides. The α chain is coded by MHC. It is estimated that the loci HLA-A, HLA-B and HLA-C contain at least 30 α genes; most of them, however, are pseudogenes which are not expressed. The first domain, α_1, is non-covalently bound to β_2-microglobulin, which forms the fourth outer domain of class I antigens. The **β_2-microglobulin** is a protein with M_r = 12 000 and is coded by a gene present on chromosome 15, *i.e.* outside MHC. All class I antigens contain an identical β_2-microglobulin indicating that their specificity is determined by the structure of the α chains.

Class II antigens consist of two non-covalently bound glycosylated polypeptide chains, designated α and β, both being encoded in MHC. Each chain has two outer domains (85–95 and 109 amino acids, respectively), a transmembrane segment (23 amino acids) and a small cytoplasmic part (—COOH end, 8–15 amino acids). The α_1 domain does not possess an intrachain disulphide bond. Both chains contain about 230 amino acid residues, although their relative molecular weights are slightly different. The relative molecular weights of the α chain varies from 31 000 to 34 000, whereas that of the β chain lies between 25 000 and 29 000. These differences are due to the different saccharide content in each chain.

The alloantigenic epitopes are expressed on the outer domain of the extracellular region. The domain close to the membrane is constant and shows similarities to class I α_3 and the constant Ig domains. In the DR β chain, hypervariable regions can be seen at amino acid portions 9–13, 26–33 and 67–74; in the DQ α chain there is only one hypervariable region (45–56) and in the DQ β chain there are two hypervariable regions (53–57 and 66–77). The genes for both chains are present in all loci of the HLA-D region and are designated A and B (*Table 5.4*). The HLA-DP locus contains two A genes and two B genes similarly to the HLA-DQ locus. The family of HLA-DR genes contains one A gene for the α chain and five B genes for the β chain.

The class I loci HLA-A, B and C show a remarkable multiple allelism: taking into account only the specificities recognized until 1989 and one allele coding for not yet detectable products, 25 HLA-A, 32 HLA-B and 11 HLA-C alleles are recognized. From these figures, 251 HLA-A, 641 HLA-B and 56 HLA-C phenotypes can be computed; the number of possible HLA-ABC phenotypes amounts to 9 009 896.

The class II molecules defined by serology are similarly polymorphic: 43 HLA-DR and 13 HLA-DQ alleles are officially recognized, corresponding to 882 DR and 68 DQ phenotypes. Because of the extremely strong linkage disequilibrium between HLA-DR and DQ alleles, the DQ factors do not increase the HLA polymorphism. According to Mayr (1990), the number of possible HLA phenotypes which can be detected by serology, taking into account the products of the loci HLA-A, B, C and DR, mount to 1.6×10^{10}. Other polymorphism can be detected by cellularly defined specificities.

5.4.1 Split genes

Until 1977, it had been accepted that nucleotides encoding a gene in the DNA molecule have a linear continuous sequence, with no interruptions. The gene structure and function, however, had previously been studied in relatively simple organisms, particularly bacteria. In 1977, Berget *et al.* observed that the DNA chain encoding a certain gene of the adenovirus is longer than the corresponding mRNA. This observation was almost simultaneously confirmed by Breathnach *et al.* (1977) for ovalbumin and by Tonegawa *et al.* (1978) for the murine immunoglobulin chain. It is therefore apparent that the DNA chain contains segments that are not transcribed into mRNA. In 1978, Gilbert called these non-transcribed segments **introns** and DNA segments which are transcribed into mRNA were termed **exons**. It was soon discovered that such a gene organization occurs not only in vertebrates, but rather in all eukaryotic cells, including unicellular organisms with differentiated nuclei, as well as in parasitic viruses of these cells. The colinearity principle, which is typical for the arrangement of nucleotides in the genes of prokaryotes (except *Archaebacteria*) does not hold true for genes of eukaryotic organisms. These genes are, therefore, called **split genes**. This term expresses the fact that the gene (a DNA molecule section encoding for one polypeptide chain) is not homogeneous, but consists of fragments transcribed into the amino acid sequence, *i.e.* exons, that are separated by non-transcribing introns of various length.

A schematic representation of the split gene is shown in *Fig. 5.3*. The gene begins with the *leading* or *signal sequence* of nucleotides (L). This is a chain of between 60 and 180 nucleotides, encoding for a segment of hydrophobic amino acids, which permits anchoring of the ribosome into the membrane of the endoplasmic reticulum during initiation of protein synthesis on the membrane-bound ribosomes. The leading sequence is also transcribed in the form of a *signal peptide* into the polypeptide chain. After its synthesis has been terminated, it is split off by a specific proteinase, before its release from the ribosome. Thus, the completed polypeptide chain does not contain the signal peptide.

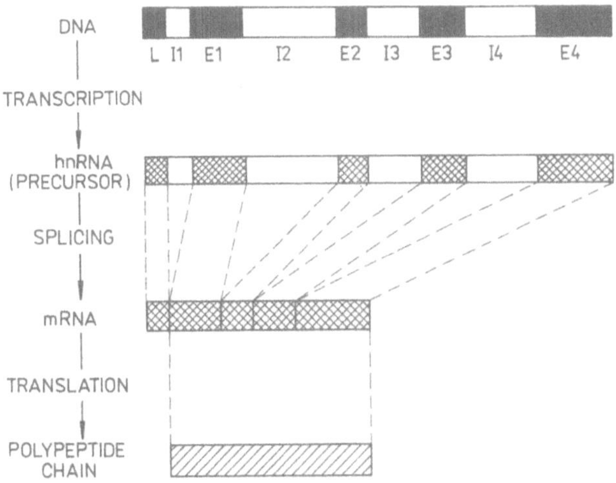

Fig. 5.3. A split gene.
L, leading sequence, I1–I4, introns, E1–E4, exons

The first exon is separated from the leading sequence by a short intron. Introns are usually longer than exons and some vertebrate genes may possess several tens of introns and exons. The complete DNA sequence (exons and introns) constituting one gene, is transcribed into precursor RNA (heterogeneous nucleic RNA, hnRNA). This if further modified in the nucleus by "**splicing**". The basis of this process is cleavage of the introns and joint exons into a linear chain of messenger RNA (mRNA), which leaves the cell nucleus. Being in the cytoplasm, it becomes a matrix for the translation of nucleotides (triplet genetic code) into the amino acid sequence in the polypeptide chain during protein synthesis.

Most genes in the genome of eukaryotic organisms, including MHC genes, are organized according to this pattern. There are only a few exceptions, the most important being genes for histones and interferons, which are organized colinearly.

Using genetic engineering methods (cross-hybridization with complementary DNA–cDNA isolated from various species), STEINMETZ *et al.* (1982) and STEINMETZ and HOOD (1983) found that all genes of class I murine H-2 antigens are divided into eight exons (*Fig. 5.4*). Individual exons correlate with the structural domains of these antigens. The first exon encodes for the signal peptide, exons 2, 3 and 4 encode for three outer domains, exon 5 for the transmembrane segment and exons 6, 7 and 8 for the cytoplasmic domain. Even the genes for class I HLA antigens possess an almost identical exon–intron organization; only the cytoplasmic domain is different: it is encoded for by two exons only and it is shorter by 10 amino acid residues. The genes for class I HLA antigens therefore have seven exons and encode 338 amino acid residues.

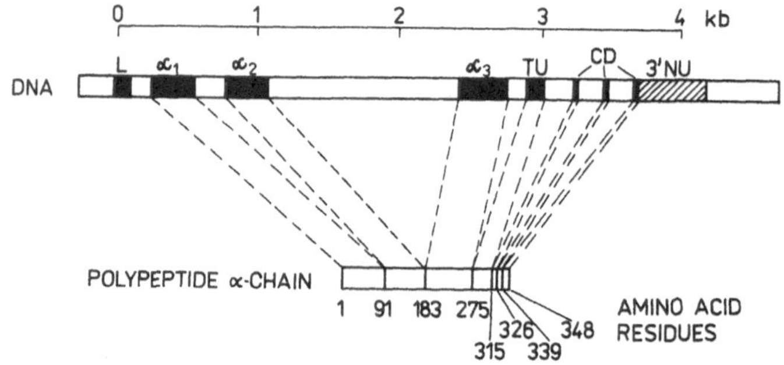

Fig. 5.4. Arrangement of the gene encoding the α chain of class I antigens of the H-2 complex (Steinmetz and Hood, 1983).
Exons are represented by black regions and introns by white regions in the DNA chain. kb, kilobases in the polynucleotide chain; L, leading sequence; α_1, α_2, α_3, exons for three outer domains; TU, exon for the transmembrane segment; CD, exons for the cytoplasmic domain; 3′NU, 3′-terminal segment, not translated into the polypeptide chain. The numbers designate corresponding exons translated into the amino acid residues sequence in the polypeptide chain.

Sequence analysis of class I antigens, as well as of DNA from individual alleles, revealed that most sequence differences occur in the first and second outer domain (α_1 and α_2). The changes in amino acid sequence in these two domains are, therefore, considered to be the main cause of polymorphism of class I antigens. Not all positions in these domains are equally variable. Three "hypervariable" segments (positions 62–83, 92–121 and 135–157) were discovered, where the differences in amino acid sequence among individual antigens are much more frequent. The functional consequences of such allele differences are as yet unknown.

The genes for class II antigens are split into between four and six exons; four exons include genes for the E_α chain from the H-2 complex and genes for DR_α chain, five exons include genes for the DR_β and DQ_β chains, whereas the genes for murine chain A_β and E_β comprise six exons (*Fig. 5.5*). Even in class II antigens, there is a direct correlation between the exon–intron gene organization and the structure of their protein products. DNA sequence analysis has shown that the murine A_α, A_β and E_β and human DQA, DQB and DRB genes are polymorphous, whereas murine E_α and human DRA genes have a conservative structure. The structural difference among individual alleles, coding for polymorphous class II antigens, are relatively high and are localized particularly in domains α_1 and β_1, whereas the domains α_2, β_2, the transmembrane segment and cytoplasmic domains are rather conservative (with a relatively constant amino acid sequence).

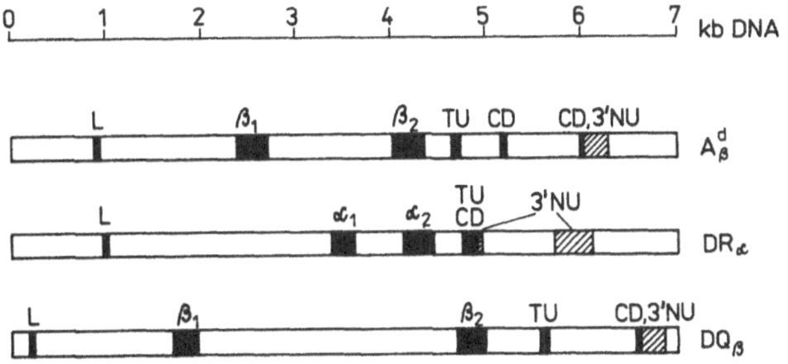

Fig. 5.5. Organization of genes for certain chains of class II histocompatibility antigens. For individual symbols see *Fig. 5.4*.

5.5 What is the biological significance of histocompatibility antigens?

Histocompatibility antigens are the marks on the cell surface which enable their mutual recognition. Such primary (primitive) self-markers have been found in simple marine organisms, living in colonies. They enable recognition of the members of their own colony from other individuals of the same species, belonging to another colony. During phylogenetic development, the self-marker system has gradually improved, and besides the recognition and signal function, the regulatory function has emerged. In mammals, the histocompatibility antigens — besides enabling mutual cellular recognition — reflect the individuality of every representative of the same species and regulate interactions between the cells of the immune system and their effector functions. In addition, in man they are somehow associated with an increased susceptibility to certain diseases. Presumably, the association of HLA with disease — at least viral diseases — results from the fact that the

HLA structures directly represent (or influence) the receptor sites for a virus on the cell host surface.

Mutual recognition and cell control in multicellular organisms are part of their basic characteristics. It is essential for maintenance of life in all organisms. It permits the junction, interaction, cooperation and coordination of physiological processes and prevents the self-destruction of the organism. During recognition of self and non-self, the histocompatibility antigens must fulfil two functions — to receive and transfer the information. Recognition is subsequently achieved by a pair of markers with complementary structure. One member of the pair plays the role of the *effector* (it provides a signal), the second operates as a *receptor* (it receives a signal). Both markers may be present in one or two different antigen molecules.

Class I histocompatibility antigens are involved in mutual cell recognition and effector functions of the immune system as they are present on the surface of all cells except erythrocytes. Once a foreign marker has occurred on the surface of a cell, it signals that the cell has suffered genetic damage, or has been infected by a virus, and should be eliminated. Such a cell can be killed by cytotoxic *T*-lymphocytes, although only if both the infected cell and the T_C-lymphocyte possess identical class I antigens on their surface (meaning that they belong to the same organism). Similarly, the class II antigens participate in the interaction of macrophages and lymphocytes and also in lymphocyte subpopulations. Macrophages, for example, can present an antigen to induce the immune response, but only to *T*- and *B*-lymphocytes possessing identical class II antigens to the macrophages. This essential agreement between two reacting cell types is called the **MHC-restriction** phenomenon or the phenomenon of *syngeneic preference*. It is based on the fact that lymphocytes can read information from another cell if this cell provides the antigenic information together with self-markers (histocompatibility antigens).

The *Ir-genes*, regulating the immune response, are present in the major histocompatibility complex. The Ir-genes are inherited dominantly. In order to react with an immune response, the individual must possess an appropriate Ir-gene in his genetic equipment. When this gene is lacking, the immune response does not take place. The Ia antigens are products of Ir-genes. It is thought that the MHC also contains the *Is-genes* (immune suppression) that can induce unresponsiveness to a given antigen, despite the fact that the given individual has the necessary Ir-gene. The intensity of the resulting immune response is then dependent on the complex activity of Ir- and Is-genes, and, in addition, on other genes localized outside the MHC. According to DAUSSET and PLA (1985), each HLA haplotype controls a characteristic genetic profile of the immune response. Every individual has a certain ability to resist various immunologically aggressive agents. The complementary action of both HLA haplotypes plays an important role in this activity.

It has been established that many diseases occur more frequently in individuals with certain genotypes in the HLA complex. Thus, for example, 96% of patients suffering from ankylosing spondylitis possess the HLA-B27 antigen. The occurrence of this antigen in a normal human population is much lower — only 9.4%. Patients with subacute viral thyroiditis have the HLA-B35 antigen in almost 80% of cases, while the frequency in the normal population is 14%. Children suffering from coeliac disease (a disease with disturbed normal resorption of nutrients from the small intestine probably due to allergic reaction to gluten), have a higher occurrence of the HLA-DR3 antigen (by 50%). Patients with insulin-dependent diabetes have a higher frequency of the HLA-DR4 antigen (by 43%) and, finally, patients suffering from multiple sclerosis were found to have a higher frequency of HLA-DR2 antigen (by 34%). The molecular basis of the association of these and other HLA antigens with diseases is so far unknown. It is not clear whether the disease-associated HLA antigens are directly involved in the pathological process, or serve as a marker (associated event) of so far unknown factors that are associated with them in a binding imbalance. Elucidation of this problem would clarify why certain individuals are more susceptible to some diseases than are other individuals.

The biological functions of histocompatibility antigens are diverse. They can be summarized, according to HAŠKOVÁ and BUC (1985), as follows:

1. Their combination on the cell surface characterizes any individual very precisely, and they are therefore useful, for example, in paternity testing, forensic medicine and criminology.
2. After allogeneic transfer of cells, tissues and organs, they are able to induce transplantation immunity. Therefore, the donors and recipients in transplantation should have the highest degree of antigen similarity.
3. They play a basic role in presenting foreign antigens to *T*-lymphocytes with help from macrophages and other accessory cells, and in binding of cytotoxic *T*-lymphocytes to target cells.
4. They clear the cell surface of foreign or altered self particles. For example, certain viruses can be bound to class I antigens (often instead of β_2-microglobulin). The complex is then engulfed by endocytosis and thereby removed from the cell's surface.
5. They provide a signal marker for cellular proliferation and maturation.
6. They possess an important immunoregulatory function even under physiological conditions. In addition, they influence the *T–B-cell* interaction and are related to receptors for various endogenous regulators, *e.g.* hormones.
7. Most probably, they also influence morphogenesis.

References

Berget, S. M., Moore, C. and Sharp, P. A. (1977) Spliced segments at the 5'-terminus of adenovirus 2 late mRNA. *Proc. Nat. Acad. Sci. USA,* **74**, 3171–75.

Billingham, R. E., Brent, L. and Medawar, P. B. (1954) Quantitative studies on tissue transplantation immunity II. The origin, strength and duration of actively and adoptively acquired immunity. *Proc. Roy. Soc.*, B 143, 58–70.

Bodmer, J. G., Marsh, S. G. E. and Albert, E. (1990) Nomenclature for factors of the HLA system, 1989. *Immunol. Today*, **11**, 3–10.

Breathnach, R., Benoist, C., O'Hare, K., Gannon, F. and Chambon, P. (1978) Ovalbumin gene: Evidence for a leader sequence in mRNA and DNA sequences at the exon-intron boundaries. *Proc. Nat. Acad. Sci. USA*, **75**, 4853–57.

Breathnach, R., Mandell, J. L. and Chambon, P. (1977) Ovalbumin gene is split in chicken DNA. *Nature*, **270**, 314–19.

Buc, M. (1989) *HLA complex in biology and medicine*. Bratislava, Veda, 188 pp. (*in Slovak*).

Buc, M. and Ferenčík M. (1993) *Immunogenetics*. Bratislava, Alfa, in press. (*in Slovak*).

Dausset, J. (1958) Isoleuco-anticorps. *Acta Haematol.* (Basel), **20**, 156–61.

Dausset, J. and Pla, M. (1985) *HLA complexe majeur d'histocompatibilité de l'homme*. Paris, Flamarion Médicine-Science, 414 pp.

Gilbert, W. (1978) Why genes in pieces? *Nature*, **272**, 581–3.

Hána, I., Ivašková, E. and Dostál, C. (1982) *The main histocompatibility system of man (HLA) and diseases*. Prague, Avicenum, 224 pp. (*in Czech*).

Hašková, V. and Buc, M. (1985) Advanced notions about the major histocompatibility complex. In: Ferenčík, M. and Štefanovič, J. (eds.), *Immunology 1985*. Bratislava, Immunol. Sect. Czechoslov. Biol. Soc., pp. 191 – 99.

Mayr, W. R. (1990) HLA 1990. *Blut*, **61**, 207–12.

Payne, R. and Rolfs, M. R. (1958) Foeto-maternal incompatibility. *J. Clin. Invest.*, **37**, 1756–61.

Snell, G. D. (1948) Methods for the study of histocompatibility genes. *J. Genetics*, **49**, 87–108.

Snell, G. D. (1958) Histocompatibility genes of the mouse. II. Production and analogs of isogenic resistant lines. *J. Nat. Cancer Inst.*, **21**, 843–55.

Snell, G. D. (1976) Histogenetic methods. In: Snell, G. D., Dausset, J. and Nathenson, S. (eds.), *Histocompatibility*. New York, Acad. Press, pp. 11–21.

Steinmetz, M. and Hood, L. (1983) Genes of the major histocompatibility complex in mouse and man. *Science*, **222**, 727–33.

Steinmetz, M., Winoto, A., Hinard K. and Hood, L. (1982) Clusters of genes encoding mouse transplantation antigens. *Cell*, **28**, 489–98.

Tonegawa, S. (1983) Somatic generation of antibody diversity. *Nature*, **302**, 575–81.

Tonegawa, S., Maxam, S. M., Tizard, R., Bernard, O. and Gilbert, W. (1978) Sequence of a mouse germ-line gene for a variable region of an immunoglobulin light chain. *Proc. Nat. Acad. Sci. USA*, **75**, 1485–92.

Van Rood, J. J., Eernise, J. G. and van Leeuwen, A. (1958) Leucocyte antibodies in sera of pregnant women. *Nature*, **181**, 1735–6.

6 The immunoglobulins (antibodies)

The immunoglobulins are officially defined by the World Health Organization (WHO, 1969) as "proteins of animal and human origin that are endowed with known antibody activity and which also include certain proteins related to antibodies in chemical structure and hence in antigenic specificity". Thus, the term "immunoglobulin" is somewhat wider than the term "antibody". The incomplete immunoglobulin molecules produced in multiple myeloma and related lymphomas (*e.g.* Bence-Jones protein and defective heavy chains), β_2-microglobulin, α_1-microglobulin and some other proteins are also thought to be immunoglobulins despite the fact that none of them has an antibody activity.

Many years ago, various observations suggested that the serum of man and animals, who had overcome an infectious disease, contained substances which could protect them from a repeated infection by the same disease. It was also observed that such an immune serum could agglutinate the infectious agents, lyse bacteria, precipitate microbial antigens, neutralize bacterial toxins *etc.* These substances were called agglutinins, lysins, precipitins or antitoxins. At the end of 1891, PAUL EHRLICH named them "antibodies" (from the German "Antikörper") and the term became popular from the beginning of this century.

It is clear that the main biological function of antibodies had been discovered well before that their chemical structure was known. This is easily understood because at the beginning of this century, when the antibody activity of these serum substances was well known, there were no biochemical methods available for the detailed chemical characterization of antibody molecules. The first breakthrough in the study of the chemical structure of antibodies was their physicochemical characterization. The main impetus for this advance came from the development of electrophoresis by ARNE TISELIUS (1937) which led to his Nobel prize and later to the Presidency of the Nobel Foundation. TISELIUS first demonstrated the separation of serum into albumin and a series of bands that he called α-, β-, and γ-globulin. In 1939, TISELIUS and KABAT showed that antibody activity was mainly confined to the γ-globulin fraction (*Fig. 6.1*). Thus it was demonstrated that *antibodies were serum proteins of the globulin type.*

Further understanding of the chemical structure of antibodies was facilitated not only by the development of numerous protein separation techniques, but also by the discovery of the myeloma proteins, and by the realization that these were related to the normal immunoglobulins. **Myeloma proteins** are produced by antibody-synthesizing cells in persons suffering from multiple myeloma and related conditions. Any patient with this disease produces a monoclonal protein unique to that individual, often in considerable amounts, and all molecules of the protein are structurally identical. On the other hand, antibodies from a healthy individual represent a heterogeneous population of molecules, differing from each other by the sequence of amino acids in their polypeptide chains. It is interesting that the first myeloma protein was discovered in 1847 by HENRY BENCE–JONES in the urine of a patient with a disease now known as multiple myeloma. This protein, which was the first tumour marker and one of the first proteins ever to be identified, was named after him — the **Bence-Jones protein**. In 1944, WALDENSTRÖM described a new disease which he called **macroglobulinaemia** because it was characterized by the presence in the serum of a great excess of a high-molecular-weight 19S immunoglobulin which is now termed IgM. Subsequently, other myeloma diseases were discovered and the myeloma proteins of the remaining classes of immunoglobulins were characterized. Due to their uniform structure, the myeloma immunoglobulins played a pivotal role in the elucidation of the chemical structure of antibody molecules. Most of the initial structural information was gained from their analysis and later confirmed on isolated antibodies.

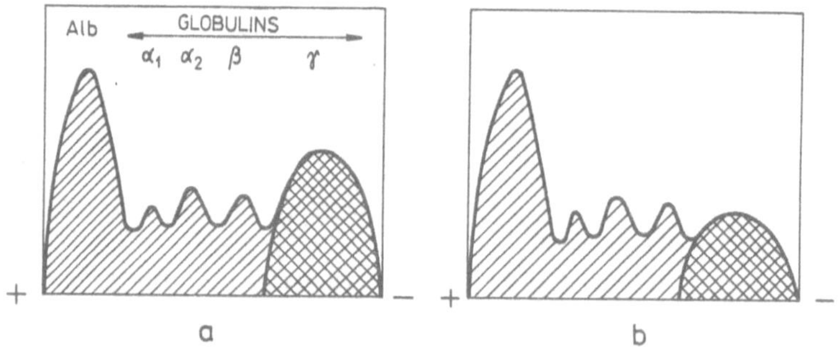

Fig. 6.1. Separation of serum proteins by free electrophoresis.
(a) Serum from an immunized individual, (b) serum from an immunized individual after adding the antigen used for immunization Addition of antigen caused antibody precipitation which resulted in decrease of the globulin fraction

In the meantime, the development of paper electrophoresis in 1948–1950 was followed by cellulose acetate and numerous other methods which were quickly adapted to the study of immunoglobulins. The powerful method of immunoelectrophoresis was introduced in 1953 by GRABAR and WILLIAMS and led HEREMANS (1960) to the discovery of the immunoglobulin class we now call IgA.

Intensive study of antibody structure began in 1950, when PORTER succeeded in splitting the IgG molecule into three fragments with the aid of the proteolytic enzyme papain. In 1960, EDELMAN *et al.* proved definitively that antibody molecules were composed of two types of polypeptide chain. According to these findings PORTER introduced the four-chain model of antibody molecules (*Fig. 6.2*). This model comprises two pairs of polypeptide chains of which each pair is always identical. The polypeptide chains are mutually linked by several dusulphide bonds. EDELMAN (1973) found that, after reduction of these disulphide bonds, antibody molecules dissociated into their individual chains, if the reduction was performed under conditions preventing repeated aggregation.

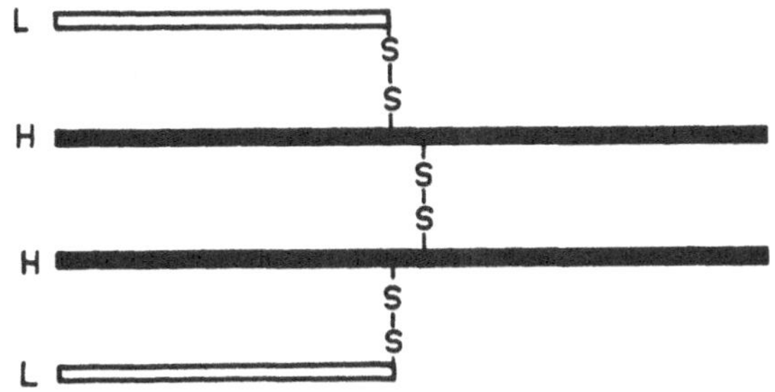

Fig. 6.2. Porter's model of the antibody molecule.
L, light chain; H, heavy chain; *N*-terminus of polypeptide chains is to the left; *C*-terminus to the right.

It was gradually shown that the basic structure of all antibodies was the same. Nevertheless, the small differences exist among them are responsible for their distinct physicochemical and biological properties. Individual investigators isolated antibodies by means of different methods and called them various names. To improve antibody nomenclature, the Commission of WHO experts met in Prague in 1964, clarified the nomenclature and recommended the use of the term **immunoglobulins (Ig)** as it reflected precisely both the biological function and chemical structure. The immunoglobulins known then were named IgG, IgM and IgA. Subsequently, ROWE and FAHEY (1965) discovered IgD, and in 1966, ISHIZAKA *et al.* showed that the long elusive allergy-reagin was IgE. This completed the five currently known immunoglobulin classes.

In the 1960s, structural studies of immunoglobulins also began to accelerate. Through the work of HILSCHMANN and CRAIG (1965), TITANI *et al.* (1965), and PUTNAM *et al.* (1966) the amino acid sequence was first established for the Bence-Jones proteins (immunoglobulin light chains or their dimers). In 1969, EDELMAN *et al.* presented the primary structure of the IgG1 heavy chain (*Fig. 6.3*), and this was followed by a description of the covalent structure of the entire IgG1 molecule. Once again, new methods such as column chromatography, gel filtration and automated amino acid analysis were essential to this progress. In 1972, PORTER and EDELMAN were awarded the Nobel prize in Physiology and Medicine for their contribution to the resolution of antibody structure. It is symbolic that the Nobel prize winners in Chemistry in the same year were MOORE and STEIN for the construction of an automatic amino acid analyser which contributed decisively to the elucidation of the primary structure of antibodies.

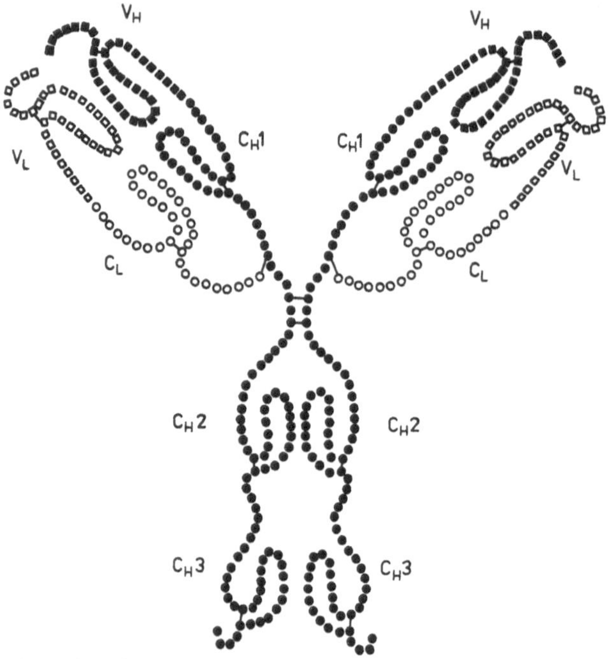

Fig. 6.3. Edelman's model of the IgG molecule
Each circle represents three amino acid residues Open circles, light chains, filled circles, heavy chains, squares, the variable portion of both light and heavy chains, short lines, disulphide bonds among peptide chains or their parts

From a chemical point of view, *the immunoglobulins belong to the glycoprotein group.* Thus, they are biopolymers made up primarily of polypeptide chains composed of amino acids, together with a carbohydrate moiety constituting 3–18% of the total molecular weight.

Antibody molecules play two major roles in the humoral system of immune defence. The first is *antibody specificity*, which consists of recognition and combination with complementary antigenic sites (determinants), generally of foreign origin. The second is a set of biological *effector functions* that are triggered by the antibody–antigen reaction and which initiate a series of events that result in elimination of the foreign antigen.

6.1 The chemical structure of immunoglobulins

The normal immunoglobulin molecule is made up of two different polypeptide chains — small light chains and large heavy chains. A myeloma immunoglobulin molecule may consist of only one type of chain.

The molecular weight of human immunoglobulin **light chains** is about 23 000; they consist of 215 amino acid residues and their abbreviation is *"L"*.

There are two types of light chains which are known by the Greek letters $\varkappa$ (kappa) and λ (lambda), both of which are present in all immunoglobulin classes. Every immunoglobulin molecule contains only one type of light chain — $\varkappa$ or λ. The proportion of $\varkappa/\lambda$ in all the immunoglobulin molecules of an individual is constant in a given species. In humans this proportion is 65/35, and in mice it is 97/3. Light chains determine the *type* of immunoglobulin molecules, which may be *K* (the molecule is composed of $\varkappa$-chains) or *L* (the molecule consists of λ-chains); *e.g.* IgGK or IgGL. Thus, the letter *"L"* may have two meanings. It may be the symbol of a light chain generally, or it may indicate one type of an immunoglobulin molecule.

The immunoglobulins are divided into five principal classes on the basis of chemical and isotypic properties. Each of the five classes has similar sets of light chains but an antigenically distinctive set of **heavy chains**, which are known by the corresponding Greek letter (γ- chain in IgG, μ in IgM, α in IgA, δ in IgD and ε in IgE; *Table 6.1*). The differences in antigenic determinants of individual heavy chains (abbreviation "H") facilitate their immunochemical determination. Proteins of the IgG class can be further divided into four *subclasses* (IgG1–IgG4) each with distinctive heavy chains termed γ_1, γ_2, γ_3, γ_4. Besides IgG there are also two subclasses in the IgA class (IgA1 and IgA2 which contain α_1- and α_2-chains respectively).

Table 6.1 Polypeptide chains in immunoglobulin molecules

Class	*Subclass*	*L-chain*	*H-chain*	*Relative molecular weight of H-chain*	*Other chain*	*Molecular formula*
IgG		$\varkappa$, λ	γ	50 000		$L_2\gamma_2$
	IgG1		γ_1			
	IgG2		γ_2			
	IgG3		γ_3			
	IgG4		γ_4			
Serum IgA		$\varkappa$, λ	α	55 000		$L_2\alpha_2$
Secretory IgA		$\varkappa$, λ	α	55 000	SC, J	$(L_2\alpha_2)_2SCJ$
	IgA1		α_1			
	IgA2		α_2			
IgM		$\varkappa$, λ	μ	70 000	J	$(L_2\mu_2)_5J$
IgD		$\varkappa$, λ	δ	65 000		$L_2\delta_2$
IgE		$\varkappa$, λ	ε	70 000		$L_2\varepsilon_2$

L, light chain,
H, heavy chain,
SC, secretory component (relative molecular weight 70 000),
J, joining chain (relative molecular weight 15 000)

The number of amino acid residues, and thus the molecular weight, is not identical in all classes of heavy chains. Gamma-chains are relatively the lightest. The size of human heavy chains in different classes increases as follows: IgG < IgA < IgD < IgE < IgM.

The first human heavy chain sequenced was the γ-chain; reported by the PORTER and EDELMAN groups in 1969. This γ-chain was composed of 445 amino acid residues with a molecular weight of 50000. In 1973, PUTNAM *et al.* determined the primary sequence of the μ-chain (576 amino acid residues), followed in 1976 (LIU *et al.*) by the sequence of the α_1-chain (472 amino acids). The molecular weight of the sequenced α_1-chain was about 54000 and it contained 17 cysteine residues. Subsequently, PUTNAM (1982) and TAKAHASHI *et al.* (1982) published the primary sequence of the human δ-chain. This chain comprised 512 amino acid residues and the molecular weight of its polypeptide portion was 56213. However, there are seven oligosaccharides attached at different sites of the chain. These contribute approximately 9000 daltons to the carbohydrate moiety, yielding a total molecular weight of about 65000. The molecular weights of the other heavy chains also include both polypeptide and carbohydrate parts.

Both heavy chains in a single immunoglobulin molecule are identical. In the IgM and IgA classes their molecules contain, besides the light and heavy chains, one additional polypeptide called *J* **(joining chain)**. In addition, the secretory IgA has a fourth polypeptide chain — the **secretory component (SC)**. The *J* and SC chains differ antigenically and in amino acid composition from each other and also from light and heavy chains.

The basic four-chain structure of the immunoglobulin molecule is Y-shaped. However, it may change from a "Y" to a "T" shape depending on the reaction conditions. This is facilitated by the existence of a **hinge region** located in the middle of the heavy chain. The polypeptide chain in the hinge region is flexible and, therefore, allows the Y-arms to bend (*Fig. 6.4*). Besides the differences in the primary structure, there are also distinctions among individual classes and subclasses in their spatial arrangement. The conformation (three-dimensional structure) of an immunoglobulin molecule is mainly influenced by the number and position of the disulphide bonds (*Fig. 6.5*), which are situated between cysteine residues of the same chain or between

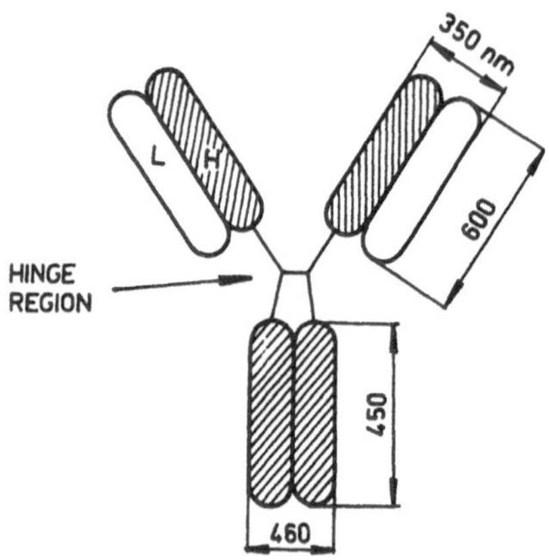

Fig. 6.4. Model of a basic immunoglobulin subunit according to electron microscopy studies.

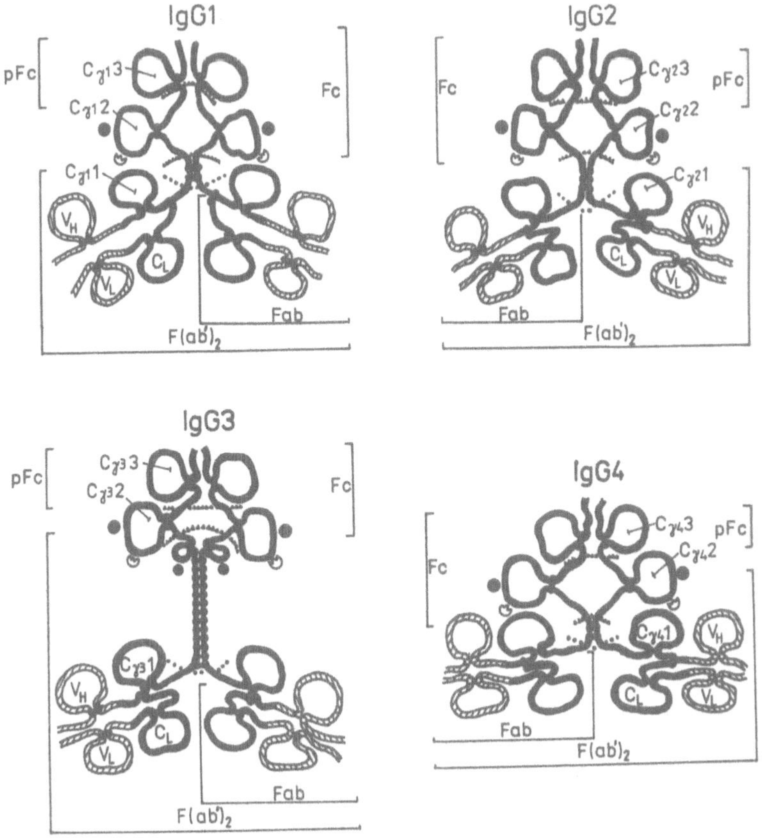

Fig. 6.5. Schematic representation of the four subclasses of human IgG (Reproduced from *Immunology Today*, June 1980).

V_L, V_H — variable domains of the *L*-chain or *H*-chain, C_L — constant part of the *L*-chain, $C_{\gamma 1}1$ — $C_{\gamma 1}3$ — constant domains of the *H*-chain. ●● — disulphide bond between both *H*-chains. IgG3 molecule contains up to 15 disulphide bonds. The figure shows the major pepsin cleavage points (▲▲▲▲▲) and the major papain cleavage points (●●●●●●●), Clq binding site (♁), and carbohydrate side chains (●).

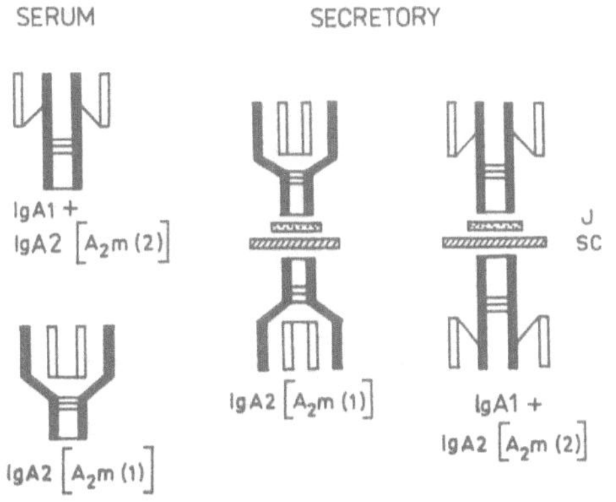

Fig. 6.6. Schematic representation of the serum and secretory IgA molecule.

J, joining chain; SC, secretory component.

different chains. The disulphide bonds are important for the biological activity of immunoglobulins. After their reduction, the three-dimensional structure is altered and this usually leads to the loss of antibody activity.

IgA exists in two forms: *serum IgA* (largely as a monomer composed of two light and two heavy chains) and *secretory IgA* (SIgA). Every SIgA

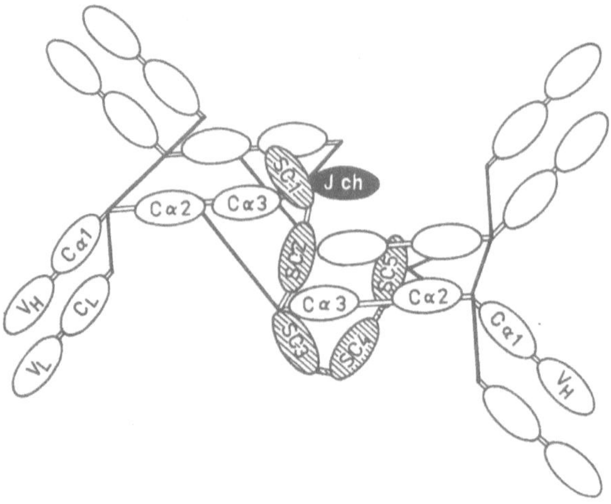

Fig. 6.7. Model of a human dimeric SIgA molecule (Mestecky *et al.*, 1988).
Each elliptical area represents an immunoglobulin (V_L, C_L, V_H, C_α1—3) or immunoglobulin-like (SC1-5, *J* chain) domain of *H*, *L* and *J* chains and SC Bold lines indicate the probable position of interchain disulphide bridges that connect the component chains The *J* chain, covalently linked to one of the monomeric IgA molecules through the penultimate cysteine residues of two α chains, together with C_α3 domains, forms a site for non-covalent interaction with the first domain of SC (SC1) The fifth domain (SC5) is disulphide-linked to the cysteine residues in the C_α2 domains of the other monomeric IgA subunit of the SIgA molecule

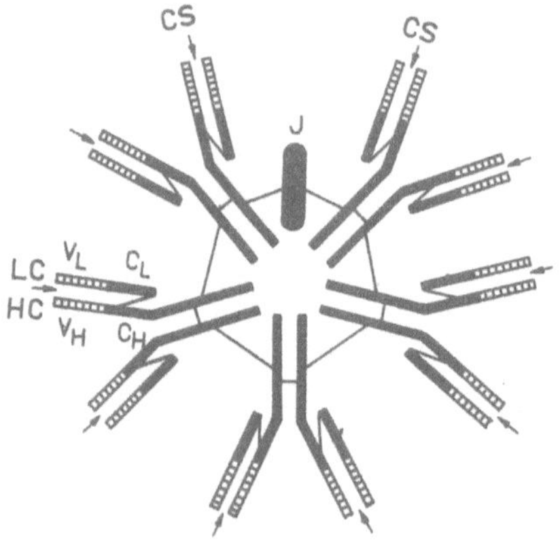

Fig. 6.8. Structure of the human IgM molecule.
Fine lines, disulphide bonds, LC, light chain, HC, heavy chain, *J*, joining chain, V_L, variable part of the *L*-chain, V_H, variable part of the *H*-chain, C_L, constant part of the *L*-chain, C_H, constant part of the *H*-chain Arrows indicate combining sites (CS)

molecule consists of two four-chain basic units, one secretory component, and one *J* chain (*Figs. 6.6 and 6.7*).

The IgM molecule is a pentamer, composed of five basic monomeric units (10 light and 10 heavy chains) and one *J* chain (*Fig. 6.8*). The molecular weight of a *J* chain is 15 000 and it consists of 119 amino acid residues. This polypeptide chain is normally found only in polymeric immunoglobulins.

6.1.1 Determination of the primary structure of immunoglobulins

The linear sequence of amino acid residues in a polypeptide is referred to as its primary structure. For the study of primary structure, the immunoglobulin molecule must initially be reduced to single polypeptide chains or fragments. These fragments may subsequently be hydrolysed to several smaller peptides in which the sequence of amino acids can be analysed. The sequencing may begin from the peptide terminus carrying an amino acid with a free amino group (*N*-terminal amino acid) or from the opposite end with a free carboxyl group (*C*-terminal amino acid). However, knowledge of the amino acid sequence of all peptides in not sufficient for the determination of complete polypeptide chain primary structures. The order in which the individual peptides are arranged in the polypeptide chain is not yet known. Therefore, every polypeptide chain must be twice cleaved into peptides. However, the second cleavage should be performed at different positions to the first and the amino acid sequence of the second set of polypeptides compared with that found in the first set. The homogeneous portions can then be identified in the peptides of both sets and the amino acid sequence of the entire polypeptide chain may be reconstituted.

The degradation of an immunoglobulin molecule can be performed enzymatically or chemically. During enzymatic digestion, the peptide bonds are hydrolysed and various **fragments** derived from one or several polypeptide chains are formed. For example, the digestion of IgG with papain leads to the formation of two distinct fragments, Fab (Fragment antigen binding) and Fc (Fragment crystallizable), whereas pepsin degradation results in formation of fragments $F(ab')_2$ and pFc′ (*Fig. 6.9*). The enzymatic cleavage is caused by the hydrolysis of peptide bonds largely in the hinge region of heavy chains. The region is localized around the interchain disulphide bridges which link both heavy chains in the basic immunoglobulin unit.

Chemical degradation leads to the cleavage of disulphide or certain peptide bonds in the immunoglobulin molecule. The disulphide bonds may be hydrolysed by reduction or sulphitolysis. The originating sulphydryl groups may be transformed with alkyl reagents to derivatives which facilitate easy identification and isolation of fragments. It is also possible to cleave the peptide bonds with cyanogen bromide (CNBr). By using this reagent the polypeptide chain is cleaved at the methionine position which is transformed in the lactone of homoserine and becomes the *C*-terminal amino acid of the fragment. Cyanogen bromide is one of the most specific agents used for the fragmentation of immunoglobulins and other proteins. The high selectivity of cyanogen bromide cleavage is based upon the rare occurrence of menthionine in proteins.

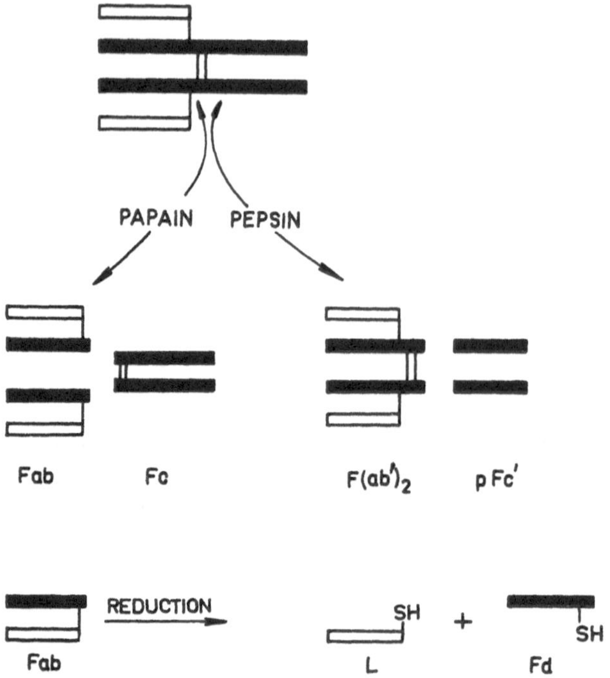

Fig. 6.9. Fragmentation of the human IgG1 molecule.

The second relatively selective method for the fragmentation of a polypeptide chain depends on the use of *S*-ethyl trifluorothioacetate which can mask the ε-amino groups of lysine. After reaction with a polypeptide, its *N*-trifluoroacetyl derivative is formed. If such a modified polypeptide chain is digested with trypsin, only the peptide bonds close to arginine are cleaved.

Using chemical fragmentation, single light and heavy chains may be obtained. Fragment Fab may be degraded by reduction to a light chain and the *N*-terminal half of a heavy chain which is named Fd. Both IgM and secretory IgA can be fragmented in a similar way. However, IgM should first be depolymerized. Secretory IgA is relatively the most resistant immunoglobulin to the enzymatic digestion.

6.1.2 Variable and constant regions of light and heavy chains

Detailed analysis of the amino acid sequence of the polypeptide chains of myeloma immunoglobulins showed that the sequence of one part of all chains of the same isotype was constant, but was variable in the other part.

The first portion was, therefore, called **the constant region** (C) and the second **the variable region** (V). These C and V regions were distinguished during the first studies on the amino acid sequence of Bence-Jones proteins (HILSCHMANN and CRAIG, 1965), and were soon extended to heavy chains.

The constant and variable regions contain all types of L chains and all classes of H chains. In the case of L chains, the variable part is represented by the amino-terminal half of the chain and is designated V_L. The carboxyl half of the chain is constant and called C_L. A similar terminology is used for the variable and constant regions of heavy chains — V_H and C_H. The V_L region consists of 108–109 amino acid residues, and the V_H region contains between 118 (γ-chains) and 129 (δ-chains) amino acid residues. The variable region of the heavy chain represents approximately one-quarter of the whole chain. In addition to the variable and constant regions, the L and H chains also contain a **junction region** which includes the last 13–17 amino acid residues of the variable part, and are called J_L and J_H, respectively (*Fig. 6.10*). The J region is situated between the V and C regions. It should be emphasized that the J region and the J chain (in polymeric immunoglobulins) are, structurally and functionally, completely different portions of immunoglobulin molecules.

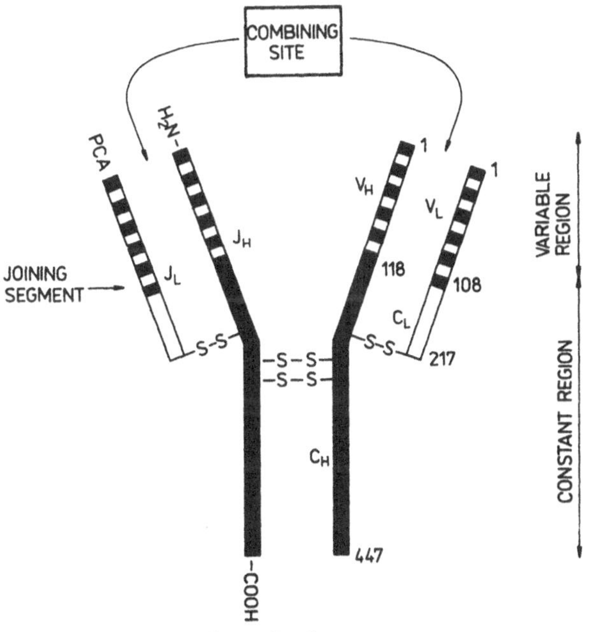

Fig. 6.10. Structure of the human IgG1 molecule.
PCA, pyrrolidoncarbonylic group

The existence of variable regions is due to the fact that in the same position of the polypeptide chain of a particular immunoglobulin molecule there is one amino acid residue, while in the other molecule there may be a different amino acid. However, not all amino acid residues in the variable region are generally changeable. For example, the V_L region contains about 65–70 positions in which amino acids may be exchanged. The other positions are constant (with the same amino acid residues in all immunoglobulin molecules). In the majority of variable positions only the alternation of two or three amino acids is observed. Changes of more than three diverse amino acids in the same position are less frequent. In the V_L region there are only about 25 such hypervariable positions located in three **hypervariable areas**. These are often called the *complementarity determining regions* (*CDR*) because, together with similar CDR regions on the heavy chains, they determine the conformation of the antibody-combining sites of immunoglobulins. The hypervariable areas of human *L* chains are localized in positions 24–34, 50–55, and 88–98 (*Fig. 6.11*), whereas the CDR regions on human *H* chains are found in positions 32–37, 50–67 and 96–110.

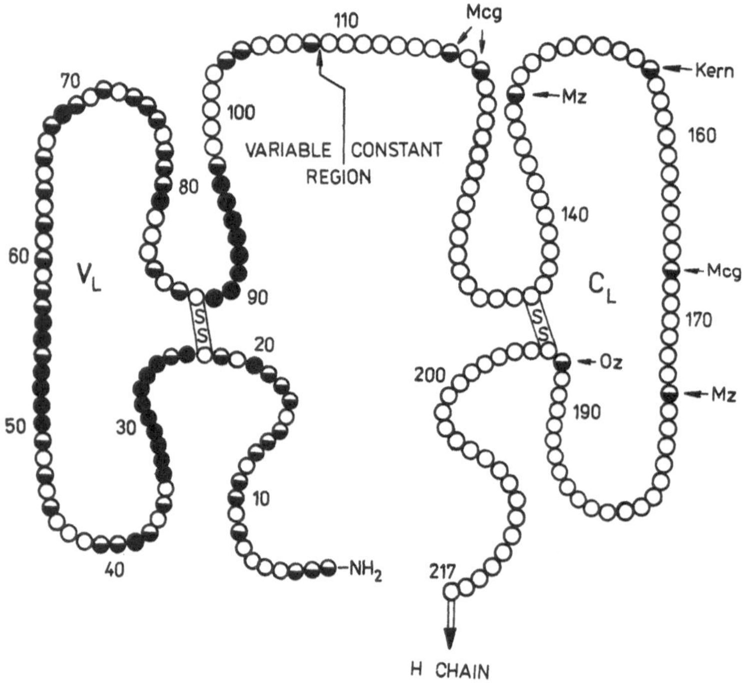

Fig. 6.11. Primary structure of the human λ-chain.
Each circle represents one amino acid residue. Open circles, invariable amino acids (identical in all the λ-chain types), filled circles, more than three different amino acids may alternate at each position (hypervariable regions), semifull-circles, in various chains only two or three amino acids alternate. Oz, Kern, Mz, Mcg, isotypic variants

6.1.3 Three-dimensional structure of immunoglobulins

Research into the primary structure of immunoglobulins yielded several findings that may be summarized in four points:

1. Variable regions of *L* and *H* chains are homologous (they have a common basis, meaning that identical amino acid residues are present at many positions). A lower structural homology was found even between the variable and constant regions of the same chain.
2. The constant region of γ-, α- and δ-chains consists of three homologous domains (C_H1, C_H2 and C_H3) exhibiting a high degree of mutual homology, and homology with the constant region of *L* (C_L) chains. Heavy chains of the μ and ε type have four such regions homologous in their constant region (*i.e.* even C_H4).
3. Every *V*-, as well as *C*-homologous region of both chain types (*L* and *H*) contains one disulphide bond, by means of which two parts of the same region are connected; at the same time, a loop is generated forming a regular and characteristic structure. The loop contains about 60 (*C*-region) or 67 (*V*-region) amino acid residues and usually begins in position 23.
4. The segment containing disulphide bonds formed among different chains, is localized in the centre of the linear sequence of *H* chains, and is not homologous with any other part of *L* or *H* chains and is called the hinge region.

On the basis of the above-mentioned facts, as well as findings suggesting that individual physiological properties of immunoglobulins may be localized in various parts of their molecule, SINGER and DOOLITTLE (1966) and EDELMAN and GALL (1969) named the segments of polypeptide immunoglobulin chains with homologous amino acid sequence, **domains**.

Immunoglobulins of the IgG class have six different domains: V_L, V_H, C_L, C_H1, C_H2 and C_H3 (*fig. 6.12*). In addition, the IgM molecule contains an additional domain in the μ-chain (C_H4). Two domains of the μ-chains are in the basic subunit associated with the *L* chain (V_H and $C\mu1$) and two are released during the limited tryptic hydrolysis from the Fc fragment ($C\mu3$ and $C\mu4$). The fifth domain ($C\mu2$) is degraded by trypsin to its constituent polypeptides. The cysteine residue participating in disulphide bonds among the subunits during the IgM pentamer formation, is localized in the $C\mu3$ domain in position 414.

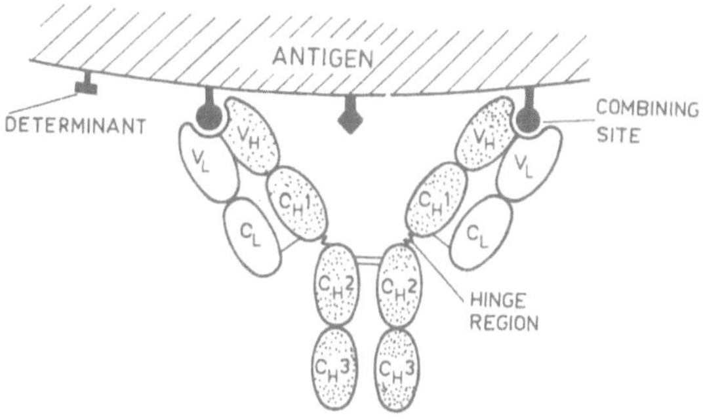

Fig 6 12. IgG domain structure.

A survey of domains, amino acid residues and oligosaccharides in the heavy chains of human immunoglobulins is given in *Table 6.2*.

Table 6.2 Number of domains, amino acid residues and oligosaccharides in heavy chains of human immunoglobulins

Chain	*Domains*	*Amino acid residues* *V-region*	*C-region*	*Hinge region*	*Oligosaccharides* *GalN (hinge region)*	*GlcN (domains)*
γ_1	4	118	314	15	0	1
γ_2	4	118	313	12	0	1
γ_3	4	118	313	62	0	1
α_1	4	119	327	26	5	2
α_2	4	119	327	13	0	4 or 5
δ	4	129	319	64	4 or 5	3
μ	5	126	450	0	0	5
ε	5	125	420	0	0	6

GalN, galactosamine-containing oligosaccharides, GlcN, glucosamine-containing oligosaccharides

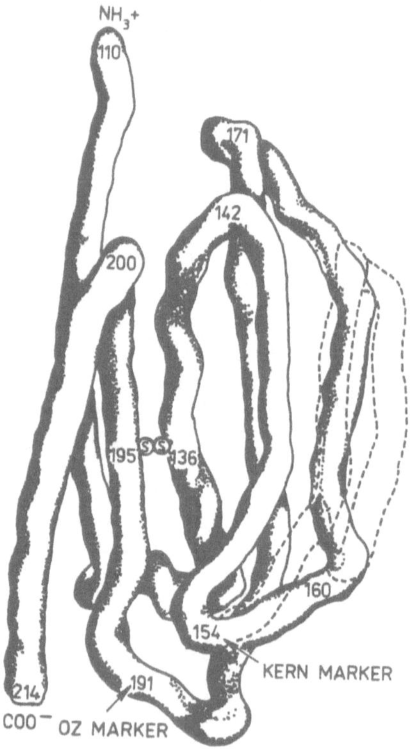

Fig. 6.13. Tertiary structure of C_L- and C_H1-domains (according to Poljak, 1975).
Dotted line, another polypeptide chain loop localized in V_L- and V_H-domains Numbers refer to the amino acid residues of the λ-chain

The tertiary structure of domains was elucidated on the basis of X-ray diffraction analysis of crystals prepared from Fc- and Fab-fragments and subsequently also from whole immunoglobulin molecules. This was accomplished by the groups of POLJAK, DAVIES, EDMUNDSON and HUBER. In 1975, EDMUNDSON *et al.* showed that each domain, regardless of whether it originated from the *L* or *H* chains, possessed an identical three-dimensional structure, termed a "fold" (*Fig. 6.13*). It consists of several long segments interconnected by arcs of various length and shape (*Fig. 6.14*). In comparison with *C* domains, *V* domains have one segment with arcs more over, which is given by a higher number of amino acid residues. Each domain is stabilized by one intrachain disulphide bond and individual domains are connected through short, less-folded chains. The adjacent segments form a folded leaf structure (β-structure).

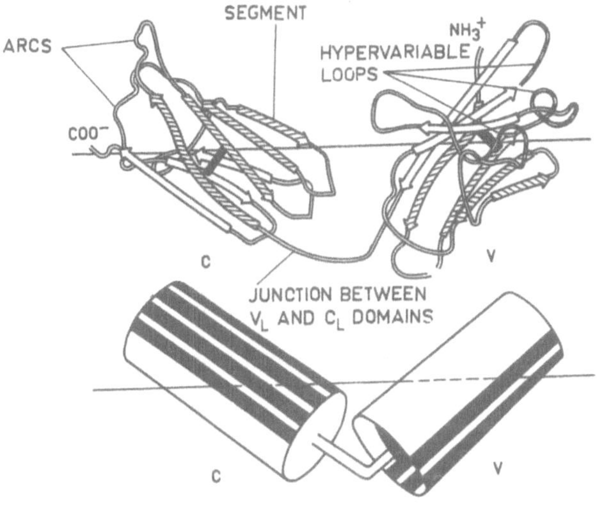

Fig. 6.14. Tertiary structure of the *L*-chain of the IgG molecule (according to Capra and Edmundson, 1977).
C, constant region; V, variable region.

As regards the spatial arrangement and mutual homology, three domain types exist: V_H and V_L; C_L, C_H1 and C_H3; and C_H2. The C_H2 domain differs from other domains by binding with branched oligosaccharide units (carbohydrate moiety).

The interaction between various parts of the immunoglobulin chain is responsible for its arrangement into domains, and the interaction among domains results in the final three-dimensional structure of the immunoglobulin molecule. There are two possible types of inter-domainal interaction: lateral and longitudinal. As a result of lateral interactions the **modules** V_L–V_H, C_L–C_H1 and C_H3–C_H3 are generated. The C_H2 domains remain isolated, because the mutual interaction is blocked by oligosaccharide chains that are bound here (*Fig. 6.15*) or, more precisely, the interaction of the C_H2 domain-forming polypeptide chain can be achieved with these oligosaccharide residues only. The domains in *V* modules overlap at different sites from the domains in *C* modules. During the module generation, the Van der Waals forces are much more important than hydrogen bonding.

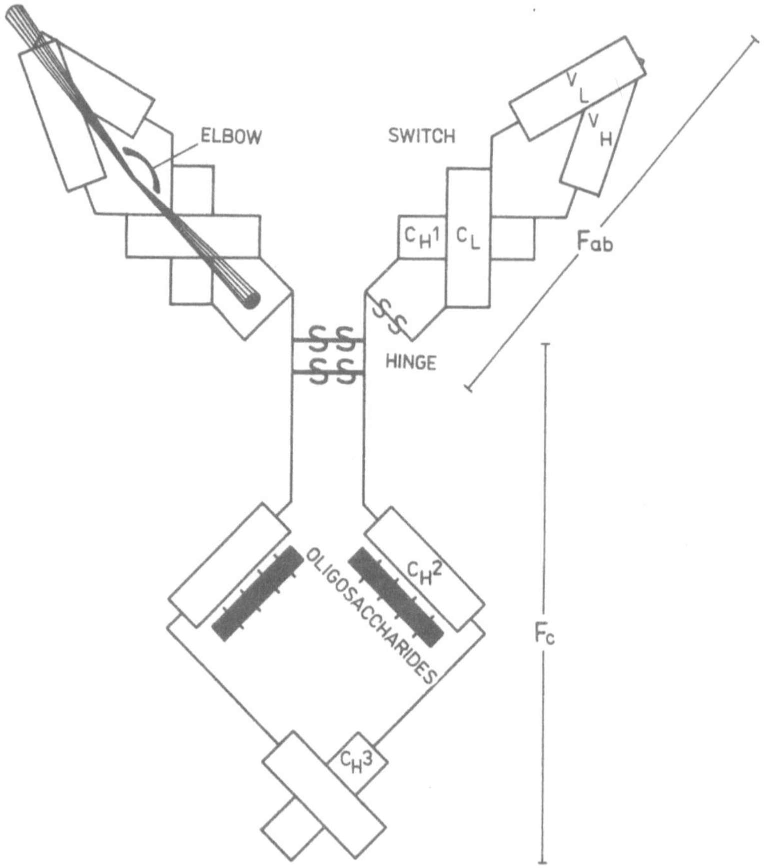

Fig. 6.15. The three-dimensional structure of the IgG1 molecule (Marquart and Deisenhofer, 1982).

The longitudinal interactions along the *H* or *L* chain are much weaker than lateral interactions. However, they are of considerable interest as they influence the antibody conformation changes. Such longitudinal contacts of V_H–C_H1 and V_L–C_L domains result in bending of V_L–V_H and C_L–C_H1 modules between 10 and 45° from the common axis. Even the C_H2–C_H3 orientation may be influenced by external forces to an angle of $\pm 6°$.

6.2 The antigen-combining site

For spatial reasons, the antigen molecule cannot react with the antibody molecule with its whole surface (imagine the contact of two spheres). The specificity requires that only a specific and not just any site of the antibody molecule can react in this way. Precise immunochemical (affinity labelling with specific hapten analogues) and recently crystallographic studies have shown that such a **combining site** is present between the variable parts of heavy and light chains.

In 1963, FRANĚK and NEZLIN showed that the generation of the combining site requires the cooperation of both *H* and *L* chains; the presence of one chain type being insufficient. It was confirmed by HABER (1964) and became generally accepted that the specificity of antibody was related to the primary immunoglobulin sequences (PRESS and PIGGOT, 1967; WAXDAL *et al.*, 1967). In 1970, WU and KABAT termed the segments with high amino acid exchange *hypervariable regions* and FRANĚK suggested that hypervariable regions participate in the formation of a binding site; a hypothesis which was later confirmed (GIVOL, 1974; DAVIES *et al.*, 1975; KABAT *et al.*, 1979). Therefore, the hypervariable segments are also called the **Complementarity Determining Region (CDR)** and the relatively non-variable segments, localized between them, were called the **Framework Region (FR)** (*Fig. 6.16*).

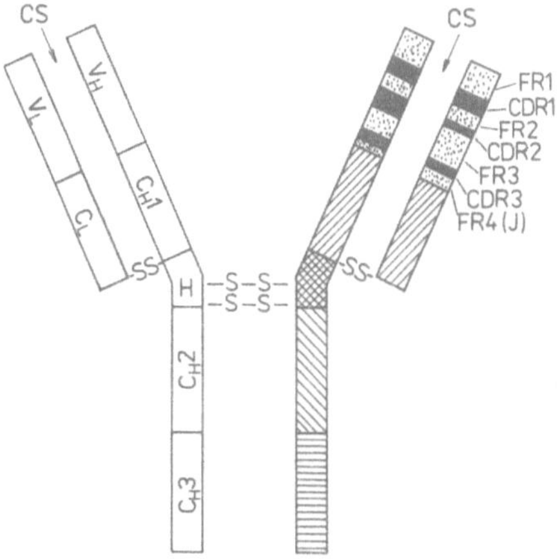

Fig 6 16 Schematic representation of domains and complementarity determining regions (CDR) in the human IgG1 molecule

CS, combining site, H hinge region, FR framework regions (relatively constant regions in V-domains) J joining segment between V_L and C_L or V_H and C_H1 domains

The location of the combining site among variable domains of *L* and *H* chains explains the vast heterogeneity (diversity) of antibodies. Any change in the primary structure is also reflected by changes of the tertiary structure. Thus, the substitution of even a single amino acid can theoretically induce a change in the three-dimensional structure of the polypeptide chain sufficient to alter specificity of the combining site.

Hypervariable regions of each *H* and *L* chain contain approximately 25 amino acid residues. Thus, one can assume that at least 50 positions may participate directly in the formation of the combining site architecture. If it is assumed that 20 naturally occurring amino acids may alternate in each position, an immense number of combinations are obtained. If one gram of each combination was prepared, their total weight would exceed that of the Earth. In reality however, all these combinations are never realized. As will be discussed later, 10^6–10^7 combinations are sufficient for the human organism to allow the formation of all potential antibodies.

Using crystallography, even the size of the combining site could be determined. In the three-dimensional structure of the *V*-region of the IgG molecule, the binding site forms a shallow trough 150×60 nm $\times$ 60 nm deep or eminence with the same height.

Every combining site may bind one antigen determinant or hapten. The strength of the combining site-determinant bond is known as the **affinity**. Antibodies usually possess two or more combining sites. The strength of the bond between the whole antibody molecule and the antigen is termed the **avidity**. Affinity is determined by the complementarity of the spatial structure of the antigen determinant and antibody combining site — the larger the complementarity, the higher the affinity. The CDR regions form loops in the tertiary structures of *V* domains (*Fig. 6.14*), which are located closely to each other on the surface of the V_L–V_H module. The cavity or convexity originating in this way between hypervariable segments of *L* and *H* chains, represents the antibody combining site (*Fig. 6.17*).

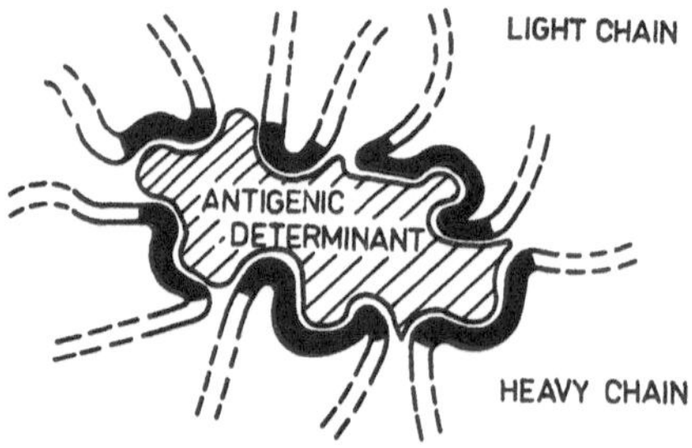

Fig. 6.17. Schematic representation of the antibody combining site.
Hypervariable (CDR) segments of *L* and *H* chains are shown in black.

Antibody molecules of class IgG, IgD, IgE and serum IgA possess two binding sites, *i.e.* they are *divalent*. Thus, every molecule can bind two antigen determinant groups. After fragmentation of the IgG molecule, one combining site remains in each Fab fragment. However, the $F(ab')_2$ fragment contains both binding sites and thus its antibody activity is equivalent to that of the whole immunoglobulin molecule. In theory, the secretory IgA (dimer) molecule has four combining sites, whereas the IgM molecule (pentamer) has ten. The number of effective valencies in the antibody molecule actually depends on the size and shape of the antigen molecule. According to METZGER (1974) the mean distance between binding sites in the IgM subunit is 1 000 nm. When the distance between two determinant groups in the antigen molecule is higher than the above value, or when the antigens are large, all the combining sites do not react. For spatial reasons, IgM antibodies only have five binding sites for most antigens.

EDBERG and coworkers prepared antibodies against dextran that belonged to the IgG or IgM class which subsequently were allowed to react with dextran fragments of various molecular weights. The results are shown in *Table 6.3*. It was found that IgG and IgM antibodies may have 2–1.1 and 10–2.3 combining sites respectively, depending on the size of the dextran molecule.

Table 6.3 Relation between the number of antibody binding sites and antigen relative molecular weight (according to Edberg *et al.*, 1976)

Relative molecular weight of dextran (antigen)	*Number of antibody binding sites*	
	IgG	*IgM*
342 (isomaltose)	2	10
2 800	2	7.5
4 300	2	6
7 100	2	5
237 000	2	5
492 000	2	4
1 870 000	1.1	2.3

6.3 Relationship between structure and function of immunoglobulins

Antibody molecules have two main roles in the humoral immune system: (1) in antibody specificity, which is the basis of recognition, and binding with complementary antigen determinants, particularly of foreign origin (the *recognition function*); and (2) in other biological functions, called *effector functions*. These are often triggered by the reaction with the corresponding antigen; parts of the immunoglobulin molecule other than the combining site are responsible for this activity and they secure the elimination of the foreign antigen and clearing of the internal environment from foreign or altered components.

Immunoglobulins were among the first proteins systematically studied with respect to the relationship between structure and biological function. Thus, relatively abundant information on this subject has accumulated. It was found that each domain possesses specific functions (*Fig. 6.18*). The Fab fragment domains determine the recognition function of antibodies, whereas the domains in the Fc fragment are responsible for the effector functions. Major or minor differences have been observed in these functions among individual immunoglobulin classes or subclasses.

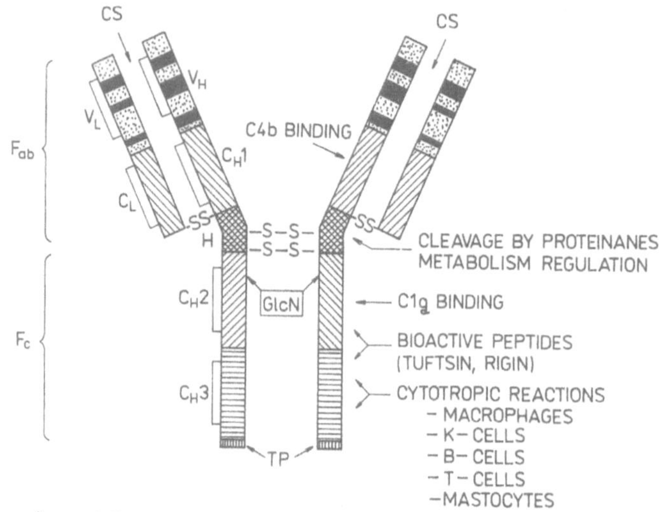

Fig 6 18 Properties and functions of domains in the immunoglobulin molecule
CS, combining site, H, hinge region, GlcN, glucosamine-containing oligosaccharides, Clq, C4b, fragments of complement components, TP, tail piece of heavy chains

It is presumed that the reaction of the combining site with a specific antigen elicits a signal that is transferred gradually by means of lateral and longitudinal interdomain interactions up to the Fc region. Such a signal induces conformation changes in the C_H2 domain which consequently results in emergence of the binding site for the first complement component — Clq. Its binding then triggers the whole complement cascade *via* the classical pathway. In the later phase of the cascade, the fragment C4b is covalently bound to the C_H1 domain.

Immunoglobulins alone are usually unable to kill microorganisms or foreign cells, which have penetrated the body. Thus, they have to work in combination with the complement system, phagocytes or other immune mechanisms. *Complement activation* is therefore an important part of immunoglobulin function. Not all immunoglobulins, however, are able to activate complement. The classical complement pathway is activated only by antibodies of the IgM, IgG1 and IgG3 classes and, to a lesser degree, also by IgG2 (*Table 6.4*). The ability to activate complement is localized in the $C\gamma2$ and $C\mu3$ domains of the Fc region. The site for complement binding *via* the alternative pathway is present in other parts of immunoglobulin *H* chains, including the hinge region. Other immunoglobulins, in addition to IgM and IgG, can also activate complement by the alternative pathway.

Domains $C\gamma2$ and $C\gamma3$ have an amino acid sequence with is responsible for *passage of whole IgG molecules across the placenta.* Antibodies of other classes lack this sequence, and therefore, do not pass across the placenta.

Table 6.4 Properties of human immunoglobulins

Property	Classes								
	IgG				*IgM*	*IgA*		*IgD*	*IgE*
	IgG1	*IgG2*	*IgG3*	*IgG4*		*IgA1*	*IgA2*		
% of all immunoglobulins			78		7	14.5		0.5	0.002
Saccharides (%)			2.6		10	5 to 10		10 to 12	13
Type of *L*-chain			$\varkappa, \lambda$		$\varkappa, \lambda$	$\varkappa, \lambda$		$\varkappa, \lambda$	$\varkappa, \lambda$
Type of *H*-chain	γ_1	γ_2	γ_3	γ_4	μ	α_1	α_2	δ	ε
Distribution (%)	68	24	7	1		90	10		
Sedimentation coefficient (S)			6.7		19	7 to 15		7	8
Relative molecular weight			150 000		900 000	160 000 390 000[a]		180 000	190 000
Mean serum concentration (g/l)			12.5		1.1	2.5		0.05	0.00015
	8.5	3.0	0.85	0.15		2.25	0.25		
Biological half-life (days)	21	20	7	21	5.1	5.9	4.5	2.8	2.4
Complement activation by the classical pathway	+	±	+	−	+	−	−	−	−
Placental transfer	+	−	+	+	−	−	−	−	−
Binding to macrophages	+	±	+	−	−	−	−	−	−
Fixation to skin:									
heterologous[b]	+	−	+	+	−	−	−	−	−
homologous	−	−	−	−	−	−	−	−	+
Number of binding sites	2	2	2	2	5	2	2	2	2

[a] Secretory IgA, sedimentation coefficient 11.45 S.
[b] Sensitization of guinea-pigs for anaphylaxis.

The hinge region transfers the signal between Fab and Fc regions, and at this particular site IgG and IgD molecules may be cleaved to Fab and Fc fragments by limited proteolysis. This is not true for IgA, because its hinge region is highly resistant to proteinases. This is because of its particular structure — it contains numerous proline residues and is coated with galactosamine-containing oligosaccharides. IgM and IgE do not possess a hinge region, but they do have an extra domain. In IgM, this extra domain is cleaved by proteolytic enzymes which results in formation of the Fab fragments and the circular Fc fragment, which retains its ability to bind Clq. It seems that the extra domain (C_H2) of the μ-chain is an evolutionary precursor of the hinge region. Proteinases may also cleave immunoglobulin chains at other sites. Such fragmentation is an initial step of normal immunoglobulin catabolism. It may induce formation of some *immunoregulatory peptides.*

Tuftsin and rigin belong to the best-known immunoregulatory peptides originating from immunoglobulin molecules. **Tuftsin** (named after the Tufts University, Boston, where tuftsin was discovered by NAJJAR and NISHIOKA in 1970) is a tetrapeptide (Thr-Lys-Pro-Arg) which splits off from the C_H2 domains of all four human IgG subclasses (positions 289–292 in the γ_1-chain). It participates in the regulation of phagocytosis. Phagocytosis is also stimulated by another tetrapeptide called **rigin** (Gly-Gln-Pro-Arg). Rigin was named by CHIPENS after the place of its discovery, Riga. It is localized in the C_H3 domains of heavy chains of all four IgG subclasses — in the γ_1-chain it is formed by the first four amino acid residues of this domain (position 340–343).

In addition to complement activation, the *cytotropic reactions* belong to the most important immunoglobulin functions, in which the IgG antibody molecule binds through the *C*-terminal part of its *H* chains (domains C_H3) to specific sites (Fc-receptors) on the surface of phagocytes, *K*-cells, *B*-cells, *T*-cells and mastocytes. Thus, the basis of the cytotropic reaction is binding of antibody to the target cell by means of the Fc region and not through the combining site. This permits the antibody molecule to bind with antigen determinants on the surface of one cell through binding sites, and through the Fc region with Fc-receptors on the surface of the second cell. In this way, a bridge is formed which permits contact and interaction of both cells. When such a bridge is formed, *e.g.* between the target cell and *K*-cell, a signal for killing the target cell by the *K*-cell is emitted.

So far, it is not clear whether only the C_H3 domain, or also the C_H2 domain participate in cytotropic reactions. It is presumed, however, that the interaction of both C_H3 and C_H2 domains is primarily involved and that the key role is played by a segment of heavy chains, called the tailpiece (TP).

The existence of **Fc-receptors (FcR)** on phagocyte membranes enables the rapid phagocytosis of antigen–antibody complexes and thus uptake and removal of undesirable antigens and molecules from the body. Phagocytosis of such immune complexes is much faster than phagocytosis of the pure antigen. In the Fc-receptor-mediated phagocytosis, the activation of the Fc-region domains by reaction with antigen is required. Only a small proportion of IgG antibodies has a high enough binding energy, to allow spontaneous binding to phagocytes even without a reaction with the antigen. These are termed *cytophilic antibodies*. Phagocytosis is facilitated by cytotropic IgG or even by IgM antibodies, although the latter act in cooperation with complement only. Macrophage membranes also bear FcR for IgA and IgE and it appears that IgE plays an important role in the defence against parasitic infections.

By using monoclonal antibodies which recognize Fc-receptors, three types of IgG binding structure — FcRI, RcRII and FcRIII — have been defined (*Table 6.5*). They are expressed on the surface of different cells and they have distinctive affinity for the individual subclasses of IgG:FcRI has the highest affinity for IgG1 and IgG3 and smaller for IgG4; FcRII and FcRIII preferentially bind IgG1 and IgG3. GERGELY *et al.* (1986) have shown that FcRIII express one $C\gamma2$ and another $C\gamma3$ domain-specific binding site while FcRI possess only one ($C\gamma2$-specific) site. Besides binding, the $C\gamma2$-specific site transfers killing signals for a target cell while the $C\gamma3$-specific site has only binding capacity.

Table 6.5 Fc receptors for human immunoglobulins

Ig isotype (ligand)	*Receptor*	*Affinity*	*Mol. wt. (kD)*	*Location on cells*	*Differentiation markers*
IgG	$Fc_{\gamma}RI$	high	70	Ma, Mo	—
	$Fc_{\gamma}RII$	low	40	*B*	CDw32
			33	Ma, Mo, Ne, Eo, Tr	
	$Fc_{\gamma}RIII$	low	50–70	*NK*, *K*, Ne, Eo, Ma	CD16
IgE	$Fc_{\varepsilon}RI$	high	100	Ba, MC	—
	$Fc_{\varepsilon}RII$	low	45	*B*, some *T* Eo, Ma, Mo, *NK*, Tr	CD23
IgM	$Fc_{\mu}R$	medium	40	Mo, Ma, Ne, *T*	—
IgA	$Fc_{\alpha}R$	medium	?	Mo, Ma, Ne, Eo, *B*, *T*	—

Ma — macrophages, Mo — monocytes, Ne — neutrophils, Eo — eosinophils, Ba — basophils, Tr — platelets, MC — mast cells, *B*, *T* — *B*- or *T*-lymphocytes, *NK* — *NK*-cells, *K* — *K*-cells

The low affinity $Fc_{\gamma}RIII$ (CD16) is a surface glycoprotein with $M_r = 50\,000–60\,000$. It is expressed on neutrophilic granulocytes, *NK*-cells and on some eosinophilic granulocytes and tissue macrophages. Through the $Fc_{\gamma}RIII$ the *NK*-cells mediate antibody-dependent cytotoxicity whereas the neutrophils are responsible for immune phagocytosis. In addition to these activities, the Fc-receptors and their ligands generally participate in the control of antibody biosynthesis and transport, lysosomal enzyme release and arachidonic acid metabolism.

The structure of the majority of Fc-receptors is similar; they form a single polypeptidic chain in which the extracellular Ig-binding portion is composed of two or more Ig-like domains. The high affinity receptor for IgE ($Fc_{\varepsilon}RI$) is an exception, and so far, is the only one known to consist of three different polypeptide chains. In addition to the IgE-binding α-chain, it is composed of a single β and two disulphide-linked γ-chains. The α-chain is similar to the functionally homologous chain of the $Fc_{\gamma}R$. It contains 227 amino acid residues (180 of them in two extracellular domains), 30% carbohydrate and has a molecular weight of 37 000. The molecular weights of the β- and γ-chains are 33 000 and 7 000 respectively and both are devoid of carbohydrate.

The high affinity $Fc_{\varepsilon}RI$ is only found on mast cells and basophils. IgE antibodies may bind to them through the Fc domains. Once the bound IgE has reacted with specific antigen, histamine and other mediators of the immediate type of hypersensitivity are released, which in turn induces the allergic (anaphylactic) reaction or even anaphylactic shock.

Because of this ability to induce *immediate-type allergic reactions*, IgE antibodies are also called **reagins**. **Anaphylaxis** can be transferred indirectly (passively) when an individual is injected with the cytotropic antibody and subsequently, with the specific antigen. Passive transfer of anaphylaxis between two individuals of the same species may be accomplished with *homocytotropic antibodies*, while an interspecies transfer can be achieved with *heterocytotropic antibodies*. Human homocytotropic antibodies belong to the IgE class, whereas heterocytotropic antibodies belong to the IgG class.

Besides the high affinity $Fc_{\varepsilon}RI$, METZGER (1988) distinguished additional low affinity $Fc_{\varepsilon}RII$ that is composed of 321 amino acid residues and is homologous to hepatic lectin.

6.3.1 The function of the carbohydrate moieties of immunoglobulins

Heavy chains of all classes of immunoglobulins possess sites which bind to oligosaccharides (the carbohydrate moiety). γ-chains only bind one oligosaccharide, whereas other *H* chains can bind several oligosaccharides. Two oligosaccharide types are present in immunoglobulins. The first oligosaccharide type contains *glucosamine* (*GlcN*), which binds to asparagine through the *N*-glycosidic bond if the polypeptide chain has the signal sequence Asn-X-Ser-Thr (where X = any amino acid residue). As a result of incomplete arrangement (post-translation processing) or later degradation, the **GlcN oligosaccharides** do not possess a uniform structure. They are formed either by a simple chain containing mainly mannose residues or a di- or tri-branched structure with a more heterogeneous saccharide composition. GlcN oligosaccharides bind to domains in the *H* chain (*Table 6.2*).

The second main oligosaccharide type contains galactosamine (GalN) which is bound through the *O*-glycosidic bond to serine or threonine residues. So far, no signal sequence has been found in the vicinity of Ser or Thr, but the **GalN-oligosaccharides** are normally bound to several sites close to each other in the hinge region of IgAl and IgD. Even GalN-oligosaccharides may have a heterogeneous structure. Their molecular weight is about 750, *i.e.* lower than GlcN-oligosaccharides, which have molecular weights of 2 500–3 000.

As the saccharide chains are hydrophilic, they must be localized on the surface of the immunoglobulin molecule or between domains. Therefore, they participate in immunoglobulin-to-cell interaction as well as in lateral and longitudinal domainal interactions.

GlcN and GalN oligosaccharides are normally found only in constant regions of immunoglobulin chains. Their number, type and position are characteristic of each immunoglobulin class. IgG of any subclass and any animal species have only one GlcN oligosaccharide (bound approximately to position 300 of the γ-chain). Other heavy chain classes have between two and five GlcN oligosaccharides. The α_1- and δ-chains have four to five GalN oligosaccharides in their hinge region. Thus, the saccharides contribute to the total relative molecular weight of immunoglobulins by between 3% (IgG) and 13% (IgE).

Despite intensive studies, the biological function of the carbohydrate moiety of the immunoglobulin molecule has not been sufficiently elucidated. Presumably, it facilitates solubility, acts as a spatial gap between chains and domains, influences signal transduction from the combining site to the effector segments in the Fc region, participates in certain effector functions such as Clq binding and cytotropic interaction with lymphoid cells and regulates catabolism both by the protection against proteases and by governing immunoglobulin uptake by hepatic receptors. According to MELCHERS and KNOPF (1967) it appears that the carbohydrate moiety might facilitate transfer of immunoglobulins across membrane systems of plasma cells. Biosynthesis of immunoglobulins takes place in plasma cells on membrane-bound polyribosomes. Immediately after biosynthesis, IgG molecules contain no saccharides at all, or only minute amounts of glucosamine and mannose. The completed IgG molecules pass into the cisternae of rough endoplasmic reticulum, and, during this process, additional monosaccharides are bound to them (the glucosamine and mannose content is completed and some galactose is added). The immunoglobulin molecule then passes to the smooth endoplasmic reticulum, where IgG acquires additional monosaccharides (except fucose, which is bound only after leaving the plasma cell). Evidence against such a carbohydrate function in the transport of immunoglobulin molecules out of the cell is the observation that *L*-chains do not contain carbohydrate, but are nevertheless excreted into the extracellular space. In patients with multiple myeloma this occurs in the form of large amounts of Bence-Jones protein. It was also found that deficient clones of plasma cells may excrete immunoglobulins with an incomplete carbohydrate moiety.

Recently (BAENZIGER, 1984), it was shown that the immunoglobulin oligosaccharide component might be important in the biological turnover of these substances. When the immunoglobulin molecule contains complete oligosaccharides, it remains in the circulation. However, when the terminal monosaccharides are split off from oligosaccharide chains, the whole immunoglobulin molecule is degraded. Degradation occurs mainly in the liver. The liver cells (hepatocytes) have specific lectin receptors on their surface that recognize the penultimate saccharide residue in the oligosaccharide chain as "foreign". The terminal saccharides, however, are not recognized by these receptors as foreign. The penultimate immunoglobulin saccharides are usually D-galactose and D-mannose, whereas the terminal saccharide is sialic acid (*N*-acetylneuraminic acid) or L-fucose. Immunoglobulins that lose terminal monosaccharides become incomplete and altered and are therefore removed from the circulation by hepatocytes and macrophages. The half-life of such physiological degradation ("uptake") depends on the immunoglobulin type, individual genetic factors and state of health. Changes in the structure of the carbohydrate moiety of immunoglobulin molecules are catalysed by specific enzymes — glycosyltransferases.

This system is also responsible for the physiological turnover of other glycoproteins. In addition, it fulfils many other biological functions, particularly during recognition, adherence and cell interaction. During immunoglobulin catabolism, however, certain limiting mechanisms are necessary. Thus, only oligosaccharides with a complex heterogeneous structure (containing various monosaccharides) can bind to the hepatocyte receptors, whereas oligosaccharides rich in mannose residues do not bind. As limited proteolysis is one of the first steps of catabolism, the biological half-life of individual immunoglobulin classes depends on the conformation of the protein component and on the function of the carbohydrate moiety during binding to corresponding receptors.

6.4 Properties of immunoglobulins

Individual immunoglobulin classes and subclasses differ in chemical, physicochemical, antigenic, metabolic and biological properties (*Tables 6.4, 6.6* and *6.7*).

The chemical properties of immunoglobulins are determined by their primary structure, *i.e.* the amino acid content and sequence and, in addition, since immunoglobulins are glycoproteins, also by the carbohydrate content. The amino acid sequence determines the type of *L* and *H* chains. The *H* chain type permits the classification of immunoglobulins into individual classes and subclasses.

Other characteristics of immunoglobulins are also determined by their structure. Among their physicochemical properties, the sedimentation coefficient is particularly important in determining the molecular weight which typifies the individual classes. However, minor differences in relative molecular weight can be observed within individual IgG subclasses both in man and other mammals. Indeed differences in relative molecular weight and other characteristics within the same immunoglobulin class can be observed even among individual animal species. For example, murine and rat IgD have lower relative molecular weights because their heavy chains are lacking one domain (C_H2).

6.4.1 Normal immunoglobulin concentrations

Immunoglobulins are present in various tissues and body fluids (blood, cerebrospinal fluid, saliva, mucous membrane secretions *etc.*). Their concentrations are not, however, uniform; the highest concentration is found in colostrum and blood serum. In addition, there are large differences in the proportion of individual classes and subclasses. IgG is present in the highest amounts (78%) and IgE in the lowest (it comprises only 0.002% of the total amount of circulating immunoglobulins). The levels of individual classes vary within certain physiological limits, reflecting the immunological reactivity of

Table 6.6 Properties of rat immunoglobulins

Property	*IgG1*	*IgG2a*	*IgG2b*	*IgG2c*	*IgM*	*IgA*	*IgD*	*IgE*
Sedimentation coefficient (S)	6.7	6.4	6.5	6.7	17–19	7	6.5	7.6
Molecular weight ($\times 10^3$)	156	156	156	156	900	163	140	190
Serum concentration (g/l)	11.1	11.0	1.3	2.2	0.97	0.15	0.008	0.1×10^{-5}
T/2 (days)	5.5	5.0	?	?	2.6	?	?	0.5
C1 activation	+	+	?	–	+	–	–	–
Placental transfer	+	+	?	?	–	–	–	–
Presence in colostrum and milk	+	+	?	?	–	+	?	+

In comparison with human IgD, rat IgD lacks $C_{\delta}2$ domain and its molecular weight is therefore lower.

the organism. The finding of immunoglobulin levels above or below normal values is of great diagnostic significance — it may indicate infectious processes, immunodeficiency or immunopathological conditions.

The mean normal immunoglobulin values in serum, colostrum, milk and saliva in man and some animals are given in *Tables 6.6–6.8*. Human immunoglobulin levels are expressed in g/l or international units (IU) deter-

Table 6.7 Characteristics of human and some animal species immunoglobulins

Species		*Classes and subclasses*								
Man		IgG1	IgG2	IgG3	IgG4	IgA1	IgA2	IgM	IgD	IgE
	a	150	150	160	150	160	160	900	180	190
	b	6.7	6.7	6.7	6.7	7	7	19	7	8
	c	8.5	3.0	0.85	0.15	2.25	0.25	1.1	0.05	0.00015
	d	21	20	7	21	5.9	4.5	5.1	2.8	2.4
Cow		IgG2	—	—	IgG1		IgA	IgM	?	IgE
	a	150	—	—	160		400	1000		200
	b	6.6	—	—	6.8		10.9	19.4		7.5
	c	9.2			11.2		0.37	3.05		?
	d	17.7	—	—	9.6					
Horse		IgGa	IgGb	IgGc	IgGT		IgA	IgM	?	IgE
	a	150	150	150	160		160	900		
	b	6.6	6.6	6.6	6.7		6.7	19		
	c		13.4		8.2		1.5	1.2		
Pig		IgG1	—	IgG2			IgA	IgM	?	IgE
	a	150		160			330	1000		
	b	6.6		6 8			9.5	18.1		
	c		21.5				2.2	2.8		
Rabbit		IgG2	—	—	IgG1	IgA1	IgA2	IgM	?	IgE
	a	160	—	—	150	160		900		190
	b	6.9	—	—	6.7	7		19		8
	c		11.1							
	d		5.6							
Mouse		IgG2a	IgG2b	IgG3	IgG1	IgA1	IgA2	IgM	IgD	IgE
	a	155	155	155	155	160		900	140	190
	b	6.6	6.6	6.6	6.6	6.7		19	6.5	8
	c	15.0	4.0	0.9	1.9	0.1		1.6	0.006	0.00001
	d	6.5	3.0	?	8	0.5		1.5	0.5	?
Chicken		IgG					IgA[e]	IgM	?	?
	a	180					160–400	950		
	b	7.1					7.17–19	16.7		
	c	5.5					0.35	2.5		
	d	4.1						1.7		
Snake		IgG					?	IgM	?	?
Amphibians		IgG					?	IgM	?	?
Fish		?					?	IgM	?	?

a, Relative molecular weight ($\times 10^3$)
b, Sedimentation coefficient (S)
c, Mean serum concentration (g/l)
d, Biological degradation half-life (days)
[e]Avian sera contain two IgA types — 7S and 17–19S Serum 17–19S IgA is identical to IgA isolated from the bile Very different immunoglobulin serum concentrations may be found in individual strains of mice and rats and in various age groups of animals

mined according to an international standard. The World Health Organization recommended the following conversion factors for individual immunoglobulins:

IgG 100 IU/ml = 8.00 g/l
IgA 100 IU/ml = 1.42 g/l
IgM 100 IU/ml = 0.83 g/l
IgD 100 IU/ml = 0.14 g/l
IgE 100 IU/ml = 0.0003 g/l

Individual laboratories and manufacturers of kits for immunoglobulin determination calibrated their own standards according to the WHO standard and determined conversion coefficients (IU/ml to g/l). Unfortunately, different manufacturers have so far also used different conversion factors. Nevertheless, immunoglobulin values, expressed in IU, are comparable even when kits produced by various manufacturers are used. Thus, when the

Table 6.8 Mean immunoglobulin concentrations in serum, colostrum, milk and saliva of man and some domestic animals

Species	*Concentration (g/l)*		
	IgG	*IgM*	*IgA*
Man			
serum	12.5	1.1	2.5
colostrum	0.45	1.5	17.5
milk	0.04	0.1	1.0
saliva	0.01	traces	0.015
Pig			
serum	21.5	2.8	2.2
colostrum	58.7	3.2	10.7
milk	3.0	0.8	3.7
Cattle			
serum	20.0	3.0	0.4
colostrum	55.0	6.5	5.4
milk	0.5	0.09	0.08
saliva	0.05	0.01	0.5
Sheep			
serum	21.0	1.2	0.25
colostrum	60.0	4.1	2.0
milk	0.3	0.03	0.06
saliva	0.1	traces	0.2
Goat			
serum	22.0	1.6	0.35
colostrum	58.0	3.8	1.7
milk	0.3	0.03	0.06
saliva	0.1	traces	0.2
Chicken			
serum	5.5	2.5	0.35
saliva	0.3	0.01	0.09
bile	0.2	0.3	3.15

immunoglobulin concentration is expressed in IU, no further explanation is required. However, when their concentration is expressed in weight units (g/l), the standard used (or manufacturer) must be given.

The mean normal immunoglobulin values in human serum are: IgG, about 140; IgA, 125; IgM, 130; IgD, 35; and IgE, 50 IU/ml. There is a wide variation in IgE values between 10 and 500 IU/ml. A similar high variance is also found in other immunoglobulin classes. For example, in 95% of all USA citizens examined, the following values were found (IU/ml): IgG, 74–234; IgA, 44–347; IgM, 56–408; IgD, 0–170. Immunoglobulin values in the remaining 5% of the population were outside these limits. The following normal ranges were found in the Central Europe population: IgG, 92–207; IgA, 54–268; IgM, 69–287 (males) and 80–322 (females); IgD, less than 100 IU/ml.

Immunoglobulin levels change with age and the above values may only be considered normal in adults between 20 and 50 years of age. The serum immunoglobulin concentration in children is lower and depends on their actual age (*Fig. 6.19*). This is due to the differences in passage of individual immunoglobulins across the placenta, and to the different onset and degree of biosynthesis during pre- and postnatal life. In elderly individuals, on the other hand, the normal levels are higher and some differences can be observed between males and females (*Table 6.9*).

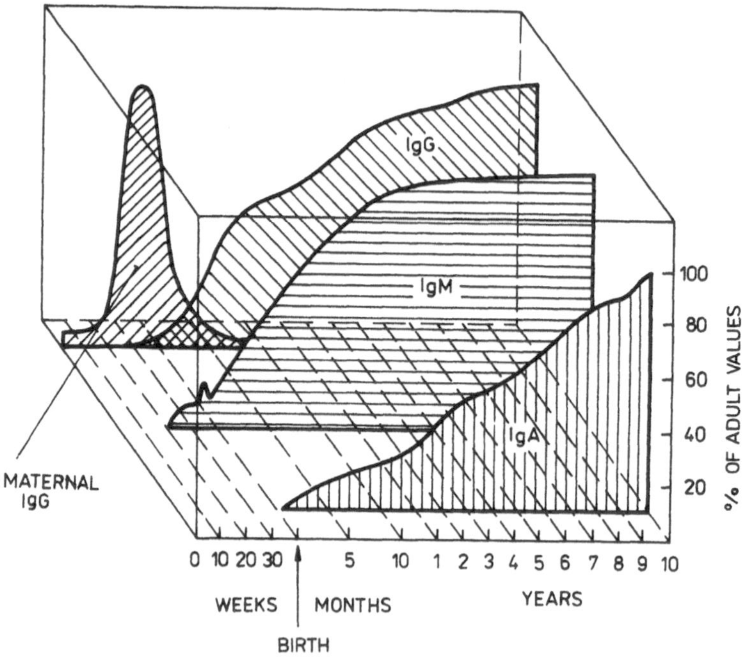

Fig. 6.19. Development of serum immunoglobulins during human ontogeny (according to Nouza and John, 1987).

Table 6.9 Normal human immunoglobulin serum levels (g/l) in adults

Age category (years)	*Mean age*	*IgG*	*IgM*	*IgA*
Males 19–46	32.5	11.7 ± 3.6	0.9 ± 0.6	2.3 ± 1.1
Females 19–46	33.9	12.9 ± 4.7	1.3 ± 0.7	2.7 ± 1.4
Males 52–92	68.5	15.4 ± 3.9	1.2 ± 0.8	2.8 ± 1.7
Females 52–92	73.5	19.1 ± 8.1	1.6 ± 1.3	3.3 ± 1.6

The values are given as means ± standard deviation.

In man, only IgG traverses the placenta. Although by the end of prenatal development, the human foetus has a developed immune system which is capable of immunoglobulin production, this synthesis is minimal because the foetus develops in a sterile environment. The difference between immunoglobulin synthesis and common protein synthesis lies in the fact that the production of large amounts of immunoglobulins that can be excreted extracellularly, requires the presence of antigens, stimulating appropriate antigen-sensitive cells. Thus, under normal conditions, the foetus does not possess significant amounts of its own immunoglobulins. In fact, newborn sera contain only IgG antibodies acquired by transplacental transfer from the maternal circulation. The synthesis of its own immunoglobulins begins only after birth when microflora colonize the mucous membranes and the neonate comes into contact with antigen. The IgG level gradually decreases after birth due to catabolism of passively transferred maternal antibodies, while the infant's own antibody synthesis is still insufficient to replace this loss. The lowest antibody level is found approximately between three and five months of age. Thereafter, the level increases because biosynthesis of the infant's own IgG predominates. Similar developmental changes occur with the IgM and IgA classes. Thus they are synthesized in significant amounts only after birth and reach normal values in adulthood.

The immunoglobulin levels increase throughout life. This is probably associated with their decreasing reactivity and specificity, or with their degradation rate (catabolism). These problems are studied by gerontologists.

If an individual is born and maintained postnatally under germ-free (sterile) conditions, its immune system produces less immunoglobulins compared with individuals living under normal conditions. Such an individual is referred to as a *gnotobiont* and the study of germ-free conditions on life processes is called **gnotobiology**. If the gnotobiont is deprived of dietary antigens (experimental animals are fed a sterile diet containing low-molecular-weight substances only) or even viral antigens, it synthesizes virtually no immunoglobulins because of the lack of adequate antigenic stimulation. Gnotobiotic animal models are advantageous for experimental studies, although gnotobiotic principles have also been successfully applied to human clinical medicine. For example, gnotobiotic isolators help to support life in infants born with serious immune defects, or are used in patients exposed to high doses of ionizing radiation or cytostatic drugs.

6.4.2 *Metabolism of immunoglobulins*

Immunoglobulin levels, as with other proteins, are determined by biosynthetic and catabolic processes that are dynamically balanced under physiological conditions. The basic metabolic characteristics of human immunoglobulins are given in *Table 6.10*.

Table 6.10 Metabolism of human immunoglobulins

	IgG	*IgA*	*IgM*	*IgD*	*IgE*
Synthetic rate (mg/kg/day)	33	24	6.7	0.4	0.02
Daily synthesis (% of total amount in the organism)	3	12	14	28	65
Catabolic rate (%/day)	6.7	25	18	37	89
Biological degradation half-life (days)	21	5.9	5.1	2.8	2.3
Distribution of circulating immunoglobulins (%):					
intravascular space	50	40	76	75	50
extravascular space	50	60	24	25	50

The amount of daily synthesized immunoglobulins decreases and their catabolic rate increases in the following order: IgG–IgA–IgM–IgD–IgE. These differences in biosynthetic and catabolic rates are shown by the decrease in normal levels of individual immunoglobulins in the above order. The most stable is IgG with a half-life of 21 days, and the most labile is IgE with a catabolic half-life of 2.3 days. These values, however, are not the same for all subclasses. Thus, the biological catabolic half-life of IgG1 — contrary to other IgG subclasses — is only 8 days; in IgA1 it is 5.9 and in IgA2 4.5 days. The rate of immunoglobulin catabolism is altered in various diseases; for example, in patients with chronic progressive polyarthritis, the half-life of IgG is only 12–14 days, and in tuberculosis patients, it is 15–17 days.

The immunoglobulin-synthesizing machinery can be imagined as follows: a man with a body weight of 70 kg synthesizes approximately 2 310 mg IgG per day. One mg contains 3.75×10^{15} IgG molecules. Thus, the daily synthesis is 8.7×10^{18} molecules, *i.e.* 10^4 new IgG molecules are produced per second. The same number of molecules are degraded.

All immunoglobulins do not circulate in the intravascular space. Some are present at other sites. IgG and IgE are roughly evenly distributed in the intravascular and extravascular space while a larger proportion of IgA is found outside the blood circulation. This ratio is reversed with IgM and IgD. This distribution is related to the function of individual immunoglobulin classes. Secretory IgA is responsible for local humoral immunity and, therefore, it is present primarily in secretions of various mucous membranes (*e.g.* nasal and intestinal), colostrum, milk and saliva.

Proteolytic enzymes, particularly intracellular proteinases, are responsible for catabolism of immunoglobulins. In the first place, fragments are produced that are further degraded to small peptides and amino acids. The carbohydrate moiety is degraded by glycosidases. The viability of individual fragments is not equal. Experiments with radiolabelled fragments have shown that *L* chains have the lowest viability with a biological degradation half-life of only 0.8–1.6 h. Small amounts of *L* chains are normally present in the urine and serum. Their amount increases in patients with kidney diseases or in patients suffering from the neoplastic growth of plasma cells. The latter condition is associated with synthesis of the Bence-Jones *proteins* mentioned previously. These are dimers or monomers of *L* chains; in any one individual, Bence-Jones proteins of either $\varkappa$- or λ-type occur, but both types are not found. Under normal conditions, both the *L* chains and Bence-Jones proteins enter the blood circulation like the products synthesized *de novo*. These proteins are thus not catabolic products formed by degradation of complete immunoglobulin molecules synthesized previously. Catabolism of *L* chains takes place in the kidney, whereas degradation of whole immunoglobulin molecules occurs at other sites, particularly the liver.

The biological half-life of IgG Fab and $F(ab)_2$ fragments is also very short, usually only 3.6–6 h. The most stable part of the IgG molecule is the Fc fragment with a biological degradation half-life of approximately 11 days. Similarly to other immunoglobulin classes, the Fc fragment has the longest survival. It is therefore responsible for the biological half-life of the whole molecule.

H chains and Fc fragments are not present in the serum of healthy individuals. They can be found, however, in sera and urine of patients with malignant growth of certain plasma cell clones (*heavy chain disease*).

6.4.3 Isotypes, allotypes and idiotypes

Although immunoglobulins possess a similar structure and share a number of characteristics, they are highly heterogeneous in other properties (particularly in antibody specificity). The molecular basis of this *heterogeneity* was elucidated in studies of their primary structure. It was shown that the division of immunoglobulins into different classes, subclasses, types and groups can be explained by a different amino acid sequence in the constant or variable regions of peptide chains. Each polypeptide chain contains characteristic antigenic determinants. Changes in the amino acid sequence may alter the specificity of these determinants, and may be used for preparation of specific antibodies. The determinant groups are present in every antibody molecule. Any antibody is thus an antigen at the same time, and with suitable immunization, an "anti-antibody" against it can be prepared; this again behaves as an antigen and an "anti-anti-antibody" can be prepared *etc*.

The existence of antibody against antibodies permits their serological or immunochemical determination, but at the same time, it makes it possible to detect relatively small structural differences among individual antibodies. Immunoglobulins can thus be divided not only on the basis of their different physicochemical and biological properties, but also on the basis of their antigenic characteristics.

Immunoglobulin antigenic determinants ↗ isotypic → allotypic ↘ idiotypic

Individual classes and subclasses can be differentiated on the basis of antigenic determinants (specific amino acid sequences) in the constant region of heavy chains. Similar differences in the amino acid sequence in the constant region of light chains serve as a basis for differentiation of immunoglobulins into types and subtypes (*Fig. 6.20*).

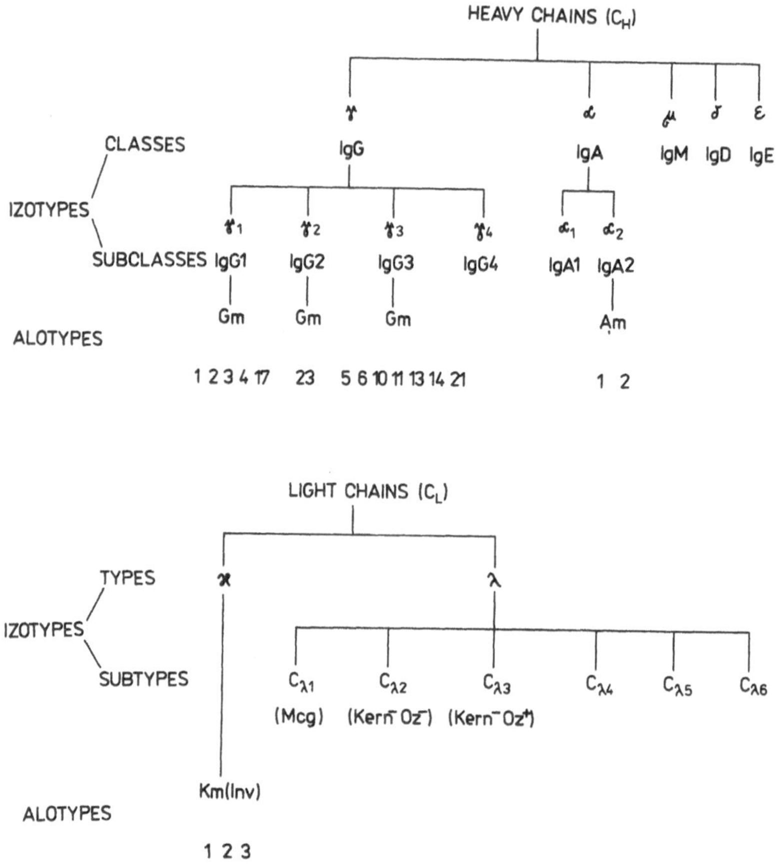

Fig. 6.20. Immunoglobulin classes, subclasses, isotypes and allotypes.

Antigenically different immunoglobulin variants that are common in all individuals of a given biological species are called **isotypes** (for example, IgG, IgA and IgM including the subclasses). They are present in many animal species. Each species, however, has its own characteristic isotypic determinants in immunoglobulin molecules of individual classes and subclasses. Isotypes are products of various structural genes and can be detected using xenogeneous antisera. For example, the rabbit antiserum against the heavy chain γ of human IgG, specifically reacts with IgG in all normal human sera, but does not react with IgG of other animal species.

The existence of several allotypic variants can be detected in many immunoglobulin isotypes. **Allotypes** are not present in all individuals of a particular species, but only in some of them. They are products of individual alleles of a certain structural gene. In a group of related individuals, they behave as genetic markers and are therefore important for the immunogenetic characteristics of an individual. Until now, 18 γ-chain-associated Gm allotypes, two Am allotypes bound to α_2-chain, one allotype Em bound to ε-chain and three Km factors associated to $\varkappa$-chain, have been found in human immunoglobulins (DE LANGE, 1989). The Km factors used to be called Inv factors according to a previous nomenclature (*Fig. 6.20*). Allotypic determinants are also usually present in the constant regions of corresponding chains. An exception to this rule can be found, for example, in rabbit immunoglobulins containing allotypic determinants even in variable domains of *H* chains.

Differences in the antigenic structure of variable portions of *H* and *L* chains serve as a basis for immunoglobulin differentiation into groups, subgroups and idiotypes. *Groups* and *subgroups* belong to isotypic and allotypic variants (*Fig. 6.21*). Thus, they are characterized by determinants specific to a particular animal species and group of individuals within one species.

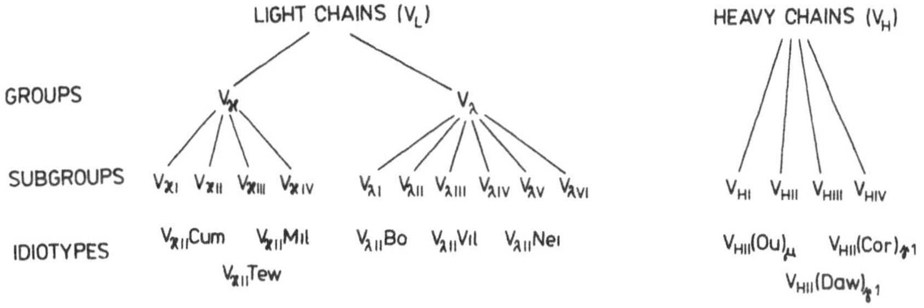

Fig. 6.21. Scheme of groups, subgroups and idiotypes of human immunoglobulins
For simplicity, some known idiotypes are shown only in one subgroup (idiotypes found in myeloma immunoglobulins)

Idiotypes are also characterized by a complex of antigen determinants present primarily in variable domains of the antibody molecule. According to JERNE (1974, 1985) any antigen determinant can be called an *epitope*. If such an epitope is localized in a variable domain of the immunoglobulin chain, it is known as an *idiotope*. This term does not designate a new quality, but only emphasizes a particular epitope location, *i.e.* that it is localized in the same region of the chain as the antigen combining site. A set of idiotopes represents the idiotype. Every individual possesses a specific complex of idiotypes which reflects his history of contact with particular antigens. For the sake of accuracy, it is necessary to note that idiotopes are present not only in hypervariable regions of H and L chains but even in relatively constant segments forming a framework (FR) of variable regions. Idiotypes are thus formed by the spatial arrangement of chains and by their mutual interaction, *i.e.* by a complex of sequential and conformational determinants of variable domains. Using this definition, one can easily understand that molecules of the same idiotype may possess a different isotype (*e.g.* they belong to different immunoglobulin classes), and that immunoglobulins of various species or individuals may share the same idiotype.

Idiotypes common to antibodies of various species and individuals are designated as dominant, public or cross-reacting idiotypes. They are determined by idiotopes outside the combining site. Idiotypes that are present only in individual antibodies or in a small number of molecules, are called individual or private idiotypes. Cross-reacting idiotypes are determined genetically and have an important regulatory role, therefore they are also termed regulatory idiotypes.

Every idiotope can recognize a complementary *paratope* (antibody combining site) in another antibody which is called the anti-idiotype antibody. The anti-idiotype antibody is a mirror image of the idiotype. These are the "anti-antibodies" mentioned above and may be of two types: α and

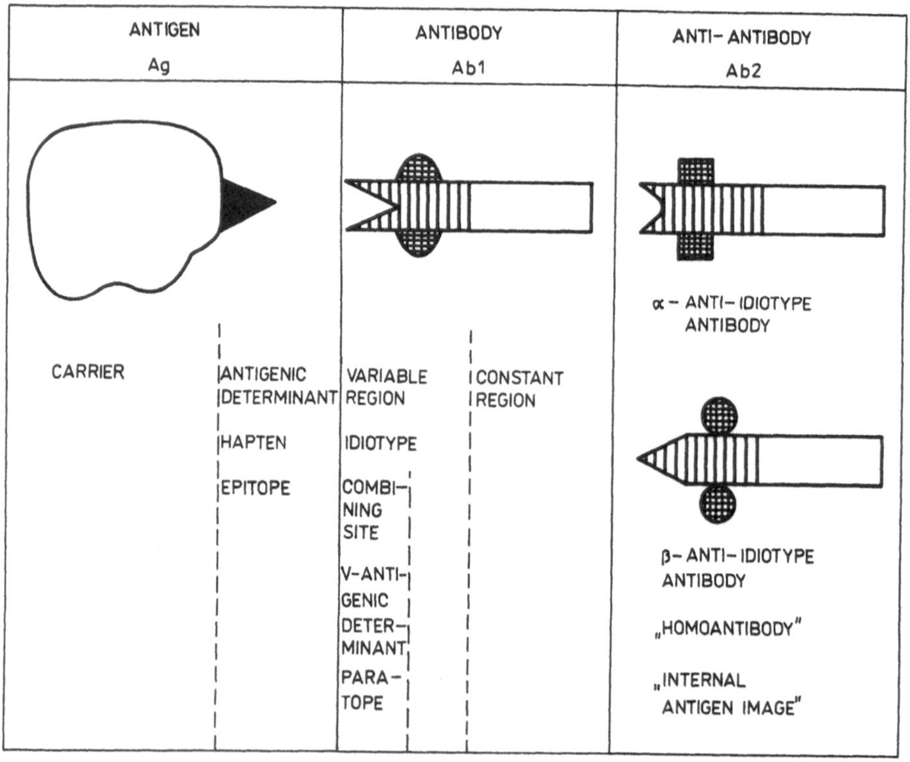

Fig. 6.22. Scheme of idiotypes and anti-idiotypic antibodies (according to Šterzl *et al.*, 1985).

β (*Fig. 6.22*). The *α-anti-idiotype antibody* has a paratope which specifically recognizes an idiotope present in another part of the variable region as the antibody site (cross-reacting idiotype). The *β-anti-idiotype antibody* has a paratope which is specifically complementary with the combining site of the first antibody site (cross-reacting idiotype). The *β-anti-idiotype antibody* has a paratope which is specifically complementary with the combining site of the first antibody (idiotype). It is, therefore, also called the internal antigen image. The β-anti-idiotype antibodies can be used, for example, for preparation of artificial vaccines when the original antigen is not available. Antibody against β-anti-idiotype antibody (*i.e.* anti-anti-antibody) has a binding site of identical specificity to the first antibody.

α- and β-anti-idiotype antibodies contain idiotopes (antigen determinants) in the variable parts of their molecules, which are localized even outside the combining sites. Thus, α-anti-idiotype antibodies of further generation may be formed against them. Interactions between antibodies and anti-antibodies form the *idiotype–anti-idiotype network* which is considered to play a key role in regulatory mechanisms of both humoral and cellular immunity. This network includes not only molecules of circulating antibodies, but also antibodies acting as antigen receptors that are incorporated into the cytoplasmic membrane of *B*-lymphocytes, as well as receptors for antigen on the surface of *T*-cells (their structure is different from that of immunoglobulins). So far, it has not been established whether anti-idiotype antibodies are formed during every immune response under physiological conditions. NIELS KAJ JERNE was awarded the Nobel prize for his discovery of the immunoregulatory idiotype network in 1984.

6.4.4 Biological properties of human immunoglobulins

IgG represents the main fraction of serum immunoglobulins and carries a major part of antibody activities. IgG antibodies are formed after repeated immunization (hyperimmunization) of the organism. Certain antibodies are preferentially present in particular subclasses. For example, anti-DNA antibodies are found most frequently in the IgG1 and IgG3 subclasses whereas antibodies against some carbohydrates, *e.g.* dextran, are present primarily in IgG2. Because of the ability of IgG to pass across the placenta, IgG antibodies form the main defence barrier against infection during the first weeks of life. The protection may be further enhanced by colostral IgG in the intestines of the neonate. IgG diffuses more easily into extracellular spaces than other immunoglobulins. IgG is the main immunoglobulin class responsible for neutralization of bacterial toxins and antigen agglutination; in addition, IgG antibodies can bind to bacterial surfaces (opsonization) and facilitate their phagocytosis. IgG–bacteria complexes activate complement which results in chemotactic attraction of polymorphonuclear leukocytes. The complexes are then bound through Fc- and C3-receptors to the surface of leukocytes; the occupied Fc-receptor provides a very effective signal for initiation of phagocytosis. In cooperation with complement, IgG antibody may produce bacteriolysis and cytolysis (cell lysis). Under pathological conditions, immune complexes may be deposited in the walls of blood vessels and thus participate in pathological processes.

IgM (M indicates macroglobulin) is the first immunoglobulin to occur in both phylo- and ontogeny. IgM antibodies are formed after the first contact of the organism with certain antigens. They are preferentially produced after immunization with cellular antigens whereas IgG antibodies are mainly formed after immunization with soluble antigens. IgM antibodies are primarily present in the blood circulation; they do not pass across the placenta. The IgM class contains, for example, natural antibodies against microorganisms and against endotoxins of Gram-negative bacteria *etc.* As IgM antibodies appear early as a response to immunization and their occurrence is limited to the bloodstream, they obviously play an important role in the defence against bacteria in the bloodstream (bacteraemia).

Polymerization of subunits into the basic IgM pentamer is dependent on the presence of the *J chain* which contains numerous cysteine and asparagine residues as well as several (7–8) oligosaccharides. The *J* chain probably stabilizes sulphydryl groups in the Fc portion during biosynthesis so that they become available for generation of transversal bonds among individual subunits to form a pentamer.

Theoretically, IgM antibodies may possess 10 combining sites. This number, however, is only found during reactions with small haptens. When larger antigens are used, the number of effective combining sites (valency) decreases to five or even less. This phenomenon is caused by the steric hindrance responsible for the loss of the molecule's flexibility. IgM antibodies have a relatively low affinity to individual determinants (haptens), but because of their higher valency they are more avid to the antigens containing a large number of determinants. This is due to *the additional effect of multivalency*. For the same reason, IgM antibodies possess a high agglutinating and, in cooperation with complement, a high cytolytic capacity. Compared to IgG, IgM antibody–antigen complexes activate the classical complement pathway significantly more efficiently.

In addition to the IgM pentamer, an **IgM monomer** with an 8S sedimentation coefficient also occurs. However, no free monomer is found in the circulation, but is incorporated into the surface of *B*-lymphocytes acting as an antigen receptor. It differs from the basic unit of the pentamer, in that the carboxy terminus of its heavy chains is extended by a segment containing mainly hydrophobic amino acid residues. This permits anchoring of the monomeric IgM through the phospholipid bilayer in the cytoplasmic membrane of *B*-cells.

IgA — as secretory IgA (SIgA) — is present in various sero-mucous secretions such as saliva, nasal fluid, sweat, colostrum, lung secretions and secretions of the urogenital and gastrointestinal tracts. Its main function is to protect mucous membranes and other external body surfaces that are in direct contact with microorganisms. However, besides SIgA, immunoglobulins of other classes (IgG, IgM, IgE) also participate in these defence mechanisms. IgA is present in external secretions and on mucous membranes in such concentrations that local biosynthesis rather than passive transfer from the serum must be assumed. In patients suffering from IgA deficiency, increased amounts of IgM with a bound secretory component are found in secretions of various mucous membranes. The *secretory component* is not synthesized in plasma cells like the *L* and *H* immunoglobulin chains, but in epithelial cells of the mucous membrane. It facilitates the passage of SIgA synthesized in subepithelial regions on the outer side of mucous membranes and, in addition, it stabilizes SIgA — particularly IgA2(1). Originally, the secretory component was assumed to confer protection of the IgA molecule against the effect of proteolytic enzymes. The secretory component is formed independently of IgA synthesis and this is why patients, lacking the ability to synthesize IgA antibodies, still have a secretory component on their mucous membranes.

Recently, the function of the secretory component was elucidated more precisely (MESTECKY *et al.*, 1988). It was found that the secretory component is a part of the so-called **poly-Ig receptor** which facilitates transfer of polymeric immunoglobulins across mucous membranes, *e.g.* into the gut lumen. After the secretory IgA or IgM has been synthesized it binds to the receptor present on the epithelial cell surface and the resultant complex is engulfed by endocytosis. It is then transferred across the cell to the plasma membrane adjacent to the lumen of the gut and there it is enzymatically cleaved. Its major portion remains bound to the immunoglobulin molecule and represents the secretory component (MOSTOV *et al.*, 1984). This Polyimmunoglobulin receptor has a structural homology with immunoglobulins and belongs to the immunoglobulin superfamily.

IgA class antibodies may possess activity against viruses, toxins, food antigens, erythrocyte membranes, nucleic acids, polysaccharides *etc.* They are generated mainly in response to oral immunization whereas after parenteral immunization, IgG or IgM antibodies are formed. It is not clear whether IgA antibodies possess opsonizing activity. However, Fc-receptors for IgA are found on the surface of phagocytes which permit their participation in bacterial opsonization and potentiation of phagocytosis. In addition, it was also found that SIgA coat the bacteria and thus prevent their adherence and penetration *via* mucous membranes into the organism. Aggregated IgA, similarly to other aggregated immunoglobulins, activates complement, but only *via* the alternative pathway. Complement activation by the alternative pathway is probably the basis of the well-known synergism between SIgA, complement and lysozyme during killing of some microorganisms.

SIgA plays an important role in preventing passage of soluble proteins from the gut into the blood. It forms SIgA–protein complexes that are degraded by local proteinases in intestinal mucosa. Recently, the role of SIgA in the bile has been discussed intensively. It seems that bile SIgA is formed by hepatocytes that remove IgA-containing immunocomplexes from the circulation.

The function of monomeric IgA which is present in considerable amounts in human serum, has not yet been firmly established. A small amount of serum IgA is present in the form of various polymers.

IgD was the last but one of the immunoglobulin classes to be discovered and the biological function of circulating IgD is almost unknown. Its antibody activity is rarely detected, *e.g.* against cell nuclei, serum albumin and insulin in some insulin-treated diabetic patients. There is no evidence for a role in the defence against pathogenic microorganisms. Besides monomeric IgM, IgD is the second most frequent immunoglobulin, incorporated as a receptor into the cytoplasmic membrane of *B*-lymphocytes. This IgD receptor differs from serum IgD by having a longer tail part of heavy chains (*Fig. 6.18*).

Of all the immunoglobulins, **IgE** is present in the lowest concentration. It has the shortest biological half-life and the highest rate of synthesis. Contrary to other immunoglobulins, IgE is thermolabile (at 56 °C an irreversible change in conformation of the *C*-terminal parts of heavy chains occurs within 30 min). In man, IgE represents homocytotropic antibodies with reagin activity.

Reagin antibodies have been known since PRAUSNITZ and KÜSTNER (1921) described a biological assay which subsequently became known as the *PK-test*. It is based on the following principle: 0.1 ml of diluted or undiluted serum containing reagin antibodies obtained from an allergic patient, is injected intradermally to a non-sensitized (non-allergic) individual. After 24 h, an antigen (allergen) is injected at the same site. In a positive reaction, itching, papula and erythema occurs at the site of injection. The reaction reaches its maximum within approximately 10 min, and starts to disappear after 20 min. In 1966, ISHIZAKA and coworkers obtained evidence that reagins do not belong to any previously known immunoglobulin class and therefore represented a new class — termed IgE.

The biological properties of IgE are based on its ability to be bound through C_H3 and C_H4 domains to high-affinity specific receptors on the surface of mast cells and basophils. Another type of Fc-receptor for IgE with a lower binding affinity is also present on eosinophils, neutrophils, mononuclear phagocytes and both *T*- and *B*-lymphocytes. When IgE is bound by its Fc region to the $Fc_\varepsilon RI$ in the cytoplasmic membrane of mast cells or basophils, the binding sites on Fab-fragments remain free and may thus react with a specific antigen (*Fig. 6.23*). Bridging of two IgE molecules on the mast cells surface by antigen causes liberation of pharmacological mediators (histamine, serotonin, leukotrienes *etc.*). The **immediate type allergic reaction** is thus generated, inducing for example, hay fever, atopic bronchial asthma, urticaria or allergic eczema. It was found that mast cells of allergic patients (with genetically conditioned enhanced susceptibility to immediate type allergic reactions) bear much higher numbers of high-affinity Fc-receptors for IgE on their surfaces than mast cells of healthy individuals. The number of leukocytes bearing low-affinity $Fc_\varepsilon RII$ is also four times higher in atopic patients than in normal individuals. In some cases, binding of IgE to Fc-receptors can be blocked by IgG4 antibodies. One cannot thus exclude the possibility that antibodies of this subclass may possess a regulatory function in the above mechanisms. The biological half-life of IgE is 2.4 days. However, this is not the case when IgE is bound to a "self" receptor on the surface of a suitable cell. In such cases the half-life may be as long as 1–2 weeks.

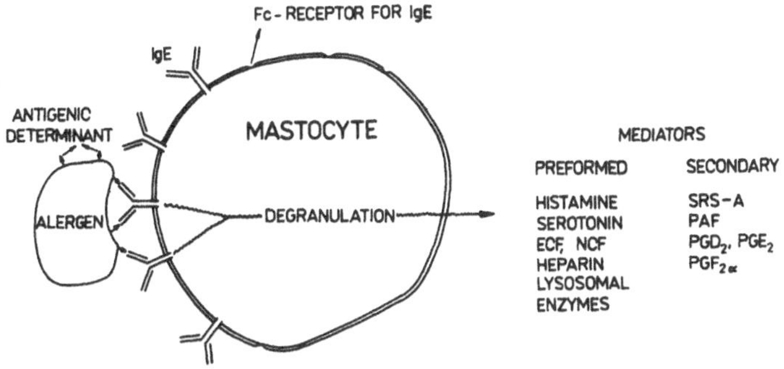

Fig. 6.23. Release of mediators from mastocytes during atopic reaction.
SRS-A, slow-reacting substance of anaphylaxis (a mixture of leukotrienes LTC_4, LTD_4, LTE_4), ECF, eosinophil chemotactic factor, NCF, neutrophil chemotactic factor, PAF, platelet-activating factor (1-O-alkyl-2-acetyl-sn-glyceryl-3-phosphorylcholine) Preformed mediators are present in the cells, secondary mediators are synthesized after stimulation

A decapeptide has been isolated from the $C_\varepsilon 4$ domain, which binds to Fc-receptors and can alone liberate histamine and other mediators from the cells. Some neuropeptides, *e.g.* substance P, also exhibit a similar ability. This may serve as another example of a direct interconnection between the nervous and immune systems. These neuropeptides may then induce certain allergic diseases even without participation of IgE. On the other hand, psychotherapy might be successful in the treatment of such allergic diseases.

The role of IgE described above concerns its participation in pathological reactions. However, one can hardly imagine the existence of a biological mechanism which is only designed to cause diseases. Obviously, the original biological function of reaginic antibodies was different. It is presumed that the homocytotropic antibodies, cells containing mediators (mast cells and basophils) and eosinophils form a system, the main role of which is defence against parasites. Thus, the ability to form IgE antibodies is, in fact, a marker of resistance against parasites. In the past, such an ability had been advantageous. Nowadays, however, since the incidence of parasitic diseases has been significantly reduced due to hygienic measures (particularly in affluent societies), this mechanism has become less important and is unfortunately manifested only by its undesirable effects, *i.e.* the induction of immunopathological (allergic) reactions of the immediate type.

References

Baenziger, J. (1984) The oligosaccharides of plasma glycoproteins: Synthesis, structure, and function. In: Putnam, F. W. (ed.), *The Plasma Proteins*, 2nd ed. Orlando, Acad. Press, vol. 4, pp. 271–315.

Bence-Jones, H. (1847) *Papers on chemical pathology*, Lecture III. *Lancet*, **2**, 88–92.
Capra, J. D. and Edmundson, A. B. (1977) The antibody combining site. *Sci. Amer.*, **236**, 50–59.
Chipens, G. I., Veretennikova, N. I., Nikiforovich, G. V. and Atare, Z. A. (1981) Elongated and cyclic analogues of tuftsin and rigin. In: Brunfeldt, N. (ed.), *Peptides 1980*. Copenhagen, Scriptor, pp. 445–50.
Davies, D. R., Padlan, E. A. and Segal, D. M. (1975) Three-dimensional structure of immunoglobulins. *Annu. Rev. Biochem.*, **44**, 639–67.
De Lange, G. G. (1989) Polymorphism of human immunoglobulins: Gm, Am, Em, and Km allotypes. *Exp. Clin. Immunogenet.*, **6**, 7–17.
Edelman, G. M. (1973) Antibody structure and molecular immunology. *Science*, **180**, 830–44.
Edelman, G. M., Cunningham, B. A., Gall, W. E., Gottlieb, P. D., Rutishauser, V. and Waxdal, M. J. (1969) The covalent structure of an entire γG immunoglobulin molecule. *Proc. Natl. Acad. Sci. USA*, **63**, 78–85.
Edelman, G. M. and Gall, W. E. (1969) The antibody problem. *Annu. Rev. Biochem.*, **38**, 415–66.
Edelman, G. M., Heremans, J. F., Heremans, M. T. and Kunkel, H. G. (1960). Immunological studies of human γ-globulin: Relation of the precipitin lines of whole γ-globulin to those of the fragments produced by papain. *J. Exp. Med.*, **112**, 203–23.
Edmundson, A. B., Ely, K. R., Abola, E. E., Schiffer, M. and Panagiotopoulos, N. (1975) Rotational allomerism and divergent evolution of domains in immunoglobulin light chains. *Biochemistry*, **14**, 3953–61.
Ehrlich, P. (1891) Experimentelle Untersuchungen über Immunität. II. *Ueber Abria. Dtsch. Med. Wochenschr.*, **17**, 1278–88.
Franěk, F. (1970) The character of variable sequences in immunoglobulins and evolutionary origin. In: Šterzl, J. and Říha, I. (eds.), *Developmental Aspects of Antibody Formation and Structure*. Prague, Academia, pp. 311–13.
Franěk, F. and Nezlin, R. S. (1963) Recovery of antibody combining activity by interaction of different peptide chains isolated from purified horse antitoxins. *Folia Microbiol.*, **8**, 128–30.
Gergely, J., Sármay, G., Rozsnyay, Z., Stanworth, D. R. and Klein, E. (1986). Binding characteristics and isotype specificity of Fc receptors in K cells. *Molec. Immunol.*, **23**, 1203–9.
Givol, D. (1974) Affinity labeling and topology of the antibody combining site. *Essays Biochem.*, **10**, 1–31.
Grabar, P. and Williams, C. A. (1953) A method permitting the combined study of the electrophoretic and immunochemical properties of a mixture of proteins: application to blood serum. *Biochim. Biophys. Acta*, **10**, 193–4.
Haber, E. (1964) Recovery of antigenic specificity after denaturation and complete reduction of disulfides fragment of antibody. *Proc. Natl. Acad. Sci. USA*, **52**, 1099–106.
Heremans, J. F. (1960) *Les globulines sériques du systéme gamma*. Brussels, Edition Arsia.
Hilschmann, N. and Craig, L. C. (1965) Amino acid sequence studies with Bence Jones proteins. *Proc. Natl. Acad. Sci. USA*, **53**, 1403–9.
Huber, R. (1976) Antibody structure. *Trends Biochem. Sci.*, **1**, 174–78.
Ishizaka, K., Ishizaka, T. and Hornbrook, M. M. (1966) Physicochemical properties of reaginic antibody. V. Correlation of reaginic activity with γE antibody. *J. Immunol.*, **97**, 840–53.
Jerne, N. K. (1974) Towards a network theory of the immune system. *Ann. Immunol.* (Inst. Pasteur), **125C**, 373–89.
Jerne, N. K. (1985) The generative grammar of the immune system. *Science*, **229**, 1057–59.
Kabat, E. A., Wu, T. T. and Bilofsky, H. (1979) *Sequences of immunoglobulin chains*. Washington, Government Print. Office. NIH 80-2008, 185 pp.
Liu, Y. S., Low, T. L. K., Infante, A. and Putnam, F. W. (1976) Complete covalent structure of a human IgA1 immunoglobulin. *Science*, **193**, 1017–20.
Marquart, M. and Deisenhofer, J. (1982) The three-dimensional structure of antibodies. *Immunol. Today*, **3**, 160–66.
Melchers, F. and Knopf, P. M. (1967) Biosynthesis of the carbohydrate portion of immunoglobulin: Possible relation to secretion. *Cold Spring Harbor Symp. Quant. Biol.*, **32**, 255–62.
Mestecky, J., McGhee, J. R. and Elson, C. O. (1988) Intestinal IgA system. *Immunol. Allergy Clin. N. Amer.*, **8**, 349–68.

Metzger, H. (1974) The effect of antigen binding on the properties of antibodies. *Adv. Immunol.*, **18**, 169–207.
Metzger, H. (1988) Molecular aspects of receptors and binding factors for IgE. *Adv. Immunol.*, **80**, 227–312.
Mostov, K. E., Friedlander, M. and Blobel, G. (1984) The receptor for transepithelial transport of IgA and IgM contains multiple immunoglobulin-like domains. *Nature*, **308**, 37–43.
Najjar, V. A. and Nishioka, K. (1970) Tuftsin, a natural phagocytosis stimulating peptide. *Nature*, **228**, 672–3.
Nouza, K. and John, C. (1987) *Immunology of health and illness*. Praha, Avicenum, 356 pp. (*in Czech*).
Poljak, R. J. (1975) Three-dimensional structure, function and genetic control of immunoglobulins. *Nature*, **256**, 373–6.
Poljak, R. J., Amzel, L. M., Chen, B. L., Phizackerley, R. B. and Saul, F. (1974) Three-dimensional structure of the Fab' fragment of a human myeloma immunoglobulin at 2.0Å resolution. *Proc. Natl. Acad. Sci. USA*, **71**, 3440–44.
Poljak, R. J., Goldstein, D. J., Humphrey, R. L. and Dintzis, H. M. (1967) Crystallographic studies of rabbit and human Fc fragments. *Cold Spring Harbor Symp. Quant. Biol.*, **32**, 95–8.
Porter, R. R. (1950) The formation of a specific inhibitor by hydrolysis of rabbit antiovalbumin. *Biochem.* J., **46**, 479–84.
Porter, R. R. (1959). The hydrolysis of rabbit γ-globulin and antibodies with crystaline papain. *Biochem. J.*, **73**, 119–26.
Porter, R. R. (1973) Structural studies of immunoglobulins. *Nobel lecture*, December 12, 1972. Prix Nobel, 174–83.
Prausnitz, C. and Küstner, H. (1921) Studien über die überempfinclichkeit. *Zentralbl. Bakteriol.* (*Orig. A*), **86**, 160–72.
Press, E. M. and Piggot, P. (1967) The chemical structure of the heavy chains of human immunoglobulin G. *Cold Spring Harbor Symp. Quant. Biol.*, **32**, 45–51.
Putnam, F. W. (1983) From the first to the last of the immunoglobulins. *Clin. Physiol. Biochem.*, **1**, 63–91.
Putnam, F. W., Florent, G., Paul, C., Shinoda, T. and Shimizu, A. (1973) Complete amino acid sequence of the mu heavy chain of a human IgM immunoglobulin. *Science*, **182**, 287–91.
Putnam, F. W., Takahashi, N., Tetaert, D., Lin, L.-C. and Debuire, B. (1982). The last of the immunoglobulins. Complete amino acid sequence of human IgD. *Ann. N. Y. Acad. Sci.*, **399**, 41–68.
Putnam, F. W., Titani, K. and Whitley, E. Jr. (1966) Chemical structure of light chains. Amino acid sequence of type $\varkappa$-Bence-Jones proteins. *Proc. R. Soc., Lond., Ser. B*, **166**, 124–37.
Rowe, D. W., and Fahey, J. L. (1965) New class of human immunoglobulin. II. Normal serum IgD. *J. Exp. Med.*, **121**, 171–84.
Singer, S. J. and Doolittle, R. F. (1966) Antibody active sites and immunoglobulin molecules. *Science*, **153**, 13–25.
Šterzl, J. (1989) *Development and induction of immune response*. Prague, Academia, 464 pp. (*in Czech*).
Šterzl, J., Tlaskalová, H., Rejnek, J., Šimečková, J. and Zikán, J. (1985) Natural and inducible idio-antiidiotype relations. In: Ferenčík, M. and Štefanovič, J. (eds.), *Immunology 1985*, pp. 31–65. (*in Czech*).
Takahashi, N., Tetaert, D., Debuire, B., Lin, L.-C. and Putnam, F. W. (1982) Complete amino acid sequence of the δ-heavy chain of human immunoglobulin D. *Proc. Natl. Acad. Sci. USA*, **79**, 2850–4.
Tiselius, A. (1937) A new apparatus for electrophoretic analysis of colloid mixture. *Trans. Faraday Soc.*, **33**, 524–31.
Tiselius, A. and Kabat, E. A. (1939) An electrophoretic study of immune sera and purified antibody preparations. *J. Exp. Med.*, **69**, 119–31.
Titani, K., Whitley, E., Jr., Avogardo, L. and Putnam, F. W. (1965) Immunoglobulin structure. Partial amino acid sequence of a Bence-Jones protein. *Science*, **149**, 1090–2.

Waldenstrom, J. (1944) Incipient myelomatosis or "essential" hyperglobulinemia with fibrinogenopenia. A new syndrome? *Acta Med. Scand.*, **117**, 216–47.
Waxdal, M. J., Konigsberg, W. H. and Edelman, G. M. (1967) The structure of a human gamma G immunoglobulin. *Cold Spring Harbor Symp. Quant. Biol.*, **32**, 53–63.
WHO (1969) An extension of the nomenclature for immunoglobulins. (Participants: Asofski, R., Binaghi, R. A., Edelman, G. M., Goodman, H. C., Heremans, J. F., Hood, L., Kabat, E. A., Rejnek, J., Rowe, D. S., Small, P. A., Jr. and Trnka, Z.). *Bull. W. H. O.*, **41**, 975–8.
Wu, T. T. and Kabat, E. A. (1970) An analysis of the sequences of the variable regions of Bence–Jones proteins and myeloma light chains and their implications for antibody complementarity. *J. Exp. Med.*, **132**, 211–50.

7 Biosynthesis of antibodies

7.1 Antibody production at the organism level

Antibody formation results from contact of the antigen with antigen-sensitive cells of the immune system. It occurs either under natural conditions during the development of the individual, or after an artificial antigen administration, *i.e.* vaccination and immunization. In man, *vaccination* is mainly performed to induce protection against infectious diseases, whereas *immunization* of animals, besides conferring protection, is primarily used to obtain antibodies for experimental purposes (immune sera, antisera).

7.1.1 Preparation of immune sera

Rabbits or other laboratory animals are usually used for antisera preparation for laboratory purposes. When large amounts of therapeutic and diagnostic antisera are required, larger animals, particularly horses, pigs, goats or sheep are used.

If just one simple antigen is used for immunization, *monospecific antibodies* are obtained; when a complex antigen or a mixture of antigens is used, *polyspecific antibodies* are formed. Polyspecific and monospecific antibodies are often incorrectly called polyvalent and monovalent antibodies respectively. The antibody valency does not designate the ability to react with one or more antigens, but rather indicates the number of binding sites on the antibody molecule to which an antigen can be bound.

The ability to form antibodies is genetically determined, at both the species and individual level. This fact should be considered when choosing a suitable animal for the immunization. Selection depends primarily on the origin and type of antigen, against which the antibodies should be prepared. For animal antigens, the phylogenetically most distant species should be used. Antibody formation can be induced even by minor antigen differences within the same biological species (*e.g.* immunoglobulin allotypes). In such a case, an individual of the same species, lacking the appropriate antigen, should be chosen as the antibody producer. If genetically diverse (outbred) animals are used for immunization, a larger number of specimens should be used since not all of them will produce antibodies to the same degree.

The efficiency of immunization depends on the route of antigen administration and on the character, dose and form of antigen. Antigens are usually administered intravenously, intradermally, subcutaneously, intraperitoneally or intramuscularly. After intravenous and intraperitoneal administration, the antigen is taken up by the blood circulation mainly in the liver and spleen. In addition, it stimulates various lymphatic tissues. The maximum immune response is achieved in the spleen. After intradermal, subcutaneous or intramuscular injection, the antigen is taken up by regional lymph nodes, where the highest immune response is observed.

A single antigen injection usually yields low titres of antibodies (the primary response). Experimental animals are therefore immunized repeatedly at various intervals for several months (up to 6 months) according to different immunization schemes. *Hyperimmune sera* with high antibody titres are obtained from hyperimmunized animals. However, possible anaphylactic shock should be taken into consideration.

The character of the antigen is important for the mode of immunization. If a cellular antigen (foreign erythrocytes, microorganisms or other cells) is used, the cell suspension is usually injected directly. For example, in order to obtain anti-erythrocyte antibodies, a 10–15% suspension of erythrocytes in a physiological buffer solution (pH 7.2–7.4) is used for immunization. Erythrocytes are injected intravenously (0.2–0.5 ml/kg of animal body weight) in four or five doses at two day intervals. The antibody titre is determined 7–10 days after the last immunization. If the titre is not high enough, another erythrocyte injection is given and the antibody titre is re-determined after a further 7–8 days.

When preparing antisera against biological fluids (*e.g.* blood serum), immunoglobulins or other antigens should be present at a concentration of at least 5%. Therefore, fluids containing low antigen concentrations (cerebrospinal fluid, urine, media after cultivation of tissues and cells) must be concentrated before immunization.

When preparing antisera against tissue and organ antigens, certain factors must be considered. For example, lipids can mask certain antigens and block their immunogenicity. It is, therefore, necessary to disintegrate the tissue by sonication, by repeated freezing and thawing, by mechanical homogenization, or to remove the lipid fraction by extraction. However, extraction with organic solvents may cause denaturation of protein antigens and thus changes in their specificity.

Another problem, connected with the preparation of antisera against organ antigens, is the possible occurrence of autoimmune processes. These may occur in immunized animals when the administered antigens possess a structure which partially corresponds to that of the host tissue and organ antigens. Autoimmune reactions cause damage to some organs and may even kill the immunized animal. Some animal species are unable to respond to certain antigens. Such unresponsiveness may be caused by the phenomenon of *immunological tolerance,* especially when administered and host antigens are similar. In such a case another animal species must be used for the antisera preparation.

Unresponsiveness (tolerance) to an antigen may also occur when either a high or low dose is used; this is particularly found with polysaccharides. The possibility of tolerance induction by protein antigens is lower; experimental animals may therefore be immunized with a wide range of antigen doses. Haptens or low-molecular-weight hapten-like substances must be conjugated with the carrier for immunization purposes.

When a highly specific antiserum to a single antigen is to be prepared, an absolutely pure antigen must be employed. However, the preparation of pure antigens is often very complicated, time-consuming and expensive and it is therefore desirable to avoid the purification procedures. This can be achieved by the stepwise *absorption of polyspecific antisera* by individual antigens, leaving antibodies agaist a single antigen. The antiserum thus becomes monospecific. The use of immobilized antigens for absorption is even more convenient as antigens are not added to antisera and the unwanted non-precipitating antibodies are also removed. It is not always necessary to use purified antigens for absorption of polyspecific antisera. Antigens or their mixtures, obtained from certain individuals or prepared by simple procedures, can often be used for that purpose. For example, newborn sera contain almost no IgA. When rabbits are immunized with an unpurified human IgA preparation, the unwanted antibodies can be removed by absorption with newborn serum. As this serum does not contain IgA, only anti-IgA antibodies remain in solution, whereas other antibodies form a precipitate with the antigen present in the serum.

Soluble antigens induce a weaker immune response than the cellular antigens. The immunization period is longer and antigen administration must be repeated many times. This is why the immunogenicity of soluble antigens (particularly proteins) is enhanced by adjuvant compounds.

7.1.2 What are immunological adjuvants?

Adjuvants enhance the immune response to soluble (colloid) antigens. The mechanism of the adjuvant effect is not exactly known. It is based on an increased antigen affinity to the cells of the lymphatic system. In addition, adjuvants protect antigens from rapid degradation, so that their effect is prolonged. At the same time, adjuvants activate mechanisms of the inflammatory reaction which in turn induce formation of important amplification compounds which enhance the intensity of the immune response.

Some inorganic compounds, *e.g.* aluminium hydroxide, potassium ammonium sulphate and potassium aluminium sulphate are among the most common adjuvants. Such mineral carriers adsorb the antigen on their surface which slows down its degradation and elimination from the organism. The mineral carrier is usually used at a 0.1% concentration and contains up to 5 mg protein antigen in 1 ml. The immunization lasts for 2–4 weeks; the animals receive four doses of antigen solution weekly; always at 0.2–0.5 ml/kg body weight. The mixture is administered intravenously or intraperitoneally. The antibody titre is determined 7–10 days after the last immunization.

Freund's adjuvant (FA) is most commonly used. This is an emulsion of the aqueous antigen solution in mineral oil and an emulsifier. The adjuvant effect of this mixture can be enhanced by the addition of killed mycobacteria or related microorganisms (*Mycobacterium tuberculosis*, *Mycobacterium butyricum*, *Mycobacterium smegmatis*, *Corynebacterium parvum etc.*). Such a mycobacteria-containing adjuvant is known as *complete* Freund's adjuvant; in addition to prolonged antigen persistence, the glycolipid component of the mycobacterial cell wall stimulates and activates cell proliferation of the reticuloendothelial system. *Incomplete* Freund's adjuvant contains no mycobacteria.

Both complete and incomplete FA are either commercially available, or the individual components must be prepared in the laboratory using: paraffin oil, Bayol F or 55, or sometimes *n*-hexadecane; lanolin, Aquaphore, Arlacel A, glycerolmono-oleate *etc.*, serve as emulsifiers. Incomplete FA is prepared by mixing 3–4 aliquots of oil with one volume of emulsifier which is then completed by adding 1–2 mg of dried, killed mycobacteria per ml of the mixture. The emulsion for immunization is obtained by adding the above mixture to an equal or double volume of a 1–5% antigen solution (pH 7.2–7.4). The antigen in the incomplete FA is usually administered subcutaneously or intramuscularly, whereas the antigen in the complete FA is given intramuscularly only. The antigen is administered at three day intervals for 3–4 weeks during a short-term immunization, and every second or third week for several months (minimum 4–5 doses) during a long-term immunization. The first antigen injection in FA is often combined with an intravenous or intraperitoneal injection of the pure antigen or antigen adsorbed to a mineral carrier.

Freund's adjuvant is *toxic* and is therefore not used for immunization of humans. However, mineral adjuvants of the aluminium hydroxide type are common in human medicine.

After obtaining the required antibody titre, blood should be withdrawn under aseptic conditions and merthiolate (0.005%) or sodium azide (0.05%) added to the serum as preservatives before storage at −20 °C.

7.1.3 Dynamics of antibody production

The term *dynamics of antibody formation* includes two types of antibody response, primary and secondary. In both cases, however, the extent of the immune response depends not only on the character immunization, but also on the sensitivity of methods used for detecting the antibodies produced. The response can be followed at the cellular level (direct determination of antibody-producing cells), or serologically (determination of antibodies in blood serum or other body fluids).

The immune response, generated after the first contact of the immune system with a given antigen, is called the **primary immune response**. It does not occur immediately, but after a specific induction (latent) period, the length of which depends on the properties of the antigen, route of immunization and the sensitivity of methods used for detection of the immune response. Thus the *inductive phase* of antibody formation is the period between the antigen administration and appearance of antibodies in the serum. The inductive phase is followed by the *productive phase* when the presence of antibodies or effector lymphocytes and their products can be determined in the immunized individual. During the productive phase, antibody production increases up to a peak and then starts to decline (*Fig. 7.1*).

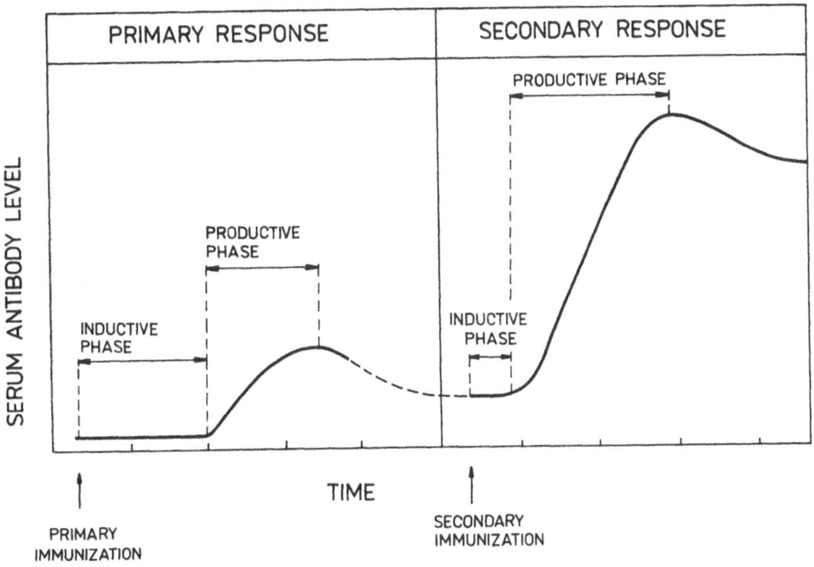

Fig. 7.1. Characteristics of antibody formation during the primary and secondary immune response.

When the organism comes into contact with the antigen for a second time, a **secondary response** occurs. The second antigen dose is frequently called a booster dose. During the secondary response the inductive phase is shorter and the immune response is stronger. This is because the immune system remembers the antigen and is able to respond more readily. The phenomenon of **immunological memory** operates in both the humoral and cellular immune response. The memory cells are usually both *B*- and *T*-lymphocytes. When the induction of antibody synthesis is elicited by thymus-independent antigens, the function of memory cells is taken over by *B*-lymphocytes.

Even during the secondary response, IgM antibodies are formed first, although production quickly switches to IgG antibodies. The mean amount of antibodies, produced by one cell, is equal both in the primary and secondary response. Thus the observed differences are due to different numbers of committed cells. During the secondary response the cell number is increased by memory cells generated during the primary response.

7.2 Antibody production at the cellular level

Antibody production and associated processes of cellular immunity represent an important form of individual *adaptation* to the external environment. Similarly to adaptation of a species requiring hereditary differences among individuals and their selective development, the adaptation of the individual immune system to external antigens requires hereditary differences among cells and their selective growth. During species adaptation, the selection unit is represented by an individual; during an individual adaptation, the selection unit is a cell. Cell selection and their mutual cooperation are the basic requirements of antibody formation. The basic principles, on which the theory of cell selection has been developed, are those of immunological memory and specificity.

7.2.1 Theories of antibody formation

The first complex theory of antibody formation was elaborated by PAUL EHRLICH (1854–1915) at the turn of this century. He found that the ability to induce antibody formation is a general property of antigens. His **side-chain**

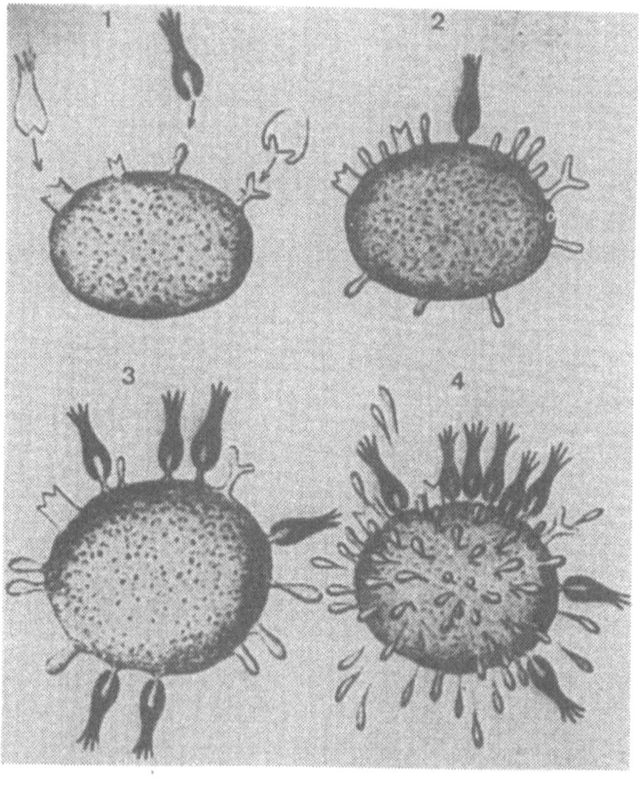

Fig. 7.2. Ehrlich's concept of side-chains (selective cell receptors) and their multiplication after binding to antigen (Ehrlich, 1900).

theory (selective cell receptors) was especially ingenious and the findings of contemporary immunology have confirmed its basic principles. EHRLICH (1900) suggested that the animal organism produces various cell types with receptors that are complementary to injected or naturally invading bacteria or other cell antigens. Antigen binding to surface receptors of these cells causes stimulation of homologous receptors that are later released into the blood as antibodies (*Fig. 7.2*). The term *"receptor"* was introduced by EHRLICH, who, together with I. I. METCHNIKOFF (1845–1916) was awarded a Nobel prize in 1908. Thus, he is not only a founder of modern chemotherapy, but he also introduced the basic concept of the *selective* role of antigens.

Theories of antibody formation ↗ selective ↘ instructive

The findings of LANDSTEINER, obtained with synthetic antigens (particularly the fact that antibodies may be formed even against synthetic compounds), were the basis of the **instructive theory**, based on the presumption that antigen reacts with antibody-forming cells; as a result the cells "know" what type of antibody is to be synthesized. The *template theory* is one of the most important instructive theories, formulated by BREINL and HAUROWITZ in 1930. This theory suggests that the antigen acts directly on the antibody molecule synthesized as a template and thus forms the complementary structure. The theory was further improved by the double Nobel prize winner LINUS PAULING — the *variable folding theory* of antibody formation (*Fig. 7.3*).

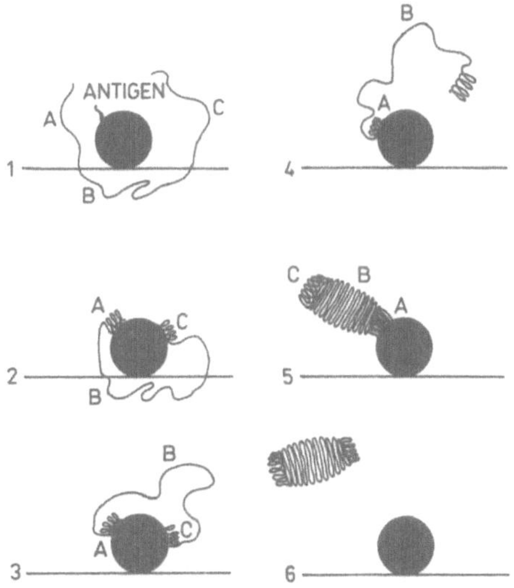

Fig. 7.3. Antibody formation according to the instructive theory (Pauling, 1940).
ABC, readily shapeable chain of the basic antibody; A and C, combining sites.

According to the classical instructive theory, there is only one basic, but very flexible antibody, whose structure can be changed by the antigen. The basis of the template theory was later modified to suggest that the antigen provides instruction indirectly by inducing formation of an enzyme or new DNA or mRNA which in turn controls further antibody synthesis.

The deeper insight into antibody structure resulted in appreciation of the EHRLICH'S original **selection theory**. The first modern contemporary theory of this type was formulated by NIELS KAJ JERNE in 1955. This was the *natural selection theory* of antibody formation. BURNET (1957, 1959) modified JERNE'S theory and formulated the clonal selection theory of antibody formation. Together with MEDAWAR, BURNET received the 1960 Nobel prize for this theory and for other basic immunological discoveries.

The **clonal-selection theory** became a basic dogma of modern immunology. It can be briefly summarized as follows. *The immune system recognizes antigens by means of lymphocytes. Every lymphocyte, circulating in the organism, can recognize (select) only a certain antigen (its determinants must be complementary with receptors on the surface of a particular lymphocyte). This specific antigen can then stimulate lymphocyte proliferation and further differentiation. Division of one lymphocyte type generates a cell clone, which produces antibodies with specific complementarity to the antigen which induced their formation.*

Thus the clonal-selection model assumes that every immunocompetent lymphocyte can only control the synthesis of one immunoglobulin which is present in its cytoplasmic membrane as a receptor. An antigen with a complementary determinant can bind to this receptor resulting in lymphocyte activation and generation of a cell clone producing a given immunoglobulin. In addition, a population of memory cells is also generated.

Contemporary immunology has not confirmed the validity of the instruction theories but has supported the clonal-selection theory. However, there is no separate gene for every antibody as originally predicted by the classical selection theory. Instead, lymphocytes possess small sets of gene subunits coding for individual parts of the antibody molecule (*split genes*). By their mutual combinations, a vast number of different antibody specificities may be generated. Similarly, the dogma "one antigen = one cell clone = one antibody" is valid only for homogeneous and comparatively simple antigens. Even then, however, antibodies against the carrier, hapten or against the whole carrier–hapten complex may be formed. This results in the well-known *antibody heterogeneity*. In addition, anti-hapten antibodies are usually not fully specific, because the combining site is usually larger than most synthetic haptens. Various cross-reactions with haptens, having a similar molecular shape, may thus occur. It follows that the hapten combining site is not ideally specific.

The situation with complex antigens such as viruses, bacteria or animal cells is rather more complicated. Such antigens contain numerous determinants that can activate lymphocytes with antigenic receptors of various specificities. The response to such a complex antigen is not due to one clone only, but a specific set of clones. It therefore follows that numerous antigen-sensitive cells must have the ability to respond to a given immunogen. It has been shown that individual cells are genetically restricted to the production of a single combining site with a certain specificity. However, they possess a potential ability to initiate its synthesis after stimulation with multiple cross-reacting immunogens. The only condition for such a cross-stimulation is a sufficient number of appropriately arranged determinants. As these incompletely complementary determinants bind more weakly to the antigen-sensitive cell receptors (they are of lower affinity), they must bind a substantially larger number of receptors; in other words, to a larger area on the lymphocyte surface. Thus a sufficiently strong signal for antibody stimulation is achieved. Specific determinants are more strongly bound to receptors and, therefore need fewer receptors. Cross-reacting immunogens use the avidity to compensate for low affinity of individual receptors for the induction of the immune response.

Affinity refers to the binding force between individual antibody combining sites or antibody receptors on *B*-lymphocytes and the antigen determinant group.

Avidity is less precisely defined; generally, this refers to the binding force between a di- or polyvalent antibody, or a whole cell with antibody receptors, and its specific antigen. Thus, avidity is a function of the number of combining sites and their individual affinities.

7.2.2 Basic mechanisms of antibody production

Antibody formation differs from biosynthesis of other proteins, particularly in the initiation mechanism. Initiation of antibody formation requires not only the genetic information and the protein synthesizing apparatus with a sufficient energy input, since an immunogen and the cooperation of two, and often three cell types, *i.e.* *B*-lymphocytes, *T*-lymphocytes and macrophages or other accessory cells, is also necessary. The thymus-dependent antigens require cooperation of all three cell types.

Thus, immunoglobulin production is more complicated than the biosynthesis of other proteins. For better understanding, this process can be divided into antigen recognition, antigen processing by accessory cells, cell cooperation, activation of *B*-lymphocytes and antibody biosynthesis on membrane-bound ribosomes. However, all these individual steps are components of one uniform process.

The most important factor in the induction of a response to all protein antigens appears to be the activation of $T_{H/I}$-cells. Such activation cannot be achieved by a free antigen, but only by an antigen presented by an accessory cell bearing syngeneic class II histocompatibility antigens (Ia-antigens) on its surface. Receptors on the $T_{H/I}$-lymphocyte recognize the presented antigen together with Ia-antigens. The activated $T_{H/I}$-cell produces a number of differentiation and growth factors. These factors then stimulate *B*-cells that had recognized the antigen, inducing their proliferation and differentiation into plasma cells. *B*-cells also recognize antigens by specific receptors although these are structurally different from the antigen receptors on the surface of *T*-lymphocytes. Together with *T*-helper cells, the suppressor *T*-

cells, which suppress the immune response, may also be activated. The final outcome of *B*-lymphocyte activation depends both on the positive cooperation with $T_{H/I}$-cells and negative cooperation with T_S-cells (*Fig. 7.4*).

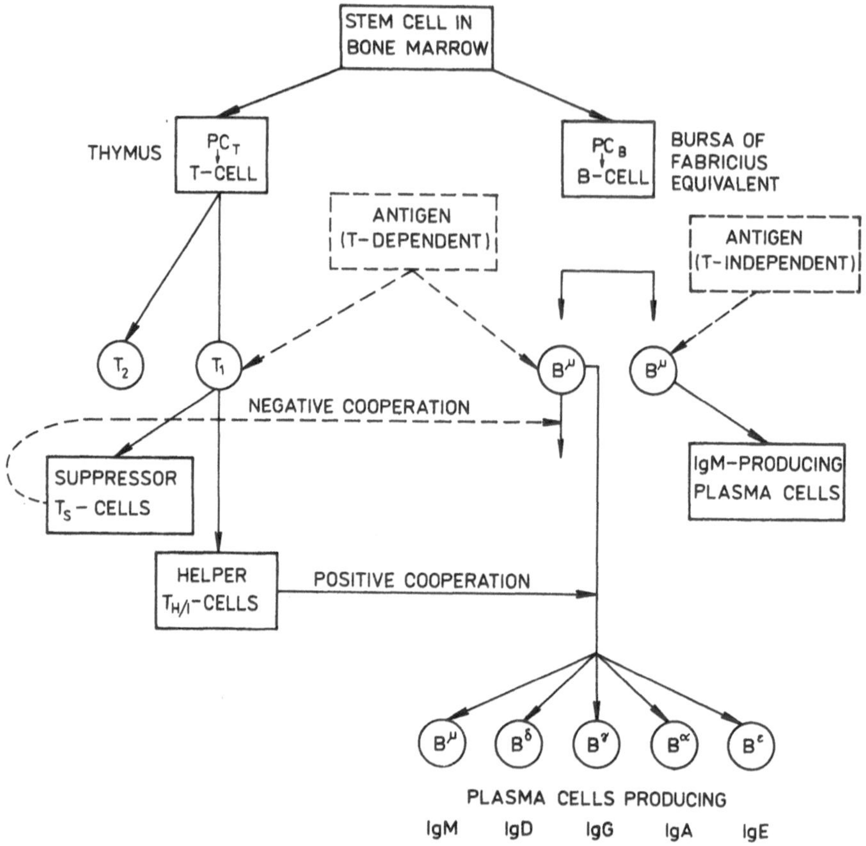

Fig. 7.4. Origin and functional relationships between *B*- and *T*-cells.
PC_T, precursor *T*-cell, PC_B, precursor *B*-cell, T_1, *T*-cell subpopulation with regulation activities ($T_{H/I}$ T_S), T_2, *T*-cells subpopulation with effector activities (T_C, T_{DH}), B^μ, IgM-producing *B*-cells, B^δ, IgD-producing *B*-cells, *etc*)

It is assumed that, among 100 000 circulating lymphocytes there are only between 1 and 10 cells which have, in a non-immunized individual (during the primary response), enough complementary receptors capable of binding a given antigen administered to the organism. During the secondary response, the number of such lymphocytes increases 5- to 100-fold. Cells that encounter the antigen for the first time are known as "virgin" cells and such interaction is metaphorically called "the first antigen sin."

7.2.3 B-cell activation

B-lymphocyte activation by thymus-dependent antigens is *restricted* by Ia antigens as the $T_{H/I}$-cells recognize the antigen determinant only together with

Ia antigens present on the accessory cells. On the surface of *B*-cells the antigen induces formation of receptors for growth and differentiation factors that are produced by $T_{H/I}$-lymphocytes and accessory cells. These **growth factors** include interleukin 1 (IL-1), IL-4 (*B*-cell growth factor) and the **differentiation factors** including IL-5 (the *B*-cell differentiating factor, BCDF). The binding of these receptors with appropriate factors provides a second signal to *B*-cells (the initial signal is antigen binding to the immunoglobulin receptor for antigen), which causes transition of the *B*-cell from the resting state to the growth cycle which may give rise to the antibody-producing cell — **the plasma cell** (*Fig. 7.5*).

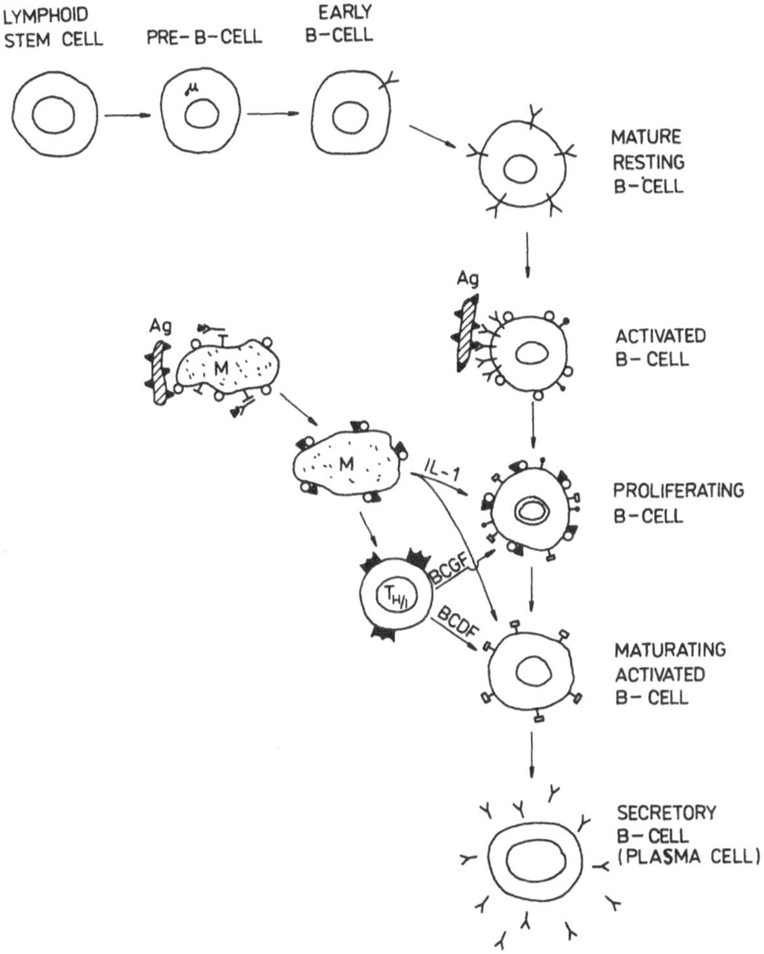

Fig. 7.5. *B*-cell development and their activation by thymus-dependent antigens (adapted from Nouza and John, 1987).
μ, heavy chain of IgM present in cytoplasm; Y, receptor for antigen or antibody molecule; Ag, antigen; ▲, antigenic determinant; O, Ia-antigen; M, macrophage; ⊢, Fc-receptor; ⚲ ⚲, receptors for growth and differentiation factors; $T_{H/I}$, helper *T*-lymphocyte; IL-1, interleukin 1; BCGF, *B*-cell growth factor; BCDF, *B*-cell differentiation factor.

It thus follows that at least two signals are required for *B*-cell activation by *T*-dependent antigens. The first signal is provided by the antigen itself (*recognition signal*) whereas the second signal (or signals) is provided by macrophages or even other accessory cells through *T*-lymphocytes (*confirmation signal*).

The results of several experiments have shown that antigen contact with a *B*-cell induces its activation or paralysis. It appears that the first contact (recognition signal) induces the state of paralysis, whereas induction of mitosis, with subsequent differentiation into the antibody-forming cell, requires another signal. The term "signal" refers to the molecular events occurring after binding the antigen determinant or other ligand to the corresponding cell receptor.

The situation is different for *T*-independent antigens. These antigens can activate *B*-cells without the aid of $T_{H/I}$-lymphocytes, independently of the class II histocompatibility antigens. However, during *B*-cell activation by thymus-independent antigens, the presence of accessory cells providing necessary growth factors (particularly IL-1) is also required. The *T*-independent antigens possess numerous repeating determinants that may form an antigen receptor network on the surface of the *B*-cell which results in their activation.

In addition to *B*-cell activation by *T*-dependent and *T*-independent antigens, *B*-lymphocytes may also be activated by anti-immunoglobulin antibodies and mitogens (*Table 7.1*). Activation by anti-immunoglobulin antibodies is most efficient when they are in an insoluble form, *e.g.* bound to an insoluble carrier (immunosorbent). This activation mechanism also requires immunoglobulin receptors for antigen and *B*-cell growth factors.

Table 7.1 Mode and conditions of *B*-cell activation

Mode of activation	*Active receptor*	*Ia antigen restriction*	*Accessory cells required*	*Lymphokines required*
T-dependent antigens	Ig	yes	yes	IL-1, IL-4, BCDF
T-independent antigens	Ig	no	yes	IL-1, IL-4, BCDF
Anti-immunoglobulins	Ig	no	yes	IL-1
Mitogens	mitogen	no	yes	IL-1

Ig, immunoglobulin, IL-1, IL-4 — interleukin 1, resp 4, BCDF, *B*-cell differentiation factor

The mechanism of mitogen-mediated activation is different. In this particular case the activation signal is not transmitted through immunoglobulin receptors, but *via* mitogenic receptors. In addition, similarly to anti-immunoglobulin antibodies, the activation is *polyclonal* in nature, *i.e.* mitogens and anti-immunoglobulins activate various *B*-cell clones.

Mitogens are substances that can induce cell activation and division; they are usually used as lymphocyte polyclonal activators. Lectins isolated from higher plants, *e.g.* phytohaemagglutinin (PHA) from red beans (*Phas-*

eolus vulgaris), or concanavalin A (Con A) from *Canavalia ensiformis* belong to the most important mitogens. **Lectins** are present not only in higher plants, but also in bacteria and animal cells. These are mainly proteins or glycoproteins, which possess a binding site that can bind specific monosaccharides on their surface. Thus, for example, PHA binds *N*-acetylgalactosamine while Con A binds α-D-mannose or α-D-glucose. The binding takes place on the basis of complementarity, similarly to binding between the antigen determinant and the antibody combining site.

This stereochemical complementarity of cell surface oligosaccharides and corresponding lectins (present as *lectin receptors* on other cells, or released from the cells and reacting with saccharides in the soluble phase), ensures a high degree of binding specificity. When a single cell possesses lectin receptors complementary with saccharide units protruding from the surface of another cell, the cells interact. This results in information transfer among the cells of the immune system as well as in other biologically significant phenomena such as symbiosis, phagocytosis *etc*. The system of saccharides and lectin receptors is one of the basic cellular recognition systems, *i.e.* the **saccharide–lectin system** present on the surface of both eukaryotic and prokaryotic cells (SHARON, 1984; LIENER *et al.*, 1986; UHLENBRUCK, 1987).

Lectins activate lymphocytes by bridging specific monosaccharide units protruding from the lymphocyte surface. This results in cross-linkage similar to that of the interaction of immunoglobulins with immunoglobulin receptors. For example, *N*-acetylgalactosamine residues are present on the surface of most lymphocytes. PHA may therefore non-specifically activate many lymphocyte clones rather than the limited number observed with certain antigens.

Contemporary experimental data suggest that lymphocyte activation following contact with an antigen or mitogen consists of several closely related events which trigger their binding to receptors. The mitogen molecule, as with the antigen molecule, contains numerous active sites that bind to lymphocyte receptors. As a result of cross-linkage by antigen or mitogen molecules, the receptors in the lymphocyte membrane are laterally rearranged (network formation) and the fluidity of membrane phospolipids increases. The cross-linked receptors aggregate on the lymphocyte surface forming *patches* or, occasionally, a polar *cap*.

Thus, the immune response of *B*-cells is regulated by antigen, by various soluble *T*-cells and accessory cell-derived molecules, and perhaps by direct *T*-cell–*B*-cell contact. The initial event in the generation of this response is the interaction of antigen or anti-immunoglobulin antibody with the immunoglobulin surface receptors. This results in the generation and transduction of intracellular signals and in the changes of both the membrane enzyme activities and transport of several substances across the membrane. Two such intracellular signals (second messengers) are *inositol-1,4,5-triphosphate* and *diacylglycerol*, which control changes in cytosol Ca^{2+} and protein phosphorylation respectively. The generation of these two second messengers occurs by hydrolysis of an inositol-containing phospholipid, phosphatidylinositol-4,5-biphosphate, present in the plasma membrane. The enzyme that catalyses this reaction, polyphosphoinositide phosphodiesterase (also known as phospholipase C), is also present on the plasma membrane.

The receptor cross-linking on the *B*-cell surface triggers the rapid induction of phosphatidylinositol-4,5-bisphosphate hydrolysis by phospholipase C yielding diacylglycerol (DAG) and inositol-1,4,5-triphosphate ($InsP_3$). DAG subsequently causes the translocation and activation of *protein kinase C* (PKC). $InsP_3$ evokes the release of Ca^{2+} from intracellular stores and perhaps extracellular Ca^{2+} influx. PKC activation and elevation of intracellular free calcium leads to increased expression of several genes including the genes encoding MHC antigens class II.

During this transduction cascade, which seems to be extremely complex, membrane (Na^+, K^+)ATPase is also activated. This enzyme increases influx of K^+, and, by opening calcium channels, the influx of Ca^{2+}. Intracellular Ca^{2+} acts through *calmodulin* which is a multifunctional modulatory protein and a primary receptor of divalent cations. In the cytosol of lymphocytes and other mammalian cells, there is an equilibrium Ca^{2+} concentration of between 10^{-8} and 10^{-7} mol/l. After stimulation of immunoglobulin or mitogen receptors it increases to 10^{-6} mol/l or more. Such a concentration is sufficient for the binding of Ca^{2+} to calmodulin and other Ca^{2+}-binding regulatory proteins. The active calmodulin–Ca^{2+} complex regulates the activity of many enzymes and physiological cell functions. In lymphocyte activation, its determining role appears to be its effect on adenylate cyclase and guanylate cyclase systems and on phosphodiesterase activity. It appears that the increase in intracellular *cAMP* level inhibits *T*- and *B*-lymphocyte proliferation and the production of lymphokines and antibodies. On the contrary, an increased intracellular *cGMP* concentration has a stimulatory effect. In this regulation, it is the cAMP/cGMP ratio (increase = inhibition, decrease = stimulation) rather than the absolute concentration of both cyclic nucleotides which is the important factor.

During lymphocyte stimulation, other biological events take place. These include interaction of occupied receptors with the cell cytoskeleton system, signal transduction to and from the cell nucleus, activation of phospholipid metabolism (which is connected with the plasma membrane fluidity and prostaglandin formation) and increased transport of saccharides, amino acids, nucleosides *etc.*

According to Jaroslav Šterzl (1988, 1989), antigen acts on antigen-sensitive cells not only specifically and in the function of a mitogen, but its effect can be likened to a membrane pump. It is analogous to the effect of certain substances that can specifically remove certain cell components (*e.g.* ionophores take up ions). The membrane-bound antigen reacts only with specific immunoglobulin receptors and withdraws them from the cell. Thus, cell metabolism is generally increased and the gene regions whose products are specific antigen-binding immunoglobulin molecules, are primarily transcribed.

7.2.4 B-cell development after activation

B-cell differentiation after antigen-induced activation into antibody-producing cells requires additional signals. It has been repeatedly shown that such signals are provided by lymphokines (immunoregulator substances originating from lymphocytes) or monokines (immunoregulators from macrophages), or by various factors released from other accessory cells (*Fig. 7.5*).

The basic **lymphokines** involved in lymphocyte development and function are listed in *Table 7.2*. These lymphokines influence various phases of *B*-cell development. For example, BCGF-1 (IL-4) influences the resting *B*-cells (phase G_0) and induces their increased response to anti immunoglobulin antibodies. BCGF-2 (IL-5) acts only at the end of the G_1-phase of the cell cycle and enables transition to the S-phase. Its effect is independent from anti-immunoglobulin antibodies. The differentiation factors (BCDF) affect cells having IgM- or IgG-receptors for antigen on their surface that differentiate into the final phase of IgM- or IgG-secreting cells. A similar factor was also shown to be involved in the stimulation of IgE formation. The IL-1, IL-2 and IFN-γ action results in *B*-cell activation associated with an increased number of surface class II histocompatibility antigens.

Table 7.2 Basic lymphokines regulating the development and function of lymphocytes

Lymphokine	*Abbreviation*	*Origin*	*Target cells*
Interleukin 1	IL-1	accessory and other cells	*T*, *B*
Interleukin 2	IL-2	*T*-cells	*T*, *B*
Growth factors	BCGF (IL-4)	*T*-cells	*B*
Differentiation factors	BCDF (IL-5)	*T*-cells	*B*
Interleukin 3	IL-3	*T*-cells	stem, *B* precursors
γ-interferon	γ-IFN	*T*-cells	*T*, *B*

Source: Tlaskalová *et al.* (1985).

The efficiency of *B*-cell regulation by non-specific factors depends on their concentration in the vicinity of cells and on the cells ability to bind these factors on their surface. The state of cells producing immunoregulatory factors (their metabolic activity, degree of stimulation *etc.*), and the presence of receptors for corresponding stimulatory factors on the *B*-cell surface also play an important role. These substances have recently attracted considerable attention for both theoretical and practical reasons. It was shown that several inborn and acquired immune deficiencies are based on abnormal immunoregulation, both at the level of lymphokine production and receptor expression on the cells. These immunodeficient patients might be treated by administering the deficient lymphokines, by stimulating their production pharmacologically, or by transferring the activated *in vitro* cell clones.

After antigen activation, not all specific *B*-cells differentiate into plasma cells. Some of them become long-living memory cells. **Memory cells** differ in some of their properties from non-activated (resting) *B*-cells. For example, the affinity of their receptors for antigen is higher. A characteristic feature of immunological memory is a stronger immune response and shorter inductive phase, as well as the production of antibodies of various isotypes after repeated stimulation.

In 1971, Šterzl and Nordin formulated a hypothesis concerning *the switch of immunoglobulin classes* during *B*-cell proliferation. According to this theory (*Fig. 7.6*), *B*-cells differentiate into IgM antibody-producing cells after initial contact with the antigen. After repeated activation there is a switch to antibodies of the IgG or other classes although their combining site remains identical. This idea has recently been confirmed experimentally even at the level of gene tandem arrangement of the constant region of immunoglobulins. The origin of isotype diversity, however, not only reflects antigen stimulation; during ontogeny, individual isotypes are generated in a specific sequence although the regulatory mechanisms involved have not yet been clarified.

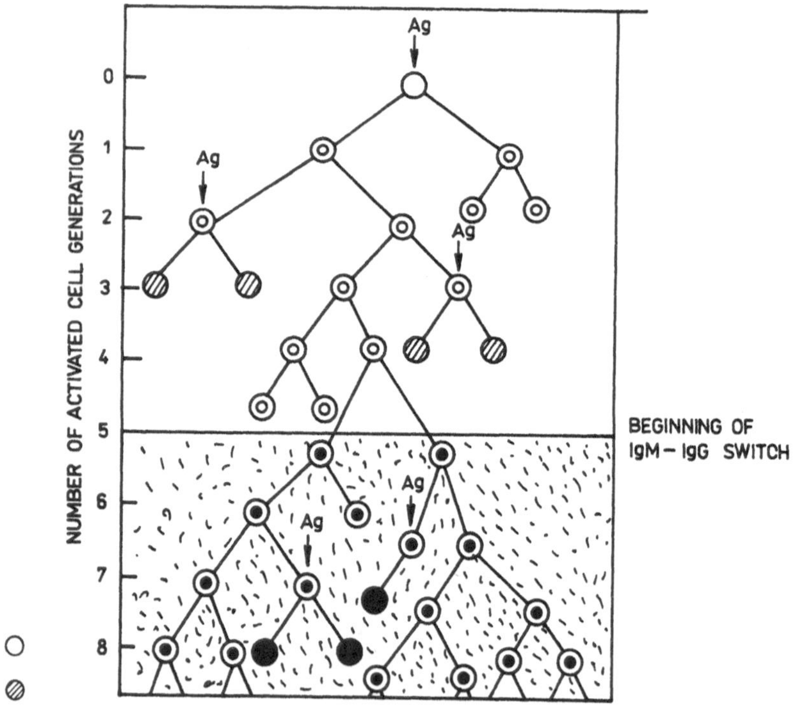

Fig. 7.6. Concept of immunoglobulin isotypes switch during *B*-cell proliferation (Šterzl and Nordin, 1971).

○, Immunocompetent cell; immunologically activated cell with IgM-receptors: ◎ or IgG-receptors: ⊙; ⊘, IgM antibody-producing cell; ●, IgG-producing cell.

7.2.5 Immunological tolerance

During ontogenic development, individuals acquire the ability to regulate the intensity of the immune response. Therefore, the antigenic stimulation which results in antibody formation and the origin of immunological memory, may also lead to suppression (unresponsiveness, immunological tolerance) under certain conditions. **Immunological tolerance** is an acquired phenomenon of total or partial unresponsiveness to antigen or cells, which would induce an immune response in normal individuals. A tolerant individual does not respond to a specific antigen (called a **tolerogen**), whereas the ability to respond to other antigens is preserved; it thus follows that tolerance is as specific as immunity.

The most typical example of immunological tolerance is *autotolerance, i.e.* the inability of the organism to stimulate immunocompetent cells by components of its own ("self") tissues and cells, which are immunogenic when transferred to the body of other individuals. When the tolerance to "self" antigens is deranged, **autoimmunization**, which causes various autoimmune diseases, may occur.

The first unifying concept to explain why the organism does not produce antibodies against the components of its own body, was proposed by BURNET. According to this theory, the organism had "learned" to recognize its own structures during embryonic development, and afterwards destroyed or removed clones of immunocompetent cells that might be able to recognize these structures (so-called "forbidden clones"). This theory was proved experimentally by PETER MEDAWAR (joint 1960 Nobel prize winner with BURNET) and MILAN HAŠEK. In MEDAWAR's basic experiments, cells isolated from *CBA* strain mouse spleens were injected into embryos or newborn mice of the *A* strain. When, after several weeks, these mice were transplanted with a skin graft obtained from spleen cells of donors, the graft was not rejected. Adult mice of the *A* strain became able to accept permanently skin grafts from *CBA* mice since their lymphocytes had learned to recognize the CBA transplantation antigens as self during development.

A similar state of tolerance was observed by HAŠEK (1953) in a different model, *i.e.* the fusion of chorioallantoic membranes of two different developing chicken embryos (the chorioallantoic membrane forms the outer foetal membrane in snakes, birds and mammals). During *embryonal parabiosis*, an intensive exchange of blood cells between each embryo occurs which results in the inability to form antibodies against each other's antigens and reject skin grafts.

It was later found that immunological tolerance could be induced not only during embryonal development, but also in adults when using both low and high antigen doses. MITCHISON (1968), in his classical experiments, repeatedly injected three groups of mice with different bovine serum albulin (BSA) concentrations: low, medium and high. Surprisingly, only the group preimmunized with the medium BSA dose responded with antibody formation. Mice immunized with either high or low BSA dose did not respond to the second antigen injection.

The induction of tolerance in adults is more difficult than in the embryonic stage of development and can be facilitated by various treatments which temporarily decrease the immunological responsiveness of the organism. Various types of **immunosuppressive treatments** can be used, including physical (irradiation with X-rays), chemical (various anti-metabolites), or biological (anti-lymphocyte serum *etc.*). Tolerance induction in adults might be important in organ and tissue transplantation. Nowadays, allotransplantation of various organs is not usually a surgical problem although suppression of the transplantation reaction which results in graft damage and rejection remains complicated. Tolerance induction to histocompatibility antigens of the donor's organ would certainly be more natural for the recipient than the contemporary long-term immunosuppressive treatment after the transplantation.

Immunological tolerance can be either natural or acquired (similar to immunity). **Natural tolerance** — spontaneous unresponsiveness to "self" antigens — develops during ontogenic maturation of the individual. **Acquired tolerance** refers to unresponsiveness to foreign antigens. This may be induced during the period of immunological maturity, but also in adults with fully developed immune system.

Tolerance induction requires certain conditions that either alter the strength of the antigenic stimulus, or allow a qualitative change in its effect. The antigen dose and form, mode of antigen administration and the age of the individual appear to be the most important factors. According to contemporary concepts, tolerance may be induced by various mechanisms that can

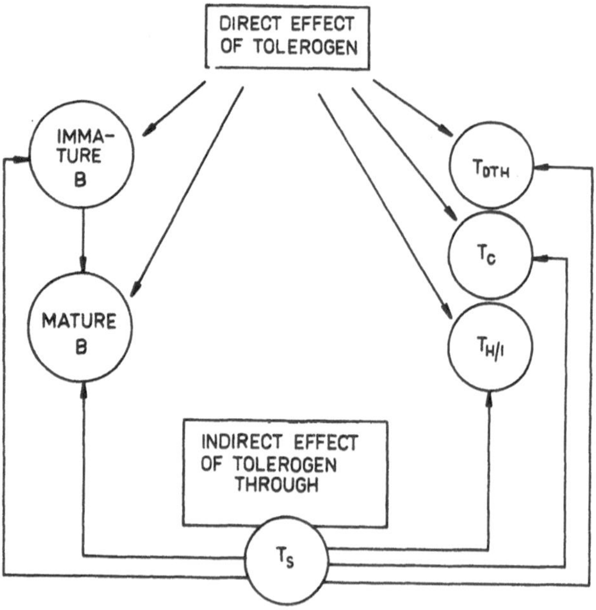

Fig. 7.7. Mechanisms of immunological tolerance induction (Strejček and Lokaj, 1985).

be divided into two main groups: passive and active. As regards passive mechanisms, the specific *T*- and *B*-lymphocytes are directly removed or inactivated by the antigen. In the active mechanisms, the antigen (tolerogen) induces formation of suppressor cells (T_S-lymphocytes), actively blocking the activity of specific lymphocytes (*Fig. 7.7*). Thus, in this case the antigens act directly through T_S-cells. Tolerance, however, need not be induced by the antigen only, but also by anti-idiotypic antibodies that may paralyse idiotypic determinants on immunocompetent cells.

Individual developmental stages of *B*-cells are not equally sensitive to tolerance induction. It appears that the early *B*-cells are most sensitive (*Fig. 7.5*). Mechanisms leading to immunological unresponsiveness to a particular antigen not only include removal (deletion) of *B*-cells, but also blockade of antigen receptors, inability of their resynthesis after endocytosis, loss or inactivation of receptors for growth or differentiation factors, interference with the Fc- or Ia-signal, inability to transmit appropriate signals between the membrane and cytoplasm and inability to process or present the antigen.

7.2.6 Cooperation of B- and T-cells in the production of antibodies

Cooperation of *B*- and *T*-cells is a very old phylogenetic mechanism which also operates in amphibians, snakes and probably fish. The *T*-cell help is not only limited to induction of antibody response, but it is also involved in its later phase; without *T*-cell help, the switch from IgM to IgG antibody formation does not occur. Furthermore, this type of cell cooperation also plays an important role in other immune mechanisms: in addition to *T*- and *B*-lymphocyte cooperation, cooperation between two *T*-lymphocyte subpopulations could also be demonstrated, *e.g.* during the graft-versus-host reaction. A similar cooperation occurs in the transplantation reaction against foreign grafts, tumours and in reactions against those parasites, fungi, bacteria and viruses where cellular immunity is of primary importance.

Cooperation of *T*- and *B*-cells during antibody formation was demonstrated in several classical experiments, performed by Claman *et al.* (1968), Rajewski *et al.* (1969), Mitchison *et al.* (1970) and Katz and Benacerraf (1972).

Mitchison (1971), using mice as a model, injected one group with NIP–CGG conjugate (NIP = 4-hydroxy-3-iodo-5-nitrophenyl acetic acid, CGG = chicken gamma globulin) before sacrifice and splenectomy. The spleen cells were then injected into the second group of mice, which had been previously X-irradiated to paralyse their own immune system. These mice received NIP bound to a heterologous carrier together with immunologically active spleen cells, obtained from the first group of mice. BSA (bovine serum albumin) was used as a heterologous carrier instead of CGG, in order to form a NIP–BSA conjugate. Under such conditions, only the NIP–CGG-sensitized implanted spleen cells yielded an immune response. However, the anti-NIP antibodies did not appear in the sera of experimental mice. These were formed only when irradiated mice were also given cells, sensitized with a pure heterologous carrier, *i.e.* spleen cells of mice immunized with BSA. Obviously, the irradiated recipients responded normally if they received the pure NIP–CGG conjugate as the antigen together with NIP–CGG-sensitized cells.

The above experiment is a classical example of the carrier effect and it further demonstrates that both the carrier and hapten are required for the induction of antibody formation. The *carrier effect* is based on the fact that in animals sensitized with hapten–carrier conjugates, the *B*-cells (as precursors of antihapten-forming cells) cannot be activated with either the free hapten, or the same hapten bound to a different (heterologous) carrier, but only with hapten conjugated with a homologous carrier.

In general, the carrier effect may be precisely defined as an immune response to a single determinant (*helper determinant*) which is present on the antigen molecule and enhances the response against another determinant (*inductive determinant*). Although the hapten and carrier determinants may be recognized by receptors of both *B*- and *T*-cells, hapten determinants are dominant primarily for *B*-cells and therefore represent inductive determinants. The carrier determinants on the other hand are dominant, particularly for *T*-cells, and may be considered helper determinants.

The existence of helper and inductive determinants was demonstrated not only in hapten–protein conjugates, but also in molecules with a subunit structure, and in complex antigens such as blood group antigens or histocompatibility antigens.

During the thymus-dependent response to a typical protein antigen, one determinant acts as a hapten and other determinants play the role of a carrier. Thus, the carrier and hapten are expressed *functionally,* rather than by specific structures. Binding of one determinant to the immunoglobulin receptor of *B*-cells provides a signal which paralyses the cells until they are activated by the second signal. The latter originates in the *T*-cell as a result of its stimulation by the carrier determinant, which is presented by the accessory cell. It thus follows that antibody response can be induced by an antigen possessing at least two determinants.

The carrier effect does not operate during induction of antibody formation with *thymus-independent antigens,* because these antigens do not require the presence of helper cells. *T*-independent antigens usually have large molecules and a high density of repeating identical determinants, *e.g.* lipopolysaccharide of Gram-negative bacteria. The direct *B*-cell stimulation is probably caused by the overall arrangement of these determinants rather than by the character of one of them. In addition, *T*-independent antigens are usually relatively non-degradable and therefore persist in the organism much longer than the thymus-dependent antigens.

It appears that *antigen degradability* may be one of the factors determining the mechanism of *B*-cell stimulation during the antibody response. SELA *et al.* (1962) found that synthetic copolymers of L-amino acids were degraded faster *in vivo* and that the humoral response against them was thymus-dependent, although their molecule contained repeating identical determinants.

Conversely, copolymers of D-amino acids were virtually non-degradable and thymus-independent.

In the past, several hypotheses concerning the *T*–*B*-cell cooperation have been proposed. According to one theory, the more motile *T*-lymphocytes transfer the antigen to less motile *B*-lymphocytes. An alternative theory claims that the antigen forms bridges between *T*- and *B*-lymphocytes and thus the *T*-lymphocytes provide the confirmatory (mitotic) second signal to *B*-lymphocytes. The contemporary hypothesis suggests that macrophages and other accessory cells participate during *T*- and *B*-lymphocyte cooperation; these cells present the antigen to $T_{H/I}$-lymphocytes and produce the growth and differentiation factors for *B*-cells.

7.2.7 *Processing of antigens in accessory cells*

Classical macrophages were originally classified as **accessory cells** (sometimes abbreviated to *A-cells*). Nowadays, the dendritic cells of lymphoid organs, interdigitating and follicular dendritic cells, Langerhans cells of the skin and some other cell types are also classified as accessory or antigen-presenting cells. In some cases, the *B*-lymphocytes themselves can behave as accessory cells. The function of accessory cells is not antigen-specific (it is not mediated through antigen receptors — apart from *B*-cells) and includes:

1. maintenance of lymphocyte viability;
2. facilitation of immune cell interactions;
3. secretion of immunoregulatory molecules;
4. antigen or mitogen presentation to lymphocytes.

Until approximately the mid-1970s, it was assumed that *A*-cell function was unrestricted. However, ROSENTHAL and SEVACH then showed that the function of these cells during presentation of *T*-dependent antigens is restricted by Ia-antigens. The accessory function can therefore only be fulfilled by cells having class II MHC surface antigens; the proper interaction between the *A*-cell and $T_{H/I}$-lymphocytes occurs only when syngeneic Ia-antigens are present on both cell types.

The expression of class II MHC antigens on macrophages is not a permanent (constitutive) feature, but is dependent on various factors. The percentage of *Ia-antigen-bearing macrophages* differs in various tissues, but is typical for a given tissue. Newborns contain a decreased number of Ia-positive macrophages in all tissues except for thymus. In general, young macrophages express Ia-antigens more efficiently than older ones. Thus, in order to maintain the basic Ia-antigen level in tissue macrophages, stem cells must permanently differentiate into macrophages. It was found that macrophages may acquire and lose Ia-antigens during their development. The ratio of Ia-positive to Ia-negative macrophages, as well as the density of Ia-antigens on their surface, can be regulated in the following ways:

1. By substances which trigger phagocytosis. These substances react with Ia-positive macrophages and inhibit their conversion into Ia-negative macrophages.
2. By products of antigen-stimulated *T*-cells. Such a lymphokine, released by $T_{H/I}$-cells, was isolated by UNANUE *et al.* (1984) and called "*macrophage Ia-positive recruiting factor*" (*MIRF*). MIRF attracts monocytes and immature macrophages to the site of antigen activity and preferentially induces Ia-antigen expression on immature Ia-negative macrophages. Interferon-γ also induces Ia-antigen formation. The Ia-antigen expression on macrophages can also be increased by drugs, *e.g.* indomethacin.
3. By several substances, which, on the contrary, decrease Ia-antigen expression on macrophages. Such substances include α-fetoprotein and prostaglandin E (PGE). PGE, however, decreases Ia-antigen expression on mouse macrophages but does not affect expression of these antigens on human monocytes and macrophages.

These findings suggest that the Ia-antigen expression plays an important role in regulation of the immune response. Unresponsiveness to certain antigens may be due to inadequate expression of Ia-antigens or inability of certain Ia-antigens to form an immunogen complex with the antigen, *i.e.* inability effectively to stimulate $T_{H/I}$-lymphocytes. It may be assumed that there is a definite quantitative relationship between Ia-antigen expression on macrophages and their accessory function. In other words, the higher the expression, the stronger the accessory function (UNANUE, 1989).

Antigens must be processed by accessory cells before proper presentation to *T*-cells. It was originally assumed that the antigen is first bound to the macrophage surface and is then mainly engulfed by endocytosis and rapidly catabolized in phagolysosomes. The products of catabolic antigen degradation are very poor antigens since most determinant groups are destroyed. However, a small portion of antigen remains bound on the macrophage surface for a relatively long period and these molecules are highly immunogenic. Presumably, they can form bridges between carrier-specific receptors on *T*-cells and hapten-specific immunoglobulin receptors on *B*-cells. The production of these bridges provides the necessary signals for the initiation of antibody formation.

This concept, however, has changed during the past decade since the discovery that **processing** of most soluble protein and bacterial antigens requires, in addition to binding, an active catabolic process within the accessory cells. Before presentation, all these antigens must therefore be processed in the phagolysosomal apparatus of accessory cells.

Antigens are bound to the surface of *A*-cells either by specific receptors, such as receptors for the Fc-domain of immunoglobulins, for the C3 complement component (CR1 to CR5), immunoglobulin receptors for antigen (*B*-cells only), or non-specifically by non-covalent interactions to unidentified structures in the plasma membrane of *A*-cells. In addition to phagocytosis, adsorption pinocytosis is involved in the ingestion of soluble antigen molecules.

GREY and CHESNUT (1985) observed that at least 45–60 min must elapse between the antigen binding to the surface of the accessory cell and the presentation of antigen to the *T*-cells. During this period, *A*-cells not only ingest, but also metabolize the antigen (Fig. 7.8). The nature of these metabolic processes is unknown, but UNANUE's experiments suggest that proteolysis is the basic event. When protein degradation by lysosomal proteinases is inhibited by lysosomotropic substances (penetrating directly into lysosomes) or by increasing intralysosomal pH, the macrophages lose the ability to present soluble protein antigens. This indicates that the basis of processing lies in limited antigen proteolysis in lysosomes of *A*-cells. As soon as the processing is completed, the antigen (more precisely its fragments) returns to the cytoplasmic membrane by exocytosis, where it becomes bound to Ia-antigens, in a form which can be recognized by $T_{H/I}$-lymphocytes. It appears that antigen degradation within phagolysosomes is mainly performed by cysteine proteinases, whereas aspartic acid proteinases are rather less significant. All larger protein antigens must be degraded. The small molecules of peptide antigens do not require metabolically active processing.

The macrophages presenting protein antigens must have the following properties. The ability to (a) bind antigen to their surface; (b) to engulf (endocytose) it; (c) to split it into fragments by limited proteolysis; (d) to excrete it back on the surface by exocytosis and (e) to associate the fragments with Ia-antigens into the immunogenic complex. Together these activities are called "processing". A defect in any of these steps results in blockade of antigen presentation and therefore also in inhibition of the immune response.

Why does the antigen require processing? To date, there is no answer to this question. The processing is required only by antigens stimulating $T_{H/I}$-lymphocytes and recognized by antigen receptors only in cooperation with class II MHC antigens. Antigens whose activity is restricted by class I MHC antigens and which are recognized by T_C-lymphocytes, do not require processing because as soon as they (*e.g.* virus antigens) occur on the surface of the target cells, they are rapidly recognized by T_C-lymphocytes which then kill the cell. Receptors on helper and cytotoxic *T*-lymphocytes possess a high degree of homology (they are very similar or even identical) and it is therefore unlikely that antigen processing would require different receptor types on $T_{H/I}$- and T_C-cells. The different requirements for antigen processing must therefore be based either on restriction elements (class I or class II MHC antigens) or on the character of antigen to be recognized.

What is the mechanism of antigen presentation by *A*-cells to $T_{H/I}$-lymphocytes? This question has not yet been answered satisfactorily. There are two basic possibilities of recognition of the processed antigen on the accessory cell surface by a single or two receptors. With the *one-receptor model*, the simplest explanation (although not the only one) of binding of the processed antigen to the receptor of $T_{H/I}$-cell would be generation of one complex, consisting of the processed antigen and Ia-antigen on the surface of the *A*-cell. The one molecular ligand thus formed can easily be recognized by one receptor. Such an interaction between the processed antigen and the Ia-antigen is not required in the *two-receptor model*, where one *T*-receptor binds the processed antigen and the second *T*-receptor binds the Ia-antigen. Recently, the existence of the processed antigen–Ia-antigen complex was demonstrated; therefore most immunologists now prefer the one-receptor model (*Fig. 7.8*).

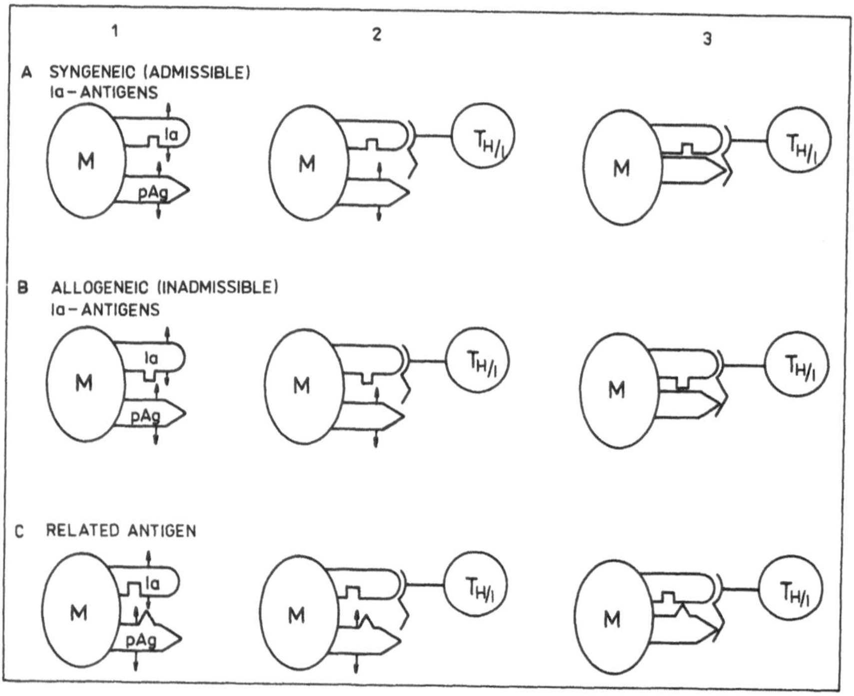

Fig. 7.8. A single-receptor model of processed antigen recognition restricted by Ia-antigens (Grey and Chesnut, 1985).

M, macrophage, $T_{H/I}$, *T*-helper cell, Ia, Ia-antigen, pAg, processed antigen 1 Processed antigen occurs on the Ia-positive accessory cell surface Ia-antigen and processed antigen do not possess any mutual affinity Both molecules may freely diffuse ("float") in the cytoplasmic membrane of *A*-cells 2 *T*-cell with receptors having two specific sites that are able to recognize separately processed Ag and Ia-antigen, they react firstly with Ia-antigens 3 Antigen diffuses into the site of interaction between Ia-antigen and the *T*-cell receptor In the case of syngeneic admissible interaction (A), processed antigen may bind to the antigen-specific site of the *T*-cell receptor which results in formation of a stable trimolecular complex that triggers the *T*-cell activity In the case of allogeneic Ia-antigen (B) or some related antigen (C), the stereochemical configuration of one of these molecules does not permit binding of processed antigen to the receptor site of the *T*-cell despite interaction with Ia-antigen This results in dissociation of the low-affinity complex Ia-antigen–*T*-cell receptor and thus the *T*-lymphocyte remains in the resting state (stimulus is lacking)

7.2.8 Are all immunoglobulins only synthesized in plasma cells?

All immunoglobulin *L*- and *H*-chains, as well as the *J*-chain, are synthesized in plasma cells although the secretory component is produced in the epithelial cells of mucous membranes. Thus, two cell types — plasma and epithelial cells participate in the **biosynthesis of SIgA** (*Fig. 7.9*). The plasma cells localized just below the mucous membrane epithelium produce IgA dimers and *J*-chains and excrete them into the intercellular space. Some of these dimers pass into the blood circulation, but most are deposited in the mucous membranes. On their way to the mucous membrane surface, IgA dimers must pass across the basal membrane located below epithelial cells of the mucous membrane, and across the epithelial cells themselves. The passage through epithelial cells is facilitated by the secretory component, which is synthesized as a part of poly-Ig receptor for IgA, or even IgM transport.

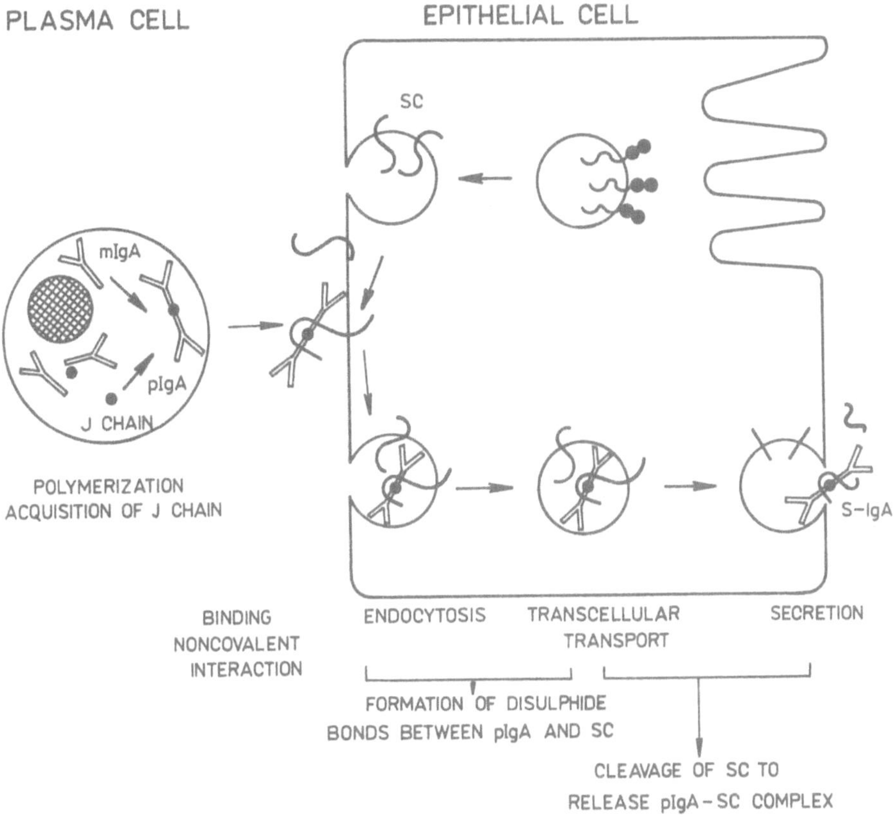

Fig. 7.9. Diagram of the selective transport of *J*-chain containing polymeric IgA (pIgA) through an epithelial cell (Mestecký *et al.*, 1988).
Polymeric IgA, assembled in subepithelial plasma cells from monomeric IgA (mIgA) with the participation of *J*-chain (●), interacts with the membrane form of SC SC is produced in the rough endoplasmic reticulum of the epithelial cell and its glycosylation is completed in the Golgi complex SC-containing vesicles fuse with the basolateral membrane, and membrane SC and interact with pIgA The SC-pIgA complex is internalized in endoplasmic vesicles, which are transported through the cytoplasm of the epithelial cell These vesicles fuse with the apical membrane of the epithelial cell, and SIgA is released into the external secretion

Some authors include **microglobulins** devoid of antibody activity in the immunoglobulin family because their structure is similar to immunoglobulin chains. However their mechanism of formation is different (an antigenic stimulus is not required) and they are produced by cells other than those producing immunoglobulins. β_2-Microglobulin and α_1-microglobulin are among the most important microglobulins. β_2-**Microglobulin** (β_2m) forms the "light" chains of class I histocompatibility antigens. It is present on the surface of all mammalian cells, except for mature red blood cells. It is also synthesized by these cells. In addition to cell surfaces, β_2m is present in blood serum and other body fluids. The values in sera of normal individuals are in the range of 0.8–3.0 mg/l. Increased levels can be detected in some pathological conditions, particularly in renal malfunction, lymphoproliferative diseases, rheumatoid arthritis, systemis lupus erythematosus (SLE), acquired immunodeficiency syndrome (AIDS) and in the sera of foetuses and newborn infants.

β_2m is a low-molecular-weight protein (M_r = 11 815) which, in contrast to other membrane proteins and immunoglobulins, does not contain a saccharide moiety. β_2m was isolated and characterized by BERGGARD and BEARN in 1968. After the primary structure (*Fig. 7.10*) had been determined, it was found that β_2m contains one polypeptide chain composed of 100 amino acid residues whose sequence is homologous with constant domains of both *L*-

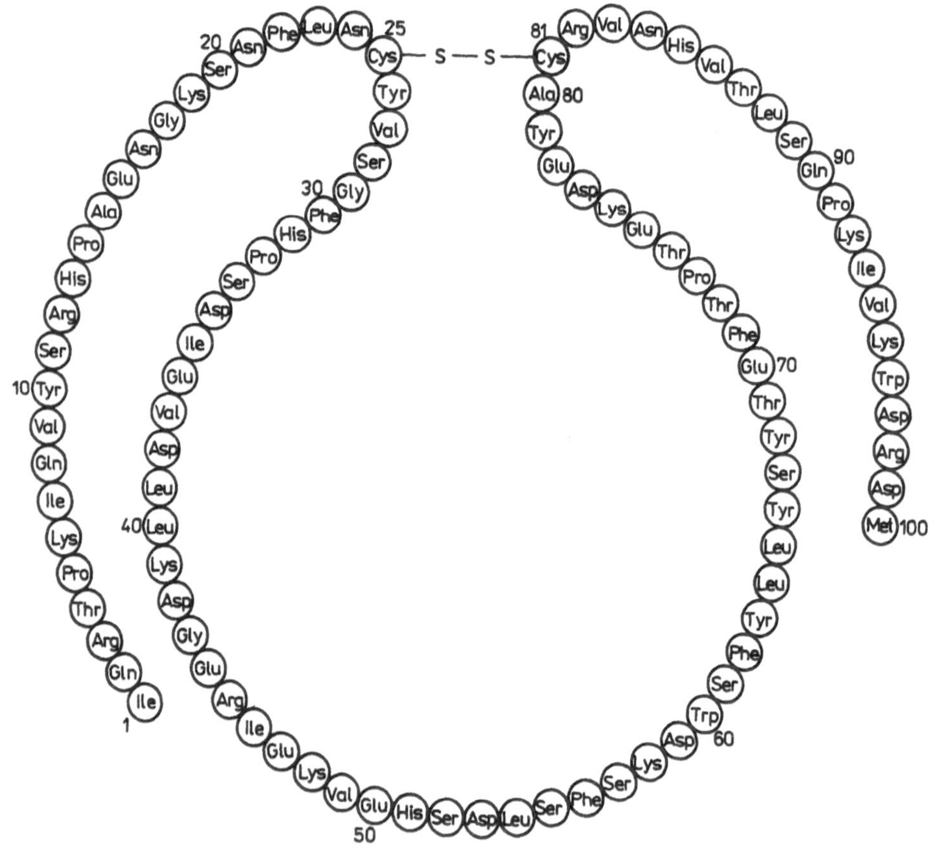

Fig. 7.10. Primary structure of human β_2-microglobulin.

and *H*-chains. It possesses the relatively highest degree of homology with $C_\gamma 3$ and $C_\varepsilon 4$ domains (*Table 7.3*). β_2m has one domain with a spatial arrangement similar to immunoglobulin domains. It is encoded by three exons including the exon encoding for the leading sequence (*Fig. 7.11*). β_2-Microglobulins of various animal species possess a rather conservative structure. Thus, for example, the amino acid sequence in human and rabbit β_2m differs only in 10 amino acid residues.

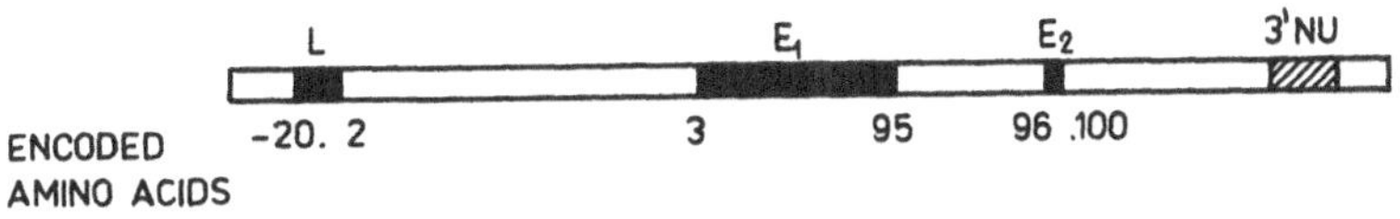

Fig. 7.11. Organization of the gene for β_2-microglobulin.

Table 7.3 Number of identical amino acids determined in studies on the primary structure of β_2-microglobulin and constant domains of human immunoglobulin light and heavy chains

	Light chains	
	C_κ-26 *C_λ-22*	
	Heavy chains	
C_γ1-21	C_μ1-16	C_ε1-20
	C_μ2-20	C_ε2-16
C_γ2-21	C_μ3-18	C_ε3-23
C_γ3-27	C_μ4-25	C_ε4-27
IgG	IgM	IgE

The biological function of β_2m has not yet been satisfactorily resolved. It shares some properties with immunoglobulins — particularly the ability to activate complement and bind to cells and tissues. These properties are associated with the fact that β_2m possesses the highest homology with the domains of the immunoglobulin Fc-regions. As the homology is only partial, these properties are not as pronounced as in the immunoglobulin Fc-fragment. The appropriately aggregated β_2m, for example, activates the alternative complement pathway much less than does the aggregated IgG.

In 1975, EKSTRÖM *et al.* isolated **α_1-microglobulin** (α_1m) from the urine of patients suffering from renal disease. Soon after this discovery, its presence was also demonstrated in normal sera (mean level 44 mg/l), urine (5.7 mg/24 h), and cerebrospinal fluid. In patients with renal disease, much higher concentrations were found: 200–250 mg/l and 100 mg/24 h in the serum and urine respectively. α_1m is a glycoprotein (M_r = 30 000 containing 20 % saccharides), present in the serum as a monomer or dimer, or even a high-molecular-weight polymer. Its biological function is unknown although some data suggest that it might possess an immunosuppressive effect. It is present on the surface of both *T*- and *B*-cells and is probably synthesized by these cells and excreted into the environment (BERGGÅRD *et al.*, 1980).

7.3 Polyclonal and monoclonal antibodies

The mechanism of antibody formation implies that every antigen determinant activates a single cell clone which synthesizes only one type of immunoglobulin molecule, all of them belonging to the same type, allotype, idiotype and subgroup. Their class and subclass can however be different. Their variable domains therefore have an identical amino acid sequence. Every individual, however, comes into contact with a number of antigens during development and many immunoglobulin-producing cell clones are therefore formed. Therefore, immunoglobulins of a healthy individual include various molecular species differing in their idiotype even within certain subclasses and subgroups. This is the basis of the well-known **immunoglobulin heterogeneity**.

Under some conditions, a clone of immunoglobulin-synthesizing plasma cells begins to multiply uncontrollably, regardless of the physiological need for these products. Such uncontrolled growth transforms normal plasma cells or their precursors into tumour cells resulting in multiple myeloma or WALDENSTRÖM's macroglobulinaemia. Each malignant clone of plasma cells produces large amounts of a single type of immunoglobulin molecule (with a homogeneous structure) that usually greatly exceeds the concentrations of other immunoglobulins of the same class. These **monoclonal immunoglobulins** can therefore be easily isolated. Analysis of their primary structure revealed only one amino acid at each position, because the molecules are identical. This finding made it possible to elucidate the structure and organization of immunoglobulin molecules.

Monoclonal immunoglobulins used to be considered pathological proteins and in clinical medicine were termed **paraproteins**. It is now known that these are not pathological immunoglobulins (with altered molecular structure) but are actually normal products which are produced in enormous quantities as a result of impaired regulation.

Monoclonal immunoglobulins cause **monoclonal gammopathies** of both benign and malignant type. Monoclonal gammopathies include the above-mentioned multiple myeloma, Waldenström's macroglobulinaemia and heavy chain disease. *Multiple myeloma* is caused by malignant tumour growth of plasma cell precursors in the bone marrow associated with the excessive production of IgG, IgA, IgD or IgE of the monoclonal type.

The occurrence of particular myeloma immunoglobulins in humans corresponds to the distribution of a normal immunoglobulin class in the peripheral serum. Thus, monoclonal IgGs are most frequent and IgEs are the least frequent. Myeloma immunoglobulins are formed either by the entire molecule or by light chains only (*Table 7.4*). In most cases, complete molecules are found in the serum of patients. In addition to complete molecules, monomers or dimers of *L*-chains (Bence-Jones *proteins*) can sometimes be found in serum and urine. Bence-Jones may sometimes occur alone. They are not degradation products of immunoglobulin molecules, but are synthesized in excess *de novo*. As Bence-Jones proteins are monoclonal products, each patient always has either $\varkappa$- or λ- type only.

Waldenström's macroglobulinaemia is caused by malignant growth of lymphoid cells which produce monoclonal IgM. Free and usually incomplete *H*-chains may be found in the serum of patients with *heavy chain disease*. They lack a certain domain (usually the variable domain) or its part. So far, three main types of this disease, having defective γ-, α-, or μ-chains, have been described. The defective monoclonal *H*-chains are not products of catabolic degradation of immunoglobulin molecules, but are synthesized by a corresponding malignant plasma cell clone.

Myeloma immunoglobulins have been found in man and also in some animals, *e.g.* mice, rats, chickens *etc.*

Table 7.4 Monoclonal gammopathies

Gammopathy	*Monoclonal immunoglobulin*	*Schematic structure of monoclonal immunoglobulin*[a]
Multiple myeloma	IgG, IgA, IgD, IgE	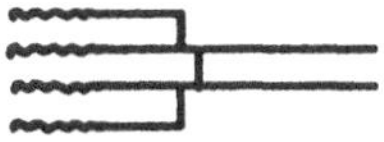
	IgG, IgA, IgD, IgE + Bence-Jones protein (BJP)	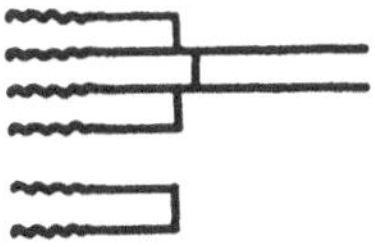
	BJP, type $\varkappa$ or λ	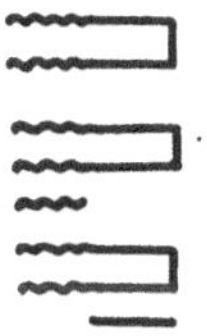
Waldenström's macroglobulinaemia	IgM	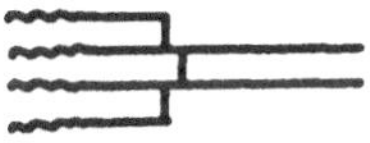
	IgM + BJP	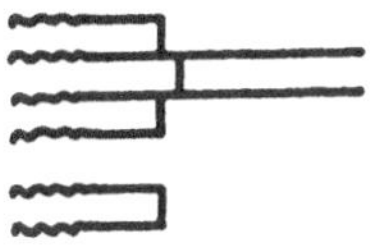
Heavy chain disease	α, γ, μ	

Points (...) mark the missing parts of the *H*-chain. [a] Only the basic IgM subunit is shown.

7.3.1 *Preparation of monoclonal antibodies*

Monoclonal antibodies are homogeneous and are therefore basically different from polyclonal antibodies. During immunization of an experimental animal, **polyclonal antibodies** are formed, because every antigen has more than one determinant group. Therefore, such serum, although highly specific, contains a mixture of antibodies with different affinities and biological functions (*e.g.* the ability to activate complement). This heterogeneity is responsible for differences in the quality of sera prepared in various laboratories which may give rise to conflicting experimental results. The antisera prepara-

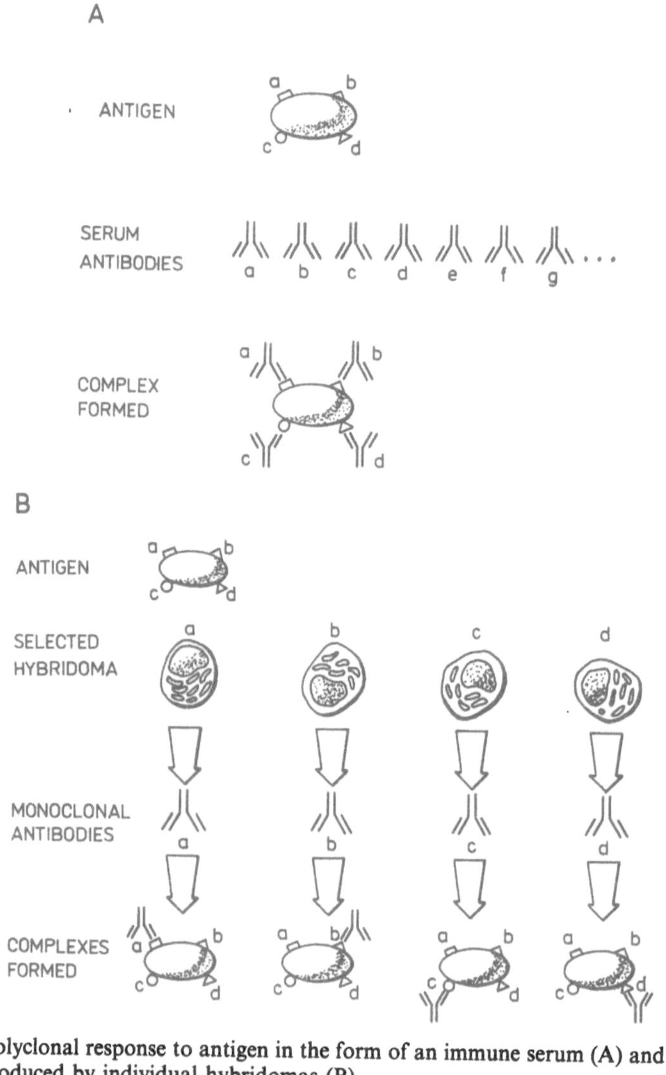

Fig. 7.12. Polyclonal response to antigen in the form of an immune serum (A) and monoclonal antibody produced by individual hybridomas (B).
a, b, c, d, antigenic determinants.

tion itself is difficult because the intensity of antibody response in a given experimental model cannot be predicted.

All these facts stress the advantage of **monoclonal antibodies**. Spontaneously produced monoclonal antibodies causing gammopathies cannot be used because the antigen (epitope) they can react with can only seldom be determined. Therefore, the possibility of generating artificial monoclonal antibodies has been intensively investigated during the past two decades. Attempts were made to obtain antigens having only one determinant in many copies, and small spleen fragments were cultured in the hope of obtaining a single cell clone producing the desired antibodies. However, the yields of antibodies, secreted into the culture medium, were very low because the plasma cells had only limited growth capacity and died after the short *in vitro* culture. The major turning point in the possibility of monoclonal antibody preparation of known specificity was the discovery by KÖHLER and MILSTEIN in 1975 (the Nobel prize in 1984).

The basis of their discovery is the fusion of myeloma (tumour) cells cultivated *in vitro* with mouse spleen lymphocytes immunized with the desired antigen. Thus, hybrid myeloma–spleen cells (**hybridomas**) were obtained which possessed properties of both parent cells. They are able to form antibodies specific for the antigen used — similarly to the spleen cells — and have the immortality of myeloma cells, *i.e.* the ability to multiply for an unlimited number of generations. Individual hybridoma cells can be cloned (multiplication of a single cell yielding many identical clones) and every clone produces large amounts of identical antibodies directed against a single determinant. The difference between monoclonal and polyclonal antibodies is schematically depicted in *Fig. 7.12*. The isolated hybridoma clones can be maintained in a frozen state for an unlimited time and can always be cultured and multiplied either *in vitro* (in tissue culture) or *in vivo* (after injecting into an appropriate experimental animal).

7.3.2 Lymphocyte hybridomas

The principle of lymphocyte **hybridization** with myeloma cells can be used for monoclonal antibody preparation against various antigens. The basic procedure is shown in *Fig. 7.13* and consists of several steps. The first step is the fusion of spleen cells from an immunized animal with myeloma cells grown in tissue culture. The resultant hybridomas are then submitted to selection in **HAT medium** (containing hypoxanthine, aminopterin and thymine). Culture media from individual hybridoma cultures are tested for the presence of secreted antibodies. An aliquot of each positive culture is frozen (a reserve in case the culture is destroyed during further treatment). The second aliquot is cloned and the presence of antibodies is re-tested. The cells producing antibodies with the desired specificity are divided into two portions; one is frozen, the other is repeatedly cloned. The selected variants are analysed and the required clones are multiplied and then maintained in the frozen state. After thawing they can be used to prepare large amounts of monoclonal antibodies.

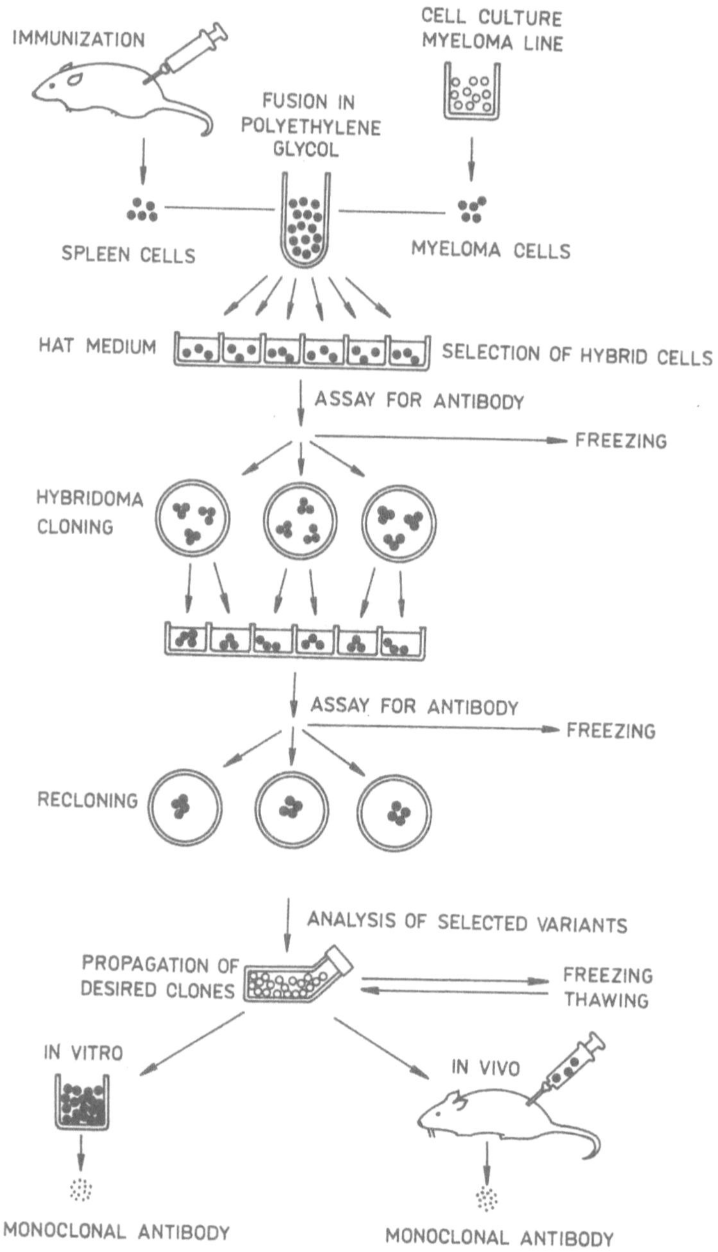

Fig. 7.13. General procedure for the preparation of monoclonal antibodies (Milstein, 1985).

In order to prepare hybridomas it is necessary to obtain appropriate *myeloma lines*. Such lines must permanently grow both *in vitro* and *in vivo*, must not produce own myeloma antibodies and must be in a differentiation stage permitting formation of specific antibodies, even after fusion with antibody-forming cells. During fusion, only a small proportion of hybridoma cells retains the ability to grow and multiply *in vitro*. Other cells remain either unfused or fuse with the same cells (*i.e.* lymphocyte–lymphocyte or myeloma–myeloma cell hybrids are generated). Thus, a mixture of cells is obtained after hybridization, and the required hybridomas must then be separated.

For this purpose, a method based on *enzyme defects* (lack of an essential enzyme) is most frequently used. It is based on the following principles. (1) Certain enzyme defects can be induced in the cultured *in vitro* cells. (2) The cells with enzyme defects are unable to grow in media containing the so-called selection components. (3) The hybridoma cells may compensate for the absence of the essential enzyme in one cell by its presence in the second partner cell. The absence of some nucleotide-synthesizing enzymes is used for selection of hybridomas.

Normal cells can synthesize nucleotides in two ways: (1) nucleotides are synthesized *de novo* from saccharides and amino acids; (2) nucleotides are synthesized from preformed purine bases and nucleosides (*Fig. 7.14*). Nucleotide synthesis by the former mechanism can be blocked by aminopterin, whereas the latter synthetic pathway depends on the activity of two enzymes — hypoxanthine(guanine)phosphoribosyltransferase (HGPRT) and thymidine kinase (TK). HGPRT catalyses the reaction of hypoxanthine or guanine with 5′-phosphoribosyl-1-pyrophosphate yielding corresponding nucleotides — IMP and GMP. Similarly, TK catalyses the conversion of thymidine into TMP.

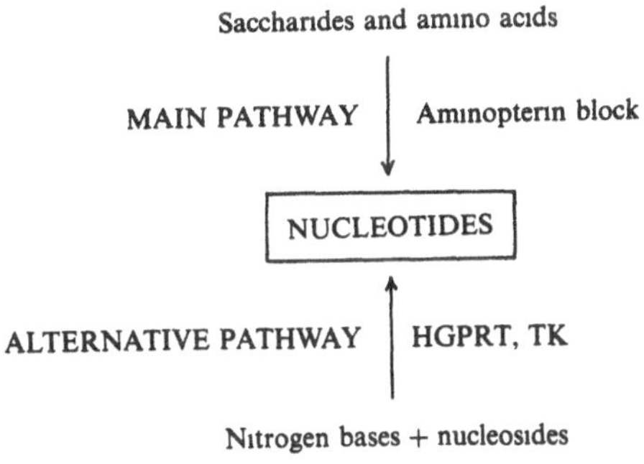

Fig. 7.14. Main and alternative pathway of nucleotide synthesis (Dráber and Vojtíšková, 1982). HGPRT, hypoxanthine(guanine)phosphoribosyltransferase, TK, thymidine kinase

Mutant cells lacking HGPRT can be obtained by culturing myeloma cells with 8-azaguanine (a guanine analogue) which is incorporated into the DNA of cells with functional HGPRT resulting in cell death (destruction of genetic information). In the presence of 8-azaguanine, only spontaneous mutants deficient in HGPRT can survive. Similarly, by culturing cells in the presence of a thymidine analogue — 5-bromodeoxyuridine — TK-deficient myeloma lines can be prepared. Under normal culture conditions, the cells synthesize nucleotides primarily *via* the main mechanism and the absence of enzymes of the alternative pathway does not interfere. However, when HGPRT- or TK-deficient cells are cultured in the presence of aminopterin (an inhibitor of the main pathway), they die. They can only survive if they fuse with cells that can compensate for the enzyme deficiency, provided the culture medium contains hypoxanthine and thymine in addition to aminopterin (*HAT-medium*). When a cell mixture is cultured in this medium after hybridization, only the hybridomas survive. The unfused myeloma cells, as well as myeloma cell–myeloma cell hybrids, die rapidly, since they cannot synthesize DNA. The lymphocytes or lymphocyte—lymphocyte hybrids do not multiply in the tissue culture and their half-life is also short.

Not all myeloma cell lines are suitable for preparation of hybridomas. Those lines that do not form immunoglobulin chains are most suitable. When the myeloma cell produces the light or heavy chain or complete imunoglobulin molecules, their chains may randomly recombine with immunoglobulin chains coming from the lymphocyte partner. This requires another step, in which a clone producing only the required variant (antibody coded by the parent lymphocyte), is selected.

The technology of monoclonal antibody preparation by cell fusion was originally limited only to mouse immunoglobulins, as only mouse myeloma lines were available and the interspecies fusions were usually unsuccessful. During interspecies fusion, many chromosomes of one partner, including the chromosomes bearing genes for immunoglobulin molecules, are lost within several weeks. Myeloma lines suitable for preparation of rat and human monoclonal antibodies are currently available. In the preparation of human monoclonal antibodies, however, the immunization and source of the spleen cells represents a serious problem.

Immunization of the spleen cell donor serves two purposes. First, it increases the number of spleen cells producing antibodies against the antigen used for immunization, and, second, it induces proliferation and differentiation of corresponding cell clones into the stage which permits generation of hybridomas after fusion with myeloma cells. The principles of immunization and the immunization protocols are similar to those used for preparation of conventional antisera. In some cases the immunization can also be performed *in vitro* by culturing the spleen cells with the antigen. At present, such techniques are limited, but if these problems are solved they might be useful in the preparation of human monoclonal antibodies.

Under normal conditions, relatively little fusion of lymphocytes with myeloma cells occurs. Therefore, fusion accelerating substances called **fusogens** must be employed. Köhler and Milstein (1975) used the inactivated virus *Sendai* as a fusogen in their first experiments. *Polyethylenglycol* (*PEG*) of molecular weight 2000–6000 is most frequently used at present. Fusion results in bi- or multinucleate cells (*e.g.* one myeloma cell may fuse with several lymphocytes). In the selective media, only the heterokaryotic cells formed by fusion of one myeloma cell and one lymphocyte can usually proliferate. Multinucleate heterokaryotic cells and homokaryotic cells formed by fusion of identical cell types die.

When selecting hybrid cells obtained by fusion, small numbers of the cells are transferred to HAT-medium and then pipetted into the wells of tissue culture plates. This facilitates the isolation of individual hybridoma clones. The growth of hybrid cells is significantly increased by the presence of supporting cells or their products.

Identification of hybrid cell clones producing antibodies of the desired specificity is the most time-consuming stage in hybridoma preparation. Hundreds or thousands of hybridoma cell cultures must be analysed before a culture with the desired antibody specificity is found. Therefore, there has been a tendency to try to detect microamounts of antibody and automate the whole procedure.

The first cloning of hybrid cells is performed immediately after fusion and the cells producing the required antibodies are cloned repeatedly. If all cell clones produce antibodies of the desired specificity after recloning, the process is terminated. When only a proportion of the clones produces the desired antibodies, the cell population is still heterogeneous and cloning must therefore continue. Cloning is performed even after long-term hybridoma culture *in vitro* in order to verify its stability. This may be performed in two ways — either by dilution or in semi-liquid media. Using the dilution cloning method, the cells are diluted to obtain a culture containing, on average, less than one cell per well of the culture plate. A semi-liquid medium is prepared from agar or agarose. The repeated cloning of hybridoma cells enhances the probability that the cell line obtained will be stable.

Multiplication of hybridomas for large-scale production of monoclonal antibodies is performed *in vitro* in culture medium or *in vivo* in mouse ascitic fluid. Cultivation can also be performed in synthetic media (without serum) which simplifies the isolation of pure immunogloblins. From 1 ml of culture medium between 10 and 200 μg of monoclonal antibodies can be obtained. In order to obtain large amounts of monoclonal antibodies, the ascitic (peritoneal inflammatory) fluid of mice, injected with hybridoma cells, is used. However, syngeneic mice must be used. One mouse yields 3–15 ml of ascitic fluid containing 2–20 mg/ml of monoclonal antibodies.

In addition to hybridomas which use plasma cells (as the terminal stage of the *B*-cell development), hybridomas can also be prepared from *T*-cells. They are produced by fusing *T*-lymphoma (a malignant disease of *T*-lymphocytes) cells with normal *T*-cells, enriched with the helper, suppressor or cytotoxic *T*-lymphocytes. The ***T*-hybridomas** then represent cell clones with specific regulatory or effector functions. In addition, they can produce various immunoregulatory and effector lymphokines.

This type of **hybridoma technology** can be used for the preparation of monoclonal antibodies and also for obtaining other biologically important substances, *e.g.* enzymes and hormones that can be isolated from natural sources only in trace amounts or with an insufficient purity. In this case, cells producing the desired substance are fused with appropriate tumour cells.

7.3.3 Properties and general application of monoclonal antibodies

The properties of monoclonal antibodies are not markedly different from those of polyclonal antibodies. However, they do possess significant advantages. They are biochemically homogeneous and are therefore monospecific and efficient at low concentrations, and can be prepared in virtually unlimited amounts even against relatively impure antigens. The majority of monoclonal immunoglobulins are *non-precipitating* antibodies, *i.e.* antibodies that do not form precipitating immune complexes after reaction with the corresponding antigen. This is because the monoclonal antibody only binds to a single antigen determinant and thus cannot form the three-dimensional lattice structure required for the generation of precipitating immune complexes. Such a lattice is generated only if the antigen is a polymer with repeating identical structural units.

To date, monoclonal antibodies have been prepared against viruses, bacteria and parasites, major histocompatibility antigens, surface cell antigens, tumour and differentiation antigens, receptors, hormones, enzymes and isoenzymes, various polysaccharides, glycoproteins and glycolipids, and nucleic acids *etc*. Therefore, they can be used in immunology, biochemistry, molecular biology and genetics, virology, microbiology and parasitology, clinical medicine and even industry (biotechnology) and other areas.

The isolation of proteins together with their use in analytical methods, both in research and practice, are the most important applications of monoclonal antibodies. Monoclonal antibodies, bound to a solid carrier, can be used in immunoaffinity chromatography for the isolation of a single antigen from a complex mixture of various antigens. Using such techniques, various enzymes, proteohormones and other bioregulatory substances, including lymphokines, interferons *etc*., can be isolated. Using monoclonal antibodies, even DNA fragments, used in genetic engineering, can be isolated. In addition, polyribosomes (ribosome aggregates) can be precipitated yielding a specific messenger RNA-enriched material. They can also be used as specific probes to localize genes for enzymes or other proteins.

Monoclonal antibodies have gradually replaced traditional polyclonal antisera in various immunoanalytical techniques, *e.g.* radioimmunoassay (RIA), enzyme-immunoassay (EIA) *etc*. Monoclonal antibodies are advantageous because: (a) they are homogeneous and therefore have uniform affinity constants which in turn allow shorter incubation times for the analysis; (b) they are highly standardized which ensures better reproducibility; (c) they exclude cross-reactions and thus allow accurate measurements even at high serum concentrations; (d) calibration curves remain linear within a broader concentration range which allows the use of fewer standards for construction of the curve.

Monoclonal antibodies have facilitated the relatively rapid and simple solution of a number of problems in biochemistry and molecular biology, which can be illustrated by the following examples.

Conformation changes play an important role in the biological activity of proteins. For example, complement activation *via* the classical pathway begins with the interaction of the Clq subcomponent with an immune complex. The mechanism of this reaction was clarified by a monoclonal antibody against the epitope generated during the conformational change of the Clq molecule. It was found that the free Clq molecule does not contain such an epitope, but it is expressed immediately after the interaction with the immune complex and initiates the complement cascade.

Many enzymes exist as inactive zymogens that have particular epitopes on their surface. When part of their chains is cleaved by limited proteolysis the proteins are transformed into active enzymes and new epitopes may be uncovered. It is now possible to prepare specific monoclonal antibodies against epitopes on the zymogen molecule, as well as against epitopes on the active enzyme, which permits their differentiation.

Monoclonal antibodies against individual epitopes make it possible to study the structure of complex biopolymers and subcellular structures as well as their developmental aspects. They are able to recognize the substitution of one amino acid in the polypeptide chain, as well as the total molecular conformation and the minute changes caused by the binding of the allosteric effector. They act as allosteric effectors and may influence the functional state of proteins or other biopolymers.

The plasma membrane contains numerous markers (antigens) characterizing the biological species, individual and tissue, as well as specific cellular functions and properties. Monoclonal antibodies allow a completely new approach to studies of surface cell antigens and their functions that are mostly unknown so far. Monoclonal antibodies, specifically recognizing surface antigens, make it possible to discriminate between *T*- and *B*-lymphocytes and their subpopulations (*e.g.* $T_{H/I}$- and T_S-lymphocytes); in addition, the developmental stage of individual cells, the presence of virus or tumour-associated antigens *etc.* can be determined.

Defects in regulation of the immune system are usually manifested by changes in the relative proportions of various lymphocyte subpopulations and their determination therefore became an important tool in the diagnosis of primary and acquired immunodeficiencies, autoimmune diseases, leukaemias, lymphomas, and infectious diseases. For example, patients with AIDS have markedly decreased numbers of helper *T*-lymphocytes, and in autoimmune diseases, a decreased number of suppressor *T*-cells is often observed. *Table 7.5* summarizes the most important monoclonal antibodies used for classification of human leukocytes.

Monoclonal antibodies became an important tool for the differential diagnosis of viral, bacterial and parasitic infectious diseases since they allow a more accurate determination of the infectious agent.

In clinical medicine, monoclonal antibodies are used particularly for the determination of tumour-associated antigens, for the accurate localization of tumours in the organism and the isolation of enzymes used in the treatment of certain diseases. They can also be used as carriers of cytotoxic drugs in the directed treatment of specific tumours ("magic bullet" therapy).

Table 7.5 Monoclonal antibodies used for classification of human leukocytes

Differentiation group	*Monoclonal antibody (symbol)*	*Characterized cell population*	*Application (identification of)*
CD1	T6, M241, SK9/Leu6	cortical thymocytes	thymocytes
CD2	T11, LFA-2, S2/Leu56	rosette-forming *T*-lymphocytes	peripheral *T*-lymphocytes
CD3	T3, BW264/56 SK7/Leu4	mature *T*-lymphocytes	peripheral *T*-lymphocytes
CD4	T4, MT 321, BW264/123	T_H-lymphocytes	$T_{H/I}$-cells
CD8	T8, BW135/80, UCHT4	T_C- and T_S-lymphocytes	T_C- and T_S-lymphocytes
CD9	BA2	monocytes	monocytes
CD10	J5, VIL-A1	pre-*B*-cells, PMN leukocytes	
CD11	Ki-M5, VIM-12, OKM1	monocytes, granulocytes, *NK*-cells, some lymphocytes	CR3 (α-chain)
CD15		leukocytes	CR3 (β-chain)
CD16	BW243/41, VEO13	granulocytes, *NK*-cells	$Fc_\gamma RIII$
CD21		*B*-lymphocytes	CR2 and receptor for EBV
CD23	MHMb	*B*-cells, monocytes, eosinophils	low affinity Fc IgE receptor
CD25	TAC, T1A	activated *T*-lymphocytes	IL-2 receptors
CDw29	4B4	*T*-lymphocytes	CD4$^+$ helper cells
CD32		granulocytes, macrophages, platelets, *B*-cells	$Fc_\gamma RII$
CD35		*B*-cells, monocytes, granulocytes, erythrocytes	CR1
CD45	2H4	*T*-lymphocytes	CD4$^+$ suppressor cells
CD56		*NK*-cells, monocytes	*NK*-cells
CD64		monocytes, Kupffer cells	$Fc_\gamma RI$

CD, "cluster of differentiation" This new nomenclature for differentiation antigens was approved by the 2nd International Conference on Human Leukocyte Differentiation Antigens (Boston, 1984) The symbols (abbreviations) of individual antibodies represent the original nomenclature of individual laboratories and manufacturers, previously used in the scientific literature

At present, monoclonal antibodies against various tumour-associated antigens are available, *e.g.* against carcinoembryonic antigen (CEA), α-fetoprotein (AFP), chorionic gonadotropin, β_2-microglobulin, ferritin, acid prostatic phosphatase. **Tumour-associated antigens** are produced in large amounts by certain tumours and are liberated into the blood at detectable levels. These antigens are produced in low amounts even by normal tissues and therefore only their elevated levels, and not simply their detection, is of diagnostic value. Monoclonal antibodies against tumour-associated antigens can also be conjugated with radioactive isotopes. When injected into the blood, they bind to tumour cells and specifically label them. By radioactivity measurement (**radioimmunoscintigraphy**) the tumour can be precisely localized. Modern radioimmunoscintigraphy allows the detection of a tumour weighing 0.1 g, with a diameter of approximately 1 cm. ^{123}I with a short half-life is usually used for the conjugation and the patient therefore receives a relatively low dose of radiation. The principle of conjugating a monoclonal antibody against a particular tumour antigen with a radioactive isotope may also be used for selective tumour irradiation.

Clinical **immunotoxins** are conjugates of antibodies and cytotoxic compounds that may bind to the target (*e.g.* a tumour cell) through a specific binding site, which subsequently kills the target cell with the cytotoxic agent without damaging other cells of the organism. Because of their higher specificity and affinity, it is advantageous to use monoclonal rather than polyclonal antibodies for this purpose (BYERS and BALDWIN, 1988).

The efficiency of immunotoxins depends on the pharmacokinetics of their degradation and on potential chemical modification in the organism. They should not themselves be immunogenic, which means that in man, human or at least **chimeric monoclonal antibodies** must be used. The term "chimeric" refers to immunoglobulin molecules, certain parts of which (*e.g.* the constant regions) are derived from one biological species and other parts (*e.g.* the variable domains) from another species.

Various toxic compounds, cytostatic drugs or radionuclides are used as the toxic components. The toxins, such as ricin, abrin, diphteric toxin, are proteins whose molecules consist of two polypeptide chains. One is responsible for its toxic activity (mainly inhibition of protein synthesis), whereas the second contains the active site which binds the toxin molecule to receptors of various cells. In order to ensure the specificity of finding through the antibody only, the toxic compounds must be cleaved so that only the chain responsible for the toxicity is used. Of low-molecular-weight cytostatic drugs (inhibiting cell growth), adriamycin, daunomycin and methotrexate are most frequently used to prepare immunotoxins. The toxic component can express its activity only if the target cell can split this component from the antibody. This takes place after endocytosis by lysosomal enzymes. It is important that conjugation of the toxin with the antibody molecule should not affect its combining site.

The therapy of tumours with immunotoxins is still in the experimental phase, and many problems remain to be solved before its routine clinical use.

Monoclonal antibodies have been successfully used in studies on the structure and function of various receptors on the cell surface. Several autoimmune diseases are caused by autoantibodies against certain receptors. For example, *myasthenia gravis* is caused by blocking autoantibodies against the acetylcholine receptors of muscle cells; *Grave's disease* is caused by stimulatory autoantibodies against the thyrotropic receptor of the thyroid gland; *insulin-resistant diabetes mellitus* is caused by blocking antibodies against the insulin receptor. Anti-idiotypic monoclonal antibodies against these autoantibodies may inactivate their combining sites and might be of clinical importance in treatment of these diseases. On the other hand, monoclonal antibodies against receptors may assume the function of autoantibodies and elicit such diseases in experimental animals in studies on pathogenesis.

For therapeutic uses human monoclonal antibodies are preferred. They minimize the problem encountered when administering foreign animal monoclonal antibodies (*e.g.* anaphylaxis, clinical manifestation of immune complex formation, and reduced efficacy secondary to anti-antibodies). In a great number of the patients treated to date with murine monoclonal antibodies, the human anti-mouse antibody response has limited their usefulness. But, only a fraction of the anti-mouse immune response is directed to the variable domain (idiotype) of the rodent immunoglobulins because the most immunogenic portion of the antibody molecule are the species-conserved constant regions. This suggests that human monoclonal antibodies will be more effective therapeutic molecules than their rodent counterparts.

Therefore, the effort has been evolved to obtain human monoclonal antibodies directly or to *"humanize"* rodent monoclonal antibodies. In recent years, reproducible reliable techniques for the *in vitro* immunization of human peripheral *B*-cells have been developed (BORREBAECK *et al.*, 1988). Another possibility represents the use of mice with *severe combined immunodeficiency (SCID)* after the reconstitution of the human immune system in their organism (MOSIER *et al.*, 1988).

Recombinant DNA technology enables the construction of chimeric rodent–human monoclonal antibodies by attaching human constant regions to the rodent variable regions (**chimeric antibodies** — *Fig. 7.15*). A gene encoding for certain antibody can be isolated, incorporated into an appropriate vector, modified and transferred to myeloma cells which will then produce recombinant molecules with new properties (BOULIANNE *et al.*, 1984; MORRISON and OI, 1989). Because the antibody-combining site is localized within the variable regions, these molecules maintain their combining affinity for the antigen and acquire the function of the substituted constant region. Humanized antibodies have been also constructed from rodent monoclonal antibodies by splicing the rodent hypervariable regions (CDR) onto the human variable framework (FR) sequences (**composite antibodies**). **Hybrid monoclonal antibodies** can be formed by attaching appropriate combining domains to effector molecules such as enzymes, toxins, *etc.*

Besides the minimized immunogenic adverse side-effects, monoclonal antibodies used for the treatment of human diseases or diagnostic purposes *in vivo* must also meet certain requirements. Such preparations must not contain bacteria, mycoplasmas or viruses. In addition, undesirable molecules, including DNA and pyrogens, must be removed and specific pharmacological and toxicological control tests must be applied.

Certain practical problems have also to be solved before the routine use of monoclonal antibodies in tumour therapy or in the regulation of the immune responses. Kilogram amounts of monoclonal antibodies would then be required; at present, however, about 100 000 litres of cell culture are required to produce 1 kg of monoclonal antibody. Preparation and isolation of antibody from such a large volume would certainly not be simple. Until now, for example, culture has been performed experimentally in tanks with a volume less than 1 000 litres. Moreover, the use of monoclonal antibodies in antitumour therapy is further complicated by the fact that, in most spontaneous human tumours, a specific tumour antigen that might be used for their preparation cannot be detected.

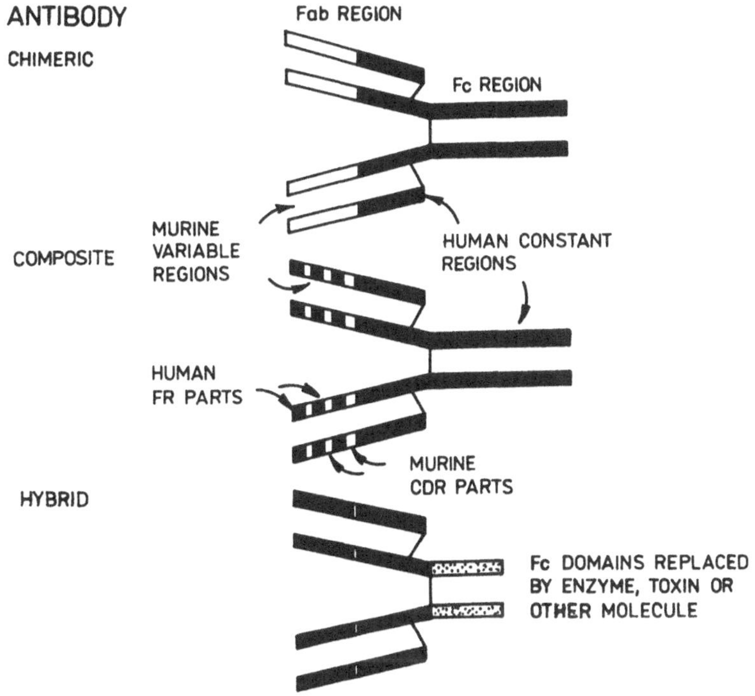

Fig. 7.15. Schematic diagram showing construction of chimeric, composite and hybrid human monoclonal antibodies.

Hybridomas are artificial cell lines that can be used for further manipulations using genetic and molecular biological techniques. This opens a completely new field of immunochemistry — the production of various antibody-based molecules. The preparation of **heterobispecific (bifunctional) antibodies** by fusion (hybridization) of two hybridomas may serve as an example.

Heterobispecific antibodies possess two combining sites with different specificity. Each of them is specific for a different epitope and originates from different immunoglobulin molecules — these are thus hybrid antibodies that do not occur in nature. Until recently, they could only be prepared by selective cleavage of disulphide bonds between two heavy chains in the immunoglobulin molecule, and subsequently combining two halves from different immunoglobulin molecules. However, this procedure is inefficient.

The modern technique, employing fusion of two hybridomas producing two different monoclonal antibodies, is much more effective.

Bifunctional antibodies were originally conceived as an aid to cancer chemotherapy, but are now perceived to have a much broader range of applications. One of the most significant advantages of bifunctional antibodies is that the antibody molecule is active in its native state and needs no chemical alteration in order to bind either the target cell or the functional agent. Likewise, agents bound by these antibodies do not require chemical modification. They have been successfully used *in vitro* for the localization of drugs, markers and cytotoxic cells to tumour cells, resulting in effective cell killing. Bifunctional antibodies have been utilized in a wide variety of immunocytochemical and enzyme-immunoassay techniques, especially in one-step immunoanalysis: one combining site is bound specifically to the antigen and the second to the marker molecule (*e.g.* enzyme) which is then used to detect the whole immune complex. It is envisaged that many of these techniques will be greatly improved by the use of bifunctional antibodies, increasing both the sensitivity and selectivity of the assay (NOLAN and O'KENNEDY, 1990).

Since antibodies can be raised against essentially any chemical structure, one can readily envisage the "engineering" of complementary antibody combining sites to appropriately chosen haptens which will provide an environment for the selective binding of reactants and promotion of specific chemical reactions. Thus, antibodies raised against structures which mimic the transition states of chemical reactions should selectively catalyse those reactions. Introducing catalytic activity into antibody combining sites should lead to a new class of enzyme-like catalysts — **catalytic antibodies** or **abzymes**. Catalytic antibodies could have considerable value as biochemical or molecular-biological tools, as therapeutic agents, or in the synthesis of pharmaceuticals and novel materials (SCHULTZ, 1989).

7.4 Genetic control of antibody formation

Since the origin of modern immunology, immunologists have been trying to answer the key question: what is the source of information for the millions of antibodies that may be synthesized by a normal individual? The clonal-selection theory, proposed on the basis of ideas of JERNE, BURNET and LEDERBERG in the late 1950s, assumes that any clone of immunocompetent cells can only synthesize antibodies of one specificity. In 1941, BEADLE and TATUM formulated the "one gene = one polypetide chain" hypothesis which became a dogma of molecular genetics for a quarter of a century. As the organism does not initially know what antigens it will meet, it must have genes encoding for antibodies of all possible specificities (**germline hypothesis**). It is estimated that the human organism may synthesize antibodies with at least 10^6 various specificities. Such a vast diversity of antibodies would require an extremely large number of genes. If one considers the IgG antibody only, 2×10^3 nucleotides would be required to code for a single molecule according to the triplet genetic code. When considering the lowest presumed number of antibody specificities, it amounts to $2 \times 10^3 \times 10^6 = 2 \times 10^9$ nucleotides. The whole human DNA molecule, however, contains only 6.6×10^9 nucleotides and thus it is impossible that it would be used only for antibody coding.

Therefore, at the beginning of the 1960s, PORTER formulated a hypothesis that the immunoglobulin molecule is encoded by two genes — one for the Fab-fragment, the other for the Fc-fragment. However, he was unable to demonstrate this hypothesis experimentally. It was in 1965, that DREYER and BENNET showed that the immunoglobulin chains must be coded for by at least two genes — one for the variable and the other for the constant region (the "two genes = one polypeptide chain" hypothesis). By cleavage of the DNA molecule with restriction endonucleases followed by hybridization, HOZUMI and TONEGAWA (1976) proved conclusively that the immunoglobulin light chains are encoded by specific genes for the *V*- and *C*-domains. A year later, BRACK and TONEGAWA found that these genes are separated by an intron, *i.e.* that the immunoglobulin-encoding genes are **split genes**. Further studies by TONEGAWA (Nobel prize winner in 1987), LEDER, RABBITS, HOOD, BALTIMORE and others, clarified the organization of immunoglobulin genes within the next 3–4 years. It was shown that the variable segments are not encoded by one joint gene, but by two to three gene segments that can be recombined and by genes for constant regions. During differentiation of *B*-lymphocytes, these genes join and form one structural gene, so that each *B*-cell clone synthesizes just one mRNA, coding the amino acid sequence of the whole immunoglobulin chain. The recombination processes guarantee generation of a large number of specificities from a small number of genes and gene segments (HONJO and HABU, 1985; BLACKWELL and ALT, 1989).

7.4.1 Organization of immnoglobulin genes

The human and murine immunoglobulin chains are encoded by three sets of genes (**translocons**) localized on three different chromosomes (*Table 7.6*). One translocon contains genes for the ϰ-light chain, the second for the λ-chain

Table 7.6 Translocons of human immunoglobulin chains

Translocon	*V-genes*	*C-genes*	*Position on chromosome no.*
L_κ	V_κI to V_κIV	C_κ	2
L_λ	V_λI to V_λV	$C_\lambda 1$ to $C_\lambda 6$	22
H	V_HI to V_HIV	$C_\gamma 1$ to $C_\gamma 4$	14
		$C_\alpha 1$ and $C_\alpha 2$	
		C_μ	
		C_δ	
		C_ε	

Until now, at least six genes for the constant part of λ-chains have been detected. They are generated by combination of subtypes Oz, Kern, Mz and Mcg (*Fig. 6.20*).

and the third for the heavy chains. The genes of all three chromosomes are arranged in the same manner — they are split (segmented) into exons coding for individual functional units of the chain. Introns of various length are localized between exons. The complete genes for the *V*-region consist of two or three germinal segments: *V* and *J* for the *L*-chains and *V*, *D* and *J* for the *H*-chains. The gene cluster for mouse L_λ-chain has the simplest organization (*Fig. 7.16*). It contains only two V_λ-segments that can combine with two pairs of J_λ-segments and C_λ-genes. Human translocon L_λ contains substantially more V_λ-segments that can be divided into five groups; in addition, it contains at least six C_λ-genes.

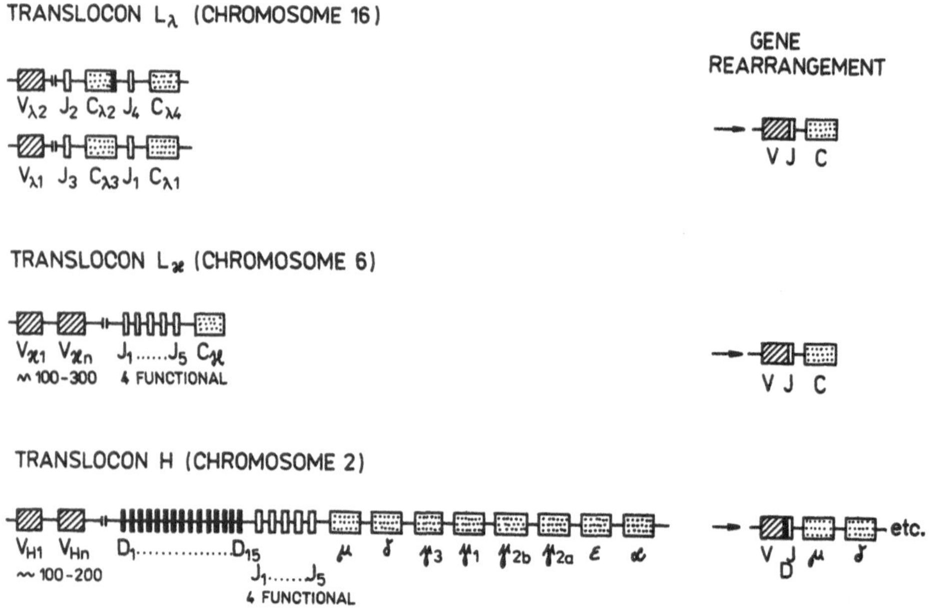

Fig. 7.16. Organization of genes for mouse immunoglobulins and the principle of their rearrangement during differentiation of *B*-lymphocytes.

Translocon $L_\varkappa$ has a similar organization in both man and mouse. It is composed of a number of $V_\varkappa$-segments, five $J_\varkappa$-segments and one $C_\varkappa$-gene. One of the mouse $J_\varkappa$-segments is inactive (*i.e.* not transcribed). Various authors, using different hybridization techniques, established that mouse genome contains up to 2000 diverse $V_\varkappa$-segments. It is generally thought, however, that their number does not exceed 100–300. The human genome probably contains fewer $V_\varkappa$-segments; only about 25. The total number of *V*-segments for both human *L*-chains is estimated to be approximately 100.

The gene system coding *H*-chains is much more complicated than systems coding the *L*-chains. In mice, it consists of 100–200 V_H-segments (*variable*), approximately 15 D_H-segments (*diversity*), five J_H-segments (*join-*

ing) and eight C_H-genes (*Fig. 7.16*). The V_H-segments are associated into at least 20 groups, each of them containing 5–10 segments. The human translocon H is generally identical with that of the mouse, from which it differs by having probably six J_H-segments and at least nine C_H-segments. The total number of human V_H-segments is unknown but it is presumably not more than 200.

The ***V*-segments** encode the first 95 amino acid residues in the immunoglobulin chains. The next part of the *V*-domains is encoded by ***D*-segments** of various length (1–15 amino acid residues), having a highly variable sequence, because they encode for the third hypervariable region of *H*-chains, or its part. The ***J*-segments** contain information for 13–17 amino acid residues and exhibit a highly homologous sequence. The highest variability is found in the first position which thus becomes the terminal part of the third hypervariable region.

Complete sets of all possible immunoglobulin gene segments and genes are present in germinal and precursor *B*-cells. Two functionally different rearrangements of immunoglobulin genes occur during differentiation of *B*-lymphocytes into antibody-producing cells. First, selection and final arrangement and assembly of *V*-genes occurs within each translocon so that the resulting gene contains one *V*-, *D*- and *J*-segment (**rearrangement of the *V*-gene**). This is followed by the joining of *V*- and *C*-genes. Thus a functional gene for the light and heavy immunoglobulin chain is generated, which determines the lymphocyte antigen-specificity. Such a rearrangement is independent of the presence of the antigen. During the second rearrangement the preformed V_H-gene recombines with various C_H-genes (**switch between *H*-chain classes**). Thus the antibody class is changed (together with the biological activity), although the antigen specificity, determined by the first rearrangement, remains unchanged. The second type of gene rearrangement requires the presence of the antigen.

During ontogenic development, the V_H-gene rearrangement precedes rearrangement of the V_L-gene. In the course of the V_H-gene rearrangement the *D*- and *J*-segments are first joined and afterwards the *V*-segments combine with the *D–J* complex. The complete *VDJ*-gene joins with the C_μ-gene, which permits coding of the complete μ-chain. Such a chain can be found in the cytoplasm of *pre-B-cells,* while the developmentally "younger" *pro-B-cells* contain only the translocated *DJ*-fragment, and in even "younger" *pre-pro-B-cells*, the rearrangement has not yet started, indicating that they contain a complete set of all genes and gene segments (*Fig. 7.17*). In the *virgin B-cells,* rearrangement of one of the V_L-genes is already completed which permits the synthesis of a complete IgM molecule which is subsequently incorporated into the plasma membrane as an antigen receptor. Thus, a mature *B*-cell is generated that may react with a specific antigen, which then induces further differentiation into the *plasma cell.*

Individual *V*-gene segments are actually exons separated by introns. The exon–intron organization of mouse germline DNA, coding heavy

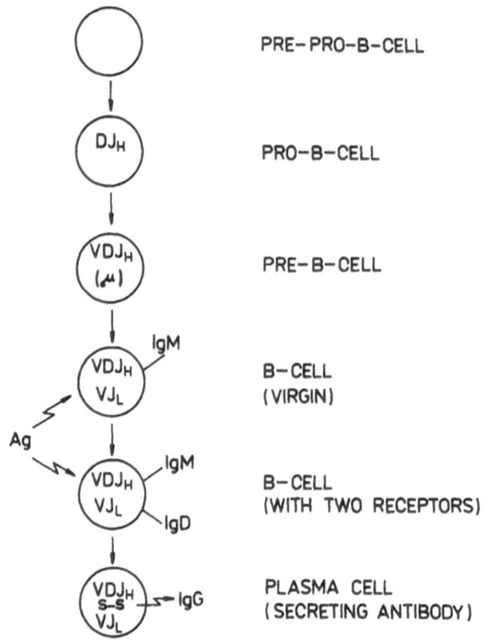

Fig. 7.17. Differentiation steps of *B*-lymphocytes (adapted from Honjo and Habu, 1985).
The degree of *V*-gene rearrangement (in circles) defines the maturation stage of *B*-cells

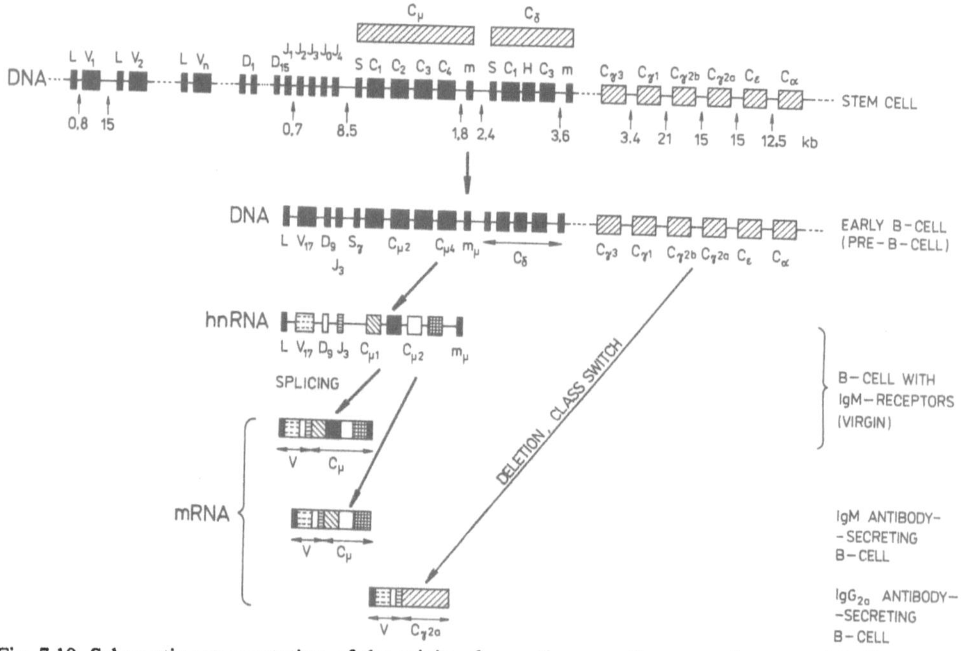

Fig. 7.18. Schematic representation of the origin of an active gene for mouse immunoglobulin *H*-chains and of its transcription into mRNA during *B*-lymphocyte development.
Exons are shown as rectangles, introns as lines L, leading sequence, S, switch site, m, membrane exon (enables anchoring of heavy chain C-end onto the *B*-cell cytoplasmic membrane), kb, kilobases (mark the size of introns in the DNA molecule), hnRNA, heterogeneous nuclear RNA, mRNA, messenger RNA

chains, is schematically depicted in *Fig. 7.18*. By rearrangement of *V*-, *D*- and *J*-segments and by their combination with the C_μ-gene, the active gene for the complete μ-chain is generated, which is transcribed into heterogeneous nuclear RNA (hnRNA) — a precursor of messenger RNA (mRNA). The active mRNA, which is translated into the amino acid sequence in the polypeptide chain, is generated from its precursor by post-transcription processing (splicing). The IgM antibody-producing cell may be generated from the virgin *B*-cell, bearing immunoglobulin receptors, after stimulation with a corresponding antigen. A further antigen stimulation induces a switch of the antibody formation to another immunoglobulin class. Special switching sites (*S*) are involved in this process. They are localized at the beginning of genes for constant domains of individual heavy chain classes, and genes between the end of the transcribing *V*-gene and the beginning of the selected *C*-gene are deleted. *Fig. 7.18* shows, as an example, deletion of C_μ-, C_δ-, $C_{\gamma 3}$-, $C_{\gamma 1}$ and $C_{\gamma 2b}$-genes during the switch of IgM synthesis to IgG_{2a} synthesis.

The switch of immunoglobulin production from one to another class always occurs on the basis of interaction between two *S*-sites; the first switch always originates from an *S*-site that is a component of a gene for the constant domains of the μ-chains ($S\mu$). The *S*-sites are composed of

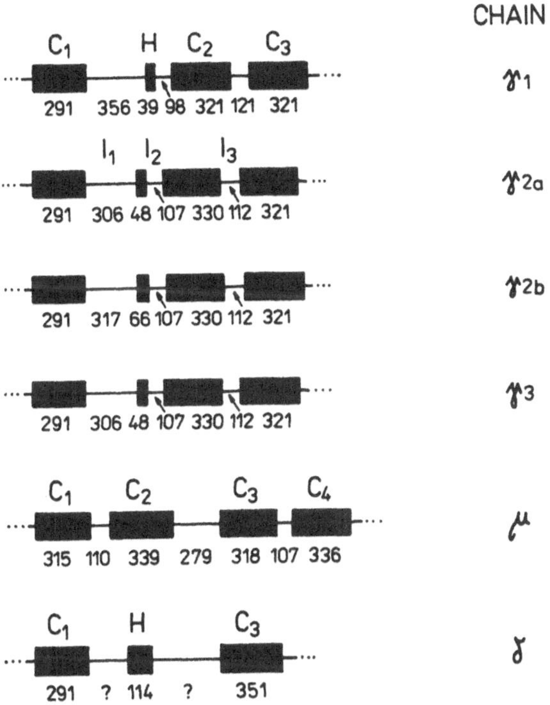

Fig. 7.19. Scheme of exon-intron structure of genes for constant regions of mouse immunoglobulin heavy chains.
C_1 to C_4, exons for constant domains; H, exon for the hinge region; I_1 to I_3, introns. The number under the introns and exons indicates the number of nucleotides.

1 500–2 000 nucleotides having different sequences for individual classes; all of them, however, contain numerous repeated sequences GAGCTG and TGGGG.

During *B*-cell differentiation, not all possible switches and formation of all immunoglobulin classes occur; usually just one or two.

The genes for the constant regions of immunoglobulin chains are also composed of exons and introns. Each domain and the hinge region have specific exons (*Fig. 7.19*).

After the immunoglobulins have been synthesized on membrane-bound ribosomes, they may be incorporated into the plasma membrane of *B*-lymphocytes as specific antigen receptors, or may be secreted by the cell into the external environment. The **membrane-bound immunoglobulins** have their last constant domain extended by one short segment composed mostly of hydrophobic amino acid residues. This enables the whole molecule to be anchored into the membrane. The secretory and membrane immunoglobulin molecule is encoded by the same gene. What is the mechanism of transcription leading to either the membrane immunoglobulin or the secretory immunoglobulin? This is due to alternative (double) processing of mRNA (schematically shown in *Fig. 7.20* for IgM). A polyadenylate chain (poly *A*–polynucleotide segment) is usually bound to the 3′-end of the mRNA molecule. Once the poly *A*-chain is bound to the end of the exon for the $C_\mu 4$-domain, where the "secretory tail" is present, gene transcription is terminated and mRNA, encoding secretory IgM, is produced. If, however, transcription of the gene for the μ-chain is not terminated at this stop signal, hnRNA is

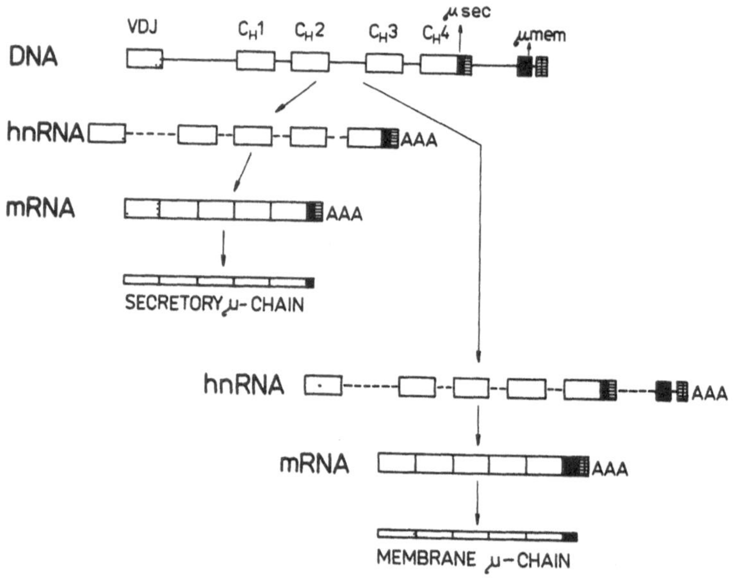

Fig. 7.20. Model of formation of two different mRNA species coding the secretory and membrane μ-chain from one gene.

μsec, gene segment for "secretory tail", μmem, exon for "membrane tail", AAA, polyadenylate chain

synthesized from which the sequence for the "secretory tail" is cut out, and mRNA translated into membrane IgM is generated. Genes for γ- and δ-chains and probably even for ε- and α-chains have a similar structure. In the case of δ- and γ-chains, the "secretory tail" is localized at the end of the exon encoding the C_H3-domain. This mechanism permits the simultaneous production of IgM and IgD by some *B*-cells. In such a case transcription is not terminated beyond the exon μ_{mem}, but proceeds through genes for the constant domains of the δ-chain (*Fig. 7.18*). Thus, a primary RNA transcript (hnRNA) is generated, from which two functional mRNA are formed — one for μ-chains, the other for δ-chains.

In summary, changes in the production of immunoglobulin heavy chains during *B*-cell development may be induced by at least two different mechanisms. One is based on rearrangement and deletion of germinal DNA, whereas the second is based on different (alternative) hnRNA splicing.

Recently it was shown that the amino acid constitution of the 25 amino acid transmembrane-spanning region of the mIgM chain also contains polar amino acids and is probably insufficient to ensure the transfer of activation signal from the antigen receptor to the intracellular molecules. It was therefore postulated that the *H* chain membrane-spanning region must interact with another membrane protein which was indeed characterized as a disulphide-linked heterodimer of 32 and 37 kD polypeptide chains. These findings indicate that ***B*-cell antigen receptor (BCR)** is not a simple Ig molecule but a complex of the heavy and light immunoglobulin chains (mainly IgD or monomeric IgM) noncovalently associated with at least one heterodimer composed of $\alpha\beta$, and in some cases $\alpha\gamma$, subunits that contain sites ensuring the intracellular transfer of activation signal after receptor ligation. These Ig-α and Ig-β or Ig-γ are members of the Ig supergene family and are composed of extracellular, transmembrane and cytoplasmic domains. Transfection studies have established that Ig-α is required for surface expression of membrane immunoglobulin (RETH *et al.*, 1991; KEEGAN and PAUL, 1992).

7.4.2 Generation of antibody diversity

Several hypotheses and theories were formulated in an attempt to explain the mechanism of genetic regulation of antibody diversity, the most important being the germinal, somatic mutation and recombination theories.

Advocates of the **germinal hypothesis**, formulated by DREYER and BENNET (1965), assume that the individual genome contains all necessary V_L- and V_H-genes and have estimated their number to be 10^4–10^5 (HOOD and TALMAGE, 1970).

The **somatic theory** (initially formulated by BRENNER and MILSTEIN in 1966) is based on the assumption that there are substantially fewer *V*-genes and their diversity arises during the ontogenetic development of antibody-forming cells, either as a series of point mutations (*somatic mutation* theory, COHN, 1970) or by mutual recombinations (*somatic recombination* theory,

GALLY and EDELMAN, 1972). Another somatic recombination theory was formulated by KABAT *et al.* (1978). According to this theory, the gene for a complete *V*-domain consists of small gene segments (*mini-genes*), coding three hypervariable segments (CDR) and four relatively non-variable segments (FR).

There are objections to all these theories; for example, the germinal theory places too high requirements for the number of genes in the individual genome while the somatic mutation theory fails to explain why the point mutations are only generated in the hypervariable regions. On the other hand, the validity of certain ideas of each theory could be proved experimentally. No single theory can therefore explain the origin of antibody diversity by itself, although individual elements of each theory, taken together, can provide a satisfactory explanation (*Table 7.7*).

Table 7.7 Mechanisms conditioning the origin of antibody diversity

Mechanism	*Its expression*
Germinal	Existence of numerous *V*-gene segments
Recombination	Various *V*-(*D*)-*J* recombinations
	Various V_H and V_L recombinations
Joining	Various possibilities of V_L–J_L, V_H–D, D–J_H segments joining
Mutation	Somatic point mutations — replacement of one nucleotide by another one in *V*-genes

It was found that the mouse genome contains approximately 100 V_L-segments and four *J*-segments which yield $100 \times 4 = 400$ V_L-genes. In addition, there are about 200 V_H-segments, 15 *D*-segments and four *J*-segments which in turn yield $200 \times 15 \times 4 = 12\,000$ V_H-gene combinations. If one assumes that the product of any V_L-gene can be joined with the product of any V_H-gene during formation of a combining site in the antibody molecule, this yields a total of $400 \times 12\,000 = 4.8 \times 10^6$ antibody specificities. This calculation took only the germinal and recombination mechanisms into consideration. The action of joining and mutation mechanisms further increases the number of possible antibody specificities 100- to 1 000-fold. It, therefore, follows that less than 350 genes and gene segments may theoretically specify hundreds of millions of different antibodies. Similar data also hold for the genome of man and of other mammals.

The mechanism of additional antibody diversity enhancement is based on the fact that recombination of *V*- and *J*-segments for the light chain as well as *V*–*D* and *D*–*J* segments for the heavy chain may be achieved in various ways (*Fig. 7.21*). During the "correct" recombination, the 95th codon (a nucleotide triplet which is a basic unit of the genetic code and governs the incorporation of one amino acid into a polypeptide chain) of the *V*-segment joins directly to the first codon of the *J*-region. This first codon of the *J*-segment then represents the 96th codon of the V_L-gene. Other joinings are inaccurate and involve the codon downstream from the 95th codon of the

V-segment which is normally a component of the intron excised during combination of both segments. One nucleotide of this extra codon can be utilized, thereby changing the triplet sequence and the amino acid which it encodes (GOUGH, 1981). Alternatively, the whole extra codon can be utilized, which results in extension of the *V*-gene by one codon and *V*-domain by one amino acid residue. In addition to such codon insertion, deletion of the first codon from the *J*-segment may also occur which reduces the polypeptide chain by one amino acid residue. *Fig. 7.21* shows that inaccuracy in joining the $V_\varkappa$- and $J_\varkappa$-segments yields five different ways in which the structural diversity of this light chain can be increased. Similar possibilities also exist in the second light chain (λ). Heavy chains have more possibilities of such inaccurate joining because up to three gene segments are recombined ($5 \times 5 = 25$). Theoretically, there are $(5 + 5) \times 25 = 250$ different possible combinations of all *V*-gene segments. If this number is multiplied by the number of antibodies that may be generated by the simple recombination of all genes and gene segments, $4.8 \times 10^6 \times 250 = 1.2 \times 10^9$ antibody specificities are obtained.

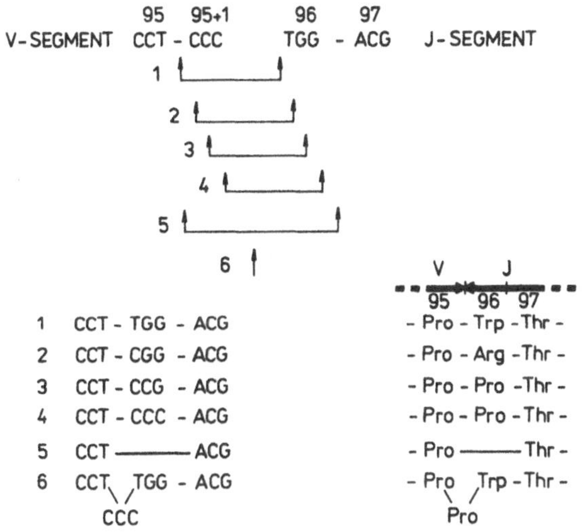

Fig. 7.21. Scheme illustrating various possibilities of $V_\varkappa$-segment and $J_\varkappa$-segment joining into the active $V_\varkappa$-gene.

Left, mRNA sequence transcribed in the form of a triplet code, right, amino acid sequence in the polypeptide product

It is generally accepted that the human *B*-cell system possesses the ability to form only about $2\text{–}5 \times 10^7$ various antibody specificities. This would mean that certain restrictions which decrease the number of theoretically possible antibody specificities should be considered (*e.g.* it has not been proved whether each heavy chain can combine with any light chain).

Another somatic mechanism enhancing the antibody diversity are mutations generated during the ontogenetic development of *B*-lymphocytes.

Such mutations occur not only in hypervariable segments but also in the whole *V*-regions. Substantially more mutations are generated during the development of IgG- than IgM-producing cells. The reasons for this difference have not yet been clarified.

All the above mechanisms responsible for the antibody diversity operate at the gene level. However, it is likely that there are other mechanisms acting at the RNA level, particularly during maturation of the functional mRNA.

7.4.3 Regulation of immunoglobulin gene expression

Regulation of immunoglobulin gene expression includes both the events that can be encountered during expression of other autosomal genes, and events that are specific for genes encoding immunoglobulin light and heavy chains. The following special requirements must be met.

1. Despite the fact that immunoglobulin genes are present in all nucleated cells of the organism, they can be expressed only in cells of the *B*-line.
2. A gene encoding expression of a chain requires *rearrangement* of gene segments in the DNA molecule.
3. Only a gene in one of two chromosomes (paired chromosomes — one maternal, one paternal) may be functionally expressed. This means that a certain plasma cell transcribes only one allele for the light chain and one allele for the heavy chain. This phenomenon is called *allelic exclusion.* Thus, during allelic exclusion the cell phenotypically expresses just one of two different alleles that are present in its genome for the given gene locus.
4. During *B*-cell development, genes for heavy chains are expressed earlier than genes for light chains.
5. When genes for light chains are expressed, expression of both light chain types must be coordinated. One cell can synthesize either $\varkappa$-chains or λ-chains, but not both (*isotypic exclusion*).
6. Production of immunoglobulin molecules is amplified (by several orders of magnitude) when the pre-*B*-cell matures into the final differentiation stage — the plasma cell.
7. Immunoglobulin molecules may be resident on the *B*-cell surface as receptors for antigen, may be secreted from the cell into the external environment, or may be both incorporated into the plasma membrane and secreted (from the same cell).
8. Genes for immunoglobulin polypeptide chains are related (structurally similar) to genes for some other glycoproteins belonging to the so-called immunoglobulin superfamily. Presumably, all these genes have been generated from one ancestral gene by gradual evolution.
9. The *B*-cell growth and maturation, together with immunoglobulin gene expression, are modulated by various signals transmitted to their surface. The signals coming from the antigen, *T*-cells and accessory cells, are of primary importance.

Although a particular *B*-lymphocyte can only express one of the allele pair for heavy and light chain genes, an individual may express both alleles — some *B*-cells express a product of one allele while other *B*-cells a product of the second allele. Thus the **allelic exclusion** occurs at the single cell level rather than at the level of the entire cell population. The allelic exclusion phenomenon significantly influences the specificity and efficiency of antibodies formed. If it did not exist, every *B*-cell could synthesize two different *L*-chains and two different *H*-chains. Theoretically, ten different antibodies could be generated by their combination, as shown in *Fig. 7.22.* Provided each allele is expressed with the same frequency, then each of these antibodies would represent just 10% of the total number of the antibody molecules formed. This would mean that the antibody response to a given antigen is "diluted" by the production of the remaining nine antibodies. In addition, most of this antibody population would have a mixed specificity (different combining sites). These antibodies could not participate in multibond reactions and would therefore be of limited efficiency. Isotype exclusion also prevents generation of undesirable heterobispecific antibodies.

	LL	Ll	lL	ll
HH	LHHL	LHHl	lHHL	lHHl
Hh	LHhL	LHhl	lHhL	lHhl
hH	LhHL	LhHl	lhHL	lhHl
hh	LhhL	Lhhl	lhhL	lhhl

Fig. 7.22. Combination possibilities of joining two different light chains (L and l) and two different heavy chains (H and h) to form an immunoglobulin molecule.
Combinations which represent identical molecules are boxed.

What is the mechanism of allelic exclusion? The simplest explanation would be that rearrangement of immunoglobulin genes is limited to an allele in one of the homologous chromosome pair. This allele is activated, while the second allele remains in the inactive germinal stage. Such a possibility, however, has only rarely been experimentally demonstrated. Rearrangement of immunoglobulin genes mainly occurs in both alleles. At present it is assumed that allelic exclusion is caused by the inaccuracy of *V*-gene rearrangement. The immunoglobulin gene rearrangement may be considered *productive* if a functional polypeptide is synthesized, and as *non-productive* if a functional polypeptide is not synthesized (Early and Hood, 1981). Non-productive translocation occurs when an extra nucleotide is inserted into one codon during recombination of the *V*- and *J*- segment (*Fig. 7.23*). Under such conditions, all codons localized upstream from this codon lose their meaning since the triplet reading frame is shifted by one position which results in synthesis of "nonsense" polypeptide. The existence of such non-functional polypeptides was experimentally proved in *B*-cells. In fact, both alleles can be transcribed, although only one of them into the functional immunoglobulin chain. It appears that the origin of inaccuracy in one allele during gene segments joining, is facilitated by generation of correct translocation on the second allele. The translocation process in the cells does not start in both alleles simultaneously. The cell provides "maximal care" to the allele where immunoglobulin gene rearrangement begins, whereas the gene segment joining in the second allele does not require such accuracy because it would not be advantageous to the function.

```
                 95   95+1      96    97
V-SEGMENT...CCA - CCC          TGG - ACG... J-SEGMENT

                 95    96    97
       1.     CCA - CCG - ACG -      PRODUCTIVE TRANSLOCATION
       2.     CCA - CCC - ACG -      PRODUCTIVE TRANSLOCATION
       3.     CCA - CCCG-ACG -       NON-PRODUCTIVE TRANSLOCATION
```

Fig. 7.23. An example of non-productive translation during *V*-segment and *J*-segment joining.

In man and mice, $\varkappa$-genes are expressed before λ-genes. It was found that the V_λ–J_λ joining does not take place in the cell where a correct $V_\varkappa$–$J_\varkappa$ translocation is generated. This might explain **isotypic exclusion**. It is not known, however, whether the preferential $V_\varkappa$-gene rearrangement is genetically predetermined or whether it is brought about randomly on the basis of probability which may be influenced in the mouse genome by having more $V_\varkappa$- than V_λ-segments. This is why murine immunoglobulins contain 95% $\varkappa$-chains and only 5% λ-chains. On the other hand, dog and horse immunoglobulins contain virtually only λ-chains which might indicate preference of V_λ-genes over $V_\varkappa$-genes.

7.5 Antigen receptors on *T*-lymphocyte plasma membranes

T-lymphocytes involved in specific cellular immunity reactions, cytotoxic reactions, delayed-type hypersensitivity reactions and regulation of antibody formation, can also specifically recognize antigens. For that purpose they are equipped with specific antigen receptors. For contrast to *B*-lymphocytes, however, the *T*-cell receptors require the presence of major histocompatibility complex products (MHC restriction). Antigen recognition by helper and inductor *T*-cells is restricted by the presence of class II MHC antigens (in man HLA-DP, HLA-DQ, HLA-DR), while recognition by receptors on cytotoxic *T*-lymphocytes requires the presence of class I MHC antigens (HLA-A, HLA-B, HLA-C).

Whereas the nature of *B*-cell receptors is clear, the character of *T*-cell receptors has only recently been unequivocally clarified. Originally, the possibility that *T*-cell receptors might constitute a separate immunoglobulin class (IgT) was considered, although evidence has since accumulated which indicates that *T*-cell receptors are coded by genes other than those for immunoglobulins.

Between 1982 and 1985, ALLISON, KAPPLER, REJNEK, SAITO and others studying receptors isolated from normal *T*-lymphocytes, cloned *T*-cell lines (clonotypes) and *T* hybridomas, showed that ***T*-cell receptors (TCR)** are glycoproteins with molecular weights of 80 000–90 000. TCR are heterodimers consisting of an acid polypeptide *α*-chain and a neutral or basic *β*-chain. The chains are joined by a disulphide bond to form a TCR molecule. The ***α*-chain** has a relative molecular weight of approximately 45 000 and contains 286 amino acid residues whereas the ***β*-chain** has a molecular weight of about 40 000 and 282 amino acid residues. Both chains contain a relatively large proportion (25–30%) of saccharides. Slight differences in relative molecular weight between human and mouse *α*- and *β*-chains are primarily due to the different saccharide content. Each chain is composed of a variable and constant domain (similarly to immunoglobulin light chains). In addition, they contain a short segment similar to the hinge region (where the interchain disulphide bond is located), a transmembrane peptide and the *C*-terminal cytoplasmic tail (*Fig. 7.24*).

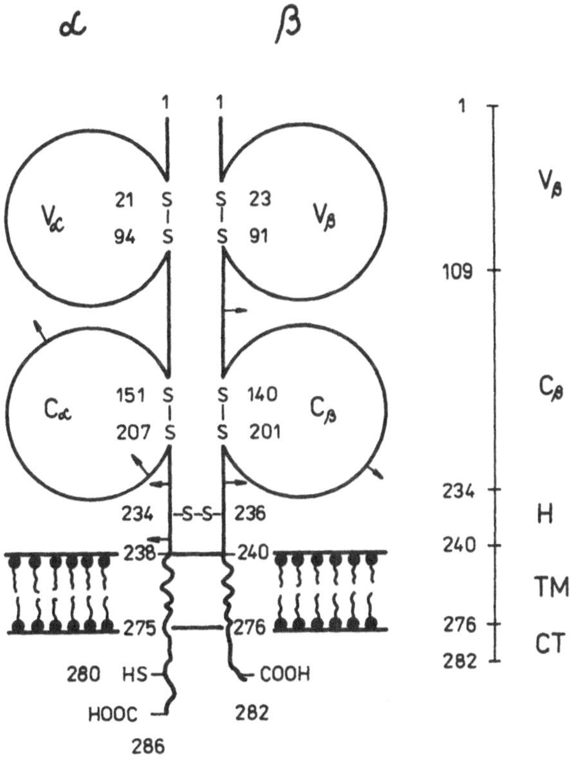

Fig. 7.24. Polypeptide structure of *T*-cell antigen receptor (Saito *et al.*, 1984; Honjo and Habu, 1985).
α, *α*-chain; *β*, *β*-chain; *V*, variable domain; *C*, constant domain; H, hinge region; TM, transmembrane segment; CT, cytoplasmic tail. Numbers refer to the amino acid position, arrows indicate points where oligosaccharide residues are bound.

When comparing the primary structure of the *V*-regions, it was found that the *V*-domain of the TCR *β*-chain exhibits a significantly higher variability than the *V*-domain of immunoglobulin chains. In addition, it contains considerably more hypervariable regions. Whereas the variable domains of immunoglobulin chains contain three hypervariable regions (two encoded by the gene *V*-segment, the third localized at the site of *V–D–J* segments joining), the *V*-domain of the *β*-chain may have up to seven hypervariable regions (*Fig. 7.25*). Three of them (designated *B, D* and *G* in *Fig. 7.25*) are localized at almost the same position as the hypervariable regions of immunoglobulin *κ*-chains. Additional segments (*A* and *F*) correspond to regions in V_H- and V_κ-domains exhibiting a higher variability than typical non-variable areas, but a lower variability than hypervariable regions. The *C*- and *E*-segments have no equivalents in immunoglobulin domains and occur in V_β-areas of TCR only. Presumably, they might be a component of the second receptor combining site on *T*-cells. Antigen is bound to one combining site and the corresponding MHC product to the second (*Fig. 7.8*).

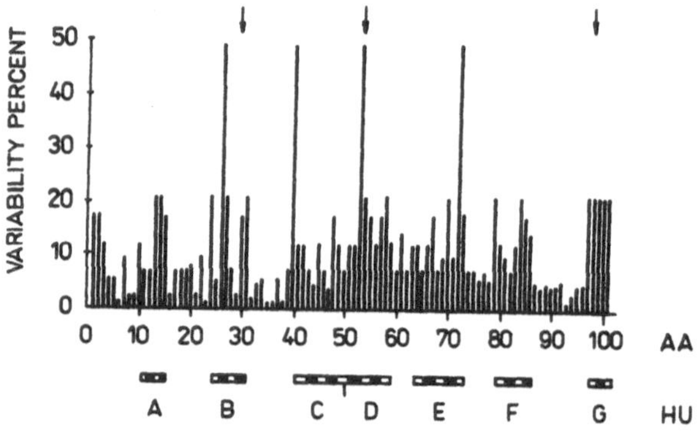

Fig. 7.25. Frequency of amino acid exchanges on individual positions of the *T*-cell receptor *β*-chain (Patten *et al.*, 1984).

The basis of the higher variability of V_β-domains is unknown. Two hypotheses were formulated in an attempt to elucidate this phenomenon. (1) From phylogenetic studies it is apparent that the specific cellular immunity reactions developed earlier than the reactions of specific humoral immunity. If the same rate of possible amino acid exchange is assumed, it takes — in the case of *β*-chains — a longer period, and thus the variability of the primary structure is also higher. (2) The V_β-domains could have developed under a higher selective pressure since the *T*-cell receptors must ensure reactions with both the antigen and with the MHC products.

T-cell receptors are expressed on the surface of human *T*-cells in association with the differentiation antigenic **CD3 complex** composed of five polypeptide chains: *γ* ($M_r = 25\,000$), *δ* (20 000), *ε* (20 000), *ξ* (16 000) and

η (28 000). γ- and δ-chains are glycosylated, while the hydrophobic ε-chain and ξ- and η-chains are not (*Fig. 7.26*).

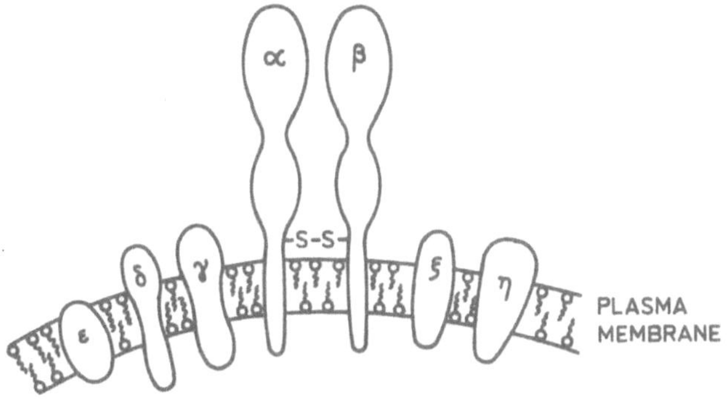

Fig. 7.26. TCR–CD3 complex on the surface of *T*-cells.
α, β, chains of *T*-cell receptor (TCR-α,β isotype), γ, δ, ε, ξ, η, chains of differentiation antigen CD3.

The classical **TCR-$\alpha\beta$** is involved in specific recognition by *T*-cells of antigens on the surface of other cells, always in the context of MHC molecules. Cells bearing this TCR-$\alpha\beta$ in peripheral lymphoid organs are either CD4- or CD8-expressing cells, the former usually recognizing class II and the latter class I MHC molecules. The CD3 complex appears to be involved in the transmission of the recognition signal received through the TCR to intracellular processes during *T*-cell activation. This process results in the display of helper (T_H) or cytolytic (CTL) effector functions (*Fig. 7.27*).

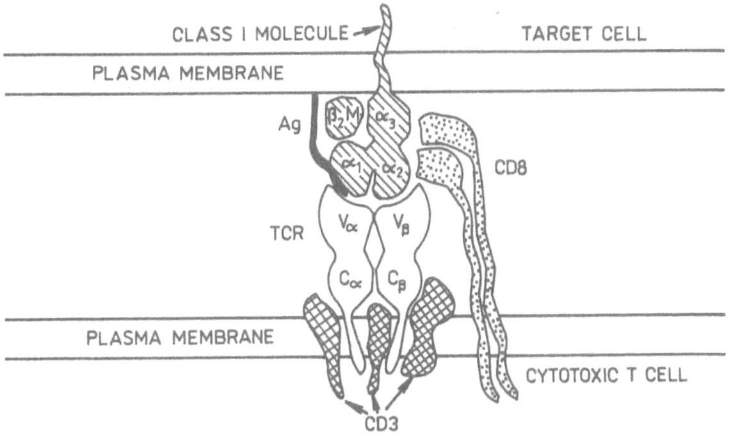

Fig. 7.27. Model for recognition by CD8$^+$ *T*-cells (Parnes, 1986).
This is a schematical model for recognition of antigen (Ag, solid area) plus class I MHC molecule (hatched) on a target cell by a CD8$^+$ cytotoxic cell. The CD8 molecule (stippled) is postulated to interact with a non-polymorphic region on a class I MHC protein. In contrast, the TCR recognizes antigen plus a polymorphic region on the class I protein.

Recently (SAITO *et al.*, 1984; KRANGEL *et al.*, 1987) a second TCR was described, named **TCR-$\gamma\delta$**, which is also expressed in association with the CD3 complex. The TCR-$\gamma\delta$ is mainly expressed on $CD4^-CD8^-$, known as *double negative (DN) T-cells,* which can be found in low numbers in the adult thymus, spleen, lymph node and peripheral blood. However, in the early stages of thymic ontogeny and in the murine epidermis, $\gamma\delta$-bearing cells account for the majority of *T*-cells (EZQUERRA and COLIGAN, 1988).

Various ligands that bind the TCR–CD3 complex induce phosphionositide (PI) degradation and Ca^{2+} mobilization, including appropriately presented antigen, monoclonal antibodies to TCR or to CD3, or mitogenic lectins (GARDNER, 1989). In light of the near universality of the **Ca^{2+} signalling mechanism** in initiation of cell growth and in light of the Ca^{2+} dependency of many cellular processes, it has generally been assumed that *Ca^{2+} is an essential second messenger for T-cell activation.* The ability of Ca^{2+} ionophores to mimic the signal of TCR ligand and the ability of Ca^{2+} chelators to block the induction process support this contention. However, it is clear that while Ca^{2+} may be a necessary signal, it is not sufficient for the full process of activation. Furthermore, the precise role that Ca^{2+} plays in subsequent activation events remains unclear.

One of the fundamental mechanisms in the transduction of a wide variety of extracellular signals into cellular responses involves phospholipase C, which can hydrolyse phosphatidylinositol-4,5-bisphosphate to yield two products: inositol-1,4,5-triphosphate (mibilizes Ca^{2+}) and diacylglycerol (activates *protein kinase C*). Together with induction of *tyrosine kinase* activity, it is an early signalling mechanism by which ligands activate resting *T*-lymphocytes (WEISS *et al.*, 1986). But, it is becoming apparent that Ca^{2+}

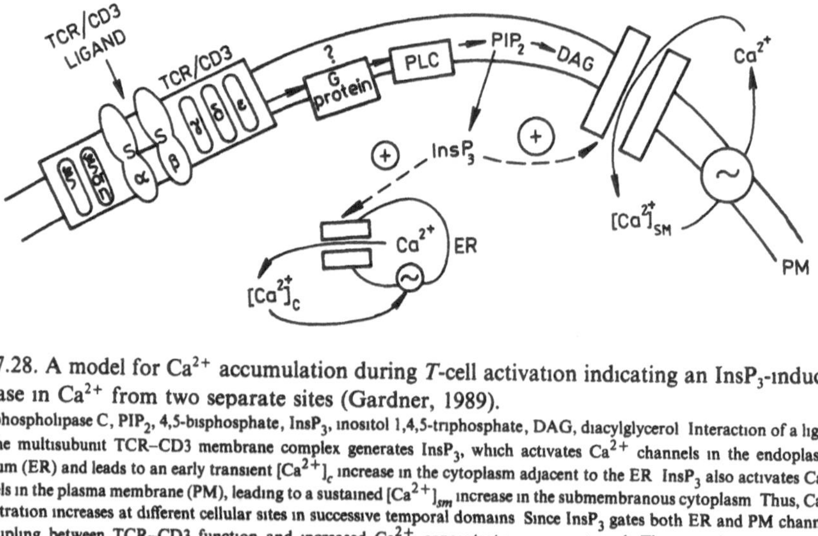

Fig. 7.28. A model for Ca^{2+} accumulation during *T*-cell activation indicating an $InsP_3$-induced increase in Ca^{2+} from two separate sites (Gardner, 1989).

PLC, phospholipase C, PIP_2, 4,5-bisphosphate, $InsP_3$, inositol 1,4,5-triphosphate, DAG, diacylglycerol Interaction of a ligand with the multisubunit TCR–CD3 membrane complex generates $InsP_3$, which activates Ca^{2+} channels in the endoplasmic reticulum (ER) and leads to an early transient $[Ca^{2+}]_c$ increase in the cytoplasm adjacent to the ER $InsP_3$ also activates Ca^{2+} channels in the plasma membrane (PM), leading to a sustained $[Ca^{2+}]_{sm}$ increase in the submembranous cytoplasm Thus, Ca^{2+} concentration increases at different cellular sites in successive temporal domains Since $InsP_3$ gates both ER and PM channels, the coupling between TCR–CD3 function and increased Ca^{2+} concentration is maintained The spatial-temporal model proposed that Ca^{2+} at the two sites has distinct molecular targets, enabling a sustained cellular response to the activating ligand signal of *T*-cell activation can be separated into at least two distinct spatial domains: one adjacent to the cytoplasmic face of the endoplasmic reticulum that is associated with intracellular stores release, and one adjacent to the inner surface of the plasma membrane that is associated with enhanced Ca^{2+} influx *via* Ca^{2+} permeable channels (*Fig. 7.28*).

7.5.1 Organization of T-cell receptor genes

Despite the fact that the organization of genes for *T*-cell receptor and immunoglobulin polypeptide chains is identical, they are coded by different genes that are mainly localized on different chromosomes (*Table 7.8*). The *T*-cell receptors are also determined by two genes (*V* and *C*) found on the same chromosome. The *V*-gene is made up of gene segments *V*, *D* and *J* — similarly to the *V*-genes for immunoglobulin heavy chains (*Fig. 7.29*).

Table 7.8 Localization of genes for immunoglobulin (Ig) polypeptide chains and *T*-cell receptors (TCR)

Chain	*Chromosome no.*	
	Human	*Murine*
H-Ig	14	12
κ-Ig	2	6
λ-Ig	22	16
α-TCR	14	14
β-TCR	7	6
γ-TCR	7	13
δ-TCR	14	14

Human and mouse genes for **TCR *β*-chains** are very similar. This gene occupies a segment of about 20 kb on mouse chromosome 6. The germline DNA contains two different genes, $C_{\beta 1}$ and $C_{\beta 2}$, separated by a polynucleotide chain of 6 kb, and having an identical transcription orientation. Each of them is associated with a separate cluster containing seven J_β-segments, six of them being functional. One J_β-segment is composed of between 46 and 49 nucleotides. Downstream from each J_β-segment cluster, there is at least one *D*-segment ($D_{\beta 1}$ or $D_{\beta 2}$) towards the 5′-end of the DNA molecule. Between the *D*-segment and the first *J*-segment, there is an intron of about 600 bp (base pairs). The D_β- and J_β-gene segments are linked in tandem. The number of V_β-segments is relatively low, approximately 20. $C_{\beta 1}$- and $C_{\beta 2}$-genes are split into four exons. The first exon contains 375 nucleotides and encodes an external domain, the second exon with 18 nucleotides encodes the hinge region, the third exon has 107 nucleotides and determines the transmembrane region, and the fourth exon contains 18 nucleotides for the cytoplasmic "tail" as well as about 200 nucleotides of the 3′-end piece which is not, however, translated into the protein chain. Both C_β-genes have a highly similar structure and their products differ only in five amino acid residues. The presence of both genes in the genome was proved using both helper and cytotoxic *T*-lymphocytes.

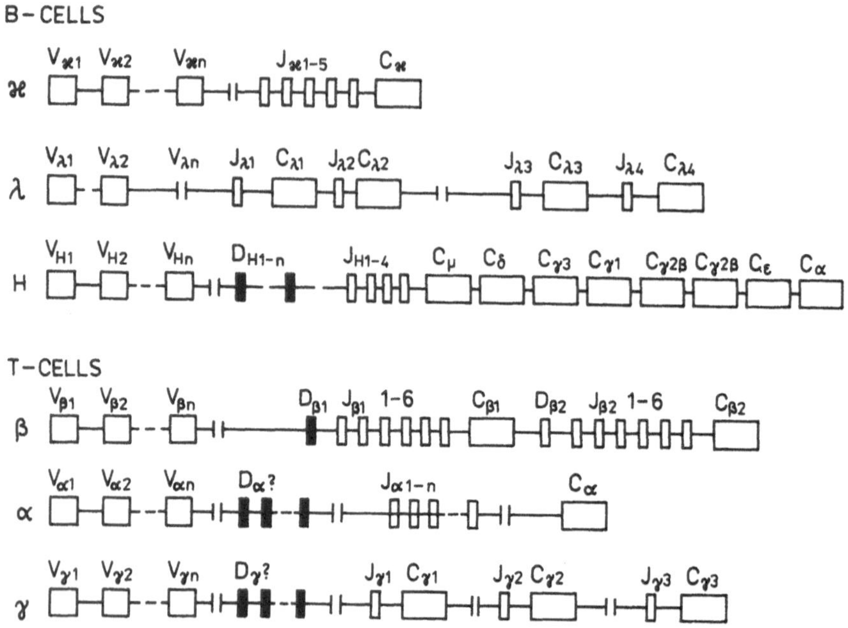

Fig. 7.29. Organization of genes for immunoglobulins and *T*-cell receptor (Yagüe and Palmer, 1985).
$\varkappa$ and λ, Ig light chains; H, heavy chains; α, β, γ, TCR chains.

Genes for human TCR β-chains have an analogous structure (*Fig. 7.30*). In comparison with the mouse β-chain genes, human C_β-genes code up to 178 amino acid residues (the mouse genes code up to 173). $C_{\beta1}$- and $C_{\beta2}$-genes have a highly similar exon structure, but completely different introns.

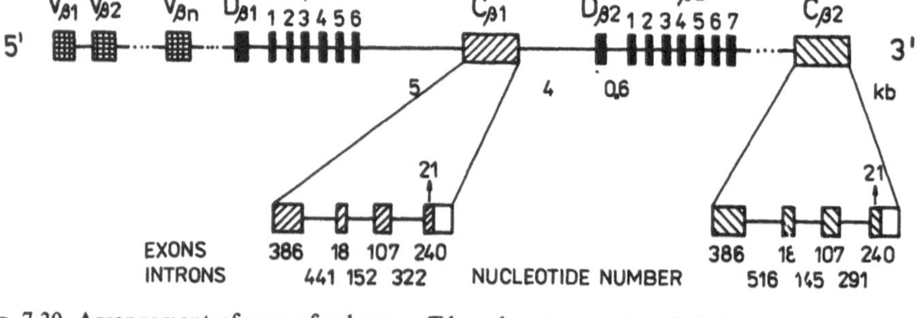

Fig. 7.30. Arrangement of genes for human *T*-lymphocyte receptor β-chains.

Germline genes for **α-chains** have a structure similar to genes for β-chains; in contrast to β-chains, however, they contain just one gene for the constant domain (consisting of four exons separated by introns of various length), one *J*-segment cluster and no *D*-segment. The polypeptide chain encoded by genes for α-chains is a bit longer than that encoded by genes for β-chains (*Fig. 7.24*).

KLEIN *et al.* (1987) have shown that the number of V_α- and J_α-segments in the human genome may be larger than estimated previously. It is composed from at least 47 V_α-gene segments and at least 37 J_α-segments, with the postulated existence of 100 J_α-segments. Thus it is possible that there are about 50 × 100 (5 000) α-chains. A similar calculation for the β-chain utilizing a germline repertoire of 60 V_β-gene segments, two D_β-segments used in all three reading frames and 13 functional J_β-segments yields about 3 500 possible β-chains. Therefore, assuming random association of α- and β-chains, the potential variability for the human TCR-$\alpha\beta$ is approximately 1.8×10^7 different heterodimers.

In 1985, HAYDAY *et al.* demonstrated a gene for the third **γ-chain** in *T*-lymphocytes (*Fig. 7.29*). In man, it is present on chromosome 7, the same as the gene for the β-chain; in mice, however, it is present on chromosome 13. The germline genome contains up to three genes for constant parts of the γ-chain — $C_{\gamma 1}$, $C_{\gamma 2}$ and $C_{\gamma 3}$, three *J*-segments and about ten *V*-segments. Because of the limited number of *V*-genes and absence of *D*-genes, the γ-gene organization resembles that of the immunoglobulin λ-chain. The γ-chains associate with δ-chains forming a γ–δ heterodimer which represents an alternative *T*-cell antigen receptor. This second TCR-$\gamma\delta$ is found on only 1% to 10% of peripheral blood *T*-cells in man or the mouse, but in the chicken these *T*-cells represent 30%.

The **δ-gene complex**, initially termed X, is located 5′ to the $J_\alpha C_\alpha$ coding region in mouse and man. The locus of the human δ-chain is composed of three *D*-region, two *J*-segments, one *V*-region and four *C*-region exons, all within 30 kb (LOH *et al.*, 1988). Rearrangements of this gene can be detected very early in thymic development, well before productive rearrangements of α- and β-genes can be detected.

The diversity of *T*-cell receptor genes, similar to those of immunoglobulin gene families, is comprised of three components — the number of germline gene segments, the nature and extent of the combination mechanisms operating at the DNA or protein level, and, finally, diversification generated by somatic mutations. Surprisingly, however, no mutations have yet been found in the origin of V_β-chain TCR diversity.

Despite the incomplete evidence, it appears that the number of *V*-segments in germline genes for TCR-chains is lower than the number of *V*-segments for immunoglobulin chains. Hybridization experiments performed so far and analysis of corresponding cDNA clones suggest that there are about 60 V_β-segments for β-chains. Such a low number of V_β-segments is compensated for by a higher number of *J*-segments and by at least two

D-segments. In addition, *T*-cells have more possibilities to combine *V*-, *D*- and *J*-segments to a complete *V*-gene, compared to *B*-cells. Whereas the corresponding segments in *B*-cells can only form *VDJ* or *VJ* complexes, in *T*-cells *VDDJ* joining is also possible. Recently, it was found (HOOD *et al.*, 1985) that D_β-segments may be translated *via* all three reading frames. This means that translation may be initiated at any nucleotide in the first triplet of the DNA *D*-segment transcription product. If, for example, the *D*-segment in the mRNA molecule has a sequence AUCGGACAUGUCA... it may be translated in triplets AUC-GGA-CAU-GUC-A..., or A-UCG-GAC-AUG-UCA..., or AU-CGG-ACA-UGU-CA, *i.e.* according to the overlapping code. Simple calculation shows that these mechanisms theoretically ensure 10^8–10^9 various specificities of *T*-cell receptors out of 120 gene segments encoding variable regions of α- and β-chains.

How can *VDJ* or *VJ* joining be generated during somatic recombination in *B*-cells and *VDDJ* joining in *T*-cells as well? This is due to recombinant recognition signals present on the margin of gene segments. The signals are represented by a sequence of seven nucleotides (heptamer) and nine nucleotides (nonamer) with a conservative structure separated by spacers, containing 12 (±1) or 23 (±1) nucleotides (approximately two turns in the DNA α-helix) (*Fig. 7.31*). The spacers do not possess a conserved structure. The *V*-segment possesses a **recognition element** at the 3′ site containing a 23-nucleotide spacer. For spatial reasons, it can only join with a 12-nucleotide recognition element at the 5′ site of the *D*-segment. The principle of joining recognition elements containing the 12-nucleotide spacer with elements whose spacer contains 23 nucleotides (the "*12–23 rule*"), operates even during recombination of $V_\varkappa$–$J_\varkappa$ and V_λ–J_λ segments during rearrangement of immunoglobulin light chain encoding genes. D_H-segments have a recognition element at their 5′ and 3′ sites with a spacer containing 12 nucleotides. This permits generation of just one $V_H D_H J_H$ recombination during gene translocation for heavy immunoglobulin chain according to the "12–23 rule".

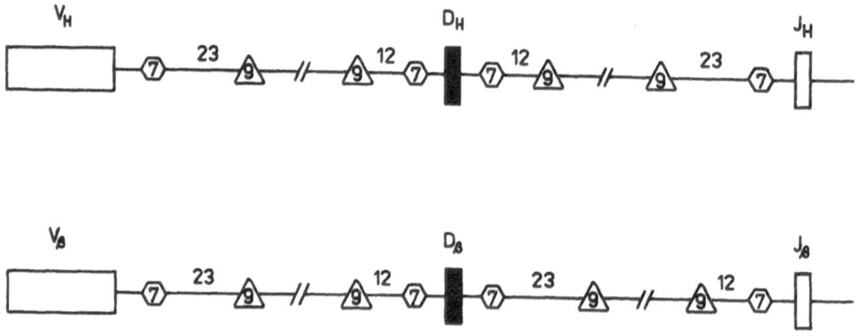

Fig. 7.31. Recombination signals for immunoglobulin heavy chains and for *T*-cell receptor β-chains.

On the margin of each gene segment (*V*, *D* or *J*) there are highly preserved heptamer (7) and nonamer (9) sequences, separated by spacers containing 12 or 23 nucleotides. These structures represent signals for mutual recombinations.

Recognition elements at the margins of *V*-, *D*- and *J*-segments encoding the variable region of *T*-cell receptor *β*-chains, have a similar structure to the recognition elements of immunoglobulin chains (*Table 7.9*). There is a basic difference, however, in D_T-segments that possess different rather than identical elements at both 5′ and 3′ sites. The "12–23 rule" is therefore valid even during recombination of two *DD*-segments which results in generation of genes of the *VDDJ* type for variable domains of *TCR*-chains.

Table 7.9 Nucleotide sequence in recognition and recombination elements of gene segments for immunoglobulin chains and *T*-cell receptor *β*-chains (Kavaler *et al.*, 1984)

3′ — end			
Gene segment	*Heptamer*	*Spacer (bp)*	*Nonamer*
V_κ	CACAGTG	12 ± 1	GGTTTTTGT
V_λ	CACAATG	23	GGTTTTTGC
V_H	CACAGTG	23 ± 1	NGTTTTTGT
$V_{T\beta}$	CACAGCA	23	AGTTTTTGT
D_H	CACAGTG	12	GATTTTTGT
$D_{T\beta}$	CACGGTG	23	CTTTTTTGT
5′ — end			
Gene segment	*Nonamer*	*Spacer (bp)*	*Heptamer*
J_κ	ACAAAAACC	23 ± 1	CACTGTG
J_λ	ACAAGAACA	12	CACAGTG
J_H	ACAAAAACC	23 ± 1	NACTGTG
$J_{T\beta}$	GCATAAACC	12 ± 2	NGCTGTG
D_H	ACAAAAACC	12	TACTGTG
$D_{T\beta}$	ACAAAAACC	12	CATTGTG

The elucidation of the *T*-cell receptor structure has raised the following question: Why has the problem of antigen recognition been solved twice during evolution of the immune system, *i.e.* recognition first by the *T*-cell system and later by the *B*-cell system? The difference in recognition capacity of both systems is probably not important. Present data suggest that each of these systems may ensure at least 10^8 different specificities, sufficient for the most complex organisms.

One possible explanation might be the lower binding capacity (lower avidity) of *T*-cell receptors compared to immunoglobulin receptors. Every TCR contains a single antigen combining site whereas the immunoglobulin receptor contains two sites. With respect to the additional effect of multivalency (p. 106), the binding strength between the antigen and immunoglobulin receptor may therefore be theoretically higher than that between the antigen and TCR. Such an assumption may also be confirmed by the fact that the TCR reacts effectively with the antigen only if the antigen is bound with an MHC product for which the TCR possesses a second combining site. In this way, the *T*-cell receptor avidity is enhanced because it reacts with the antigen–MHC product by a double bond, similarly to the immunoglobulin receptor with the pure antigen.

However, it has recently been found that not only TCR but also certain other *T*-lymphocyte surface structures participate in antigen recognition, particularly a complex designated CD3, present on all human *T*-cells.

In addition to CD3, other membrane glycoproteins — CD4 (*T*4), CD8 (*T*8) and CD2 (*T*11) — are involved in antigen recognition by *T*-cells. **CD4 glycoprotein** is present on helper *T*-cells and reacts with the constant part of class II MHC antigens, while **CD8 glycoprotein** is localized on the surface of cytotoxic *T*-cells and reacts with the constant part of class I MHC antigens. Thus, *T*-cells bear two types of structures which participate in antigen recognition. The TCR–CD3 complex recognizes the antigen bound with host MHC products, whereas CD4 and CD8 recognize MHC products only. This suggests that the process of antigen recognition by *T*-cells is much more complex than recognition by *B*-cells. Therefore, the requirements for coordination of individual recognition factors are higher; conversely, it is more vulnerable because of the probability of defects. This may be another reason why the immune system developed a simple *B*-cell recognition system during phylogenetic development.

Other possibilities should be considered, *e.g.* the fact that *T*-lymphocytes are primarily designed for antigen recognition or elimination on the cell surface. This system is probably sufficient for lower organisms. In higher organisms, however, the need also arose for recognition and efficient elimination of soluble antigens (products of microorganisms, foreign biopolymers, toxic substances from the external environment *etc.*) — this is primarily the role of antibodies. This might be the case only if the assumption, that *T*-cell-mediated specific immunity is phylogenetically older than antibody immunity, is valid.

7.6 The immunoglobulin superfamily

By extending the number of protein molecules with a known (defined) primary structure and by using computers to determine possible homologies among individual polypeptide chains, it has become possible to establish that the immunoglobulins belong to the "superfamily" which, according to NEZLIN (1987), includes three families of highly polymorphous glycoproteins (immunoglobulins, class I and II histocompatibility antigens and their functional counterparts — antigen receptors on *T*-cells) as well as multiple products of individual (non-polymorphic) genes — Thy-1, the receptor for transport of polymeric IgA and IgM, CD2, CD3, CD4 and CD8 glycoproteins on the surface of *T*-lymphocytes, *C*-reactive protein, lectins of some invertebrates *etc.* (*Table 7.10*). It was recently found that some adhesive glycoproteins of nerve cells and Fc-receptors on the leukocyte surface also belong to this superfamily (*Fig. 7.32*).

Table 7.10 Immunoglobulin superfamily

Member	*Similarity to imunoglobulin amino acid sequence*
β_2-microglobulin	homology with *C*-regions
MHC antigens	
class I	homology with *C*-regions
class II	homology with *C*-regions
T-cell receptors	homology with *H*-chains of low vertebrates
CD2, CD3, CD4, CD8	homology with V_L-domains
Thy-1	weak homology with $V_\varkappa$
Poly-Ig-receptor	homology with $V_\varkappa$-domain
N-CAM	homology with *C*-regions
C-reactive protein (CRP)	weak homology with various regions
Lectin from invertebrates *Limulus* and *Didemnum*	homology with various regions, stronger with CRP

MHC, major histocompatibility complex; CRP, C-reactive protein.

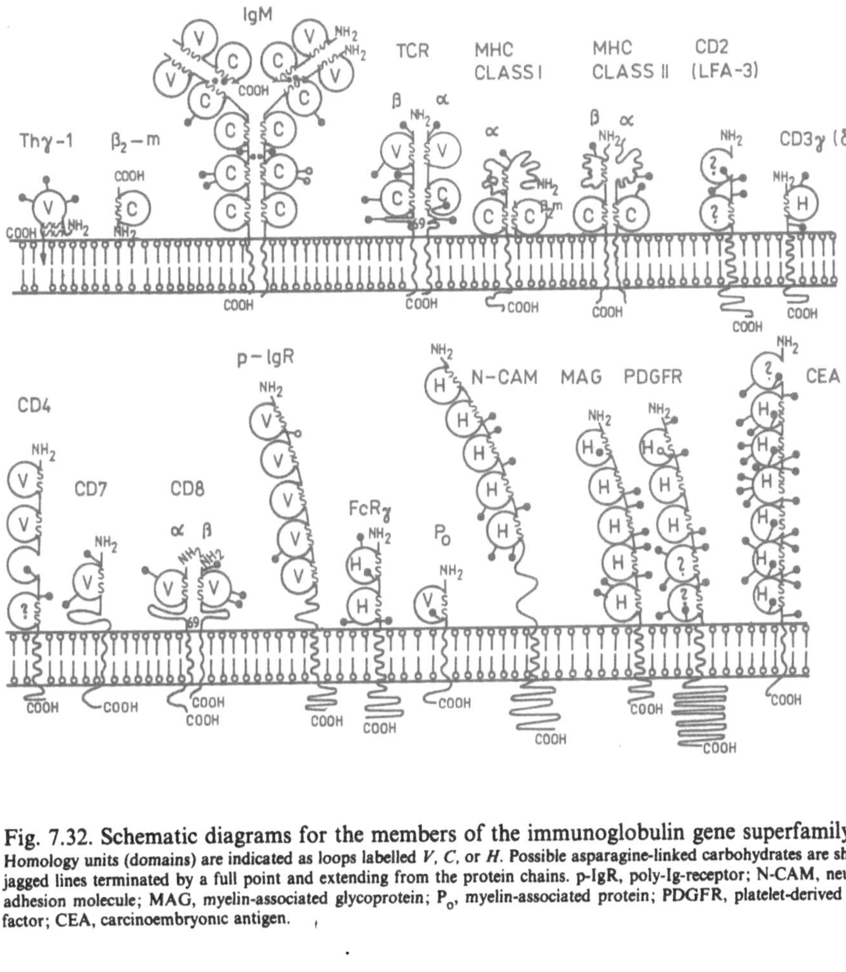

Fig. 7.32. Schematic diagrams for the members of the immunoglobulin gene superfamily. Homology units (domains) are indicated as loops labelled *V*, *C*, or *H*. Possible asparagine-linked carbohydrates are shown as jagged lines terminated by a full point and extending from the protein chains. p-IgR, poly-Ig-receptor; N-CAM, neural cell adhesion molecule; MAG, myelin-associated glycoprotein; P_0, myelin-associated protein; PDGFR, platelet-derived growth factor; CEA, carcinoembryonic antigen.

At present, more than 30 distinct immunoglobulin-related structures are known. These also include molecules that are absent from immune cells and which are particularly abundant in the brain. The main functions of the Ig-related molecules include: cell activation and mitogenesis (immunoglobulins, TCR–CD3 complex, CSF-1, Thy-1, CD28), regulation of cell–cell interactions (TCR–CD3, MHC, CD4, CD8), cell adhesion (LFA-1, N-CAM) and cell surface receptors (FcR, poly-Ig receptor). In several cases recognition occurs between pairs of molecules that are both Ig-related.

A **superfamily** is defined as a series of genes that share an evolutionary homology (*i.e.* common ancestor), but do not necessarily share function, genetic linkage, or coordinate regulation. Members of the **immunoglobulin gene superfamily** have been defined by the presence of one or more regions homologous to the basic three-dimensional structural unit of immunoglobulin. These units are named **domains** and they are characterized by a primary sequence about 70–110 amino acid residues in length with an essentially invariant disulphide bridge spanning 50–70 residues and several other relative conserved residues involved in establishing a tertiary structure referred to as an *antibody fold.* Three basic homology domain types have been defined from crystallographic analysis of the variable (*V*) and constant (*C* and *H*) regions (*Fig. 7.32*). The tertiary structure of a *V*-**region** is dominated by a series of nine antiparallel *β*-strands, connected by variable-length loop sequences, that assume a characteristic barrel or sandwich-like structure with two *β*-sheets, stabilized by the disulphide bridge. There are four *β*-strands in one sheet and three in the other. The extra pair of *β*-strands is essentially situated between the faces of the sandwich. The **C-region** domains lack the pair of internal *β*-strands. The third unit, denoted *H*, is similar by its tertiary structure to the *C*-unit. Because both *V*- and *C*-units are each more closely similar to *H*-units than to each other, the ***H*-unit** probably reflects a more primordial motif, suggesting that the original members of the superfamily arose in early metazoa and clearly carry out cell-surface recognition functions unrelated to vertebrate immunity (Hunkapiller and Hood, 1989). It is now clear (Williams, 1987) that the Ig-folding represents a stable platform that allows specific recognition between molecules with huge diversity possibly due to differences in amino acids displayed on the faces of the *β*-sheets and in the loops connecting *β*-strands. It is accepted that during evolution the immunoglobulins evolved from a primordial single domain. This primordial domain was probably involved in homophilic adhesion between primitive cells and this evolved into many sets of homophilic and heterophilic recognition pairs that function in cell–cell recognition. From these recognition systems the vertebrate *T*- and *B*-lymphocyte immune system evolved.

All molecules belonging to the immunoglobulin superfamily are therefore of common phylogenetic origin. Marchalonis *et al.* (1984) suggested that the basis of their similarity are the minigenes (gene segments) that encode homologous segments and cause serological cross-reactions of antibodies against one protein with other proteins of this family.

The **Thy-1 antigen** is present on thymocytes and on the surface of thymus-derived lymphocytes in mice, but not in man. Thus, it may be considered a surface marker of thymus-derived leukocytes. Its molecular weight is 17 000. The molecule contains 142 amino acid residues and approximately 30% saccharides.

The **C-reactive protein** is an inflammatory acute phase protein. It occurs at increased concentration in the blood serum during infection of the host or after tissue damage.

Newborn mammals are protected against infection by IgA and IgM antibodies, present in the maternal milk. These antibodies are secreted through **receptors for polymeric immunoglobulins** that are localized on the internal surface of glandular epithelial cells such that the receptor—IgA, or receptor–IgM complex is endocytosed (in the form of endosomes — membrane-coated cell organelles). The complex then diffuses across the cytoplasm of the epithelial cell and is exocytosed on the outer cell surface into the milk or other secretions (*e.g.* in the intestinal lumen). The receptor is enzymatically degraded during this process and its larger fragments represent the **secretory component**. Thus, the receptor for polymeric immunoglobulins is for single use only. It is a glycoprotein containing 733 amino acid residues and composed of five extracellular domains (they have the highest homology with variable domains of $L_{\varkappa}$-chains and each contains 100–115 amino acid residues), one transmembrane domain (it lost its disulphide bond during evolution) and a relatively long cytoplasmic tail (with 103 amino acid residues). None of these domains has a variable structure.

CD3 forms a complex with the antigen receptor of *T*-cells (TCR–CD3) which is responsible for antigen recognition by *T*-lymphocytes. CD3 molecule is composed of five non-covalently bound polypeptide chains (γ, δ, ε, ξ and η). Their amino acid sequence is invariable.

T-cells recognize antigen on stimulatory or target cells in association with membrane-bound glycoproteins encoded by genes of the MHC. There is some evidence that the mutually exclusive differentiation antigens CD8 and CD4 may contribute to MHC-restricted recognition: CD4 expression correlates closely with MHC class II restriction and CD8 expression with MHC class I restriction. Direct binding between CD4 and/or CD8 and the MHC-encoded molecules is thought to facilitate and augment *T*-cell contact with the antigen-presenting cell. However, recent experiments suggested that an additional role for CD4 and CD8 molecules may be to provide an activation signal to resting *T*-cells as soon as TCR is in close proximity to CD4 or CD8 during ligand binding.

The analogue of CD4 and CD8 on mouse lymphocytes is called Lyt2,3. Both CD4 and CD8 are present on the surface of immature human *T*-cells. The differentiated $T_{H/I}$-cells only carry CD4, whereas T_C- and T_S-cells only have CD8. The **CD4 molecule** is formed by a single polypeptide chain, consisting of 435 amino acid residues, with a molecular weight of approximately 60 000. Its extracellular portion (with 374 amino acid residues) has four domains, the first two being homologous with immunoglobulin V_L-domains (96 amino acid residues). The other two domains are not homologous with any part of immunoglobulin chains. The extracellular part is followed by a transmembrane domain (with 21 amino acid residues homologous with the analogous domains of class II MHC antigens) and, finally, a cytoplasmic tail (40 amino acids) protrudes into the cell.

The **CD8 molecule** is similar except that it is only half the length of the CD4 molecule, with a molecular weight of 34 000 and containing 214 amino acid residues. The CD8 molecule is a dimer formed by two such chains; each is composed of the *N*-terminal outer domain containing 96 amino acid residues, a hinge segment rich in Pro (65 amino acid residues), a transmembrane region (24 amino acid residues) and an internal cytoplasmic domain, containing mainly basic amino acid residues (29).

Presumably, the CD4 glycoprotein also acts as a receptor of some retroviruses, *e.g.* HIV-1, the causative agent of AIDS.

An important cell property is adhesiveness, *i.e.* the ability to adhere to certain surfaces (*e.g.* other cells, host tissues, glass, plastics *etc.*). Special glycoproteins termed **adhesins** are responsible for adhesive reactions. In 1986, EDELMAN found that some adhesins belong to the immunoglobulin superfamily. One well-defined adhesin is the glycoprotein **N-CAM** which is present on the surface of nervous system cells. The N-CAM molecule has five outer domains, a transmembrane and a cytoplasmic portion. It occurs in two forms, with molecular weights of 160 000 and 130 000; the latter form lacks the cytoplasmic tail. Both molecular forms of glycoprotein N-CAM are coded by one structural gene, although they arise as a result of different processing of precursor mRNA. **ICAM-1** is a similar adhesion glycoprotein and also acts as the major rhinovirus receptor (WHITE and LITTMAN, 1989).

7.7 Antibody deficiency disorders

Antibody disorders may be due to abnormal function of *B*-cells, *T*-cells or both cell types. They may be manifested by inadequate activity (hypogammaglobulinaemia), abnormal activity (malignant growth of *B*- or *T*-cells), immunoglobulin hypercatabolism (excessive rate of degradation), loss of immunoglobulin or lymphocytes (*e.g.* excessive bleeding) and lymphocyte damage (induced by drugs, lymphotropic viruses *etc.*). Such disorders may affect lymphocytes at various developmental stages, from precursor up to terminal effector cells. Thus, a large variety of clinical symptoms and nosological units occur and are largely of unknown pathogenetic mechanism. Defects of anti-infectious immunity are the most common clinical symptoms of antibody deficiency. These patients have increased susceptibility to recurrent severe infections resistant to common anti-infectious treatment.

In their origin, immunoglobulin deficiencies may be either *primary* (hereditary, congenital) or *secondary* (acquired during the ontogenetic development of the individual). Basic antibody disorders are listed in *Table 7.11.*

Table 7.11 Primary antibody formation deficiencies (adapted from WHO Technical Reports, 1978)

Designation	*Functional deficiency*	*Defect*
B-cell deficiency:		
Hypogammaglobulinaemia and agammaglobulinaemia	Ab	↓*B*, ↓PC
Selective IgA deficiency	IgA Ab	↓IgA ↓PC
Selective deficiency of other Ig classes and subclasses	Ab	↓PC
Deficiency of IgA secretory component	SIgA Ab	↓intestinal IgA PC
Transient infant hypogammaglobulinaemia	Ab	↓PC
Transcobalamin II deficiency	Ab and phagocytosis	↓PC
Bruton-type agammaglobulinaemia (X-linked)	Ab	↓pre-*B*
T-cell deficiency:		
Di George syndrome (thymic hypoplasia)	CMI and Ab	↓T
Purine nucleoside phosphorylase deficiency	CMI ± Ab	↓T
Combined deficiency *B*- and *T*-cells:		
Severe combined immunodeficiency (SCID)		
(a) reticular dysgenesis	CMI, Ab and phagocytes	↓*T*, *B* and phagocytes
(b) "Swiss-type"	CMI and Ab	↓*T* and *B*
(c) adenosine deaminase deficiency	CMI and Ab	↓*T* and *B*
Ataxia-telangiectasia	CMI and Ab (partially)	↓*T* and PC (namely IgA, IgE)
Wiskott–Aldrich syndrome	CMI and Ab, sometimes even phagocytosis	↓glycoprotein GP 150

Ab, antibodies; *B*, *B*-lymphocytes; PC, plasma cells; *T*, *T*-cells; CMI, cell-mediated immunity; ↓, decreased function.

7.7.1 B-cell disorders

Decreased antibody formation is observed in **hypogammaglobulinaemia**. Such a deficiency may involve various immunoglobulin classes. It is mainly associated with a functional defect of *B*-cells and plasma cells; rarely of *T*-cells. The term **agammaglobulinaemia** is used when the total serum immunoglobulin level is less than 1 g/l. Higher levels, although below the lower normal limit, are classified as hypogammaglobulinaemia. **Hypergammaglobulinaemias** are the opposite of hypogammaglobulinaemias; most of them being monoclonal gammopathies (p. 140). In immunoglobulin deficiencies, low levels of one or two (selective deficiencies) or even more classes may occur.

Selective IgA deficiency is most common. Approximately 0.2% of clinically healthy blood donors have significantly decreased IgA levels in the serum and mucous membrane secretions. The situation may become critical if the serum IgA concentration decreases below 1 mg/l. These patients may suffer from recurrent infections of the respiratory and intestinal tract. In addition, they are prone to allergic (hypersensitivity) reactions of the immediate (first) type, *e.g.* atopic bronchial asthma. Usually, high IgE levels can be observed, which contribute to clinical symptoms of atopy. The IgA secretory component deficiency has similar effects to IgA deficiency.

Selective deficiencies of other immunoglobulin classes, *e.g.* IgM, IgG, IgA + IgM, IgA + IgG *etc.* may also occur. In selective IgG deficiency an imbalance in production of individual subclasses can additionally be observed.

Transient infant hypogammaglobulinaemia. Full-term newborn serum has a maternal IgG concentration of approximately 10 g/l; but has almost no IgM and IgA. Due to biological turnover, the maternal IgG level gradually decreases and disappears by 6–12 months of age. Immediately after birth, the newborn is colonized by bacterial microflora which provide the necessary antigen signal to initiate production of the infant's own IgG, IgM and IgA (p. 98).

Inborn (congenital) **agammaglobulinaemia of Bruton's type** is an X-linked hereditary disease. Young males, suffering from this deficiency, lack immunoglobulins of all classes except IgE. The disease usually appears after 6 months of life, when all IgG of maternal origin disappears from the infant's blood circulation. Without administration of immunoglobulin preparations, antibiotics and chemotherapeutics, the patient usually dies from recurrent pyogenic (purulent) infections. However, even with successful treatment of infections, the psychosomatic development of the child is retarded. The disease is relatively rare, occurring in 1–13 per million of the population.

Transcobalamin II (a specific transport protein for vitamin B_{12} — cobalamin) **deficiency** gives rise to hypogammaglobulinaemia due to the inability of *B*-cells to differentiate into plasma cells, although the *B*-cell number is normal. It is a hereditary autosomal-recessive disease in which the patient is unable to respond by antibody formation even after intensive antigenic stimulation. In addition, disorders of phagocytic function, particularly in the intracellular killing of staphylococci, may be observed. Besides the immune mechanism disorders, transcobalamin II deficiency results in defective red blood cell function (*megaloblastic anaemia*). Most of these adverse symptoms can be reversed by regular vitamin B_{12} administration. However, acute respiratory infections occur, even in treated patients.

7.7.2 T-cell disorders

Individuals with the **Di George syndrome** have an inadequately developed or absent thymus gland. It is a hereditary disease associated with other anatomical malformations, particularly parathyroid glands, facial structures (ear, mandible, upper lip) and large blood vessel anomalies. Acute hypocalcaemia develops due to inadequate parathyroid function in the first week of life. After suitable treatment infants survive this crisis, but a series of immunodeficient episodes will develop within a few months. The number of circulating *T*-lymphocytes is normal, but they are unable to respond to antigenic and mitogenic stimulation. The immunoglobulin concentration is almost normal, but there are no specific antibodies; a finding which can be explained by altered helper function of *T*-cells which are then unable to instruct the *B*-cells correctly. Although all these defects can be diminished by administration of appropriate immunoglobulin preparations and thymic hormones, the patient soon dies from multiple cell disorders.

Purine nucleoside phosphorylase (EC 2.4.2.1, PNP) **deficiency** is associated with inhibition of DNA synthesis, particularly in *T*-lymphocytes, caused by defects of guanine and hypoxanthine metabolism. PNP is responsible for conversion of inosine into hypoxanthine and of guanosine to guanine. It is an autosomal-recessive disease associated with normal or increased immunoglobulin concentrations but decreased numbers of peripheral *T*-lymphocytes which respond poorly to mitogen stimulation. Patients suffering from this enzyme deficiency are highly susceptible to infections with DNA-viruses.

7.7.3 Stem cell disorders and combined B- and T-cell deficiency

The most serious of the immunodeficiency diseasses is **severe combined immunodeficiency (SCID)**, because both humoral and cellular immunity are almost completely lacking. It may be inherited as an autosomal-recessive or X-linked-recessive disease, as well as a non-hereditary condition. The mechanism of non-hereditary inheritance is based on the passage of maternal lymphocytes into the developing foetus, and, because the foetal immune system is underdeveloped, they induce a graft-versus-host reaction. This results in destruction of the foetal immune system. Prenatal rubella infection exhibits a similar effect.

Infants suffering from SCID have a profound lymphocytopenia (decreased number of peripheral blood lymphocytes). In addition, the small number of available lymphocytes are not functionally active — they do not respond to antigen or mitogen stimulation. Total agammaglobulinaemia develops within the first months of life. The thymus is either absent or only thymic residues can be found at anatomically atypical sites. The spleen is extremely small and the intestinal Peyer's patches are lacking. Multiple phagocyte disorders are also observed.

Severe recurrent infections occur soon after birth. The most frequent are lung infections caused by both pathogenic and opportunistic microbes that are not harmful to individuals with a normal immune system. These infants also succumb to common virus infections and *Candida* infections occur frequently. Patients suffering from SCID usually die within the first two years of life as a result of repeated infections. The only way to prevent infections is to deliver the infant under sterile conditions, followed by isolation in a sterile environment for the rest of its life. In some cases, transplantation of compatible bone marrow or foetal thymus liver fragments may be successful.

SCID occurs in several forms. The type characterized by blockade of stem cell differentiation and the type due to enzyme defects (mainly adenosine deaminase) are the most common. The most severe is the form associated with *defects in stem cell differentiation* (10–20% cases). In these patients, the following cells may be absent: (a) stem cells common for both lymphocyte and myeloid line (p. 390) — no lymphocytes, granulocytes or macrophages develop; (b) stem cells for the lymphocyte line — *B*- and *T*-cells do not develop (*"Swiss-type" agammaglobulinaemia*); (c) stem cells for *T*-lymphocytes — *B*-cells are normally developed, but cannot function properly because the instruction from $T_{H/I}$-lymphocytes is absent.

Multiple enzyme activity defects may be the pathogenetic basis of SCID; these enzymes belong to metabolic pathways which play an important role in the differentiation of active cells. Of particular importance are the enzymes of purine metabolism, *e.g.* purine nucleoside phosphorylase mentioned above, hypoxanthine (guanosine) phosphoribosyl transferase, 5′-nucleotidase *etc.* (*Fig. 7.33*). These enzymes participate in purine transformation (reutilization) and they form a pathway preferentially used by lymphocytes to synthesize essential purine, while other cells usually prefer the *de novo* synthesis of purines from glycine, 2-phosphoribosylamine, formic acid, CO_2 and aspartic acid.

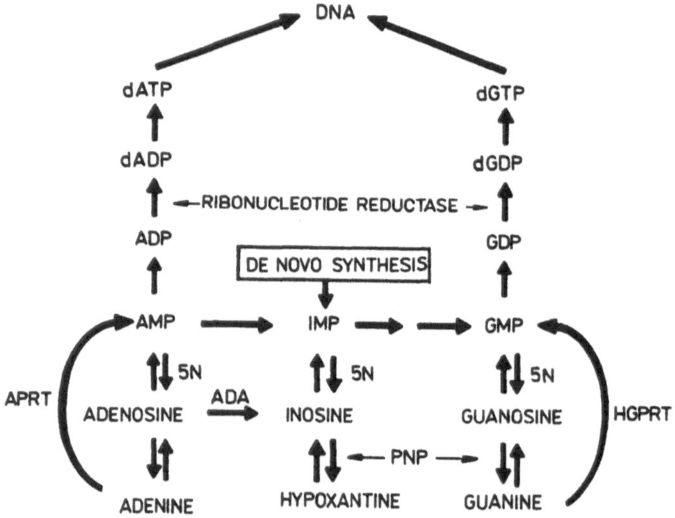

Fig. 7.33. Schematic representation of purine reutilization metabolism, preferentially used by lymphocytes for DNA synthesis.

For simplicity, only enzymes which may be assocaited with a deficiency state are shown: APRT, adenine phosphoribosyl transferase; ADA, adenosine deaminase; 5N, 5′-nucleotidase; PNP, purine nucleoside phosphorylase; HGPRT, hypoxanthine (guanine) phosphoribosyl transferase; AMP, adenosine monophosphate; IMP, inosine monophosphate; GMP, guanosine monophosphate.

Adenosine deaminase (ADA, EC 3.5.4.4) degrades adenosine (deoxyadenosine) to inosine (deoxyinosine). Its *deficiency* results in accumulation of adenosine, AMP and dATP in lymphocytes causing an inhibitory effect on DNA synthesis similar to that of PNP deficiency. Inhibition of DNA synthesis is not the only biochemical consequence of ADA deficiency. Enhanced dATP levels cause a decreased neutrophil motility that may be observed in a number of SCID patients. ADA deficiency is responsible for 20–30% of SCID cases. The prognosis of patients suffering from this defect is very poor — they survive for only about three years.

Deficiency of 5′-nucleotidase (EC 3.1.3.5), the enzyme catalysing hydrolysis of 5′-ribonucleotides and deoxyribonucleotides to the corresponding nucleosides, has much milder consequences. A low 5′-nucleotidase activity is found in non-hereditary hypogammaglobulinaemia, X-linked hypogammaglobulinaemia and in selective IgA deficiency. Normal 5′-nucleotidase activity in mature *B*-lymphocytes is four times higher than in mature *T*-lymphocytes.

Ataxia telangiectasia is a disease that is manifested by cerebellar ataxia (disordered motor coordination), cutaneous and mucous membrane telangiectasia (pathological dilatation of the capillaries), recurrent infections (particularly of the respiratory tract) and frequent occurrence of malignant tumours. The disease occurs in children of pre-school or school age and is of an autosomal-recessive hereditary character. It is associated with multiple immune mechanism disorders. Thus, final *B*-cell differentiation into plasma cells is defective, as is lymphocyte function, particularly of *T*-cells. In most cases, the IgA and IgG2 levels are markedly decreased, or completely absent, and the IgE level is also decreased. In contrast, there is an increased α-foetoprotein level, possessing an immunosuppressive effect. Even the repair mechanisms which ensure the repair of defective DNA molecules are defective, and this results in an increased occurrence of chromosomal abnormalities and enhanced susceptibility to ionizing radiation.

The **Wiskott–Aldrich syndrome** is a recessive X-linked hereditary defect. It is manifest during the first year of life in males and produces serious damage to *T*- and *B*-cell function. The IgM level is usually decreased, IgG is normal and IgA, IgD and IgE are often increased. In addition, an increased immunoglobulin and lymphokine catabolism is observed; however, this is masked by their enhanced biosynthesis. Both the chemotaxis and microbicidal activity of phagocytes is decreased. Patients' lymphocytes also lack the surface glycoprotein GP 150, which plays a key role in transmitting regulatory signals.

7.7.4 Acquired immunodeficiency

Secondary (acquired) immunodeficiencies originate during the ontogenetic development of an individual, and can be caused by various adverse physical, chemical, biological and psychosocial factors or inadequate nutrition. Of the physical factors, various types of ionizing radiation are of particular importance because they negatively influence the immune mechanisms by genotoxic effects and formation of free radicals.

Various chemicals and drugs may possess immunotoxic effects. **Immunotoxicity** can be manifest not just by suppression of antibody formation and other immune mechanisms — including the impairment of immunological surveillance leading to malignancy — but also by a pathological increase of these functions resulting in hypersensitivity reactions and autoimmune diseases. Substances with the highest immunotoxicity include halogenated aromatic carbohydrates (polychlorodibenzodioxines, hexachlorobenzene, polychloro and polybromo biphenyls), polycyclic aromatic carbohydrates (3-methylcholanthrene, 1,2-benzanthracene), phorbol esters (phorbol myristate acetate), insecticides (chlorinated — DDT, Lindan, organophosphates — parathion, malathion, *etc.*), diethylstilboestrol (a non-steroidal compound with oestrogen activity) and heavy metals (lead, cadmium, mercury, organic tin compounds). Defects in immune system functions are also induced by various industrial gas and solid pollutants. Their effects are studied by immunotoxicologists and ecological immunologists.

In addition to immunosuppressive drugs, all cytostatic drugs, some antibiotics and several other drugs, particularly those which inhibit DNA replication, RNA synthesis and protein synthesis, show immunosuppressive activity.

Parasitic, fungal, bacterial and viral infections are biological agents which may cause secondary immunodeficiencies. The immune system may become exhausted during long-lasting or recurrent, severe infections. Particularly dangerous are lymphotropic virus infections (virus replication in lymphoid cells) that can destroy certain lymphocyte populations, for example in the **acquired immunodeficiency syndrome (AIDS)**. This syndrome is caused by the *HIV-1* or *HIV-2 virus* (type 1 human immunodeficiency virus, originally termed LAV/HTLV III) which multiplies mainly in helper lymphocytes ($T_{H/I}$-cells) and completely destroys immune homeostasis. Consequently, a severe secondary immunodeficiency with high mortality develops.

AIDS occurs primarily in "high-risk" groups, *e.g.* homosexuals and intravenous drug addicts (about 90% of cases). In addition, AIDS has also occurred in normal individuals including infants who received blood transfusions or blood products. The infection may be transmitted by sexual contact, by infected blood and blood products and from the mother to the foetus. The virus may also be isolated from saliva and tears. Between 1981, when the first cases were recorded, and 1988, over 100 000 AIDS infections were reported (85% in the USA). Severe clinical symptoms including malignancy (*Kaposi sarcoma*) and infections with opportunistic organisms which are harmless in healthy individuals, occur after a long incubation period (six months to seven or more years) in about 5–10% of AIDS virus-infected individuals. The prognosis for AIDS patients is very poor; 60–90 % of patients die within 5 years. In about 20 % of infected patients, the course of the disease is milder and the mean survival is much longer. This condition is called **AIDS-related complex (ARC)** and is generally asymptomatic. These individuals, however, may become HIV-1 virus carriers and a source of infection.

Secondary antibody immunodeficiency can be found in some leukaemias; it may also be caused by low protein and calorie nutrition. Various types of malnutrition are particularly associated with cell-mediated immunity disorders. Deficiency of certain essential nutritional components, *e.g.* vitamins (B_{12}, folic acid) and microelements (Zn, Fe) *etc*. has similar consequences.

References

Allison, J. P. and McIntyre, B. W. (1983) *The murine T cell antigen receptor: a proposed structure.* In: Yamamura Y. and Tada, T. (eds.), New York, Acad. Press, pp. 755–62.

Beadle, G. W. (1945) Genetic control of biochemical reactions. *Harwey Lect.*, **40**, 179–94.

Beadle, G. W. (1948) Physiological aspects of genetics. *Annu. Rev. Physiol.*, **10**, 17–42.

Berggård, I. and Bearn, A. G. (1968) Isolation and properties of a low molecular weight β_2-globulin occurring in human biological fluids. *J. Biol. Chem.*, **243**, 4095–101.

Berggård, B., Ekström, B. and Åkerström, B. (1980) α_1-Microglobulin. *Scand. J. Clin. Lab. Invest.*, **40**, Suppl. 154, 63–71.

Blackwell, T. K. and Alt, F. W. (1989) Mechanism and developmental program of immunoglobulin gene rearrangement in mammals. *Annu. Rev. Genet.*, **23**, 605–36.

Borrebaeck, A. C. K., Danielson, L. and Moller, S. A. (1988) Human monoclonal antibodies produced by primary *in vitro* immunization of peripheral blood lymphocytes. *Proc. Natl. Acad. Sci. USA*, **85**, 3995–9.

Boulianne, G. L., Hozumi, N. and Shulman, M. J. (1984) Production of functional chimaeric mouse/human antibody. *Nature*, **312**, 644–6.

Brack, C. and Tonegawa, S. (1977) Variable and constant parts of the immunoglobulin light chain gene of a mouse myeloma cell are 1250 nontranslated bases apart. *Proc. Natl. Acad. Sci. USA*, **74**, 5652–6.

Breinl, F. and Haurowitz, F. (1930) Chemische Untersuchung des Präzipitates aus Hämoglobin und Anti-Hämoglobin-Serum und Bemerkungen über die Natur der Antikörper. *Z. Physiol. Chem.*, **192**, 45.

Brenner, S. and Milstein, C. (1966) Origin of antibody variation. *Nature*, **211**, 242–5.

Bruton, O. C. (1952) Agammaglobulinemia. *Pediatrics*, **9**, 722–9.

Burnet, F. M. (1957) A modification of Jerne's theory of antibody production using the concept of clonal selection. *Austr. J. Sci.*, **20**, 67–9.

Burnet, F. M. (1959) *The clonal selection theory of acquired immunity.* Cambridge, Cambridge Univ. Press.

Burnet, M. F. (1969) *Self and not self.* Cambridge, Cambridge Univ. Press, 318 pp.

Byers, V. S. and Baldwin, R. W. (1988) Therapeutic strategies with monoclonal antibodies and immunoconjugates. *Immunology*, **65**, 329–35.

Claman, H. N., Chaperon, E. A. and Triplett, R. F. (1966) Thymus-marrow cell combination. Synergism in antibody production. *Proc. Soc. Exptl. Biol. Med.*, **122**, 1167–71.

Cohn, M. (1970) Selection under a somatic model. Cell. *Immunol.*, **1**, 461–7.

Di George, A. M. (1968) Congenital absence of the thymus and its immunologic consequence: Concurrence with congenital hypoparathyroidsm. In: Bergsma, D. and Good, R. A. (eds.), *Immunologic Deficiency Diseases in Man. Birth Defects*, **4**, 116–28.

Dráber, P. and Vojtíšková, M. (1982) Monoclonal antibodies produced by hybridomas. *Biol. Listy*, **47**, 136–52. (*in Czech*).

Dreyer, W. J. and Bennett, J. C. (1965) The molecular basis of antibody formation: A paradox. *Proc. Natl. Acad. Sci.* USA, **54**, 864–9.

Early, P. and Hood, L. (1981) Allelic exclusion and non-productive immunoglobulin gene rearrangements. *Cell*, **24**, 1–2.

Edelman, G. M. (1986) Cell adhesion molecules in the regulation of animal form and tissue pattern. *Annu. Rev. Cell Biol.*, **2**, 81–116.

Ehrlich, P. (1900) On immunity with special reference to cell life. *Proc. R. Soc. (Biol.)*, **66**, 424–36.

Ekström, B., Peterson, P. A. and Berggård, I. (1975) A urinary and plasma α_1-glycoprotein of low molecular weight: isolation and some properties. *Biochem. Biophys. Res. Commun.*, **65**, 1427–35.

Ezquerra, A. and Coligan, J. E. (1988) *T* cell receptors: structure and genetics. *Curr. Opin. Immunol.*, **1**, 77–83.

Gally, J. A. and Edelman, G. M. (1972) The genetic control of immunoglobulin synthesis. *Ann. Rev. Genet.*, **6**, 1–46.

Gardner, P. (1989) Calcium and *T* lymphocyte activation. *Cell*, **59**, 15–20.

Gough, N. (1981) Gene rearrangement can extinguish as well as activate and diversify immunoglobulin genes. *Trends Biochem. Sci.*, **6**, 300–2.

Grey, H. M. and Chesnut, R. (1985) Antigen processing and presentation to *T* cells. *Immunol. Today*, **6**, 101–6.

Hašek, M. (1953) Vegetative hybridization of animals by joining of blood circulations in embryonal development. *Čs. Biologie*, **2**, 265–74. (*in Czech*).

Hašek, M. and Hraba, T. (1955) Immunological effect of experimental embryonal parabiosis. *Nature*, **175**, 764–5.

Hayday, A. C., Saito, H. and Giles, S. D. (1985) Structure, organization, and somatic rearrangement of *T* cell gamma genes. *Cell*, **40**, 259–68.

Honjo, T. and Habu, S. (1985) Origin of immune diversity: genetic variation and selection. *Ann. Rev. Biochem.*, **54**, 803–30.

Hood, L., Kronenberg, M. and Hunkapiller, T. (1985) *T* cell antigen receptors and the immunoglobulin supergene family. *Cell*, **40**, 225–9.

Hood, L. and Talmage, D. W. (1970) Mechanism of antibody diversity: germ line basis for variability. *Science*, **168**, 325–34.

Hozumi, N. and Tonegawa, S. (1976) Evidence for somatic rearrangement of immunoglobulin genes coding for variable and constant region. *Proc. Natl. Acad. Sci. USA*, **73**, 3628–32.

Hunkapiller, T. and Hood, L. (1989) Diversity of the immunoglobulin gene superfamily. *Adv. Immunol.*, **44**, 1–63.

Jerne, N. (1955) The natural selection theory of antibody formation. *Proc. Natl. Acad. Sci. USA*, **41**, 849–57.

Kabat, E. A., Wu, T. T. and Bilofsky, H. (1978) Evidence supporting somatic assembly of the DNA segments (minigenes), coding for the framework, and complementarity-determining segments of immunoglobulin variable regions. *J. Exp. Med.*, **149**, 1299–313.

Kappler, J., Kubo, R., Haskins, K., White, J. and Marrack, P. (1983) The major histocompatibility complex-restricted antigen receptor on *T* cells in mouse and man. V. Identification of constant and variable peptides. *Cell*, **35**, 295–302.

Katz, D. H. and Benacerraf, B. (1972) The regulatory influence of activated *T* cells on *B* cell responses to antigen. *Adv. Immunol.*, **15**, 1.

Kavaler, J., Davis, M. M. and Chien, Y. H. (1984) Localization of a *T* cell receptor diversity-region element. *Nature*, **310**, 421–3.

Keegan, A. D. and Paul, W. E. (1992) Multichain immune recognition receptors: similarities in structure and signaling pathways. *Immunol. Today*, **13**, 63–8.

Klein, M. H., Concannon, P., Everett, M., Kim, L. D. H., Hunkapiller, T. and Hood, L. (1987) Diversity and structure of human *T*-cell receptor α-chain variable region genes. *Proc. Natl. Acad. Sci. USA*, **84**, 6884–8.

Kohler, G. and Milstein, C. (1975) Continuous cultures of fused cells secreting antibody of predefined specificity. *Nature*, **256**, 495–7.

Krangel, M. S., Band, H., Hata, S., McLean, J. and Brenner, M. R. (1987) Structurally divergent human *T* cell receptor γ proteins encoded by distinct C_γ genes. *Science*, **237**, 64–7.

Landsteiner, K. (1933) *Die Spezifizität der serologischen Reaktionen*. Berlin, Springer.

Liener, I. E., Sharon, N. and Goldstein, I. J. (1986) *The lectins: properties, functions, and applications in biology and medicine*. Orlando, Acad. Press, 600 pp.

Loh, E. Y., Cwirla, S., Serafini, A. T., Phillips, J. H. and Lanier, L. L. (1988) Human *T*-cell--receptor δ chain: genomic organization, diversity, and expression in populations of cells. *Proc. Natl. Acad. Sci. USA*, **85**, 9714–8.

Marchalonis, J. J., Vasta, G. R., Warr, G. W. and Barker, W. C. (1984) Probing the boundaries of the extended immunoglobulin family of recognition molecules: jumping domains, convergence and minigenes. *Immunol. Today*, **5**, 133–42.

Medawar, P. B. (1958) *The uniqueness of the individual*. New York, Basic Books.

Mestecký, J., McGhee, J. R. and Elson, C. O. (1988). Intestinal IgA system. *Immunol. Allergy Clin. North Amer.*, **8**, 349–68.

Milstein, C. (1985) From the structure of antibodies to the diversification of the immune response (Nobel lecture). *Angew. Chem. Int. Ed. Engl.*, **24**, 816–26.

Mitchison, N. A. (1968) The dosage requirements for immunological paralysis by soluble proteins. *Immunology*, **15**, 509–18.

Mitchison, N. A. (1971) The carrier effect in the secondary response to hapten-protein conjugates: I, II. *Eur. J. Immunol.*, **1**, 10–19.

Mitchison, N. A., Rajewsky, K. and Taylor, R. B. (1970) Cooperation of antigenic determinants and of cells in the induction of antibodies. In: Sterzl, J. and Ríha, I. (eds.), *Developmental Aspects of Antibody Formation and Structure*. Prague, Academia, 547 pp.

Morrison, S. L. and Oi, V. T. (1989) Genetically engineered antibody molecules. *Adv. Immunol.*, **44**, 65–92.

Mosier, D. E., Gulizin, R. J., Baird, S. M. and Wilson, D. B. (1988) Transfer of a functional human immune system to mice with severe combined immunodeficiency. *Nature*, **335**, 256–9.

Nezlin, R. S. (1987) Superfamily of proteins related to immunoglobulins. *Biol. Membr.*, **4**, 341–65 (*in Russian*).

Nolan, O. and O'Kennedy, R. (1990) Bifunctional antibodies: concept, production and applications. *Biochim. Biophys. Acta*, **1040**, 1–11.

Nouza, K. and John, C. (1987) *Immunology of health and illness*. Prague, Avicenum, 356 pp. (*in Czech*).

Parnes, J. R. (1986) Structure and function of *T* lymphocyte differentiation antigens. *Trends Genet.*, **2**, 179–83.

Patten, P., Yokota, T., Rothbard, J., Cien, Y. H., Arai, K. I. and Davis, M. M. (1984) Structure, expression and divergence of *T*-cell receptor "beta"-chain variable regions. *Nature*, **312**, 40–6.

Pauling, L. (1940) A theory of the structure and process of formation of antibodies. *J. Amer. Chem. Soc.*, **62**, 2643–57.

Rajewsky, K., Schirrmacher, V., Nass, S. and Jerne, N. K. (1969). The requirement of more than one antigenic determinant for immunogenicity. *J. Exp. Med.*, **129**, 1131–40.

Rejnek, J., Tučková, L., Zikán, J., Říhová, B. and Pospíšil, M. (1983) Participation of VH and I region in the formation of antigen-specific *T* cell receptors. *Develop. Comp. Immunol.*, **7**, 757–65.

Rejnek, J., Tučková, L., Zikán, J., Říhová, B. and Kostka, J. (1985) Antigenic properties of *T* cell antigen-specific receptors isolated from the surface of rabbit and mouse spleen and lymph node cells. *Folia Microbiol.*, **30**, 212–22.

Reth, M., Hombach, J., Wienands, J., Campbell, K. S., Chien, N., Justement, L. B. and Cambier, J. C. (1991) The *B*-cell antigen receptor complex. *Immunol. Today*, **12**, 196–201.

Rosenthal, A. S. (1978) Determinant selection and macrophage function in genetic control of the immune response. *Immunol. Rev.*, **40**, 136–52.

Rosenthal, A. S. and Sevach, E. M. (1973) Function of macrophages on antigen recognition by guinea pig *T* lymphocytes. I. Requirement for histocompatible macrophages and lymphocytes. *J. Exp. Med.*, **138**, 1194–212.

Saito, H., Kranz, D. M., Takagaki, Y., Hayday, A. C., Eisen, H. N. and Tonegawa, S. (1984a) Complete primary structure of a heterodimeric *T* cell receptor deduced from cDNA sequences. *Nature*, **309**, 757–62.

Saito, H., Kranz, D. M., Takagaki, Y., Hayday, A. C., Eisen, H. N. and Tonegawa, S. (1984b) A third rearranged and expressed gene in a clone of cytotoxic *T* lymphocytes. *Nature*, **312**, 36–40.

Schultz, P. G. (1989) Catalytic antibodies. *Angew. Chem. Int. Ed. Engl.*, **28**, 1283–95.

Sela, M., Fuchs, S. and Arnon, R. (1962) Studies on the chemical basis of the antigenicity of proteins. 5. Synthesis, characterization and immunogenicity of some multichain and linear polypeptides containing tyrosine. *Biochem. J.*, **85**, 223–31.

Sharon, N. (1984) Carbohydrates as recognition determinants in phagocytosis and in lectin-mediated killing of target cells. *Biol. Cell*, **51**, 239–46.

Šterzl, J. (1988) Antigen processing and presentation as a part of the immune response induction. *Bratisl. Lek. Listy*, **89**, 403–14 (*in Czech*).

Šterzl, J. (1989) *Development and induction of immune response*. Prague, Academia, 464 pp. (*in Czech*).

Šterzl, J. and Nordin, A. (1971) The common cell precursor for cells producing different immunoglobulins. In: Mäkelä, O., Cross, A. and Kosunen, T. V. (eds.), *Cell Interaction and Receptor Antibodies*. New York, Acad. Press, pp. 213–30.

Strejček, J. and Lokaj, J. (1985) *Immunology in clinical practice*. Prague, Avicenum, 294 pp. (*in Czech*).

Tlaskalová, H., Šterzl, J., Hofman, J., Holub, M., Říha, I., Říhová, B., Zikán, J., Rejnek, J., Pospíšil, M., Trebichavský, I. and Bártová, J. (1985) Factors regulating the development, differentiation and function of *B* cells. In: Ferenčík, M. and Štefanovič, J. (eds.), *Immunology 1985*. Bratislava, pp. 67–92 (*in Czech*).

Tonegawa, S. (1983) Somatic generation of antibody diversity. *Nature*, **302**, 575–81.

Uhlenbruck, G. (1987) Bacterial lectins: mediators of adhesion. *Zbl Bakt. Hyg.* A, **263**, 497–508.

Unanue, E. R. (1989) Macrophages, antigen-presenting cells, and the phenomena of antigen handling and presentation. In: Paul, W. E. (ed.), *Fundamental Immunology*. 2nd edn. New York, Raven Press, pp. 95–115.

Unanue, E. R., Beller, D. I., Lu, C. Y. and Allen, P. M. (1984) Antigen presentation: comments on its regulation and mechanism. *J. Immunol.*, **132**, 1–5.

Weiss, A., Imboden, J., Hardy, K., Manger, B., Terhost, C. and Stobo, J. (1986) The role of the T3/antigen receptor complex in *T*-cell activation. *Annu. Rev. Immunol.*, **4**, 593–619.

White, J. M. and Littman, D. R. (1989) Viral receptors of the immunoglobulin superfamily. *Cell*, **56**, 725–8.

Williams, A. F. (1987) A year in the life of the immunoglobulin superfamily. *Immunol. Today*, **8**, 298–303.

Yagüe, J. and Palmer, E. (1985) Antigen specific *T* cell receptor: isolation, structure and genomic organization. *Immunologia*, **4**, 89–98.

8 Preparation of pure immunoglobulins

Immunoglobulins are usually isolated from blood serum and only rarely from other body fluids. Different methods, yielding immunoglobulins of various purity and immunochemical activity, may be used (*Table 8.1*).

Table 8.1 Methods used for the isolation and characterization of immunoglobulins

Method	*Example*
Non-specific	
1. Precipitation:	
Neutral salts	Ammonium sulphate
Organic solvents	Ethanol
Metallic ions	Zn^{2+}
Organic cations	Rivanol
Simple polymers	Polyethyleneglycol
2. Electrophoresis	
Free electrophoresis	
Carrier electrophoresis	Paper, agarose, polyacrylamide
3. Isoelectric focusing	Carrier ampholytes
4. Ion-exchange chromatography	DEAE-cellulose, CM-Sephadex
5. Gel filtration	Sephadex G-100
6. Hydrophobic chromatography	Phenyl-Sepharose CL 4B
7. Ultracentrifugation	
Specific	
Immunoaffinity chromatography	Affinity chromatography using immuno-adsorbents

Methods used for the isolation of pure immunoglobulins can be divided into two groups: non-specific and specific. In general, **non-specific methods** are those used in classical preparative biochemistry to isolate pure proteins. **Specific methods** are based on the affinity of the binding site of immunoglobulins for complementary determinants on an antigen molecule immobilized in the solid phase (on an insoluble carrier). Thus, specific methods are based on various types of affinity chromatography — immunoaffinity chromatography or immunosorption.

The fundamental difference between the non-specific and specific methods lies in the character and properties of antibodies purified by these methods. The non-specific methods are used for the isolation of immunoglobulins of a particular class or subclass but yield, however, a heterogeneous population of molecules with various binding specificities. This mixture of polyclonal antibodies contains various idiotypes produced as a response to antigens with which the serum donor was in contact. On the other hand, the specific methods use a single pure antigen which permits the isolation of polyclonal antibodies with specificity limited to determinants present on the antigen used. Thus compared to the non-specific methods, *immunoaffinity chromatography* yields only a minor fraction of immunoglobulins, but with a substantially lower heterogeneity.

The non-specific methods for the isolation of immunoglobulins can be further subdivided into fractionation precipitation, zonal electrophoresis, isoelectric focusation, chromatofocusation, ion-exchange chromatography, gel filtration chromatography, hydrophobic chromatography and ultracentrifugation. The methods of fractionation precipitation are relatively crude and do not usually yield pure immunoglobulins in a single operation, and the whole procedure must therefore be repeated several times. In addition, these methods use reagents which may denature some antibodies and decrease their activity. The fractionation methods include precipitation with neutral salts, fractionation with organic solvents and precipitation with some organic compounds.

8.1 Non-specific methods

8.1.1 Fractionation precipitation

The solubility of proteins in aqueous solutions depends on their solubility envelope, *i.e.* on the layer of molecules and ions surrounding their molecules. The solubility envelope is due to electrostatic forces formed between electrically charged parts of the protein molecule and water dipoles. The addition of a neutral salt (*e.g.* ammonium sulphate, sodium sulphate, sodium chloride) decreases the thickness of this solubility envelope or decreases the effective water concentration, resulting in a decreased solubility of the protein in question. At a certain salt concentration the protein becomes insoluble and precipitates. Different proteins precipitate from aqueous solutions at different salt concentrations and this phenomenon is used for their fractionation.

Ammonium sulphate and *sodium sulphate* are most frequently used for the **precipitation of immunoglobulins**. The IgG fraction is precipitated at room temperature and pH 7.3. Ammonium sulphate is added to the serum to yield a 33% concentration or sodium sulphate is added to a final concentration of 18 %. The precipitated proteins are centrifuged, dissolved in phosphate buffered saline (PBS) and reprecipitated with 33% $(NH_4)_2SO_4$ or 12–15% Na_2SO_4. If required, the precipitation can be repeated once or twice. The final precipitate is dissolved in PBS and excess salt removed by filtration through a Sephadex G-25 column or by dialysis against PBS. Using this procedure, relatively pure IgG preparations can be isolated from human and rabbit serum (impurities commonly include the α-globulins). Relatively pure IgG can also be isolated from sera of other mammals and chickens. However, a different salt saturation may sometimes be recommended. For example, up to 40% saturation with ammonium sulphate is recommended for the isolation of IgG from guinea-pig serum.

A crude fraction of all serum immunoglobulins can be prepared by saturating the serum with ammonium sulphate to 40–50% at pH 7.0. The precipitated immunoglobulins are separated by centrifugation and redissolved in PBS. The resulting solution is not stable. On storage, the euglobulin fraction is particularly prone to precipitation. The main IgM fraction precipitates at an ammonium sulphate saturation of 50–60%.

Immunoglobulins can also be precipitated by some organic compounds, *e.g.* rivanol (3-ethoxy-6,9-diaminoacridinelactate), caprylic acid or polyethy-

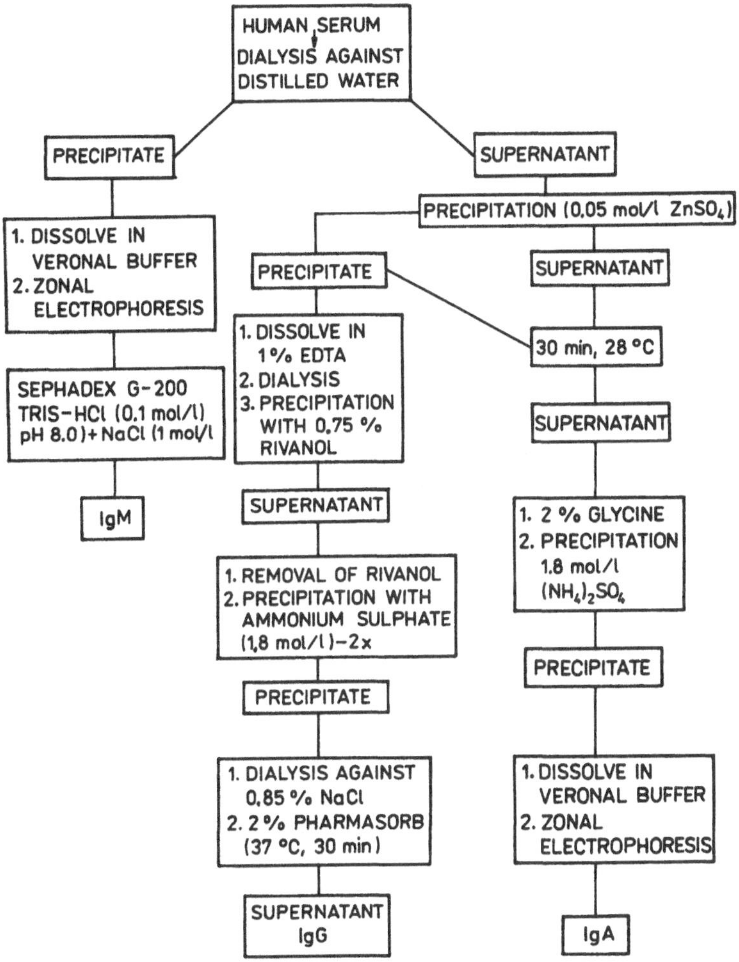

Fig. 8.1. A method for simultaneous isolation of IgG, IgM and IgA from one serum sample.

leneglycol. When using *polyethyleneglycol* (*PEG*) of M_r 6000 at 7% saturation, a precipitate largely composed of IgM in addition to other proteins is formed, whereas at 14% saturation, IgG and IgA are obtained.

Pure IgM, IgG and IgA from one serum can only be simultaneously isolated by using combinations of numerous steps and methods. An example is presented in *Fig. 8.1*.

The electric charge of proteins and thus also their solubility envelope depends on the pH of the medium (ionization of free carboxyl and amino groups). Therefore, when salting out immunoglobulins, two variables, *i.e.* salt concentration and pH, can be used.

Fractionation with organic solvents miscible with water (*e.g.* ethanol and acetone) relies on the different solubility of individual serum proteins in these solvents and in water. Increasing the concentration of the organic solvent decreases the solubility envelope around protein molecules and causes their precipitation. Fractionation with organic solvents can be combined with changes of salt concentration and pH enabling relatively selective separation of some plasma proteins. So-called **Cohn fractionation** of serum with ethanol (*Fig. 8.2*) yields a pure γ-globulin fraction containing almost only IgG and is widely used commercially. However, the temperature of ethanol fractionation must be strictly controlled. It is, therefore, necessary to use cold rooms and refrigerated centrifuges.

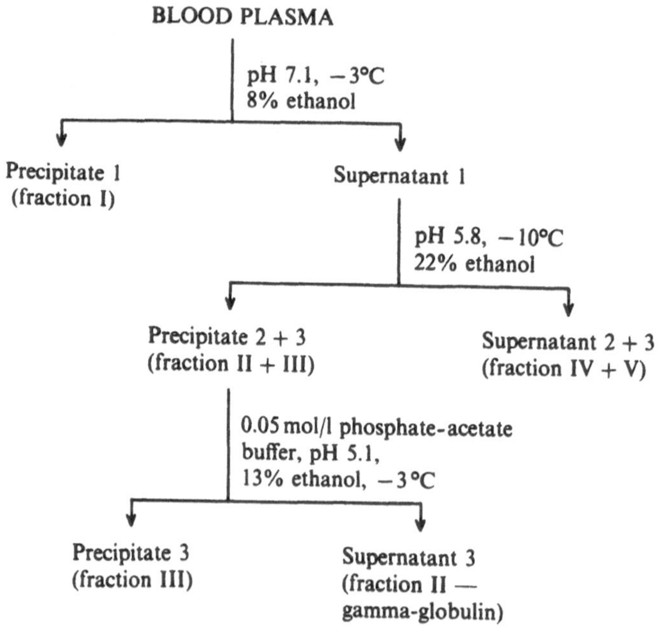

Fig. 8.2. Method of blood plasma fractionation with ethanol (according to Cohn, 1945).

8.1.2 Electrophoretic methods

Electrophoretic methods are used to separate compounds, cells and other particles bearing electric charges. Compounds must be ions or ampholytes. **Ampholytes** are compounds whose molecules carry positive and negative charges due to internal ionization. Immunoglobulins as well as other glycoproteins and proteins also belong to this group of compounds. When a mixture of such compounds is subjected to an electric field in a specific medium, molecules of the compounds (particles) begin to migrate. Their migration depends on the size of their charge, molecular size and shape, conditions of the medium and electric field strength. The size of charge of the molecule or particle is influenced by the degree of ionization, pH and ionic strength of the medium.

Electrophoresis can proceed in a free electrolyte (**free boundary electrophoresis**) or in porous carriers soaked in the electrolyte. TISELIUS and KABAT (1939) were the first to use free boundary electrophoresis and separated serum proteins into albumin, α-, β- and γ-globulins and showed that antibodies are present in the γ-globulin fraction. However, free boundary electrophoresis does not give a perfect spatial separation of the individual fractions possible and, in addition, free boundary electrophoresis equipment is relatively costly. Therefore, mainly **carrier electrophoresis** is widely used at present. As compounds separated on a carrier form easily detectable zones, the technique is often called **zonal electrophoresis**.

Paper was originally used as the carrier for electrophoresis of immunoglobulins and other serum proteins, followed by cellulose acetate and agar, agarose and polyacrylamide gels, *etc.*

Electrophoretic devices generally consist of an electrophoretic chamber, and a DC power supply. The electrophoretic chamber is composed of cathode and anode compartments with respective electrodes and electrolytes, and a carrier compartment. The carrier is connected to the electrolytes by conductive bridges or "wicks" (usually filter paper). A DC power supply serves as a source of stabilized (constant) current or voltage. Potential carrier gradients of <20 V/cm, 20–50 V/cm and >50 V/cm are known as low-, medium- and high-voltage electrophoresis respectively.

Buffers with a specific salt concentration, pH, ionic strength, conductivity, temperature, viscosity and dielectric constant usually serve as electrolytes. The main function of the buffer is to conduct current and maintain constant pH, ensuring that molecules of a particular compound maintain an identical charge during electrophoresis. An electrolyte solution is a second-class conductor and, therefore, exhibits resistance whose reciprocal value is conductivity. Increasing conductivity during electrophoresis produces heat. This increases the temperature of the carrier which may lead to denaturation of the compounds to be separated or other changes. Therefore, heat must be reduced by decreasing the ionic strength of the electrolyte (and thus conductivity), by decreasing the potential gradient or by cooling the carrier (the most efficient and most advantageous method).

Modern power supplies can provide constant current or constant voltage and the separation conditions during electrophoresis therefore remain constant. During electrophoresis on paper or cellulose acetate relatively little heat is produced and, therefore, it is not important whether constant voltage or current are maintained. In gel electrophoresis, where the carrier depth may be several mm, heat production becomes a problem. In this case it is advantageous to apply constant current since voltage decreases with time and, therefore, heat production also decreases.

Properties of the carrier. Mobility of charged particles on carriers is usually smaller than during free electrophoresis. Their migration is influenced by other factors, such as electroendosmosis, flow potential, spatial effects of the carrier, adsorption *etc.* Electroendosmosis depends on the type of carrier, potential gradient, electrolyte ionic strength and pH. Electroendosmosis forces the particles to migrate towards the cathode, together with water. The resulting particle movement is thus a sum of electrophoretic and electroendosmotic mobility. For example, with paper electrophoresis, electroendosmosis causes γ-globulins to migrate towards the cathode, even though they have a negative charge.

Flow potential is produced during water movement through the carrier and has an opposite direction with respect to the voltage applied. It thus slows down the migration of the particles.

Particle migration is also influenced by the spatial structure of the carrier. For instance, the electrophoretic migration of a particle during paper electrophoresis is not straight but "zigzag". In fact, the particle moves a greater distance than the real paper length between the start and end of migration. This phenomenon has been explained by pore curvature or by a barrier effect of the cellulose fibres. The barrier effect is based on the fact that the migrating particles strike structures of the carrier which slow its movement. Gel carriers are cross-linked to various degrees and their pore sizes are also different. In addition to the barrier effect, molecular sieving as well as ion-exchange effect may be involved. Adsorption of particles occurs in all carriers. Reversible adsorption is usually involved, but some of the material separated can also adsorb irreversibly (particularly in paper electrophoresis). Spatial effects of the carrier account for the fact that migration of numerous particles in carriers are lower than in free electrophoresis, even when the electroendosmotic movement is subtracted.

After electrophoresis, the carrier containing the separated compounds is dried, the proteins are denatured or fixed in other ways and stained in order to visualize them. Solutions of Amido Black 10B, Coomassie Brilliant Blue B 250, Ponceau S and Nigrosin WS may be used for staining. Stains are chosen according to the type of carrier used. Lipids, polysaccharides, glycoproteins and lipoproteins are stained with other dyes.

Electrophoresis of serum immunoglobulins in various carriers is used for diagnostic purposes in clinical medicine or for large-scale preparative isolation of individual classes. Cellulose acetate membranes or agarose gels are most frequently used for the electrophoretic analysis of human serum. On agarose gel, human serum proteins can be separated into nine fractions (*Table 8.2*). Immunoglobulins are found in both γ-globulin fractions. Analytical electrophoresis enables a preliminary and rapid diagnosis of agammaglobulinaemia (absence of the γ-globulin fraction) or of monoclonal gammopathies (p. 140). In the latter case, a narrow, distinct, sharp band is observed on the electrophoreogram, within γ_1, γ_2 and occasionally also β_2 ranges. A respective monoclonal immunoglobulin is found in the band.

Table 8.2 Protein fractions detected after electrophoresis of human serum in agarose gel

Fraction	*Proteins*
Anode	
Prealbumin	Prealbumins
Albumin	Albumin
α_1-globulin	α-lipoprotein, α_1-antitrypsin
α_2-globulin	Antichymotrypsin, α_2-macroglobulin, haptoglobin, ceruloplasmin
β_1-globulin	Fibronectin, haemopexin, transferrin
β_2-globulin	β-lipoprotein, C3, C4
γ_1-globulin	IgG, IgM, IgA, IgD, IgE, immune complexes
γ_2-globulin	IgG, IgM, IgA, immune complexes, CRP
Post-γ-globulin	lysozyme
Cathode	

C3, C4 — complement components, CRP — C-reactive protein.

Starch gel was previously used most frequently for the preparative electrophoretic separation of immunoglobulins. At present, Pevikon (copolymer of polyvinylchloride and polyvinylacetate) and agarose gel are applied. Gel carriers, 0.5–1.5 cm thick, are prepared, which makes it possible to fractionate a larger amount of the sample. Preparative electrophoresis is particularly used to separate molecules having a similar size but different electric charge, *e.g.* α_2-macroglobulin and IgM. Both compounds are eluted in a single peak on Sephadex G-200 chromatography and preparative electrophoresis is one of the most useful procedures in their separation.

When sodium dodecyl sulphate (SDS) is added to a protein mixture, it binds to peptide bonds and basic protein groups. This procedure gives almost all proteins a negative charge proportional to their molecular size. Electrophoresis is usually performed in thin layers of **SDS-containing polyacrylamide gel (SDS–PAGE)**. This technique is used to measure the relative molecular weights of proteins as their migration is proportional to molecular weight. A mixture of proteins with known molecular weights serves as a standard. SDS can also be used to separate light and heavy immunoglobulin chains after their hydrolysis as it prevents their reassociation. The principles of zonal electrophoresis, SDS–PAGE electrophoresis and of analytical isoelectric focusing are illustrated in *Fig. 8.3*.

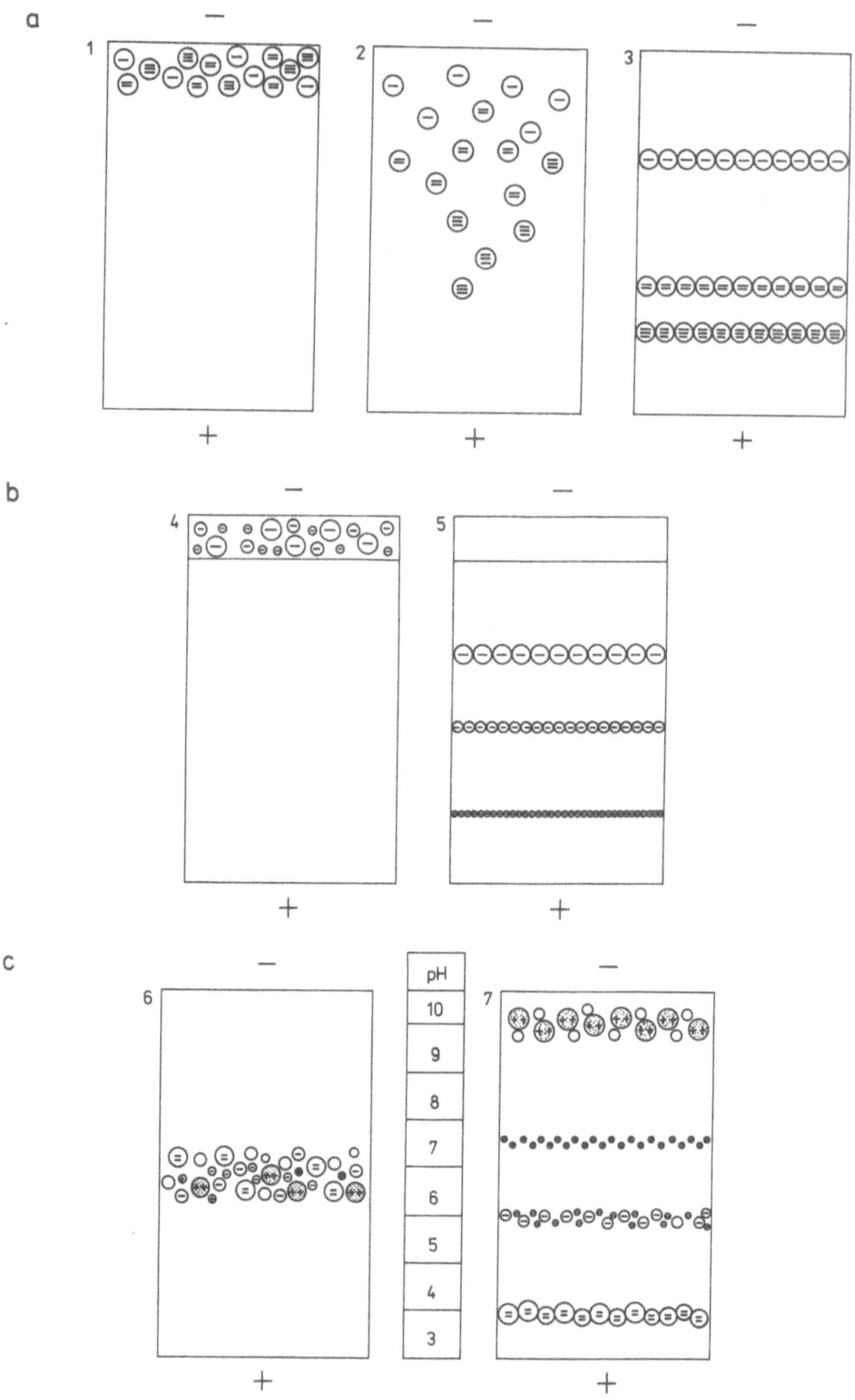
a
1
2
3
b
4
5
c
6
7
pH
10
9
8
7
6
5
4
3

8.1.3 Isoelectric focusing

During zonal electrophoresis, compounds are separated in a carrier according to their electric charge. Isoelectric focusing is, in fact, electrophoresis in a pH gradient. However, in this case individual compounds are separated not only according to charge but also according to their **isoelectric points**. During isoelectric focusing in a carrier, immunoglobulins migrate to sites which have a pH identical to their isoelectric points. They then remain stationary and their net charge becomes zero. Thus, irrespective of its application site, any protein moves to its own "focus" (hence "focusing"), where it reaches its isoelectric point and gradually concentrates. The isoelectric points of proteins with a different primary structure are different and, therefore, during isoelectric focusing they migrate to different sites of the pH gradient. It is therefore possible to separate even closely related molecular species. For instance immunoglobulin, which gives a single narrow zone on electrophoresis, separates into 2–5 components during isoelectric focusing. Isoelectric focusing is a highly discriminatory technique and is widely used to study heterogeneity of immunoglobulins and other proteins.

During electrolysis of water, H^+ ions discharge at the cathode and OH^- ions discharge at the anode. The pH, therefore, increases towards the cathode and decreases towards the anode, resulting in a pH gradient across the carrier. Buffers prevent the formation of such a gradient. However, the gradient can be stabilized by adding so-called **ampholytes** which have isoelectric points within a certain pH range. The carrier ampholyte migrates during electrophoresis to a site with a pH identical to its isoelectric point. There it remains uncharged and suppresses all possible changes which might occur. Carrier ampholyte mixtures which facilitate the formation of a stable gradient within different pH ranges (*e.g.* 3–10, 3–5, 4–6 *etc.*) are available commercially. Molecules of carrier ampholytes consist of a skeleton to which different numbers of $—COO^-$ and NH_3^+ groups are bound.

Isoelectric focusing can also be both analytical and preparative. Thin layers of polyacrylamide, agarose or tubes of polyacrylamide gel may be used as carriers for analytical isoelectric focusing. Preparative isoelectric focusing is performed on columns of Ficoll, sucrose, dextran or glycerol serving as density gradients or in thick layers of Sephadex or Bio-Gel.

Interpretation of the results of isoelectric focusing of immunoglobulins is difficult, and is only used in certain cases. For example, even a relatively pure fraction of normal serum IgG yields 10 or more separate bands. Myeloma monoclonal immunoglobulins, in spite of being synthesized by a single cell clone, appear as several bands within one hour of contact with the serum.

Fig. 8.3. Principles of zonal electrophoresis (a), SDS electrophoresis (b) and analytical isoelectric focusing (c) in polyacrylamide or agarose gels.

1, 4, 6, start; 2, run; 3, end of electrophoretic separation in a gel with sufficiently large pores; compounds separated only according to their charge; 5, end of electrophoretic separation in the presence of SDS, individual compounds separated according to their molecular size; 7, end of isoelectric focusing, individual compounds separated according to their charge and isoelectric points.

This is apparently due to their post-translation modification in the serum. Only a single band is observed when the monoclonal immunoglobulin taken directly from myeloma cells is analysed. As only proteins with a molecular weight up to 500 000 can penetrate polyacrylamide gels, isoelectric focusing of IgM in such a gel is impossible. Special types of agarose are used instead.

8.1.4 Ionex chromatography

Ionex (an ion-exchange resin) consists of a basic (chemically inert or only poorly reactive) matrix and covalently bound groups of atoms carrying electric charge and free mobile (exchangeable) ions of opposite charge. When the atom groups firmly bound to the ionex matrix have a positive charge, the exchangeable ions are negative (anions) and we speak about an **anion exchanger (anex)**. When the exchangeable ions are cations, we speak about a **cation exchanger (catex)** (*Fig. 8.4*). Polysaccharides (cellulose, dextran, agarose), synthetic bitumens or certain polymers (polyacrylamide, polystyrene, hydroxyethylmethacrylate) are usually used as the matrix. Charged groups responsible for the type and strength of the ionex are chemically bound to the above macromolecules. Cation exchangers usually contain hydroxyl (phenol), carboxyl or sulphate groups, whereas anion exchangers contain aliphatic or aromatic amino groups. Ion exchangers carrying sulphate and quarternary amino groups are considered strong, while ion exchangers carrying other groups belong to medium or weak ion exchangers (*Table 8.3*). The capacity of an ion exchanger is determined by the total number of charged groups on the ion exchanger matrix and their accessibility for exchangeable ions. The capacity of an ion exchanger thus depends on the amount of ions that can be bound and exchanged. This is usually expressed as the amount of a separated compound (mol/l) which can exchange 1 g of a dry, or 1 l of a swollen ion exchanger.

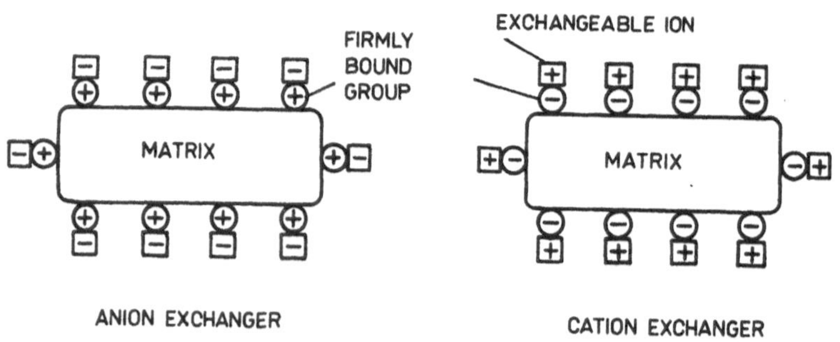

Fig. 8.4. Principle of ion exchangers.

In practice, a dry ion exchanger is first suspended in buffer with a specific ionic strength and pH, or is occasionally activated with a diluted acid or hydroxide and thus converted to a suitable "cycle". A suitable ion with an opposite charge is then bound to the functional groups. Cl^- and Na^+ are the

Table 8.3 Some ion exchangers and their functional groups

Abbreviation	*Functional group*	*Formula*	*Ion exchanger type*	*Example*
TEAE	Triethylaminoethyl	$—O—CH_2—CH_2—N^+(C_2H_5)_3\ Cl^-$	Strongly basic anion exchanger	Spheron TEAE 1000, Cellex T
Q	Quaternary amine	$—CH_2—N^+(CH_3)_3\ SO_4^{2-}$	Strongly basic anion exchanger	Q-Sepharose Fast Flo
QAE	Diethyl(2-hydroxypropyl) aminoethyl	$—C_2H_4N^+(C_2H_5)_2CH_2CH(OH)CH_3\ Cl^-$	Strongly basic anion exchanger	QAE-Sephadex, Cellex QAE
DEAE	Diethylaminoethyl	$—O—C_2H_4N^+(C_2H_5)H\ Cl^-$	Medium basic anion exchanger	DEAE-cellulose, DEAE Sephadex, Cellex D
PAB	Para-aminobenzyl	$—O—CH_2—C_6H_4—N^+H_3\ Cl^-$	Weakly basic anion exchanger	Cellex PAB
CM	Carboxymethyl	$—CH_2—COO^-\ H^+$	Weakly acidic cation exchanger	CM-Sephadex, Cellex CM
P	Phosphoryl	$—O—PO_3^{2-}\ 2Na^+$	Medium acidic cation exchanger	Spherophosphate 1000 Cellex P
SP	Sulphopropyl	$—O—C_3H_6—SO_3^-\ H^+$	Strongly acidic cation exchanger	Spheron S 1000, SP-Sephadex
SE	Sulphoethyl	$—O—C_2H_4—SO_3^-\ H^+$	Strongly acidic cation exchanger	SE-Sephadex
S	Sulpho	$—CH_2—SO_3^-\ Na^+$	Weakly acidic cation exchanger	S-Sepharose Fast Flo

most common counter ions used in anion and cation exchangers, respectively.

Columns of precycled ion exchangers are prepared in glass tubes and may be used to separate an immunoglobulin mixture or a mixture of other compounds. Separation is based on reversible adsorption. Immunoglobulins, like other compounds, contain both cationic and anionic functional groups. Thus, when an anion exchanger is used for their separation, their anion (carboxyl) groups are bound. When a cation exchanger is used, adsorption is mediated by cationic (amino) groups. A suspension of the ion exchanger in the equilibration buffer is poured into the column and the buffer is allowed to flow from the column outlet. The immunoglobulin sample (*e.g.* serum) is applied to the column on top of the ion-exchanger bed. Immunoglobulins thus enter the bed and exchange the original adsorbed ions. When the whole sample volume has been applied, a suitable buffer (usually the equilibration buffer) is added and the immunoglobulins and other adsorbed compounds are eluted. Molecules of individual immunoglobulins and of other proteins have different numbers of ionized (charged) groups, through which they have been bound to the ionex and, therefore, have different binding affinities. Compounds bound weakly to the ion exchanger are eluted first, whereas the most firmly adsorbed compounds are eluted last. The column effluent is collected in a series of test tubes and fractions containing individual immunoglobulins or their simpler mixtures are obtained. Buffers with increasing ionic strength or with increasing pH (gradient elution) are often used for desorption and elution of compounds from the ion exchanger, thereby improving separation of the test compounds.

Quantitative analysis of the individual fractions is achieved with various specific and non-specific analytical methods. For example, proteins are determined by measuring absorbance at 280 nm. Absorbance is measured in individual fractions using an ordinary spectrophotometer or by continuous flow spectrophotometry of the column effluent. In addition to the chroma-

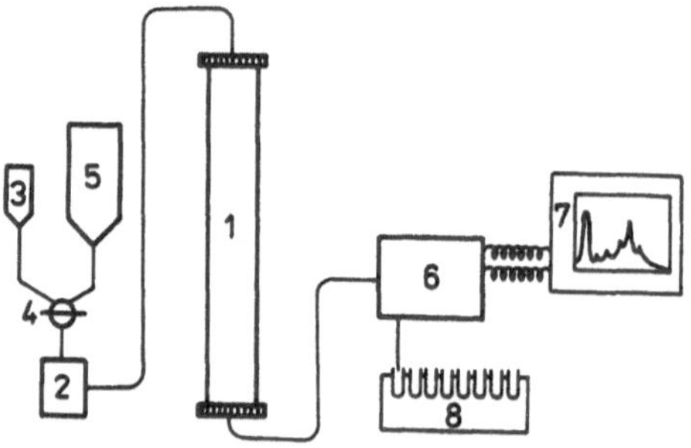

Fig. 8.5. Equipment used in ion-exchange chromatography.
A solution of compounds to be separated (3) is applied to an ion-exchange column (1) with a micropump (2) After the sample has been applied, a two-way valve is switched over (4) and the column is eluted with buffer supplied from the reservoir (5) The eluent containing the separated compounds flows through a spectrophotometer (6) in which its absorbance is measured When immunoglobulins or other proteins are to be separated, their concentrations may be determined from their absorbance at 280 nm Absorbance is automatically recorded in a chart recorder (7) Fractions of the eluent are finally collected in a fraction collector (8)

tographic column and the spectrophotometer, a sample applicator, a fraction collector to collect individual effluent fractions and a recorder are used (*Fig. 8.5*).

Ion-exchange chromatography is frequently used for the isolation of immunoglobulins from blood serum. DEAE-cellulose, DEAE-Sephadex and QAE-Sephadex are commonly used to purify IgG. A single chromatographic separation on these ion exchangers yields a virtually **pure IgG** (*Fig. 8.6*). It is not even necessary to pack a column with the ion exchanger. It is often sufficient to mix DEAE-Sephadex equilibrated with a suitable buffer (0.015 mol/l phosphate, pH 6.5) with serum in a beaker. After several minutes of mixing, the mixture is filtered and almost pure IgG is found in the filtrate. This is an example of "batch" procedure.

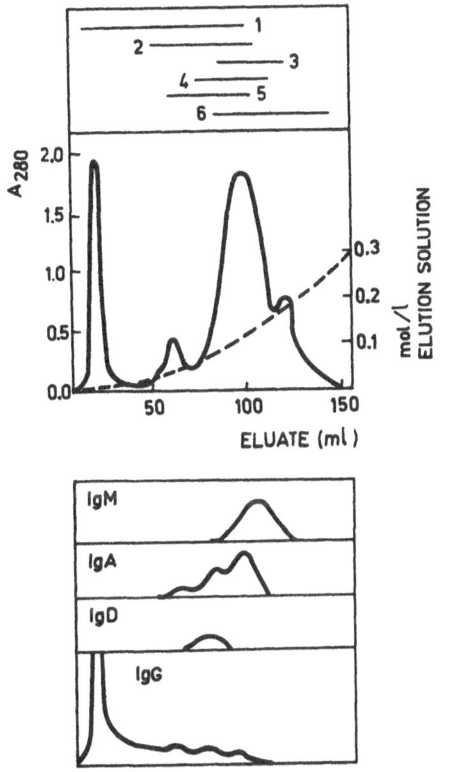

Fig. 8.6. Isolation of immunoglobulins from human serum by DEAE-cellulose chromatography. 1, IgG; 2, IgA; 3, IgM; 4, α_2-macroglobulin; 5, transferrin; 6, albumin. Serum (5 ml) was dialysed against 0.015 mol/l phosphate buffer, pH 8.0 and applied to a DEAE-cellulose column (2 × 30 cm, containing 10 g of dry DEAE-cellulose). The column was equilibrated with the above buffer and eluted with a concentration of the same buffer (0.015–0.3 mol/l). Gradient elution is shown by the dashed line.

Human IgM binds more strongly to the anion exchanger than IgG. Since solutions with higher ionic strength are required for their elution, IgM is contaminated by other serum proteins. Therefore, ion-exchange chromatography is not suitable for the isolation of IgM. A similar situation is observed with IgA which is eluted from a DEAE-cellulose column by phosphate concentrations as high as 0.10–0.15 mol/l. Using DEAE-cellulose chromatography it is only possible to obtain **pure IgA** when the test serum contains myeloma IgA, which is usually present at much higher concentrations than normal IgA. From a DEAE-cellulose column, IgD is usually eluted between IgG and IgA. Again, using ion-exchange chromatography it is only possible to isolate IgD from myeloma serum containing supranormal levels. In the preparation of ion-exchange chromatography columns, only an amount of the ion exchanger appropriate for the concentration of immunoglobulins present in the test sample should be used. With lower amounts of ion exchanger, some molecules of immunoglobulins remain unbound (*Fig. 8.7*), whereas with too high amounts, non-specific adsorption of immunoglobulins may occur, resulting in a less-efficient resolution. Ion-exchange chromatography is also used to purify monoclonal immunoglobulins which occur at high concentrations in the serum of patients with multiple myeloma or Waldenström's macroglobulinaemia.

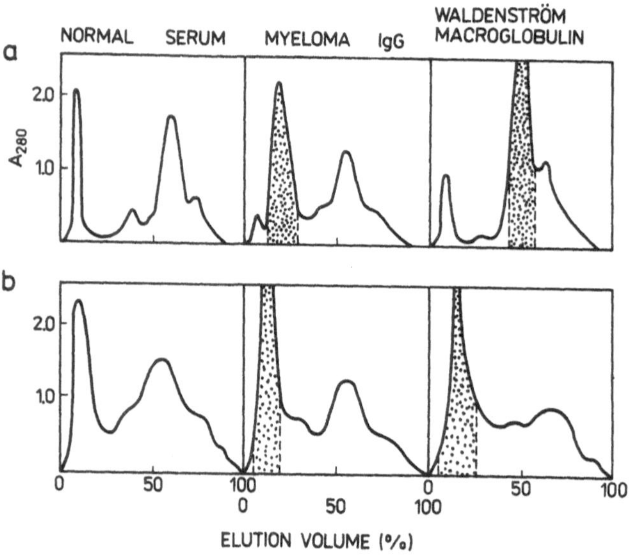

Fig. 8.7. Chromatography of immunoglobulins on an (a) adequate and (b) inadequate DEAE cellulose column.
The inadequate column has insufficient capacity to achieve a good separation of the immunoglobulins applied. Neoplastic (tumour) immunoglobulins are in the shaded region.

This method can also be used for the isolation of **animal immunoglobulins**. From rabbit serum IgG can be prepared using procedures similar to those used for the isolation of the human IgG fraction. However, rabbit serum contains two subclasses which have different electrophoretic mobility and chromatographic properties. The more slowly migrating IgG_2 is eluted from a DEAE-cellulose column with 0.015 mol/l phosphate buffer, pH 6.5, and can be isolated in a virtually pure form. The more rapidly migrating IgG_1 fractions are eluted only with 0.2 mol/l phosphate buffer, pH 8.0, and are therefore contaminated by other serum proteins. In order to purify them it is necessary to re-chromatograph the sample using the same DEAE-cellulose column.

Guinea-pig serum also contains two chromatographically different IgG subclasses. Concentrations of IgG_1 are substantially lower than those of IgG_2. Therefore, it is relatively difficult to prepare an IgG_1 subclass which does not also contain IgG_2. Both subclasses can only be purified by DEAE-cellulose chromatography on a column eluted with both concentration and pH gradients.

When isolating IgG from sera of other animals, the procedure used for the isolation of human serum IgG must be suitably modified (FERENČÍK *et al.*, 1981).

8.1.5 Gel filtration chromatography

This method is also called gel chromatography, gel filtration, exclusion chromatography, permeation chromatography and molecular sieve chromatography. The term **molecular exclusion chromatography** is technically the most accurate, since the method uses porous gels and separates individual compounds according to their molecular size. The use of a hydrophilic gel and aqueous elution solution is referred to as **gel filtration chromatography**, whereas the use of a hydrophobic gel and an organic solvent is known as **gel permeation chromatography**.

Gel filtration chromatography is performed on columns or thin layers prepared from various synthetic or natural compounds. Certain organic polymers, such as dextran (different *Sephadex types*), agarose (*Sepharose, Bio-Gel A*), polyacrylamide (*Bio-Gel P*), copolymers of styrene with divinylbenzene (*Bio-Beads S* and *Bio-Beads SM* are also suitable for work with organic solvents), copolymers of hydroxyethylmethacrylate and ethylenedimetacrylate (*Spheron* that can be swollen both in water and in organic solvents), agarose and polyacrylamide (*Ultrogel*) or porous glass (*Bio-Glas*) are most widely used.

Such compounds are known as **molecular sieves** and increase in volume, *i.e.* swell, in contact with liquids (water, organic solvents) producing gels. Gel particles have a three-dimensional structure and usually appear as small beads. After swelling they bind large volumes of the liquid phase (some of them as much as 30-fold their dry mass). The surface of the particles is not smooth and contains pores. Small molecules of the liquid can enter the pores without restriction, causing swelling of the particles and gel formation. Penetration of larger molecules is limited by the size of the pores. When a mixture of compounds with different molecular weights passes through the gel, only the molecules that can penetrate the pores are retained. Molecules larger than the gel pores flow freely with the elution solution as they cannot enter the pores (*Fig. 8.8*).

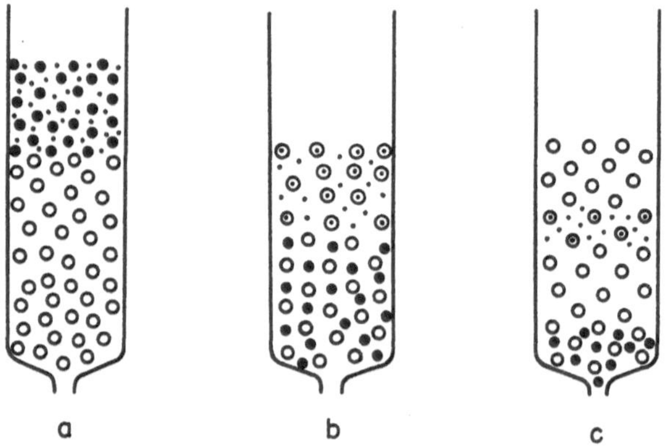

Fig. 8.8. Principle of gel filtration chromatography. Open circles, gel particles; closed circles and dots, large and small molecules respectively.
(a) A solution of small and large molecules is applied to a gel column in a glass or plastic tube; (b) the flow rate depends on the size of the molecules; small molecules are retained in gel particles while large molecules flow through the gel much faster; (c) the end of separation. Small molecules are retained in the upper band, whereas large molecules pass through the column bed.

Molecules smaller than the pore size remain in the liquid phase forming an equilibrium with the number of molecules inside the particles. As fresh elution solution flows through the gel, the equilibrium is continuously impaired and re-established. This results in a gradual elution of the compounds retained in the gel. Individual molecular species then appear in the eluate according to their size. Thus, molecules with the highest relative molecular weight are eluted first and molecules with the lowest relative molecular weight are eluted last. In addition to molecular size the flow rate of individual compounds is also influenced by the shape of the molecules to be separated. Gel column systems identical to those used for ion-exchange chromatography can be used (*Fig. 8.5*).

Gel particle pore size can be modified to retain molecules of a certain size. Different types of particles have been prepared for gel filtration chromatography and are now available commercially. For example, Sephadex G-25 retains compounds with a relative molecular weight of 1 000–10 000, Sephadex G-100 retains molecules with a relative molecular weight of 5 000–150 000 and Sephadex G-200 retains molecules with a relative molecular weight of 5 000–600 000 (*Table 8.4*). Compounds with a relative molecular weight higher than 600 000 therefore move through the Sephadex G-200 gel without entering the particles and are eluted first from the column.

When an immunoglobulin mixture is separated on a **Sephadex G-200** column, IgM is eluted first. However, a buffer with a sufficiently high salt concentration (*e.g.* 0.5 mol/l NaCl in 0.02 mol/l phosphate buffer, pH 7.3) must be used. Three characteristic peaks are obtained when separating human serum on a Sephadex G-200 column (*Fig. 8.9*). In addition to IgM,

Table 8.4 Molecular sieves used in gel filtration chromatography

Name	*Chemical composition*	*Separation range* (M_r)	*Solvent*	*Manufacturer*
Sephadex G	Dextran	$0–6 \times 10^5$	Water	Pharmacia, Sweden
Sephadex LH	Hydroxypropyl dextran derivative	$0–5 \times 10^3$	Polar organic	Pharmacia, Sweden
Sepharose	Agarose	$10^4–4 \times 10^7$	Water	Pharmacia, Sweden
Sephacryl	Agarose + polyacrylamide	$5 \times 10^3–1.5 \times 10^6$	Water	Pharmacia, Sweden
Bio-Gel P	Polyacrylamide	$10^2–4 \times 10^5$	Water	Bio-Rad Laboratories, USA
Bio-Gel A	Agarose	$10^4–1.5 \times 10^8$	Water	Bio-Rad Laboratories, USA
Bio-Beads S	Styrene–divinyl benzene copolymer	$4 \times 10^2–1.4 \times 10^3$	Organic	Bio-Rad Laboratories, USA
Bio-Glass	Porous glass	$3 \times 10^3–2 \times 10^7$	Water and organic solvents	Bio-Rad Laboratories, USA
Segavac	Agarose	$10^4–1.5 \times 10^8$	Water	Segavac Laboratories, UK
Styragel	Polystyrene	$0–4 \times 10^7$	Organic	Waters Ass. Inc., USA
Aquapak	Polystyrene	$0–1.5 \times 10^8$	Water	Waters Ass. Inc., USA
Porasil	Silicate	$0–2.5 \times 10^8$	Water	Waters Ass., Inc., USA
Ultrogel AcA	Agarose + polyacrylamide	$10^3–1.2 \times 10^6$	Water	LKB, Sweden
Ultrogel A	Agarose	$2.5 \times 10^4–2.3 \times 10^7$	Water	LKB, Sweden
Spheron	Polymethacrylate	$2 \times 10^4–1 \times 10^8$	Water and organic solvents	Lachema, CSFR; Koch-Light, UK

α_2-macroglobulin and some lipoproteins are eluted in the first peak. The last part of this fraction can also contain IgA. IgM can be separated from α_2-macroglobulin by agarose electrophoresis — IgM migrates towards the cathode and α_2-macroglobulin towards the anode. The largest IgG portion is found in the second peak, together with IgA. Albumin and other serum proteins are found in the third peak. Resolution of the first two peaks is not always satisfactory. This is due to the fact that IgG molecules spontaneously aggregate forming dimers and higher polymers. This aggregated IgG contaminates the first peak and prevents separation of IgM from IgG.

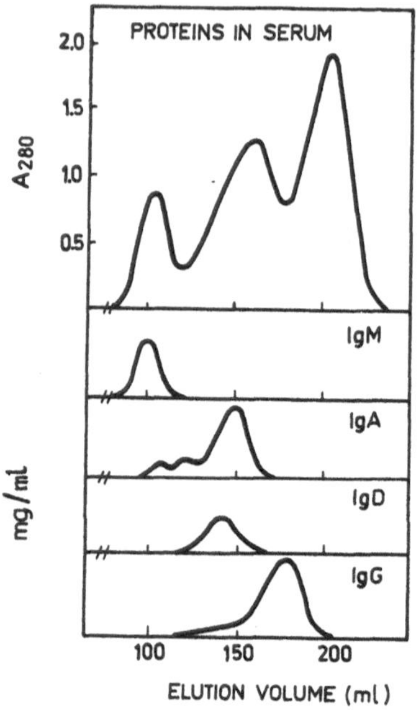

Fig. 8.9. Separation of human serum immunoglobulins by gel filtration chromatography on a Sephadex G-200 column.

In addition to separating compounds with different molecular weights, gel filtration chromatography can also be used to determine relative molecular weights, the amount of complexes (immunocomplexes, IgG aggregates) and to detect fragments of macromolecules (*e.g.* Fc- and Fab-fragments, light and heavy chains of immunoglobulins).

8.1.6 Hydrophobic chromatography

Hydrophobic chromatography is based on the interaction of hydrophobic groups of the compounds to be separated with hydrophobic groups of a chromatographic carrier, usually packed in a column. This interaction is influenced primarily by ionic strength (stability of the hydrophobic bond increases with increasing ionic strength), by the type of ions responsible for the ionic strength (ions which have desalting effects, such as SO_4^{2-}, PO_4^{3-}, NH_4^+, increase the interaction), temperature, pH *etc*. Proteins bind most firmly to a hydrophobic adsorbent at their isoelectric points. By suitable selection of these factors it is possible to modify the affinity of individual components of the mixture for the hydrophobic carrier and thus also their separation.

Agarose is the most widely used matrix for preparation of hydrophobic adsorbents. The hydrocarbon radicals octyl- or phenyl- bound to the basic matrix by ether bonds become sites of the hydrophobic interaction. *Octyl-Sepharose* and *Phenyl-Sepharose* are examples of hydrophobic adsorbents.

The chromatographic separation process is similar to that of affinity chromatography. A sample dissolved in a high ionic strength solution with a pH close to the isoelectric point of the protein is applied to a column of a hydrophobic carrier. After the required protein fraction has been bound and the unwanted components have been eluted, the adsorbed protein is released from the column by decreasing the hydrophobic interaction. This can be achieved by replacing with an ion with a lower salting out ability (most frequently), by decreasing the ionic strength of the eluent or by decreasing the polarity of the eluent.

Hydrophobic chromatography has been used for the isolation of **human IgA**. When human serum in 1 mol/l $(NH_4)_2SO_4$ is applied to a Phenyl-Sepharose CL 4B column, IgA is adsorbed. Other serum proteins are not usually bound and pass to the eluate. The adsorbed IgA is then eluted with 0.8 mol/l $(NH_4)_2SO_4$. A relatively pure IgA preparation is thus obtained which can be repurified by a single gel filtration step.

8.2 Specific methods

Affinity chromatography is the basic specific method used for the preparation of immunoglobulins. It is based on pairs of compounds reacting biospecifically, *e.g.* enzyme–substrate, enzyme–inhibitor, polynucleotide–complementary DNA or RNA region. When such a pair includes an antigen (hapten) and antibody, we speak about **immunoaffinity chromatography** (*Fig. 8.10*).

Affinity chromatography is based on immobilization (cross-linkage with an insoluble carrier) of one member of this pair. When such an immobilized component comes in contact with a solution containing a mixture of different compounds, only molecules of the second member of the pair will adsorb. The compound bound to the insoluble carrier is known as a **ligand** (affinant). Molecules of the second compound, *i.e.* the only components of the mixture which can be adsorbed by the ligand, are subsequently released by desorption and can therefore be obtained in the pure state.

In **immunoadsorption**, an antigen serves as a ligand for the isolation of a specific antibody, or alternatively an antibody acts as the ligand when a pure antigen is to be prepared. A suitable carrier with a bound ligand (antigen or antibody) is known as an *immunoadsorbent*. In practice, when isolating antibodies with immunoadsorbents, an antigen or hapten which reacts specifically with those antibody molecules (immunoglobulins) with a complementary binding site is bound to the insoluble carrier. Polyacrylamide, polymethacrylate, dextran and particularly agarose gels (Sepharose 4B or Sepharose 6B, Pharmacia) are most widely used as carriers. The antigen-adsorbent conjugate is equilibrated in a buffer which provides suitable conditions for interaction of the ligand with a specific antibody. The immunoadsorbent is packed in a glass column (this usually being more advantageous than the batch procedure) and the serum or other solution is applied. Only specifically reacting immunoglobulins remain on the column. Column length and concentration of the immunoglobulin to be isolated are not important. However, it is important to use conditions which achieve adsorption equilibrium between the immobilized antigen and the antibody. This means using

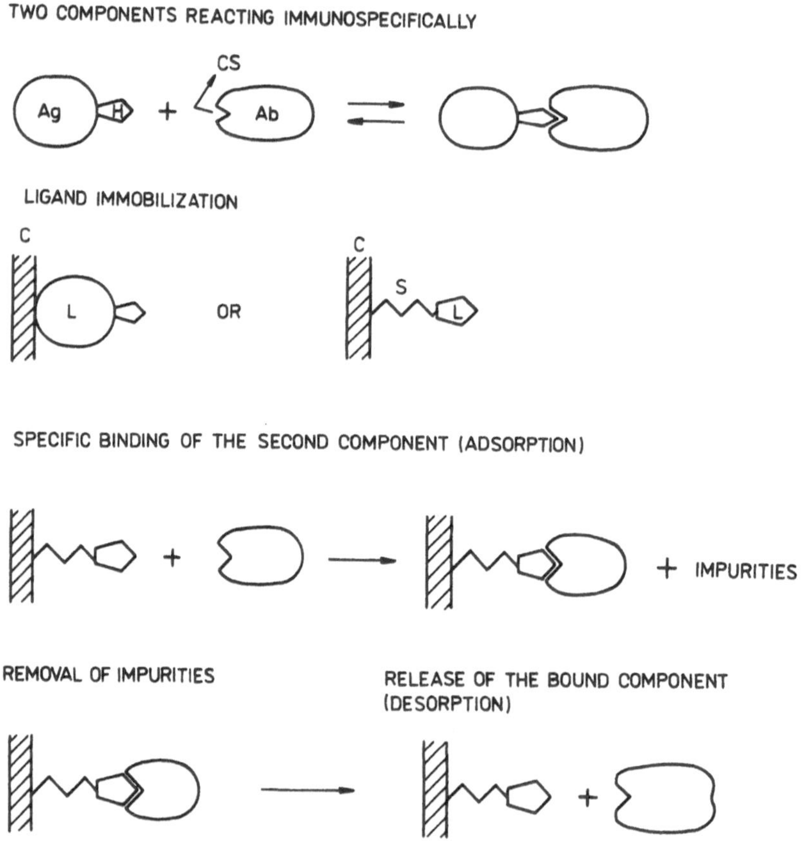

Fig 8 10 Principle of immunoaffinity chromatography
Ag, complete antigen, H, hapten (determinant), Ab, antibody, CS, combining site, C, carrier, L, ligand, S, shoulder

a low flow rate or, sometimes, leaving the solution to be separated (serum) on the column with the column outlet closed. When using highly diluted solutions, a higher flow rate can be used and the solution passed through the column several times. Since the rate of binding to the immunoadsorbent decreases with increasing temperature the process can be accelerated by cooling the column.

After the specific protein has been bound, the immunoadsorbent column is thoroughly washed with the original buffer to remove non-specifically adsorbed components. Immunospecifically bound antibody molecules are released from the sample by non-specific or specific elution. Non-specific forms of elution are usually more drastic and may employ solutions of various acids, buffers with acid pH or with high salt concentration. Specific elution is milder and is based on displacement of the bound antibody by a high concentration of the free ligand (hapten or antigen) in the eluent. Therefore, this procedure is known as *competitive elution*. Solutions used for non-specific elution should be removed as soon as possible (by gel filtration or dialysis) to avoid decreasing the activity of the isolated antibodies.

The immunoadsorbent is thoroughly washed with the eluent and, after repeated equilibration with the original buffer, can be re-used. If microbial contamination is prevented, the immunoadsorbent can be used repeatedly for several months.

The advantage of immunoaffinity chromatography is that it makes possible to isolate specific antibodies against a given antigen, rather than a heterogeneous population of a certain immunoglobulin class, as is the case with non-specific methods. From a particular serum it is possible to isolate several specific antibodies with a different affinity to the antigen that has served as ligand. In addition, immunoglobulins of individual classes and subclasses (isotypes) can be isolated. In the latter case, it is necessary to use antibodies specific for a certain class or subclass. Such antibodies are linked as a ligand to a suitable carrier forming an immunoadsorbent which selectively binds immunoglobulins of a required class or subclass.

Immunoadsorbents may be classified according to the linkage of the ligand. In **"true" immunoadsorbents** the antigen or antibody is bound to the carrier by covalent bonds, whereas in other immunoadsorbents, bonds other than covalent are involved. The latter types of immunoadsorbent include certain types based on glass and other inorganic materials or polystyrene or latex beads of well-defined size. "True" immunoadsorbents bind ten times as much ligand and have a proportionally higher capacity to bind the compound to be isolated. Therefore, "true" immunoadsorbents are more commonly used at present.

A carrier (support) used for preparation of the immunoadsorbent must fulfil certain requirements. It must have good flow properties, it must be resistant to chemical and mechanical effects, it should not exhibit gel filtration, ion-exchange or non-specific adsorption effects, it must have enough active groups to bind an antigen or antibody and carrier particles must have a permanent shape. On the other hand, the ligand must bind the target

Table 8.5 Some adsorbents used in immunoaffinity chromatography

Carrier	*Carrier functional group*	*Ligand functional group*	*Product*
Agarose	—O—C≡N	$—NH_2$	CNBr-activated Sepharose 4B[a]
Agarose	N-hydroxysuccinimide ester	$—NH_2$	Affi-Gel 10, Affi-Gel 15[a]
Agarose	$—NH(CH_2)_5—COOH$	$—NH_2$	Activated CH-Sepharose 4B[a]
Agarose	$—NH(CH_2)_5—COOH$	$—NH_2$	CH-Sepharose 4B, ECH-Sepharose 4B[b]
Agarose	—COOH	$—NH_2$	Affi-Gel 202, CM Bio-Gel A[b]
Agarose	$—NH_2$	—COOH	Affi-Gel 102[b]
Agarose	$—NH(CH_2)_6—NH_2$	—COOH	AH-Sepharose 4B, EAH-Sepharose 4B[b]
Polyacrylamide	$—NH_2$	—COOH	Aminoethyl Bio-Gel P-2[b], Aminoethyl Bio-Gel P-150[b]
Agarose	—DEAE and Cibacron Blue F3GA	Binds all serum proteins, except IgG and transferrin	DEAE Affi-Gel Blue[c]
Agarose	—CM and Cibacron Blue F3GA	Binds albumin, serum proteinases and complement	CM Affi-Gel Blue[d]

[a] The ligand binds to the adsorbent spontaneously.
[b] The carbodiimide method is used for the ligand binding.
[c] Suitable for isolating IgG from serum in a single-step procedure; IgG contains no proteinases.
[d] Suitable for preparation of a globulin fraction depleted of albumin and proteinases.

molecules and must also include a reactive group able to mediate binding to the carrier.

Procedures used to bind a ligand to a carrier include:

1. physical adsorption (this is not very advantageous since it does not produce "true" immunoadsorbents);
2. polymerization of the ligand in the gel;
3. cross-linkage with glutaraldehyde or other compounds;
4. covalent binding.

In practice, covalent bonds and, occasionally, cross-linkage with glutaraldehyde are most commonly used.

8.2.1 Agarose immunoadsorbents

Agarose is a carrier widely used for the preparation of immunoadsorbents (*Table 8.5*). As it contains numerous reactive alcohol groups (agarose is a linear polymer of the disaccharide agarobiose), it must be activated before ligand binding. Cyanogen bromide (CNBr) is usually used for the activation:

$$\text{agarose}(\text{—OH})_n + n\,\text{CNBr} \rightarrow \text{agarose}(\text{—O—C}{\equiv}\text{N})_n + n\,\text{HBr}$$

Such an agarose derivative reacts readily with amines, and therefore also with a proteinaceous antigen or antibody, *via* a free terminal amino group or ε-amino group of lysine residue.

$$\text{agarose—O—C}{\equiv}\text{N} + H_2N\text{—R} \longrightarrow \text{agarose—O—}\underset{\underset{\text{NH}}{\|}}{\text{C}}\text{—NH—R}$$

Cyanogen bromide-activated agarose is available commercially (*e.g. CNBr-activated Sepharose 4B*). It can be used for preparation of an immunoadsorbent using the procedure shown in *Table 8.6*.

Table 8.6 Method for ligand binding to CNBr-activated Sepharose 4B (Pharmacia)

Step	*Procedure*
Weighing a required amount of CNBr-activated Sepharose	1 g of dry powder forms 3.5 ml of gel after swelling
Gel swelling	15 min in 1 mol/l HCl
Washing on a glass filter	Washing solution 1 mol/l HCl — about 200 ml per 1 g of dry Sepharose
Dissolving the ligand in buffer	0.1 mol/l $NaHCO_3$ buffer pH 8.3 containing 0.5 mol/l NaCl
Mixing the ligand solution with the gel suspension (mild stirring without a magnetic stirrer)	2 h at room temperature or overnight at 4 °C
Blockade of non-reacted carrier free hydroxyl groups	Transfer of gel to buffer with a blocking agent, *e.g.* 1 mol/l ethanolamine, pH 9.0 or Tris–HCl buffer, pH 8.0
Washing of non-reacted ligand and blocking amine	Fivefold washing with 0.1 mol/l acetate buffer, pH 4.0 containing 1 mol/l NaCl and borate buffer, pH 8.0, also containing 1 mol/l NaCl

Agarose immunoadsorbents are particularly useful for the isolation of antibodies against glycoprotein and protein antigens, hapten–protein conjugates and glycosaminoglycans *etc.*

In some cases (involving small molecules or antigens consisting of subunits or proteins with a high relative molecular weight) reactive groups of the carrier cannot provide a sufficiently firm linkage between it and such a ligand. Similarly, when large molecules are bound to a ligand, steric hindrance prevents contact of the compound to be bound with the ligand. This situation occurs particularly when the ligand is bound directly to the carrier (support) matrix. It is then necessary to increase the accessibility of the reactive group. A spacer arm usually formed by between two- and ten-carbon aliphatic chains inserted between the support and the ligand is used (*Fig. 8.10*). A compound which has reactive groups at both ends of its molecule is usually used, *e.g.* 1,6-diaminohexane or 6-aminocapronic acid. Agarose with bound 1,6-diaminohexane or agarose with bound 6-aminocapronic acid are also available commercially, *i.e.* AH-Sepharose 4B and CH-Sepharose 4B, respectively. *CH-Sepharose 4B* has free carboxyl groups and, in the presence of carbodi-imide, it binds ligands with free primary amino groups:

$$\text{agarose—NH(CH}_2)_5\text{—COOH} + \text{H}_2\text{N—ligand} \xrightarrow[\text{(carbodi-imide)}]{} \text{agarose—NH(CH}_2)_5\text{—CO—NH—ligand} + \text{H}_2\text{O}$$

AH-Sepharose 4B contains free —NH_2 groups and will therefore bind ligands carrying carboxyl groups:

$$\text{agarose—NH(CH}_2)_6\text{—NH}_2 + \text{HOOC—ligand} \xrightarrow[\text{(carbodi-imide)}]{} \text{agarose—NH(CH}_2)_6\text{—NH—CO—ligand} + \text{H}_2\text{O}$$

In these reactions it is advantageous to use a water-soluble carbodi-imide which can produce urea, *i.e.* also a water-soluble compound, during the reaction.

8.2.2 *Other immunoadsorbents*

Glutaraldehyde immunoadsorbents are not based on covalent antigen or antibody binding to the carrier matrix but rather on immobilization in the form of a polymer using glutaraldehyde as a cross-linking agent. In the presence of glutaraldehyde, proteins with free α-amino groups and ε-amino groups polymerize and become insoluble when the pH of the solution is near their isoelectric points. The immobilized proteins are even stable in the presence of urea or sodium dodecylsulphate and can be used as immunoadsorbents for the isolation of corresponding antigens or antibodies. For this purpose the glutaraldehyde gel polymer may be homogenized into particles for the batch procedure (fine homogenization) or used as a bed packed in a glass column (coarse homogenization — larger gel particles). When the amount of polymerized ligand is insufficient to pack a column, it can be mixed with an appropriate quantity of an inert small-pore gel (*e.g.* Sephadex G-25) that does not retain large protein molecules.

Adsorbents without bound specific antibodies but with another ligand can also be used to isolate immunoglobulins. Protein A is an example. **Protein A** has a relative molecular weight of 41 000 and is covalently bound in the cell wall layer of almost all plasma-coagulating staphylococci. It reacts specifically with the Fc portion, and sometimes also with the Fab part of the immunoglobulin molecule. One molecule of protein A can bind two molecules of IgG, giving rise to pseudoimmune complexes. In contrast to true immune complexes, however, the binding site of IgG is not involved in the binding. It was originally assumed that protein A reacts only with IgG of different animal species. It was later found that it can also react, albeit with a much lower affinity, with IgA and IgM. Differences among individual IgG subclasses were also detected. For example, for human IgG protein A reacts only with IgG1, IgG2 and IgG4.

The covalent binding of protein A to Sepharose yields **a pseudoimmunoadsorbent** suitable for affinity chromatography of immunoglobulins, *e.g.* *Protein A–Sepharose CL-4B* (Pharmacia). This adsorbent can be packed in a column and used for the separation of human serum components. Only IgG1, IgG2 and IgG4 are adsorbed, whereas other serum proteins are eluted by PBS. One ml of the gel binds approximately 20 mg of IgG. The bound IgG is eluted from the column with 0.1 mol/l glycine/HCl buffer, pH 2.4, or with 0.5 mol/l acetic acid. It is thus possible to obtain 95% of the total serum IgG. By combining affinity chromatography on Protein A–Sepharose CL-4B with ion-exchange chromatography on DEAE-cellulose the IgG3 subclass can be isolated from the human serum.

The procedure is also suitable for the isolation of the IgG subclasses from the serum of rabbits, mice, rats, guinea-pigs and other animals.

8.3 Preparation of therapeutic immunoglobulin preparations

Three generations of purified human IgG preparations have so far been prepared for clinical use (McCue *et al.*, 1946). The first was *immune serum gamma-globulin* prepared by alcohol precipitation by Cohn *et al.* (1946) at the end of World War II (*Fig. 8.2*). This was used as a 16% solution in 0.3 mol/l glycine buffer, pH 6.8, and contained 70–80% monomeric IgG. Intramuscular injection was safe and efficient in the prophylaxis of viral diseases, such as measles and infectious hepatitis. The first generation preparations could only be given intramuscularly. This was inappropriate for treating immunodeficient states (failing to achieve a high IgG level, even when using high doses) and the injections were often painful and induced undesirable local reactions. It was therefore necessary to produce an immunoglobulin preparation which could be administered intravenously.

Intravenous preparations (Barandun and Isliker, 1986) represent the second and third generation. The second generation preparations were prepared by removing IgG aggregates and other components responsible for spontaneous anti-complement activity and undesirable anaphylactoid reactions. This treatment consisted of two general procedures — enzymatic degradation and chemical modification. The enzymatic degradation involves removal of Fc-domains by pepsin or plasmin from 30 to 80% of the IgG molecules. The chemical modification involves treatment of the IgG molecule with β-propiolactone, resulting in hydrolysis of disulphide bonds by reduction and alkylation or, sometimes by sulphonation. The second generation preparations have a decreased anti-complement activity (the ability to activate complement spontaneously) but compared to native IgG, Fc-fragment function is also considerably suppressed.

These facts led to the development of the third generation of normal human immunoglobulin preparations which do not induce anaphylactoid reactions and whose molecules are not modified chemically. IgG aggregates are removed from these preparations by selective precipitation with polyethyleneglycol, adsorption on DEAE-Sephadex, treatment with trace amounts of pepsin at pH 4.0 and by isolation in the native form using a modified Cohn's fractionation procedure. The spontaneous formation of IgG aggregates is prevented by various stabilizers, *e.g.* hydroxyethylated starch, albumin, maltose, sucrose *etc.* The available IgG intravenous preparations are listed in

Table 8.7 Preparations of normal human immunoglobulin for intravenous use

Name	*Procedure*	*Manufacturer*
Endobulin	PEG precipitation, stabilization with hydroxyethylated starch	Immuno AG, Vienna, Austria
Gamimmune N (Polyglobin N)	Native at pH 4.25	Cutter Biological, Emoryville, California
Gammagard	PEG, adsorption on DEAE-Sephadex	Hyland, Glendale, California
Gamma-Venin	Pepsin degradation	Behring-Werke AG, Marburg, Germany
Gammonativ	Adsorption on DEAE-Sephadex, stabilization with albumin and saccharides	Kabi-Vitrum AB, Stockholm, Sweden
Globulin N	Precipitation with hydroxyethyl starch and PEG	Armour, Tarrytown, New York
Intraglobin F	β-propiolactone treatment	Biotest Pharma GmbH, Frankfurt, Germany
Iveegam	Immobilized trypsin treatment, PEG	Immuno AG, Vienna, Austria
Ivega	Ultrafiltration (Amicon)	Imuna, Šarišské Michaľany, Czechoslovakia
Sandoglobulin	Treatment with pepsin traces at pH 4	Sandoz AG, Basel, Switzerland
Veinoglobuline	Plasmin degradation	Mérieux, France
Venilon	Sulphonation	Teijin Institute, Japan
Venimmune	Sulphonation	Behring-Werke AG
Venoglobulin-I	PEG, DEAE-Sephadex adsorption	Alpha Therap. Cor., Los Angeles, California

Table 8.7. An example of the fractionation scheme used for *Gamimmune* preparations of all three generations is shown in *Fig. 8.11*.

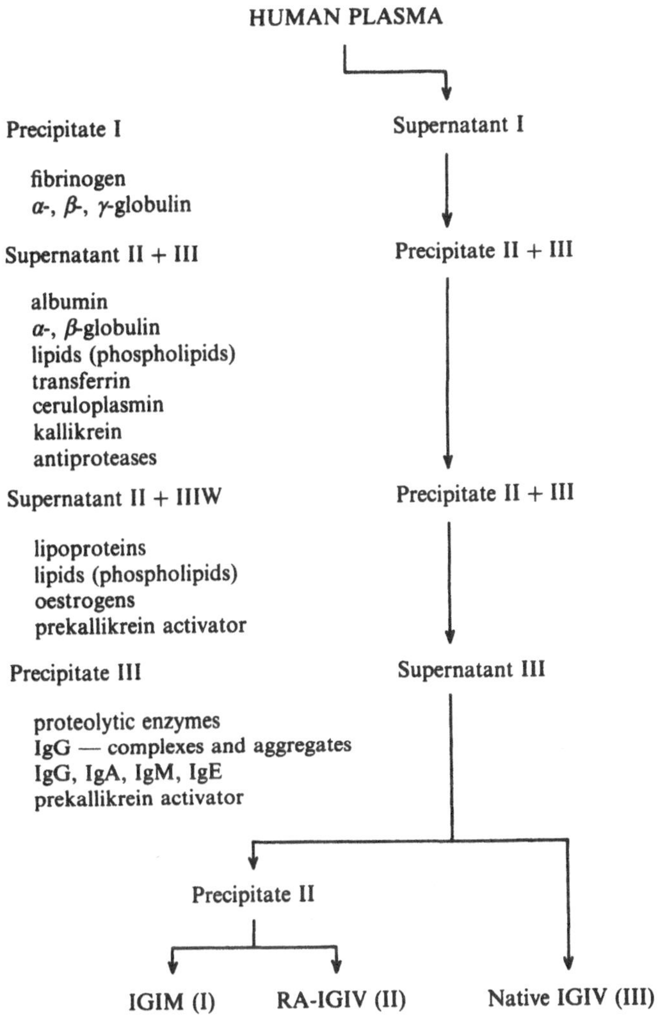

Fig. 8.11. Production of Gamimmune preparations based on Cohn's fractionation of human plasma (McCue *et al.*, 1986).
IGIM, classical the first generation gammaglobulin (for intramuscular administration only). RA-IGIV, reduced and alkylated human immunoglobulin of the second generation (for intravenous administration). Native IGIV, native normal human immunoglobulin of the third generation (for intravenous administration).

Intravenous preparations are administered as 3–5% solutions by slow infusion. High-quality second and third generation preparations are 96–100% IgG; of which at least 95% is in the monomeric form.

8.3.1 Clinical use of immunoglobulin preparations

Immunoglobulin preparations are used for three purposes: for prophylaxis (prevention of diseases), substitutive (replacement) therapy for patients with primary antibody deficiency and for the treatment of some diseases. For **prophylactic** purposes they are administered to individuals who have been, or could have been, subjected to a particular infection and have not yet developed appropriate immunity. For example, normal human immunoglobulins or immunoglobulin preparations containing specific antibodies against tetanus toxin, diphtheria toxin, hepatitis B virus (infectious hepatitis) or rabies virus *etc.* are injected intramuscularly or intravenously. Such hyperimmune immunoglobulins are obtained from serum of individuals who have successfully overcome the particular disease. Hyperimmune animal sera have also been used for prophylactic purposes (especially horse serum) and are isolated after active immunization of animals with a specific infectious agent or toxin. At present, prophylactic administration of hyperimmune immunoglobulin preparations is usually combined with active immunization.

Substitution therapy with immunoglobulin preparations is used in patients who do not produce antibodies because of an inborn or acquired deficiency. In addition to primary agammaglobulinaemia, hypogammaglobulinaemia and selective immunoglobulin deficiencies, it is also used in individuals with immunoglobulin deficiencies caused by serious infections, ionizing radiation, immunosuppressive or cytostatic treatment, following bone marrow transplantation, kidney transplantation and in immature newborns.

Substitution immunoglobulin preparations contain pure IgG and must comply with the safety and efficiency criteria recommended by the World Health Organization (WHO). Therefore, they are prepared from the plasma or serum of at least 1 000–2 000 healthy blood donors, ensuring that the composition of the resulting antibody preparation is comparable to that which occurs in the average human population.

Following the reports of transmission of infectious viruses with other blood products, in particular factor VIII, there has been some concern that this may also be the case with IgG preparations. The main worry concerns the retrovirus Human Immunodeficiency Virus (HIV) causing AIDS, but also covers hepatitis B and non-A non-B hepatitis. Therefore, all donors must be free from HBsAg (marker of hepatitis B virus) and antibodies against HIV, and the majority of manufacturers include virus inactivation steps. These include the use of β-propiolactone and ultraviolet light or the use of pepsin at pH 4, the latter initially adopted to reduce adverse reactions during intravenous administration (Newland, 1988).

Such preparations are called **normal human immunoglobulin** (instead of the previous term: *"gammaglobulin"*). The efficiency and safety of normal human immunoglobulin preparations were established by the Commission of WHO experts in 1982 and confirmed during congresses in Interlaken (Switzerland) and Chicago (Illinois) in 1985.

1. The structural and functional properties of the immunoglobulins in the preparations must be identical to those of the original serum immunoglobulins. Preparations for intravenous application (**IVIgG**) must be at least 90 % monomeric IgG, the amount of dimers and higher polymers must not exceed 5 % and the amount of fragments must not be higher than 10 %.
2. The preparation must contain the entire spectrum of IgG antibodies which occurs in the population from which the donors were selected. In particular, it must exhibit the ability to neutralize viruses and bacterial toxins (diphtheria and tetanus toxins), opsonize microorganisms and facilitate their phagocytosis.
3. The preparations should contain all four IgG subclasses in physiological ratios. However, this condition does not fully apply to intravenous preparations, as it was shown that the IgG3 subclass can also be involved in unfavourable anaphylactoid reactions (GRONSKI *et al.*, 1986) and its removal may therefore be considered.
4. The immunoglobulin preparations must not contain microorganisms and other harmful compounds.
5. They must retain their effector functions for which Fc-domains are responsible, including the activity to activate complement (p. 255) after immune complexes with specific antigens have been formed.
6. The immunoglobulins in the preparations must have an approximately normal biological half-life (applicable to IVIgG in particular).
7. Intravenous immunoglobulin preparations must not contain prekallikrein activator or other proteases which exhibit vasoactive effects and could therefore increase blood pressure.
8. Ideal IVIgG preparation should not contain IgA or IgG aggregates which cause hypersensitivity or anaphylactoid reactions respectively.

When a normal human immunoglobulin preparation does not conform to the above requirements, it may induce unfavourable side effects after intramuscular but particularly after intravenous administration. Their intensity also depends on the amount and rate at which the immunoglobulins enter the circulation of the recipient. The following side effects can be induced:

1. **Anaphylactoid reactions** occur almost exclusively after repeated administration of the immunoglobulin preparation in agammaglobulinaemic or hypogammaglobulinaemic patients. They are induced by IgG aggregates spontaneously (even in the absence of respective antigens) activating complement and stimulating secretion of various mediators from mast cells, granulocytes and other host cells. Anti-allotype antibodies synthesized after infusion of immunoglobulin preparations to hypogammaglobulinaemic patients and the production of immune complexes after repeated administration of immunoglobulins can have similar effects. It appears that the anaphylactoid reaction is directly mediated by the release of histamine.
2. **Hypersensitivity reactions** are induced by anti-immunoglobulin antibodies

of the recipient. They occur particularly after infusion of IgA-containing IVIgG preparations to patients with selective IgA deficiency and, therefore, manufacturers usually take care to remove this class from therapeutic products because about one per 800 persons has IgA deficiency (BERKMAN *et al.*, 1990). Anti-IgA antibodies of the recipient may be IgE (when they induce heavy hypersensitivity reactions), IgG, or occasionally IgM (when they induce anaphylactoid reactions).

3. Complications caused by undesirable elements in the preparations, such as pyrogens (compounds which induce fever), bacteria and their products (endotoxin) *etc.*
4. Further amplification of an ongoing inflammatory process as a reaction of the organism to the passively administered antibody.

Immunoglobulin preparations, extracted from human blood, were first used to treat immune deficiency diseases in 1952 (BRUTON). Now, it is clear that a wide spectrum of human diseases are associated with decreased or abnormal immunoglobulin levels. There are already substantial data indicating a useful role for IVIgG in persons with primary hypogammaglobulinaemia, neonates predisposed to group B streptococcal infections, selected persons with idiopathic thrombocytopenic purpura, children with Kawasaki disease, bone marrow and renal transplant recipients at risk for developing cytomegalovirus infection and pneumonia, persons with chronic lymphocytic leukaemia, and possibly persons with multiple myeloma. Areas for future study include AIDS, autoimmune disorders, and viral disorders other than cytomegalovirus. *Table 8.8* provides a list of diseases in which intravenous immunoglobulin may be useful.

Neonates have lower levels of immunoglobulins than adults. This is particularly true of premature infants in whom infections with group B streptococci are common (BAKER, 1977). The incidence and severity of streptococcal infections in premature infants and neonates is reduced by treatment with IVIgG alone or combined with antibiotics (BAKER, 1986; CHIRICO *et al.*, 1987). But, not all intravenous immunoglobulin preparations contain antibodies opsonic for group B streptococci, and antibody levels vary considerably.

Many viral infections, particularly the lymphocytotropic viruses, are accompanied by transient impairment of cell-mediated immunity. *Cytomegalovirus* (CMV) and the *Epstein Barr virus* (EBV) are the best known examples. EBV infection may also cause transient hypogammaglobulinaemia. Following infection with HIV there is damage to the *T* helper lymphocytes by the virus directly. IVIgG has been shown to have beneficial effects in diseases mediated by these viruses (NEWLAND, 1988).

There are three distinct risk periods following bone marrow transplantation: early mortality and morbidity are related to acute graft-versus-host disease, bacterial or fungal infection or a combination of these. The use of IVIgG is particularly applicable following the ablative therapy (immunosuppression) used in preparation for bone marrow transplantation and in its second risk period beginning after two months. At that time, there is the highest incidence of interstitial pneumonia and haemorrhagic gastroenteritis with CMV and rotavirus. Results with either normal pooled IgG or high titre CMV-specific IVIgG are encouraging both as prophylaxis and as treatment but are not totally satisfactory (ROGERS *et al.*, 1986).

Table 8.8 Diseases in which intravenous immunoglobulin preparations may be useful (Berkman *et al.*, 1990)

Probable efficacy
Primary immune deficiency
X-linked hypogammaglobulinaemia
Common variable immune deficiency
Severe combined immune deficiency
Secondary immune deficiency
Transplant recipients
Chronic lymphocytic leukaemia
Neonates at risk for infection
Paediatric AIDS
Other
Kawasaki disease
Idiopathic trombocytopenic purpura (acute)
Haemophilia; antibodies to factor VIII
Possible efficacy
Burns
Multiple myeloma
Autoimmune disorders
Juvenile rheumatoid arthritis
AIDS
Myasthenia gravis
Rheumatoid arthritis
Refractiveness to platelet transfusion

In humans, IgG is divided into four subclasses and they have unique biochemical, structural and metabolic features. Antibodies to specific antigens typically arise from distinct IgG subclasses. For example, antibodies to carbohydrate antigens are usually from subclass IgG2, whereas antibodies to tetanus and diphtheria toxoids and other protein antigens are usually from subclass IgG1 (YOUNT *et al.*, 1968). Also, the predominant IgG-subclass response to specific antigens changes with age. For example, antibodies to *Streptococcus pneumonia* switch from IgG1 in children to IgG2 in adults (FREIJD *et al.*, 1984).

Several reports have documented the association of hypogammaglobulinaemia and IgG-subclass deficiency with frequent infections. There are reported cases in which patients had normal total IgG level but markedly depressed levels of IgG2 or IgG4, or both. Association between specific IgG-subclass deficiencies and specific types of infections were recently reported. For example, recurent otitis media and chronic pulmonary infections have been associated with IgG2 deficiency (SMITH and BARN, 1986). Intravenous immunoglobulin has been used to prevent infections in persons with IgG-subclass deficiency (BERGER, 1987).

8.3.2 Pharmacology and mechanism of action

Peak serum immunoglobulin levels measured immediately after intravenous administration correlate with the dose. For example, a dose of 100 mg/kg body weight results in an average increment of about 2 g/l, whereas a dose of 500 mg/kg results in an average increment of about 10 g/l (BUCKLEY, 1982; MONTANARO and PIROFSKY, 1984). Serum immunoglobulin levels decrease rapidly after infusion; by 24 hours, levels are about 70–80% of peak levels and decrease to about 50% by 72 hours (PIROFSKY, 1984). Then there is a steady, exponential decrease until the baseline level is reached, usually by 21–28 days after infusion.

The half-life of most intravenous immunoglobulin preparations is similar to that of native IgG. There is, however, considerable individual variability, which reflects several factors, including the immunoglobulin level after infusion, the presence of infection or burns, the reliability in determining immunoglobulin levels, and others factors (BERKMAN *et al.*, 1990).

The proposed mechanisms of IVIgG action range from simple clearing of infective agents through blockade of the mononuclear phagocyte system to modulation or control of the immune response, either by a direct effect on the lymphocytes or *via* the mechanism of the idiotype/anti-idiotype response. Possible effects of IVIgG are summarized in *Table 8.9*.

Table 8.9 Possible mechanisms of IVIgG action (adapted according to Newland, 1988)

Enhanced elimination of circulating immune complexes formed with microbial antigens.
Competitive inhibition or steric hindrance of absorption of antibody or circulating immune complexes to the target antigen.
Non-specific blockade of Fc-receptor-mediated phagocyte function.
Increased activity of natural killer cells.
Immune modulation influencing both *T* and *B* cell numbers and function.
Suppression of antibody synthesis.
Neutralizing effect on the autoantibodies.

In immune deficiency, the major effect seems to be replacement of deficient immunoglobulins. In immune regulatory disorders, several mechanisms of action have been postulated, including blockade of mononuclear phagocytes *via* Fc receptors (JUNGI *et al.*, 1986), an increase in *T* suppressor cells with reversal of the helper/suppressor ratio (DELFRAISSY *et al.*, 1985) or natural killer cells (ENGELHARD *et al.*, 1986) and a decrease in antibody synthesis (TSUBAKIO *et al.*, 1983). The latter effect may be mediated by anti-idiotype antibodies in the immunoglobulin preparation. Common idiotypes are known to be cross-reactive intra-species and are likely to be well represented in pooled plasma. This has been clearly shown in patients with acquired antibodies to factor VIII (SULTAN *et al.*, 1984) and may be extrapolated to other conditions. IVIgG has a direct neutralizing effect *in vitro* and *in vivo* on the autoantibodies (ETZIONI and POLLACK, 1989) and may also directly influence underlying lymphocyte activity. It is likely that such multiple effects are responsible for the clinical response described in the many conditions now known to be influenced by such therapy (NEWLAND, 1988).

The mode of action of intravenous immunoglobulin in bacterial and viral infections is unknown but most likely involves antibacterial or antiviral antibodies in the preparation. It is also possible that infused antibodies block recognition of infected cells by cytotoxic *T* lymphocytes, thereby preventing immune-mediated cell damage (BERKMAN *et al.*, 1990).

References

Baker, C. J. (1977) Summary of the workshop on perinatal infections due to group B streptococcus. *J. Infect. Dis.*, **136**, 137–52.

Baker, C. J. (1986) Group B streptococcal infection in newborns: prevention at last? *New Engl. J. Med.*, 314, 1702–4.

Barandun, S. and Isliker, H. (1986) Development of immunoglobulin preparations for intravenous use. *Vox Sang.*, **51**, 157–60.

Berger, M. (1987) Immunoglobulin G subclass determination in diagnosis and management of antibody deficiency syndromes. *J. Pediatr.*, **110**, 325–8.

Berkman, S. A., Lee, M. L. and Gale, R. P. (1990) Clinical uses of intravenous immunoglobulins. *Ann. Intern. Med.*, **112**, 278–92.

Bruton, O. C. (1952) Agammaglobulinaemias. *Pediatrics*, **9**, 722–7.

Buckley, R. M. (1982) Long term use of intravenous immune globulin in patients with primary immunodeficiency disease: inadequacy of current dosage practices and approaches to the problem. *J. Clin. Immunol.*, **2**, 155–215.

Chirico, G., Rondini, G., Plenbani, A., Chiara, A., Massa, M. and Ugazio, A. R. (1987) Intravenous gammaglobulin therapy for prophylaxis of infection in high-risk neonates. *J. Pediatr.*, **110**, 437–42.

Cohn, E. J. (1945) Blood proteins and their therapeutic value. *Science*, **101**, 51–6.

Cohn, E. J., Strong, L. E., Hughes, W. L., Mulford, D. J., Ashworth, J. N., Melin, M. and Taylor, H. L. (1946) Preparation and properties of serum and plasma proteins. IV. A system for the separation into fractions of the protein and lipoprotein components of biological tissues and fluids. *J. Amer. Chem. Soc.*, **68**, 459–75.

Delfraissy, J. F., Tchernia, G., Laurian, Y., Wallon, C., Galanaud, P. and Dormont, J. (1985) Suppressor cell function after intravenous gammaglobulin treatment in adult chronic idiopathic thrombocytopenic purpura. *Brit. J. Haematol.*, **60**, 315–22.

Engelhard, D., Waner, J. L., Kapoor, N. and Good, R. A. (1986) Effect of intravenous immune globulin on natural killer cell activity: Possible association with autoimmune neutropenia and idiopathic thrombocytopenia. *J. Pediatr.*, **108**, 77–81.

Etzioni, A. and Pollack, S. (1989) High dose intravenous gammaglobulins in autoimmune disorders: mode of action and therapeutic uses. *Autoimmunity*, **3**, 307–15.

Ferenčík, M., Škárka, B. *et al.* (1981) *Biochemical laboratory methods*. Bratislava, Alfa, 856 pp. (*in Slovak*).

Freijd, A., Hammarstrom, L., Persson, M. A. and Smith, C. I. (1984) Plasma antipneumococcal antibody activity of the IgG class and subclasses in otitis prone children. *Clin. Exp. Immunol.*, **56**, 233–8.

Gronski, P., Kanzy, E. J., Ronneberger, H. J., Geursen, R. and Seiler, F. R. (1986) Quality criteria for i.v. immunoglobulins: importance of tests and product properties. *Behring. Ins. Mitt.*, **80**, 16–30.

Jungi, T. W., Eiholzer, J., Lerch, P. G. and Barandun, S. (1986) The capacity of various types of immunoglobulin for intravenous use to interact with Fc receptors of human monocytes and macrophages. *Blut*, **53**, 321–32.

McCue, J. P., Hein, R. H. and Tenold, R. (1986) Three generations of immunoglobulin G preparations for clinical use. *Rev. Infect. Dis.*, **8**, suppl. 4, S374–81.

Montanaro, A. and Pirofsky, B. (1984) Prolonged interval high-dose intravenous immunoglobulin in patients with primary immunodeficiency states. *Amer. J. Med.*, **76**, 67–72.

Morell, A. and Nydegger, U. E. (eds.) (1986) *Clinical use of intravenous immunoglobulins*. London, Acad. Press, 460 pp.

Newland, A. C. (1988) Clinical use of intravenous immunoglobulin in blood disorders. *Blood Rev.*, **2**, 157–67.

Pirofsky, B. (1984) Intravenous immune globulin therapy in hypogamma-globulinaemia. A review. *Amer. J. Med.*, **76**, 53–60.

Rogers, T. R., Riches, P. G., Walker, S. A. and Joshi, R. (1986) Changes in immunoglobulin levels and implications for immunoglobulin therapy to prevent infection following bone marrow transplantation. In: Morell, A. and Nydegger, U. E. (eds.), *Clinical Use of Intravenous Immunoglobulins*. London, Acad. Press, pp. 107–16.

Smith, T. F. and Barn, R. P. (1986). IgG subclasses in children with chronic chest symptoms. *Monogr. Allergy*, **20**, 119–27.

Sultan, Y., Kazatchkine, M. E., Maisonneuve, P. and Nydegger, U. E. (1984) Anti-idiotype suppression of auto-antibodies to factor VIII by high dose intravenous gammaglobulin. *Lancet*, **1**, 765–8.

Tiselius, A. and Kabat, E. A. (1939) An electrophoretic study of immune sera and purified antibody preparations. *J. Exp. Med.*, **69**, 119–31.

Tsubakio, T., Kurata, Y. and Katagiri, S. (1983) Alteration of *T* cell subsets and immunoglobulin synthesis *in vitro* during high dose gammaglobulin therapy in patients with ITP. *Clin. Exp. Immunol.*, **53**, 697–702.

WHO Expert Committee on Biological Standardization (1982) Report of an informal meeting on intravenous immunoglobulins (human), Geneva, *Bull. WHO*, **60**, 43–7.

Yount, W. J., Dorner, M. M., Kunkel, H. G. and Kabat, E. A. (1968) Studies on human antibodies. VI. Selective variations in subgroup composition and genetic markers. *J. Exp. Med.*, **127**, 633–46.

9 Endogeneous immunomodulators (immunohormones)

The immune system possesses self-regulatory factors, generated in the lymphoid tissue, that influence the development, differentiation and functional activity of individual cells of the immune system or even other tissues and organs. Some of these have a typical hormonal character (*e.g.* thymic hormones), others act mainly as local hormones (*e.g.* lymphokines and other cytokines). **Hormones** are secreted by *endocrine* mechanisms, and the target cells which receive their signal may be localized in any body compartment (sometimes far from the site of secretion). Secretion of **local hormones** is achieved by *paracrine* mechanisms and their effect is usually limited to surrounding cells. Although a proportion of local hormones also passes into the circulation, their concentration is higher in the microenvironment at the site of generation than in the circulation. Lymphoid cells are able to migrate through the organism and therefore their immunoregulatory products (lymphokines), despite being of local hormone character, are actually systemically active (*i.e.* at the level of the whole organism), which is highly effective and economical. With regard to their function, these substances were recently called "hormones of immunity". The immunohormones include hormones of the thymus and bursa of Fabricius, lymphokines, interleukins, *etc.* Chemically, these substances are peptides and glycoproteins with a short biological half-life. Such substances (either synthetic or isolated from biological material) may be administered to the organism as immunomodulators or modifiers of the immune response for treatment of various diseases.

9.1 Thymic hormones

The **thymus** is a primary endocrine organ of the immune system. The epithelial stromal cells secrete over 40 peptides that are responsible for thymic function. During foetal and neonatal development, they control the origin of the *T*-dependent immune cell system. During the life of an individual, they contribute to the maintenance of the fine balance among various *T*-cell subpopulations that are usually defined on the basis of their function (helper, suppressor, killer, memory, *etc.*). Thymic hormones regulate *T*-cell differentiation in all three lymphoid compartments — the bone marrow, the thymus and peripheral lymphoid system. The absence of the thymus during prenatal development causes a serious defect of cell-mediated immunity (the **Di George syndrome**, p. 182), whereas defects in thymus activity that occur after birth are usually associated with selective immunodeficiencies, variously manifest as malignant tumours, autoimmune diseases, infectious diseases *etc.*

Table 9.1 Chemical and biological properties of some thymus factors and hormones

Name	*Polypeptide structure*	*Biological effects*
Thymosin α_1 ($T\alpha_1$)	28 AR, M_r 3 108, pI 4.2	Enhances mitogenic response, production of MIF and TdT, stimulates antiviral, antifungal and antitumour immunity
Prothymosin α	112 AR, M_r 12 600, pI 3.55	The same as thymosin α_1 which constitutes its *N*-terminal part
Thymosin α_7 ($T\alpha_7$)	Acid polypeptide, M_r 2 200, pI 3.5	*In vitro* enhances activity of suppressor *T*-cells
Thymosin β_1 ($T\beta_1$)	74 AR, M_r 8541, pI 5.5	Isolated from different animal tissues and plants
Thymosin β_4 ($T\beta_4$)	43 AR, M_r 4 963, pI 5.1	Induces TdT formation and MIF production, stimulates release of LH-RH and LH, produced mainly by monocytes
Thymopoietin II	49 AR, M_r 5 562, pI 5.7	Induces *T*-cell differentiation, possesses immunonormalizing effect, influences neuromuscular transmission, enhances expression of complement receptors
Thymopentin (TP-5)	Arg-Lys-Asp-Val-Tyr, a sequence 32–36 from the thymopoietin molecule	Similar to thymopoietin II
Thymulin (FTS)	Glu-Ala-Lys-Ser-Gln-Gly-Gly-Ser-Asn, M_r 857, pI 7.5, contains zinc	*In vivo* and *in vitro* enhancement of T_S-cell formation, inhibition of antibody production against *T*-independent antigens
Thymus factor X (TFX)	M_r 4 200	*In vivo* increases *T*-lymphocyte number in the blood
Thymus humoral factor (THF)	30 AR, M_r 3 220, pI 5.7–5.9	Induces differentiation of young thymus cells, potentiates mitogenic response of peripheral lymphocytes
Thymostimulin (TS)	Polypeptide mixture	Induces specific *T*-lymphocyte functions in immunosuppressed animals and immunodeficient patients
Thymosin fraction 5	Mixture of thermostable polypeptides, M_r 1 000–15 000	Induces *T*-cell differentiation, stimulates immune functions in man and animals; stimulates release of ACTH, β-endorphins and glucocorticoids and enhances lymphokine production

AR, amino acid residues; TdT, terminal deoxynucleotidyl transferase; LH-RH, leuteinizing hormone-releasing hormone.

The best-defined thymus hormones and factors are listed in *Table 9.1*. The primary structure of most of them is well characterized (*Fig. 9.1*) and they have been prepared synthetically. The first fraction of active substances was prepared from calf thymus by GOLDSTEIN *et al.* (1982) and termed **"thymosin fraction 5" (F5)**. The F5 fraction could correct various immunodeficiencies due to functional thymus defects (*e.g.* after thymectomy) in experimental animal models as well as in patients with immunodeficiency diseases, malignant tumours and after immunosuppressive cytostatic therapy. Thymosin fraction 5 contains about 30 peptides, polypeptides and glycoproteins with molecular weights between 1 000 and 15 000. The individual components of this fraction were named after their isoelectric points (pI) which can easily be determined by isoelectric focusing. Those peptides having a pI lower than 5.0 were termed thymosins α (α_1, α_2 ... *etc.*), peptides with pI between 5.0 and 7.0 were called thymosins β (β_1, β_2 ... *etc.*) and peptides with pI above 7.0, thymosins γ. Significant homology in the primary structure was found among individual α- as well as β-thymosins. However, no homology could be found

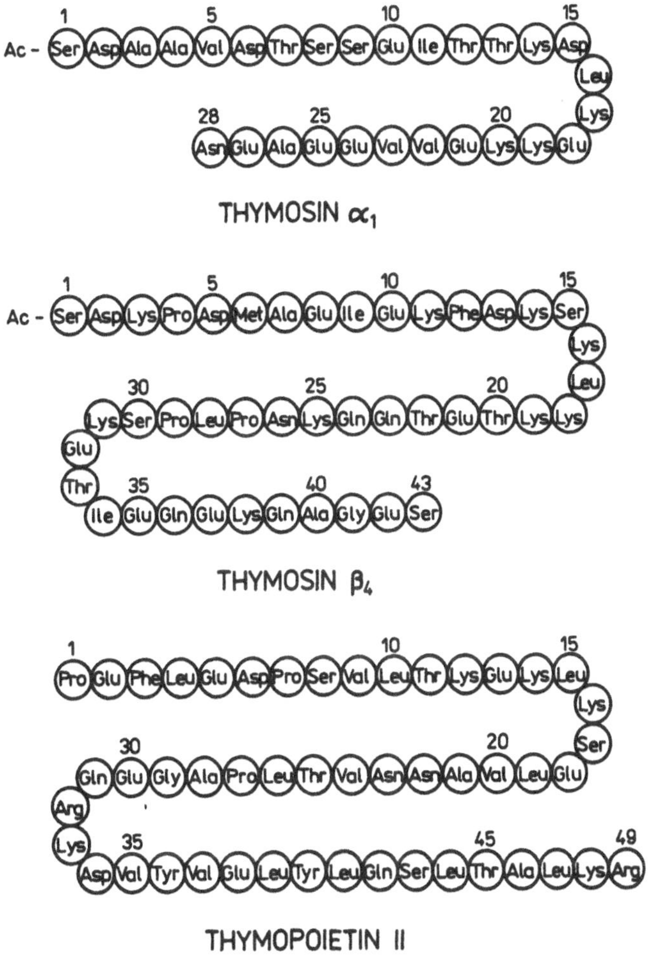

Fig. 9.1. Primary structure of some thymic hormones.

between α- and β-thymosins and thymopoietin. **Thymosin** α_1, generated from the larger precursor *prothymosin* α (*Table 9.1*), was also isolated from thymosin fraction 5. Among the β-thymosins, $T\beta_4$ has the best-characterized biological activity. In males, the normal serum levels of α_1-thymosin were found to be 670 ± 163 pg/ml and in females 652 ± 162 pg/ml, whereas the normal values of **β_4-thymosin** are 974 ± 400 ng/ml in males and 889 ± 345 ng/ml in females (NAYLOR *et al.*, 1986). No biological activity has yet been demonstrated in β_7 up to β_{11}-thymosins.

Thymopoietin has been prepared in three isomorphic variants, I, II and III, the primary structure of which differs in only three amino acid residues. GOLDSTEIN and AUDHYA (1985) observed that the effect of thymopoietin may be replaced by the thymopoietin pentapeptide (**thymopentin**) that represents the amino acid sequence between positions 32 and 36 in the thymopoietin molecule. Thymopoietin and thymopentin, besides inducing differentiation of early *T*-cells and immunoregulation, also inhibit neuromuscular transmission (and presumably participate in the pathogenesis of myasthenia gravis, p. 151).

Recently (AUDHYA *et al.*, 1986), the tripeptide **bursin** (lysyl-histidyl-glycylamide, originally called *bursopoietin*) was isolated from the chicken bursa of Fabricius; it selectively induces differentiation of chicken and mammalian early *B*-cells, but does not influence *T*-cell differentiation. Thymopoietin and bursin play different roles — the former stimulates *T*-cell development in the thymus, whereas the latter stimulates *B*-cell development in the bursa of Fabricius or its equivalent.

The immunoregulatory signal, following thymosin binding to corresponding *T*-cells, is transmitted into their nuclei through the second messengers: (1) thymosins activate guanylate cyclase and increase the intracellular cGMP level; (2) intracellular Ca^{2+} transport inside the cells is induced with subsequent binding to calmodulin; (3) arachidonic acid metabolism is modified (production of PGE_2 and some leukotrienes is stimulated). Presumably, maturation and activation of committed lymphocytes in the thymus or other compartments of the lymphoid system is intracellularly mediated largely by Ca^{2+}-dependent enhancement of relative cGMP concentration.

As thymic hormones modulate the number and function of *T*-lymphocytes, changes in their levels may be of diagnostic significance in the functional thymus defects. Thus, for example, patients suffering from AIDS or individuals belonging to AIDS high-risk groups, have elevated α_1-thymosin levels. In addition, enhanced α_1-thymosin values were also found in some *T*-cell leukaemias, tumours of brain, head and neck, and in patients with multiple sclerosis.

It seems that thymosins and thymus also participate directly in the ageing process (ZATZ and GOLDSTEIN, 1985). The thymus is one of the first glands to become atrophic before reaching adult age. Similarly, the α-thymosin level starts to decline significantly before the beginning of puberty. This may be a signal for initiation of processes leading to ageing of the immune system. On the other hand, thymopoietin serum levels of healthy individuals do not change significantly until the fourth decade of life; thereafter they gradually decline to undetectable levels after the 70th year of life.

The presence of α_1- and β_4-thymosins can also be detected in the central nervous system, and this may serve as further proof of mutual interconnection between the neuroendocrine and immune systems (p. 7). Thymosins may regulate production of the luteinizing hormone, adrenocorticotropic hormone, β-endorphins and glucocorticosteroids.

9.2 Cytokines

In the past, it was assumed that antigens were solely responsible for intracellular events leading to the immune response. Now it is clear that in its regulation not only antigens, antigen receptors, idiotypic network, helper and suppressor *T* cells are involved but antigen non-specific soluble glycoproteins or proteins take part as well. They are called cytokines, interleukins, lymphokines *etc.* and realize intercellular communication both in the immune system and between the immune and other systems.

Cytokines comprise a broad class of protein cell regulators variously termed monokines, lymphokines, and interleukins which are produced by different cells. They possess typical hormonal activities: (1) they are secreted by a single cell type, react specifically with other cell types (target cells) and regulate specific vital functions that are controlled by feedback mechanisms; (2) they generally act at short range in a paracrine or autocrine (rather than endocrine) manner, and (3) they interact first with high-affinity cell surface receptors (distinct for each type or even subtype) and then regulate the transcription of a number of cellular genes by little understood second signals. This altered transcription (which can be an enhancement or inhibition) results in changes in cell behaviour (Balkwill, 1988).

Lymphokines are cytokines secreted mainly by activated lymphocytes and the term **"monokines"** refers to analogous immunoregulatory glycoproteins produced by activated monocytes and macrophages. In order to unify the terminology of these factors, the term **"interleukins"** ("between leukocytes") was accepted at the Second International Lymphokine Workshop in 1979 (de Weck *et al.*, 1980) firstly for interleukin 1 (IL-1). This range of substances includes effector molecules that affect lymphocytes, macrophages, polymorphonuclear leukocytes, and other cell types. Besides the term expressing their origin, cytokines may also be named according to their function, as are interferons, growth and differentiation factors, colony-stimulating factors, *etc.*

Cytokines represent one broad class of agents interacting to regulate not only the immune system and inflammation, but also tissue repair, differentiation, and embryonic development. The majority of the cytokine activities described in the literature so far have been upregulatory (with the exception of cell growth inhibition and cytotoxicity) and they have *pleiotropic* nature.

The multiple, overlapping and sometimes contradictory functions of cytokines are dependent on the local concentration, the cell type they are acting on, and the other cytokines and regulatory influences to which the cell is exposed. The multiplicity of action of so many of these regulatory glycoproteins suggests that cytokines form a complex cellular signalling language with the individual proteins equivalent to characters in an alphabet code. Thus the information which an individual cytokine conveys depends on the pattern of regulators to which a cell is exposed, and not on one single cytokine. It means that all cytokines form the specific system or *network* of communication signals between cells of the immune system, and between the immune system and other organs.

The majority of cytokines are not synthesized in competent cells continuously (they are not constitutive products) and their biosynthesis is usually induced by antigens, mitogens or other cytokines. The **cytokine system** is a very potent force for good when activation of the network is local and transient, but when cytokine production is sustained and/or systemic there is no doubt that cytokines contribute to the signs, symptoms, and pathology of infections, autoimmune, and malignant disease. Under normal conditions there are usually low concentrations of cytokines in the blood but under pathological conditions, and during cytokine therapy, significant blood levels may be reached. However, the cytokines are usually eliminated rapidly, or rendered inactive, after entering the circulation.

The cytokine information network may be disturbed when an abnormality in the production of individual cytokines occurs or cytokine functions are inhibited by specific antibodies. This may lead to various diseases. Naturally occurring or therapeutically induced **antibodies to cytokines** are generally thought to inhibit cytokine activities, and the appearance of such antibodies should therefore result in various degrees of "cytokine deficiency". There is a large body of evidence that the presence of low circulating concentrations of specific cytokine autoantibodies is more common than was previously appreciated. These antibodies may not always impair the biological function of the cytokines. On the contrary, such antibodies may act as beneficial and specific carriers of cytokines in the circulation. Therefore, BENDTZEN *et al.* (1990) propose that a major role of naturally occurring autoantibodies to cytokines is to facilitate rather than neutralize functions of cytokines in the body.

9.2.1 Lymphokines and monokines

Lymphokines produced by activated lymphocytes may affect activities of macrophages, polymorphonuclear leukocytes and non-sensitized lymphocytes. Among them the most important is macrophage activation factor (MAF) or macrophage migration inhibitory factor (MMIF), macrophage chemotactic factor (MCF) and tumour necrosis factors (*Table 9.2*).

Macrophage activation factor induces various morphological, biochemical and functional changes in macrophages which enhance tumouricidal

Table 9.2 Properties of some lymphokines

Property	*MAF/MMIF*	*MCF*	*LMIF*	*TNF-α (cachectin)*	*TNF-β (lymphotoxin)*
Molecular weight	23 000–67 000	12 000–25 000	68 000	17 000	20 000 and 25 000
AR number in the chain	Various	Various		157	148 and 171
pI	3.2–6.0	5.6–10.1		3.9	5.8
Cell source	Activated *T* and *B*-lymphocytes	Activated *T* and *B*	Activated *T* and *B*	Macrophages, monocytes	Lymphocytes
Inducers of biosynthesis or secretion	Antigens, mitogens	Antigens, mitogens	Antigens, mitogens	Antigens, mitogens, LPS	Antigens, mitogens, proteases
Receptor determinant group	α-L-fucose			·	β-D-galactose
Biological functions	Inhibition of macrophage migration and functional activation	Stimulation of macrophage chemotaxis	Inhibition of granulocyte migration	Antitumour activity, participation in the development of endotoxin shock and cachexia	Cytostatic and cytolytic effect on various tumours

Abbreviations: MAF, macrophage activation factor; MMIF, macrophage migration inhibition factor; MCF, macrophage chemotactic factor; LMIF, leukocyte migration inhibition factor; TNF, tumour necrosis factor.

activity and intracellular killing of bacteria. Closer analysis has shown that MAF and MMIF belong to a single heterogeneous lymphokine group with molecular mass between 23 000 and 67 000, and are therefore designated MAF/MMIF. Enhanced tumouricidal macrophage activity is also induced by IFN-γ together with the **macrophage cytotoxicity-inducing factor** (MCIF2). Therefore, many workers consider IFN-γ to be a member of the MAF/MMIF group.

Antigen- or mitogen-activated lymphocytes also elaborate a chemotactic substance that selectively attracts macrophages or monocytes. This **macrophage chemotactic factor** (MCF), like MAF, is heterogeneous, as indicated by isoelectric focusing showing peak activities with pI of 10.1 and 5.6. MCF is heat stable at 56 °C with a molecular mass of 12 000 to 25 000 in humans. MCF induces accumulation of mononuclear phagocytes at the site of inflammation where MCF is released and MMIF prevents the accumulated macrophages from leaving this site.

Polymorphonuclear leukocytes are inhibited in their migration by a soluble material called **leukocyte migration inhibition factor** (LMIF or LIF), which is produced by sensitized lymphocytes. LMIF inhibits polymorphonuclear migration but not that of macrophages. It appears to be a protease with approximate molecular mass of 68 000.

Separately from LMIF, there are chemotactic factors for neutrophils, eosinophils and basophils produced by activated lymphocytes. They are similar in molecular mass, ranging from 24 000 to 55 000, and it is presently unclear whether these factors are all distinct molecular entities or the same substance with chemotactic activity for multiple cell types.

Sensitized lymphocytes also contain a substance that is released by either disruption of the cells or stimulation of them with a specific antigen. This material was first described by LAWRENCE (1955) and named by him **transfer factor** because in human beings it was possible to transfer delayed-type hypersensitivity (DTH) to previously unreactive recipients with extracts of these sensitized cells obtained from skin-test-positive donors. Transfer factor (TF) is defined as a dialysable material or family of materials that can be extracted from lymphoid cells of humans and certain animals. Its molecular mass is between 4 000 and 6 000 and in spite of its partial purification no one has produced a homogeneous preparation so far. TF is probably a polypeptide, possibly with two chains that are linked with disulphide bridges. The presence of nucleic acids, ribose, and phosphodiester groups, as has been proposed, is not ruled out (KIRKPATRICK, 1988).

There are two types of **tumour necrosis factors (TNF)**: α and β. TNF-α is also known as **cachectin** and TNF-β as **lymphotoxin**. Cachectin is mainly secreted by activated macrophages and monocytes whereas lymphotoxin is primarily produced by activated lymphocytes. Both possess pleiotropic activity (determined by a single gene) including cytostatic and cytotoxic effects on various tumour cells; on the other hand, normal cells are not usually influenced by these immunoregulators.

In humans, the genes for cachectin and lymphotoxin are present on the short arm of chromosome 6 and in the mouse on chromosome 17 in the region of the major histocompatibility complex (MHC). The primary transcript of human TNF-α consists of 2 762 base pairs and has three introns (De Forge *et al.*, 1990).

Cachectin is a cytotoxin that has been implicated in tumour regression, septic shock, and cachexia. Human TNF-α is a protein of molecular mass 17 000 (by SDS–PAGE) with a known sequence of 157 amino acid residues (*Fig. 9.2*). The mature mouse protein is 156 residues long. This secretory form of TNF-α is produced from a larger precursor (mol. wt. = 26 000) which contains an unusually long 76 amino acid signal peptide. Presumably, the

Fig. 9.2. Primary structure of human TNF-α (cachectin).

26 kD TNF/cachectin is a membrane-associated glycoprotein that can be clipped to generate the 17 kD secretory component. Both membranous and secretory forms of TNF-α may have cytotoxic activity. Human cachectin contains no potential glycosylation site, while the amino acid sequence of native murine cachectin contains one *N*-linked glycosylation site, an asparagine located at position 7 (PENNICA *et al.*, 1984; SHERRY and CERAMI, 1988; KRIEGLER *et al.*, 1988).

TNF acts through specific receptors which are expressed by all somatic cell types tested, with the exception of red blood cells. Receptors bind both TNF-α and TNF-β, but no other cytokines or growth factors. Recently (SPRANG, 1990), it has become evident that there are at least two molecular species of **TNF receptor** (TNFR), a high-affinity ($K_d = 0.07$ nmol/l) 75–80 kD myeloid cell-type receptor and 55–60 kD receptor of epithelial origin with an affinity content (K_d) of 0.3 nmol/l. The two molecules differ in glycosylation and amino acid sequence.

TNF/cachectin is an endogenous mediator of Gram-negative bacteria endotoxin (TRACEY *et al.*, 1988). The toxic effects of endotoxin are only manifested after cachectin release. Thus, endotoxin-resistant mouse strain C3H/HeJ do not produce cachectin. Glucocorticoids that protect experimental animals against endotoxin shock, if given before endotoxin administration, inhibit TNF-α synthesis. Cachectin is one of the cytotoxic factors produced by *NK*-cells and T_C-lymphocytes and participates in the acute phase of the inflammatory reaction. It may induce fever by exerting a direct effect on the hypothalamic hormones or by inducing IL-1 formation, which then acts as an endogeneous pyrogen. In addition, it activates granulocytes and enhances their adherence and phagocytic activity. It closely cooperates with other cytokines, especially IL-1 and IFN-γ. Cachectin presumably inhibits basic human metabolism, resulting in cachexia. **Cachexia** is a potentially lethal syndrome of unknown aetiology characterized by anorexia, weight loss and protein wasting that frequently complicates the treatment of chronic inflammation and cancer.

Lymphotoxin (LT) is one of the original lymphokines first discovered more than 20 years ago (GRANGER and WILLIAMS, 1968). Murine TNF-β has a 33 amino acid signal peptide and a mature protein of 169 amino acids. There is one potential glycosylation site at residue 60 and a cysteine residue at position 84. Human TNF-β has a 34 amino acid signal peptide, a 171 amino acid mature protein, an *N*-linked glycosylation site at residue 62 and no cysteine. LT has a molecular mass of 60 000–70 000 (determined by gel filtration) and 20 000–25 000 (determined by SDS–PAGE). The pI is 5.8. The 25 kD form is the monomeric glycosylated 171 amino acid molecule, which aggregates to produce the higher molecular weight form. The 20 kD form is missing the first 23 amino acids. Human TNF-α and TNF-β are 28% homologous and show conservation on the *C*-terminus (PAUL and RUDDLE, 1988).

The human and murine LT genes have been cloned and sequenced. Their structure consists of four exons and three introns. The genes for human and murine cachectin are also arranged similarly to LT and are also divided into four exons and three introns. The most extensive sequence homology between TNF-α and TNF-β is seen in the fourth exon (GARDNER *et al.*, 1987).

T-cells secrete LT after activation by antigen or *T*-cell mitogen; the same signals induce *T*-lymphocyte proliferation. The production of LT is induced in an antigen-specific MHC-restricted fashion from class I and class II restricted *T*-cells. Viral infection is also associated with TNF-β production by lymphoid cells.

Lymphotoxin has several effects on target cells including killing, growth stimulation and induction of differentiation. The mechanism of LT effects involves receptor binding and internalization, and several other sequelae including changes in prostaglandin production and chromosome integrity. LT probably plays several biological roles; it can contribute to immunoregulation, defence against viral and parasitic infections and to rejection of tumours.

9.2.2 *Interleukins*

The interleukins (IL) form a group of cytokines which can affect various cell functions thereby enabling communication between different cell types, especially between leukocytes. The interleukins can be released by lymphoid cells in response to antigen but in contrast to the chemical composition of anti-

Table 9.3 Human interleukins and their properties

Interleukin	*Molecular weight*	*AR number in the chain*	*Cell source*	*Target cells*	*Biological functions*
IL-1α IL-1β	17 500 17 000	159 153	MO, MA and others	*T*, *B*, MO, NK, PNM, and others	Mediator of host response to inflammatory, infectious or immune stimulus, increase the expression of IL-2 receptors
IL-2	15 000	133	Mainly T_H	*T*	Growth and differentiation factor of *T*-cells
IL-3	28 000	152	Mainly T_H	PHSC and others	Haematopoietic growth factor, multi-CSF
IL-4	15 000	129	T_H, MC	*B*, *T*	Stimulator of *B*-cell proliferation and differentiation, growth factor of *T* and MC
IL-5	15 000	134	T_{H2}	*B*, EO	Growth and differentiation factor of *B* and EO
IL-6	26 000	184	Fibroblasts, MA, *B*, *T* and others	Various	Growth and differentiation factor of the haematopoietic and immune systems
IL-7	17 400	152	Stromal cells	pre-*B*	Growth and differentiation factor of pre-*B*-cells
IL-8	8 400	72	MO, MA	NE	Activation and chemotactic factor of neutrophils

Abbreviations: AR, amino acid residues; PHSC, pluripotent haematopoietic stem cells; EO, eosinophils; MA, macrophages; MC, mast cells; MO, monocytes; NE, neutrophils.

bodies their chemical composition is not determined by that of the stimulating antigen. However, it is now clear that non-antigenic stimuli also induce synthesis of interleukins and that many non-lymphoid nucleated cells can be producers of interleukins and/or express functional interleukin receptors. At the present time, the interleukin family comprises eight (IL-1 — IL-8) well-characterized members (*Table 9.3*) and some additional members that are not so well characterized.

Interleukin 1 (IL-1) is represented by a family of polypeptides and glycoproteins with highly diverse activity. Originally, IL-1 was called LAF (lymphocyte-activating factor), LEM (leukocyte endogenous mediator) or EP (endogenous pyrogen). Under physiological conditions, IL-1 is synthesized and released primarily by mononuclear phagocytes; however, the capacity to secrete this cytokine is possessed by virtually all nucleated cells. Large amounts are liberated in almost all infections, inflammatory and immune processes. IL-1 is able to transmit an information signal, not only among leukocytes, but also to other cells and organs (*Table 9.4*).

Table 9.4 The main biological effects of IL-1 (arranged according to Malkovský *et al.*, 1988)

Cells that respond to IL-1	*Role*
T-lymphocytes[a]	Early *T*-lymphocyte development, chemotaxis, cytokine synthesis (IL-2, CSF, IFN-γ), enhancement of cytotoxic activity, IL-2 receptor expression
B-lymphocytes[a]	Chemotaxis, proliferation, enhancement of antibody production
NK-cells[a]	IL-2 receptor expression, enhancement of cytotoxic activity
Neutrophils[a]	Chemotaxis, degranulation, oxidative burst, facilitation of extravascular infiltration
Monocytes and macrophages[a]	Chemotaxis, prostaglandin synthesis, synthesis of TNF-α, generation of oxygen radicals, stimulation of cytotoxic activity
Endothelial cells[a]	Leukocyte adhesion, secretion of granulocyte-macrophage stimulating activity
Epithelial cells[a]	Proliferation, collagen synthesis
Fibroblasts[a]	Proliferation, collagen synthesis, prostaglandin synthesis
Hypothalamus	Prostaglandin synthesis, fever, somnolescence, anorexia
Pituitary gland	ACTH release
Hepatocytes	Increase of acute reactants synthesis (fibrinogen, CRP, serum amyloid, α_1-antitrypsin), decrease of albumin synthesis
Muscles	Proteolysis (amino acids release), prostaglandin synthesis
Osteoclasts	Bone resorption
Osteoblasts	Prostaglandin synthesis, proliferation
Synovial cells[a]	Proliferation, prostaglandin synthesis, secretion of collagenase and plasminogen activator
Tumour cells (some)[a]	Cytostasis, cytotoxicity, cellular differentiation
Tumour cells (some)[a]	Proliferation

[a] These cells not only respond, but also produce IL-1.

IL-1 obviously possesses highly diverse biological activities (MALKOVSKÝ *et al.*, 1988). However, closer examination reveals that all the activities are transmitted and coordinated with a single aim, *i.e.* protection of the host organism against infection and other harmful agents. Thus, for example, at the onset of a localized bacterial infection, blood monocytes and macrophages become activated during phagocytosis of invading microbes or their products. They secrete IL-1 which influences cells at the infection site and also enters the blood circulation, facilitating interaction with distant cells and tissues. It further stimulates prostaglandin synthesis in the hypothalamus which alters the "setting" of the thermoregulatory centre, thereby generating fever which, in turn, contributes to the host defence response. In addition, IL-1 influence further defence mechanisms, *e.g.* the increase in circulating immature neutrophils (neutrophilia), decrease of iron serum levels (iron is a growth factor for several bacteria), enhancement of acute phase protein production in the liver, increase in production of some hormones (insulin, glucagon, growth hormone, vasopressin), enhancement of amino acid release from the muscle tissue (required for lymphocyte proliferation, formation of acute phase proteins, immunoglobulins and collagen needed for the repair of damaged tissue). IL-1 also induces endothelial cells to secrete colony stimulating factors (CSFs) which induce proliferation of monocyte and granulocyte precursors and thus contribute to leukocytosis. CSFs are also secreted by other cells (including *T*-lymphocytes) in response to IL-1. Neutrophils respond with an oxidative burst and degranulation; activated macrophages generate oxygen radicals. It was found recently that IL-1 can induce a low--wave sleep which is also of functional importance. Diseased individuals need more sleep since the sleeping organism has a lower energy requirement and the energy saved can be utilized in defence and reparatory processes.

IL-1 is phylogenetically an ancient molecule; substances with similar activity have also been isolated in fish and amphibians.

The number of members of the IL-1 family remains unknown. On the basis of cDNA cloning it was found that human macrophages contain at least two different genes for IL-1; one encoding **IL-1α**, and the other **IL-1β** (MARCH *et al.*, 1985). Human IL-1α and IL-1β genes are present on chromosome 2 and are split into seven exons. Both IL-1α and IL-1β are synthesized in the form of precursor molecules, with a molecular mass of around 31 000 and containing 270 amino acids. The biological activity resides in the carboxy-terminal half of both precursor molecules. The precursors have no leader sequence and it is, therefore, unclear how IL-1 precursors are processed. The "normal" route of "exporting" proteins *via* endoplasmic reticulum, Golgi apparatus and vesicles is apparently not used. This explains why there is no glycosylation of IL-1 molecules at potential glycosylation sites. Both IL-1α and IL-1β molecules, secreted from cells, have a molecular mass of only 17 400. They differ, however, in pI (maximum activity is at pI 5.0 and 7.0 for IL-1α and IL-1β respectively) as well as in the primary structure.

The mature IL-1α protein consists of 159 amino acids, IL-1β is 153 residues long. Most of IL-1α remains cell-associated, while most of IL-1β is released (DINARELL ›, 1986).

IL-1α and IL-1β share the same receptor (IL-1R) on many cells. **IL-1R** is a transmembrane glycoprotein with a molecular mass of about 80 000. This is "type I" receptor. A second "*B*-cell" or "type II" receptor has been found recently (Durum *et al.*, 1991). This 60–65 kD protein has 28% homology to the type I receptor in the intracellular portion, and like the type I receptor has three Ig-like extracellular domains. The transfected IL-1RI-cDNA product devoid of the carbohydrate moiety is a 65 kD protein fully capable of binding either IL-1α or IL-1β, respectively. Analysis of the amino acid sequence shows that this protein is 576 residues long. From the amino-terminus there appears to be an extracellular part of 320 amino acids containing seven glycosylation sites and composed of three domains with homology to members of the immunoglobulin superfamily. The extracellular domains are followed by a highly hydrophobic transmembrane region, leading to a *C*-terminal cytoplasmic domain of approximately 220 residues. Resting peripheral blood *T*-lymphocytes express relatively few (about 40) receptors per cell. However, after stimulation, IL-1 receptors are rapidly expressed reaching a maximum of about 350 receptors per cell.

With respect to its multifactorial effects, one assumes that there are numerous "active sites" on the IL-1 molecule, which are each responsible for a specific activity. Thus, for example, the nonapeptide Val-Gln-Gly-Glu-Glu-Ser-Asn-Asp-Lys was isolated from the IL-1β molecule. This nonapeptide is located between positions 163 and 171 of the polypeptide precursor. It effectively stimulates *T*-lymphocyte proliferation, but does not possess pyrogenic activity (Antoni *et al.*, 1986).

To date, no specific disease states have been attributed to either a specific or non-specific deficiency in IL-1 production or response. It is of interest

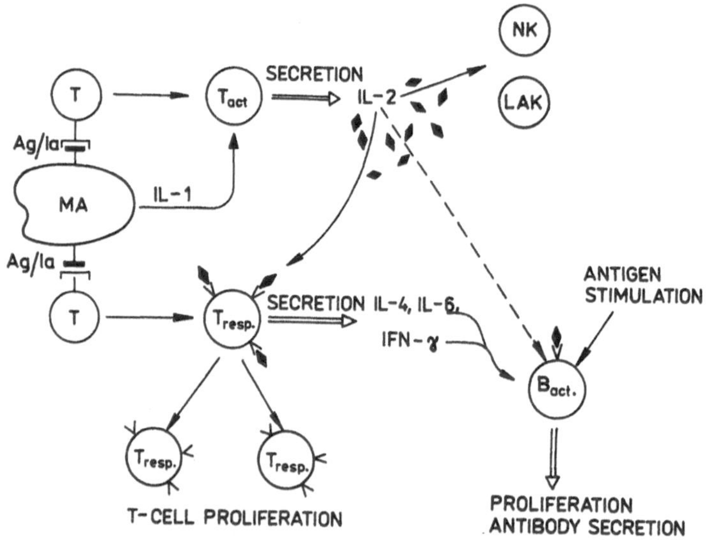

Fig. 9.3. Basic immunoregulatory function of IL-2 (Robb, 1984).
Ag/Ia, antigen—Ia-antigen complex; T_{act}, activated *T*-cell; T_{resp}, *T*-cell responding to IL-2 signal (possesses IL-2 receptors); MA, macrophage or other accessory cell; B_{act}, activated *B*-cell; NK, natural killer cell; LAK, lymphocyte-activated killer cell; IL-4, *B*-cell growth factor (BCGF); IL-6, *B*-cell differentiation factor (BCDF).

that the urine of febrile patients and pregnant women contains **IL-1 inhibitors**; however, their pathophysiological significance remains to be elucidated.

The expression of the IL-2 gene is a pivotal event in *T*-cell activation. **Interleukin 2** is produced by activated *T*-cells in response to two signals, provided by accessory cells. The first signal is delivered by antigen together with class II MHC antigens while the second signal is mediated by IL-1. Even though all the *T*- lymphocyte populations may release IL-2 under appropriate conditions, it seems that the main source are *T*-helper cells. IL-2 was originally called the *T-cell growth factor* (TCGF). In addition, IL-2 may act as a differentiation signal (*Fig. 9.3*).

The structure of human, simian and rat IL-2 is similar. It comprises a single polypeptide chain of molecular mass 15 000–17 000 (by SDS–PAGE) or 19 000–22 000 (according to gel filtration chromatography). Human IL-2 contains 133 amino acid residues and one intramolecular disulphide bond (*Fig. 9.4*). About 50% of the molecule has the structure of an α-helix. In human IL-2, microheterogeneity of molecules accounts for the variability of the *N*-terminal octapeptide, as well as the variability in the number and proportion of monosaccharide residues in the oligosaccharide that bind by *O*-glycosidic bond to threonine residue in position 3. The disulphide bond between the cysteine residues in positions 58 and 105 plays an important role in the biological activity of IL-2 (ROBB, 1984; SMITH, 1988).

The gene for human IL-2 is localized on chromosome 4q and is composed of four exons, separated by three introns.

Similarly to other growth factors, IL-2 also supports proliferation of *T*-cells through binding to their specific surface receptors. The receptor (**IL-2R**) consists of a large (75 kD) protein chain and a smaller (55 kD) chain, each of which bind separately to the IL-2 molecule (SMITH, 1989). The interaction of IL-2 with specific human *T*-cell receptors is characterized by high affinity (dissociation constant $K_d = 4 \times 10^{-12}$ mol/l). Resting *T*-cells have relatively few receptors for IL-2 binding on their surface. Their number increases following antigenic or mitogenic stimulation. One activated *T*-lymphocyte possesses 4 000–12 000 receptors.

It is supposed that there are only three parameters important for regulating *T*-cell proliferation after antigen activation: the concentration of IL-2, the density of IL-2 receptors on the cell surface and the duration of the IL-2–IL-2R interaction.

However, IL-2 is not only a growth factor of activated *T*-cells, but also possesses other activities. For example, it may induce secretion of IFN-γ and *B*-cell growth factor (BCGF-I, also called IL-4) by *T*-cells, increases cytotoxic *T*-lymphocyte activity, monocyte cytotoxicity and increases activity of *NK*-cells. IL-2 has no direct cytotoxic or cytostatic effect on most neoplastic cells. Its anti-tumour effect is mainly derived from its ability to stimulate the cytotoxic activity of *LAK*-cells, which mediate destruction of a broad range of neoplastic and transformed tissues as well as of some normal tissues.

The ***LAK*-cells** (lymphokine-activated *K*-cells) require at least three signals to trigger their cytotoxic activity, the first being antigen or mitogen, while the second and third signals provide lymhokines (IL-2 and CCDF — cytotoxic cell differentiation factor).

Fig. 9.4. Primary structure of human IL-2.
At the side marked-CHO the oligosaccharide chain is joined.

There are several immune reactions where control of the IL-2 related response could have a significant clinical effect. They include the immune responses to infectious organisms, to normal tissues in autoimmune diseases, to allogeneic transplanted tissues as well as to neoplastic tissues. Indeed, defective IL-2 production has been reported in patients with severe combined immunodeficiency (SCID) or Nezelof's syndrome, AIDS, type I diabetes mellitus, systemic lupus erythematosus and could be one possible mechanism for the abnormal immune function in patients after bone marrow transplantation. Disorders in IL-2 receptor expression have been described in patients with AIDS, hypogammaglobulinaemia, multiple sclerosis and adult *T*-cell leukaemia (MALKOVSKÝ *et al.*, 1988).

Interleukin 3 (IL-3) is a growth factor, previously known by various names (most frequently as haematopoietic growth factor or *multi-CSF* — "colony-stimulating factor"). Its molecule consists of 152 amino acid residues and is encoded by a gene composed of five exons and four introns. Target cells for IL-3 are pluripotent stem cells, megakaryocytes, erythrocytes, neutrophils, eosinophils, basophils, mastocytes and macrophages that receive the first inevitable mitogenic signal from IL-3. IL-3 also participates in *T*-lymphocyte differentiation. The main physiological source of IL-3 are the $T_{H/I}$-lymphocytes.

Interleukin 4 (IL-4), previously called "*B*-cell stimulating factor–1" (BSF-1) or "*B*-cell growing factor-1" (BCGF-1), has been shown to be a product of activated *T*-cells and mast cells and to influence the growth and differentiation of a wide spectrum of haematopoietic cell lineages. IL-4 possesses multiple biological activities. It acts on resting murine *B*-cells to increase the expression of the class II MHC molecules and CD23, the low-affinity receptor for IgE. The activity of IL-4 is not restricted to *B*-cells, since it can induce the proliferation of mast cells and activated *T*-cells and also macrophages for increased tumouricidal activity and expression of Ia antigens. Similar activities have also been reported in the human. Human IL-4 readily induces maturation of *B*-cell progenitor into mature *B*-cells.

On the basis of cDNA sequence, human IL-4 is a 14 991 dalton protein, although it has not yet been purified from non-recombinant sources. It has a single binding site for specific receptor (IL-4R). Human *B*-cells have on their surface about 1 100 IL-4Rs per cell and their molecular weight is 124 000 (SOLARI *et al.*, 1989).

Interleukin 5 (IL-5), originally termed *T*-cell-replacing factor (TRF) or *B*-cell growth factor II (BCGF II), acts as a growth and differentiation factor of eosinophils and also regulates *B*-cell activation and differentiation, particularly the switch of synthesis of membrane IgM to serum IgM, or synthesis of secretory IgM to another immunoglobulin class. IL-5 is a "late acting" factor whose receptors are not present on resting cells. It enhances the IL-2R expression on *T*-cells (SIDERAS *et al.*, 1988). The human IL-5 gene consists of four exons and three introns, and encodes the protein with 134 amino acid residues (TANABE *et al.*, 1987).

Interleukin 6 (IL-6), originally known as interferon β_2 or *B*-cell stimulating factor-2 (BSF-2), is produced by *T*-lymphocytes upon mitogen stimulation and by many other cell types, including fibroblasts, epithelial cells, monocytes and macrophages. It acts in the late stages of *B*-cell differentiation (*Table 9.5*), leading to the biosynthesis of a secretory type of immunoglobulin; as a *T*-cell activation factor as well as a cytolytic *T*-cell differentiation factor. IL-6 displays some anti-viral activity, stimulates hepatic production of acute phase proteins, increases body temperature, activates differentiation of germinal haematopoietic cells into effector cells and is a potent growth factor for myeloma cells.

Table 9.5 Interleukin signals in *B*-cell development and differentiation (Malkovský *et al.*, 1988)

B-cell maturation stages	*Induced by*
Activation	IL-1, IL-4, IL-5
Proliferation	IL-2, IL-4, IL-5
Differentiation	IL-2, IL-5, IL-6

The human IL-6 gene is located on the short arm of chromosome 7 p15–p21, and the murine IL-6 gene maps to the proximal region of chromosome 5. Both human and mouse IL-6 genes consist of five exons and four introns.

The 1.3 kb mRNA obtained by cDNA cloning is translated into a 212 amino acid precursor protein with a molecular mass of 26 000. After removal of a 28 amino acid signal peptide, the resulting 184 amino acid and 20 781 kD protein, containing two possible *N*-glycosylation sites, is glycosylated and subsequently secreted. The fibroblast and monocyte IL-6 is phosphorylated at several serine residues. Murine IL-6 contains 211 amino acids, including signal peptide of 24 residues. No *N*-glycosylation, but several *O*-glycosylation sites were identified in this sequence (Le and Vilček, 1989; Heinrich *et al.*, 1990).

The *IL-6 receptor* cDNA encoded a protein consisting of 468 amino acids including a signal peptide of about 19 amino acids and a domain of 90 amino acids that resembles a domain of the immunoglobulin superfamily. The cytoplasmic part of 82 amino acids lacks a tyrosine kinase domain, unlike other growth factor receptors.

Interleukin 7 (IL-7) is a cytokine capable of supporting the growth of both pre-*B*-cells and pro-*B*-cells *in vitro* in the absence of any stromal elements. It means that IL-7 is involved in the development of *B*-cells together with IL-1, IL-3, IL-4, IFN-γ and transforming growth factor type β (TGF-β). Nucleotide sequence analysis indicated that cDNA for IL-7 was capable of encoding a protein of 177 amino acids with a signal sequence of 25 amino acid residues and a calculated mass of 17.4 kD for the mature protein (Goodwin *et al.*, 1989).

Two alternative spliced transcripts of *IL-7 receptor* are made by human cells: one gives a membrane form and the other, which lacks a transmembrane portion, yields a soluble form that could, like the soluble tumour necrosis factor receptors, act as a soluble inhibitor for the ligand. The high affinity IL-7R is expressed on lymphoid and myeloid cells. IL-7R belongs to the **haematopoietin receptor superfamily**. This family, characterized by cysteine repeats in the extracellular domain, includes several other cytokine receptors, such as the IL-2Rβ, IL-3R, IL-5R, IL-6R, IL-7R and GM-CSFR.

Interleukin 8 (IL-8), termed also neutrophil attractant/activation protein-1 (NAP-1) or monocyte-derived neutrophil chemotactic factor (MDNCF), is an 8 400 D protein that is a chemoattractant and granule release stimulus for neutrophils. IL-8 belongs to a family of host defence small proteins, which have a degree of sequence and structural similarity. They are known as the *"small cytokine (scy) superfamily"*.

IL-8 is a typical monokine which has no significant homology with other cytokines produced by mononuclear phagocytes. It is generated as a 99 amino acid precursor with a characteristic signal peptide of 20 amino acids. IL-8 is secreted as a 79 residue protein. The commonly isolated 72 and 77 residue forms are probably extracellular cleavage products (BAGGIOLINI *et al.*, 1989; LEONARD and YOSHIMURA, 1990a).

The gene for IL-8 maps to chromosome 4 and consists of four exons and three introns (MUKAIDA *et al.*, 1989).

The biological profile of activity of IL-8 is very similar to that of the classical chemotactic peptides C5a and f-Met-Leu-Phe. It is able to induce the full pattern of responses observed in chemotactically stimulated neutrophils, *i.e.* activation of the motile apparatus and directional migration, expression of surface adhesion molecules, release of storage enzymes, and production of reactive oxygen metabolites. In contrast to many chemoattractants, IL-8 does not attract monocytes. IL-8 acts *via* a selective *receptor* with a molecular mass of approximately 58 000. Steady-state binding experiments indicate K_d values of 4 and 0.5 nmol/l and receptor number of 75 000 per a single human neutrophil (GROB *et al.*, 1990). NAP-1/IL-8 is not species-specific, and its effects *in vivo* could thus be studied in several laboratory animals.

Interleukin 9 (IL-9), in combination with erythropoietin, selectively supports the proliferation of erythroid progenitors and has no measurable effects on other haematopoietic cell lineages. Originally IL-9 was isolated through its ability to stimulate a human IL-3-dependent leukaemic cell line and named P40 *T*-cell growth factor. Thus, IL-9 is another cytokine with the potential to serve as a regulator in both the lymphoid and erythroid systems (DONAHUE *et al.*, 1990).

Interleukin 10 (IL-10) stimulates *T*-cells, but inhibits the antigen-presenting function of macrophages (including the inhibition of IL-1, IL-6 and TNF), however antigen-presenting function of *B*-cells is not inhibited by IL-10.

Interleukin 11 (IL-11) acts synergistically with IL-3 in stimulating megakaryocyte colony formation and is proposed to be an important inducer of megakaryocytopoiesis. IL-11 is the 24 kD protein lacking disulphide bridges or *N*-linked glycosylation and is produced by bone marrow stromal cell lines (DURUM *et al.*, 1991).

9.2.3 The interferon system and transforming growth factor β

In 1957, ISAACS and LINDEMANN observed that chicken cells, incubated with heat-inactivated influenza virus, produced a factor that could protect against infection with other viruses. They called this phenomenon *"interference"* and the active substance *"interferon"*. Interferon prevented virus replication, not only in the cells producing interferon, but also in surrounding cells into which interferon penetrated by diffusion. Interferon has an extraordinary activity; in cell culture, 10 pg/ml of interferon (about 5×10^{-13} mol/l) produces a 50 % inhibition of virus replication. Subsequently, it was shown that interferon formation may also be induced by other substances, and that interferon is not a homogeneous substance, but a heterogeneous population of molecules possessing various bioregulatory activities. Thus, the antiviral activity is only one of numerous activities. Borecký (1983) considers interferon to be a complex system that regulates various animal (and probably even plant cell) functions by activating other effector systems and mechanisms.

The interferon system consists of three interferon types: interferon α (IFN-α), IFN-β and IFN-γ. According to the old nomenclature, IFN-α and IFN-β belong to type I (interferon induced by viral or bacterial stimulation) and IFN-γ was considered type II (immune interferon). **IFN-α** is a main type produced by leukocytes and is induced by foreign cells, virus-infected cells, tumour cells and bacteria. **IFN-β** is formed primarily by fibroblasts; induced by nucleic acids of viral and other origin. **IFN-γ** is a product of activated lymphocytes responding to specific antigenic or mitogenic stimulation and probably of *NK*-cells. It is of interest that the same *T*-lymphocyte clone may produce both IFN-γ and IFN-α in response to virus infection. This means that the regulation of gene transcription for IFN-γ is independent from that of IFN-α. IFN-γ production by activated *T*-cells (helper, suppressor, cytotoxic) during the response to specific antigen, may possess all characteristics of the primary and secondary immune response. The production of IFN-γ is mediated by other cytokines released by cells of the immune system. For example, IL-1 and IL-2 enhance IFN-γ release.

Hybridization experiments and sequence analysis of cloned cDNA revealed that IFN-α was coded by a gene family composed of at least 13 non-allelic and eight allelic members. Their products are designated IFN-α1, IFN-α2, IFN-α3 *etc.* To date, about 16 products have been characterized. For IFN-β and IFN-γ, only one gene has so far been found. Genes for IFN-α and IFN-β are present on human chromosome 9, do not contain any introns and encode polypeptide chains containing 165 or 166 amino acid residues (lacking asparagine that can bind an oligosaccharide through an *N*-glycosidic bond). The gene for IFN-γ is located on the long arm of human chromosome 12, contains three introns and a DNA repeating segment. It encodes a polypeptide chain with 146 amino acid residues, out of which only 127 to 134 residues are translated into the mature protein (*Fig. 9.5*). IFN-γ is composed of two molecular forms, having molecular weights of 20 000 and 25 000. However, these are products of a single gene and differences in molecular weight are generated only after post-translation processing of the primary protein (mol. wt. = 17 147). Molecules of molecular weight 20 000 have one glycosylation site (Asn in position 28) and therefore contain one oligosaccharide unit, whereas 25 kD molecules possess two oligosaccharide chains (the glycosylation sites are Asn residues in positions 28 and 100). The number of monosaccharide residues in each chain is variable, and does not influence the biological activity. IFN-γ maintains its biological activity even after removal of all saccharide residues from the molecule (IJZERMANS and MARQUET, 1989). IFN-γ molecules are highly hydrophilic and possess no disulphide bridges, whereas IFN-α and IFN-β subtype molecules are hydrophobic and usually contain two disulphide bonds, one of them being essential for antiviral activity.

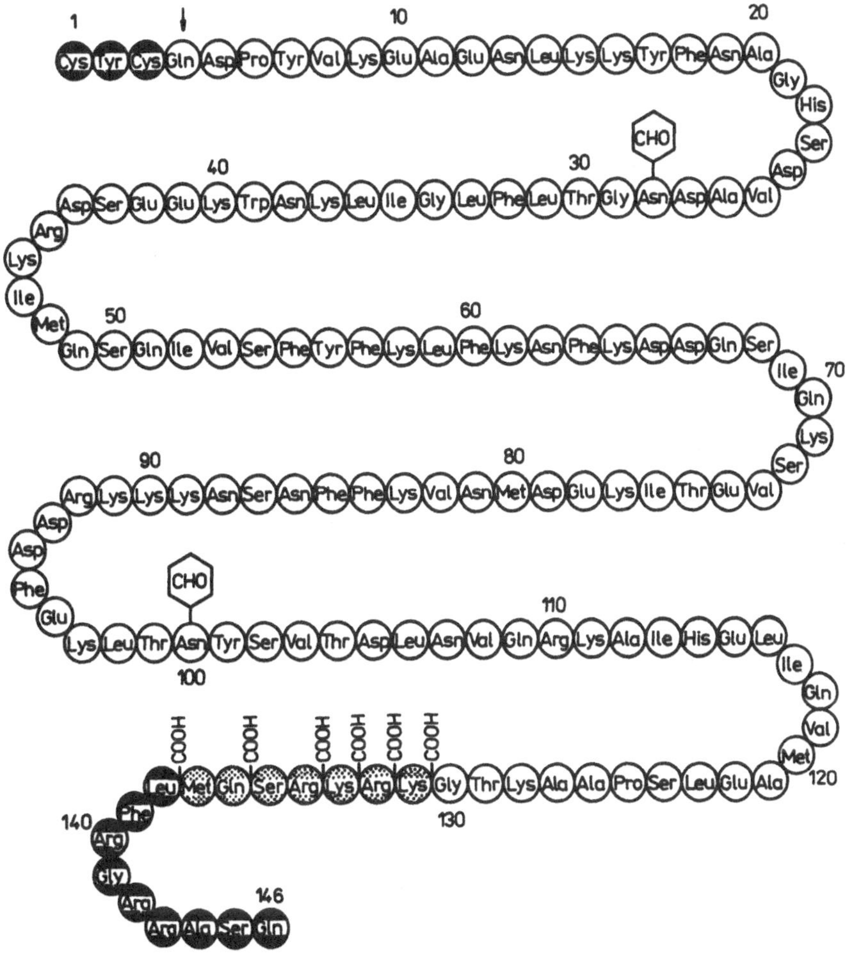

Fig. 9.5. Primary structure of human IFN-γ.
Amino acids in black circles are encoded by cDNA but are absent in the mature IFN-γ molecule. *N*-terminal Gln residue is in the form of pyroglutamate. Amino acid residues in cross-hatched circles may be present in some IFN-γ molecules, but are absent in others. COOH-groups mark the possible *C*-terminal amino acid residues of these IFN-γ variants. The —CHO groups represent sites of glycosylation where the oligosaccharide units are bound.

Interferons act by binding to receptors in the cytoplasmic membrane of corresponding cells. Receptors for IFN-α and IFN-β are different from the receptors for IFN-γ.

The basic biological properties of interferons are listed in *Table 9.6*. Antiviral and antiproliferative effects are predominant in IFN-α and IFN-β

subtypes, whereas IFN-γ primarily participates in numerous immunoregulatory effects. Even if cells are only exposed to interferons for several minutes, the resultant antiviral state usually persists for several days. Maintenance of the antiproliferative effect requires continual exposure of cells to interferon. Interferons inhibit the growth of normal and tumour-transformed cells. It seems, however, that inhibition of transformed cells may be realized to a larger extent. Intensive clinical trials are therefore in progress with the aim of using interferons for treatment of some infections and tumours. These require the successful production of recombinant interferons in order to yield enough of this substance. The most promising approach appears to be the combination of IFN-γ with TNF-α and TNF-β when the antitumour effect of IFN-γ is synergistically potentiated.

Table 9.6 Principal biological effects of interferons

Inhibition	*Stimulation*
Virus replication	Further interferon production
Cell proliferation	Macrophage activity
Tumour growth	*NK*-activity
Cell differentiation	T_C-lymphocyte cytotoxicity
Motility of cultivated cells	MHC markers expression on lyphoid cells
Origin of delayed-type hypersensitivity	Fc-receptor expression
Suppressor *T*-cells	Normal phenotype restoration of transformed cells
Antibody formation	Cell adhesiveness
	PGE_2 production
	Excitability of cultivated neurons

Among the immunoregulatory effects of interferons are the further inhibition of T_S-cells, inhibition of specific and non-specific antibody synthesis and stimulation of various cells of the immune system, *e.g.* macrophages, T_C-cells and *NK*-cells. Interferons also enhance expression of MHC antigens on many cells, and expression of high-affinity Fc-receptors for monomeric IgG (FcR1) on monocytes and macrophages.

The **transforming growth factors** were initially described as products of virally transformed cells that induce the transformed phenotype in non-neoplastic cells. Subsequently, transforming growth factor α (TGF-α) and TGF-β were identified in normal and neoplastic tissue. **TGF-β** is found in human platelets and monocytes, in bone and kidney and in other tissues. It is a 25 kD protein consisting of two identical 12.5 kD subunit chains joined covalently by disulphide bonds. TGF-β exists in several isoforms referred to as TGF-β_1, TGF-β_2 and TGF-$\beta_{1,2}$ (the latter being a heterodimeric protein consisting of one chain of TGF-β_1 and one chain of TGF-β_2). All forms of TGF-β exhibit a variety of biological effects that are consistent with their proposed role in tumourigenesis, embryonic development and wound healing (Sporn *et al.*, 1987; Sporn and Roberts, 1988). In addition, TGF-β is involved in the regulation of immune responses. For example, TGF-β_1 plays a role in autocrine growth control of activated *T*-lymphocytes, members of the TGF-β family are potent inhibitors of acquisition of immunoglobulin light chains during pre-*B*-cell maturation, TGF-β_1 can inhibit IL-2-induced human lymphokine-activated killer (LAK) cell activity *in vitro*. Many effects of TGF-β could be primarily due to its antiproliferative action.

In contrast to the other cytokines, there is an additional level of control of the TGF-β action, possibly reflecting its potency and importance. The majority of TGF-β secreted is biologically inactive and fails to bind to cell surface receptors. This latent form can be activated *in vitro* by transient acidification or alkali, but *in vivo* it is probably activated by exogeneous proteases. Moreover, as with *T*-cell activation, the TGF-β message can be present at high levels in some cells, but not secreted until a further stage of activation or differentiation is reached. Alveolar macrophages express TGF-β mRNA, but only release the protein upon activation.

Thus, TGF-β secreted at an inflammatory or immune site may be beneficial in diminishing lymphocyte function while promoting fibrosis and tissue repair. However, TGF-β is produced by various cancers, particularly glioblastomas, and therefore may contribute to immunosuppression seen in some malignancies.

There may be other cytokines that exert an inhibitory/limiting influence on immune reactions. Among them two other obvious candidates would be IFN-α and IFN-β (BALKWILL, 1988).

9.2.4 Small cytokine superfamily

The small cytokine superfamily (*scy family*) is represented by a large number of newly discovered proteins expressed by a variety of activated and transformed cell types with a molecular mass usually not higher than 10 kD. The range of known biological activities displayed by scy family members is extensive, including cell-specific chemotaxis and activation, regulation of cell growth and differentiation, modulation of immune responses, tissue remodelling and wound healing. These small cytokines can be separated into two discrete subgroups based on whether the first two of the four position-invariant cysteine residues common among the various primary sequences of their molecules are adjacent (C—C subgroup) or separated by an intervening amino acid (C—X—C subgroup) (SHERRY and CERAMI, 1991).

To the first subgroup belong monocyte chemoattractant protein-1 (MCP-1), macrophage inflammatory protein-1 (MIP-1) and others. Among the members of C—X—C subgroup are IL-8/NAP-1, MCP-2, MIP-2, platelet factor 4 (PF4), platelet basic protein *etc.*

Human **monocyte chemoattractant protein-1** (MCP-1) is a relatively basic protein migrating on SDS–PAGE in two molecular forms, as 15 kD MCP-1α and 13 kD MCP-1β. However, the molecular mass of MCP-1 based on amino acid sequence and cloning data is only 8.7 kD. Thus the SDS–PAGE results are anomalous. The cDNA open reading frame codes for a 99 residue protein. The last 76 residues correspond to MCP-1. Hydrophobicity of the first 23 residues is typical of a signal peptide, which is consistent with the fact that MCP-1 is a secreted protein. There is a single sequence (Asn-Phe-Thr) for *N*-linked glycosylation targeting amino acid 38 (LEONARD and YOSHIMURA, 1990b).

MCP-1 can be produced by leukocytes of both lymphocyte and monocyte lineages. It is a chemoattractant for human monocytes with the optimal agonist concentration of 10^{-9} mol/l and may play a role in cellular immune reactions, especially in the accumulation of monocytes over a period of 24–48 hours after interaction of antigen and sensitized lymphocytes. In addition to its role in cellular immune reactions, MCP-1 may also be important in host responses to acute tissue injury. PDGF, released by platelets in response to the loss of vascular integrity characteristic of most tissue injury, causes fibroblasts to secrete MCP-1. Monocytes attracted to the site may participate in host defence against microorganisms and in tissue remodelling.

In sharp contrast to the chemotactic and activating properties of MCP-1 and most of the *scy* family members, which may well enhance normal wound healing, another *scy* peptide, **platelet factor 4,** may interfere with tissue repair. PF-4 inhibits angiogenesis, the formation of new blood vessels by proliferation and migration of vascular endothelial cells, which is essential to normal wound repair (MAIONE *et al.,* 1990). It suggests that interplay among *scy* family members with individual profiles of activity may help to control inflammatory processes more precisely.

Macrophage inflammatory protein-1 can cause a localized inflammatory reaction characterized by a rapid influx of polymorphonuclear cells (PMNs) when injected subcutaneously. It has the ability to activate human PMNs *in vitro*, and can act as an endogenous pyrogen, which works in a prostaglandin-independent manner, and synergize with some haematopoietic factors to enhance granulocyte-macrophage colony formation.

Electrophoresis on SDS-PAGE revealed two forms, MIP-1α and MIP-1β. The cDNA for MIP-1α predicts a mature peptide of 69 amino acids in length with a molecular mass of 7889 daltons and no apparent sites for *N*-glycosylation. The cDNA for MIP-1β predicts a mature peptide also of 69 amino acids in length, with a molecular mass of 7832 daltons. However, in this case there is one potential *N*-glycosylation site (Asn-Pro-Ser) at position 53. The two cDNA are 57% identical and the predicted polypeptide sequences are 60% identical over their entire length (including the signal sequences). It is not known at the present time whether MIP-1α and MIP-1β represent two chains of a single protein or are distinct products that coaggregate during purification (WOLPE and CERAMI, 1989).

Both MIP-1 and MIP-2 are produced by activated macrophages, *T*-cells and fibroblasts. They affect neutrophils, monocytes, haematopoietic cells, fibroblasts and melanoma cells.

In contrast to MIP-1, **MIP-2** is a basic protein. MIP-2 is an extremely potent chemotactic agent for human PMNs, being more active on a molar basis than fMet-Leu-Phe. It does not induce an oxidative burst in parallel assays in which MIP-1 is active but it does cause selective degranulation of specific granules in PMNs with the release of lysozyme.

Besides the above-mentioned activity, **PF-4** is chemotactic for neutrophils, monocytes and fibroblasts. It is a 7 800 dalton protein with pI 7.6. PF-4 is released from platelet α-granules as a complex with a proteoglycan carrier weighing 350 kD. The purified protein exists as a tetramer at physiological ionic strength and has a high affinity for heparin.

β-Thromboglobulin is also contained in platelet α-granules and is initially synthesized as a 15 kD precursor called *platelet basic protein* (PBP). The cDNA encodes for a protein with predicted mol. wt. of 12 378 daltons, the message for which is increased in macrophages, endothelial cells, keratinocytes, and fibroblasts treated with IFN-γ (WOLPE and CERAMI, 1989).

9.2.5 Colony-stimulating factors

The colony-stimulating factors (CSFs) are a family of glycoprotein cytokines that regulate the proliferation and maturation of haematopoietic progenitor cells. In addition to their effects on haematopoiesis, CSFs modulate the function of fully mature cells and therefore play an important role in regulating inflammatory responses vital to host defence (WEISBART and GOLDE, 1989).

There are four CSFs directly affecting granulocyte-macrophage production: G-CSF, GM-CSF, M-CSF, and multi-CSF (IL-3). Additional cytokines affect production of other haematopoietic progenitor cells (*Fig. 9.6*).

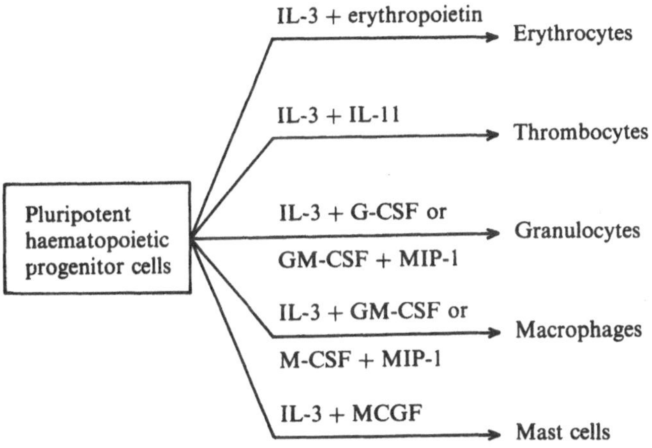

Fig. 9.6. Cytokines mainly involved in the differentiation of haematopoietic progenitor cells. MCGF, mast cell growth factor; MIP-1, macrophage inflammatory protein-1.

Granulocyte colony-stimulating factor (G-CSF) produced by endothelial cells and macrophages selectively stimulates mainly the production of neutrophils, and mature neutrophils retain receptor for G-CSF. **Granulocyte-macro-**

phage colony-stimulating factor (GM-CSF) is produced by activated vascular endothelial cells, fibroblasts and macrophages, as well as by *T*-lymphocytes responding to immune stimuli. GM-CSF stimulates the production and maturation of neutrophils, eosinophils, and macrophages which retain GM-CSF receptors when fully mature. **Macrophage colony-stimulating factor (M-CSF** or CSF-1) is elaborated by endothelial cells, fibroblasts, macrophages, and placental tissue. M-CSF is selective for macrophage differentiation, and mature cells retain M-CSF receptors. IL-3 or **multi-CSF** is produced only by activated *T*-lymphocytes and stimulates the production of neutrophils, eosinophils, basophils, monocytes, and platelets. Although mature monocytes have IL-3 receptors and respond to IL-3, mature neutrophils do not express these receptors and are therefore not responsive to IL-3.

In addition to haematopoietic activity, CSFs have multiple effects on phagocytic cells and modulate each phase of their inflammatory response, including adherence, chemotaxis, immobilization, phagocytosis, oxidative metabolism, and ADCC.

The gene for human G-CSF is composed of five exons and four introns and encodes a precursor protein with 207 amino acids (29 of them representing signal peptide).

The gene for human GM-CSF consists of four exons and is capable of encoding a protein with 144 amino acids including 17 residues of signal peptide. Molecules of both G-CSF and GM-CSF contain five glycosylation sites.

Human M-CSF is encoded by a single gene that spans about 21 kilobases on the long arm of chromosome 5 at band q33.1. The gene contains ten exons of which only the first eight include polypeptide coding sequences. Complementary DNAs encode different forms of biologically active M-CSF which are homodimeric glycoproteins generated by proteolytic cleavage of larger membrane-bound precursors. The largest human monomeric precursor consists of 554 amino acids. Alternative splicing of coding sequences within exon 6 is responsible for formation of at least two other biologically active M-CSF precursors. The smallest precursor is composed of 256 amino acids and the third precursor of 438 amino acids. Each of them have an amino-terminal signal peptide of 32 residues (Rettenmier and Sherr, 1989).

References

Antoni, G., Presentini, R., Perin, F., Tagliabue, A., Ghiara, P., Censini, S., Volpini, G., Villa, L. and Boraschi, B. (1986) A short synthetic peptide fragment of human interleukin 1 with immunostimulatory but not inflammatory activity. *J. Immunol.*, **137**, 3201–4.

Audhya, T., Kroon, D., Heavner, G., Viamontes, G. and Goldstein, G. (1986) Tripeptide structure of bursin, a selective *B*-cell-differentiating hormone of the bursa of Fabricius. *Science*, **231**, 997–9.

Baggiolini, M., Walz, A. and Kunkel, S. L. (1989) Neutrophil-activating peptide-1/interleukin 8, a novel cytokine that activates neutrophils. *J. Clin. Invest.*, **84**, 1045–9.

Balkwill, F. (1988) Cytokines — soluble factors in immune responses. *Curr. Opin. Immunol.*, **1**, 241–9.

Bendtzen, K., Svenson, M., Jønsson, V. and Hippe, E. (1990) Autoantibodies to cytokines — friends or foes? *Immunol. Today*, **11**, 167–9.

Borecký, L. (1983) *Viruses, immunity and interferon*. Martin, Osveta, 476 pp. (*in Slovak*).

DeForge, L. E., Nguyen, D. T., Kunkel, S. L. and Remick, D. G. (1990) Regulation of the pathophysiology of tumor necrosis factor. *J. Lab. Clin. Med.*, **116**, 429–38.

de Weck, A., Kristensen, F. and Landy, M. (eds:) (1980) *Biochemical characterization of lymphokines*. Proceedings of the Second International Lymphokine Workshop. New York, Academic.

Dinarello, C. A. (1986) Interleukin-1: amino acid sequences, multiple biological activities and comparison with tumor necrosis factor (cachectin). In: Cruse, J. M. and Lewis, R. E., Jr. (eds.), *The Year in Immunology 1985–86*. Basel, S. Karger AG, pp. 68–89.

Donahue, R. E., Yang, Y. C. and Clark, S. C. (1990) Human P40 *T*-cell growth factor (interleukin 9) supports erythroid colony formation. *Blood*, **75**, 2271–5.

Durum, S. K., Quinn, D. G. and Muegge, K. (1991) New cytokines and receptors make their debut in San Antonio. *Immunol. Today*, **12**, 54–7.

Gardner, S. M., Mook, B. A., Hilgers, J., Huppi, K. E. and Roeder, W. D. (1987) Mouse lymphotoxin and tumor necrosis factor: Structural analysis of the cloned genes, physical linkage and chromosomal position. *J. Immunol.*, **139**, 476–83.

Goldstein, A. L., Low, T. L. K., Thurman, G. B., Zatz, M. M., Hall, N. R., McClure, J. E., Hu, S. and Schulof, R. S. (1982) Thymosins and other hormone-like factors in the thymus gland. In: Mihick, E. (ed.), *Immunological Approaches to Cancer Therapeutics*. New York, John Wiley, pp. 137–169.

Goldstein, G. and Audhya, T. K. (1985) Thymopoietin to thymopentin: experimental studies. *Surv. Immunol. Res.*, **4**, suppl. 1, 1–10.

Goodwin, R. G., Lupton, S., Schmierer, A., Hjerrild, K. J., Jerzy, R., Clevenger, W., Gillis, S., Cosman, D. and Namen, A. E. (1989) Human interleukin 7: Molecular cloning and growth factor activity on human and murine *B*-lineage cells. *Proc. Natl. Acad. Sci. USA*, **86**, 302–6.

Granger, G. A. and Williams, T. W. (1968) Lymphocyte cytotoxicity *in vitro*: Activation and release of a cytotoxic factor. *Nature*, **218**, 1253–4.

Grob, P. M., David, E., Warren, T. C., De Leon, R. P., Farine, P. R. and Homon, C. A. (1990) Characterization of a receptor for human monocyte-derived neutrophil chemotactic factor/interleukin 8. *J. Biol. Chem.*, **265**, 8311–16.

Heinrich, P. C., Castell, J. V. and Andus, T. (1990) Interleukin-6 and the acute phase response. *Biochem. J.*, **265**, 621–36.

Ijzermans, J. N. M. and Marquet, R. L. (1989) Interferon-gamma: A review. *Immunobiol.*, **179**, 456–79.

Isaacs, A. and Lindemann, J. (1957) Virus interference. The interferon. *Proc. R. Soc.* B147, **1**, 258–68.

Kirkpatrick, C. H. (1988) Transfer factor. *J. Allergy Clin. Immunol.*, **81**, 803–13.

Kriegler, M., Perez, C., DeFay, K., Albert, I. and Lu, S. D. (1988) A novel form of TNF/cachectin is a cell surface cytotoxic transmembrane protein; ramification for the complex physiology of TNF. *Cell*, **53**, 45–53.

Lawrence, H. S. (1955) The transfer in humans of delayed skin sensitivity to streptococcal M substance and to tuberculin with disrupted leukocytes. *J. Clin. Invest.*, **34**, 219–32.

Le, J. and Vilček, J. (1989) Interleukin 6: A multifunctional cytokine regulating immune reactions and the acute phase protein response. *Lab. Invest.*, **61**, 588–602.

Leonard, E. J. and Yoshimura, T. (1990a) Neutrophil attractant/activation protein-1 (NAP-1 [Interleukin 8]). *Amer. J. Resp. Cell Mol. Biol.*, **2**, 479–86.

Leonard, E. J. and Yoshimura, T. (1990b) Human monocyte chemoattractant protein-1 (MCP-1). *Immunol. Today*, **11**, 97–101.

Maione, T. E., Grey, G. S., Petro, J., Hunt, A. J., Donner, A. L., Bauer, S. I., Carson, H. F. and Sharpe, R. J. (1990) Inhibition of angiogenesis by recombinant human platelet factor 4 and related peptides. *Science*, **247**, 77–9.

Malkovský, M., Sondel, P. M., Strober, W. and Dalgleish, A. G. (1988) The interleukins in acquired diseases. *Clin. Exp. Immunol.*, **74**, 151–61.

March, C. J., Mosley, B., Larsen, A., Cerretti, D. P., Breadt, G., Price, V., Gillis, S., Henney, C. S, Kronheim, S. R., Grabstein, K., Coulon, P. J., Hopp, T. P. and Cosman, D. (1985) Cloning, sequence and expression of two distinct human interleukin-1 complementary DNAs. *Nature*, **315**, 641–3.
Mukaida, N., Shiroo, M. and Matsushima, K. (1989) Genomic structure of the human monocyte-derived neutrophil chemotactic factor IL-8. *J. Immunol.*, **143**, 1366–71.
Naylor, P. H., Friedman-Kien, A., Hersh, E., Erdos, M. and Goldstein, A. L. (1986) Thymosin α_1 and thymosin β_4 in serum: comparison of normal, cord, homosexual and AIDS serum. *Int. J. Immunopharm.*, **8**, 667–76.
Paul, N. L. and Ruddle, N. H. (1988) Lymphotoxin. *Ann. Rev. Immunol.*, **6**, 407–38.
Pennica, D., Nedwin, G. E., Hayflick, J. S., Seeburg, P. H., Derynck, R., Palladino, M. A., Kohr, W. J., Aggarwal, B. B. and Goeddel, D. V. (1984) Human tumour necrosis factor: precursor structure, expression and homology to lymphotoxin. *Nature*, **312**, 724–29.
Rettenmier, C. W. and Sherr, C. J. (1989) The mononuclear phagocyte colony-stimulating factor (CSF-1, M-CSF). *Hematol./Oncol. clin. N. Amer.*, **3**, 479–93.
Robb, R. J. (1984) Interleukin 2: the molecule and its function. *Immunol. Today*, **5**, 203–9.
Sherry, B. and Cerami, A. (1988) Cachectin/tumor necrosis factor exerts endocrine, paracrine, and autocrine control in inflammatory responses. *J. Cell Biol.*, **107**, 1269–77.
Sherry, B. and Cerami, A. (1991) Small cytokine superfamily. *Curr. Opin. Immunol.*, **3**, 56–60.
Sideras, P., Noma, T. and Honjo, T. (1988) Structure and function of interleukin 4 and 5. *Immunol. Rev.*, **102**, 189–218.
Smith, K. A. (1988) Interleukin 2: Inception, impact, and implications. *Science*, **240**, 1169–76.
Smith, K. A. (1989) The interleukin 2 receptor. *Annu. Rev. Cell Biol.*, **5**, 397–425.
Solari, R., Quint, D., Obray, H., McNamee, A., Bolton, E., Hissey, P., Champion, B., Zanders, E., Chaplin, A., Coomber, B., Watson, M., Roberts, B. and Weir, M. (1989) Purification and characterization of recombinant human interleukin 4. *Biochem. J.*, **262**, 897–908.
Sporn, M. B. and Roberts, A. B. (1988) Peptide growth factors are multifunctional. *Nature*, **332**, 217–9.
Sporn, M. B., Roberts, A. B., Wakefield, L. M. and De Crombrugghe, B. (1987) Some recent advances in the chemistry and biology of transforming growth factor-beta. *J. Cell Biol.*, **105**, 1039–45.
Sprang, S. R. (1990) The divergent receptors for TNF. *Trends Biochem. Sci.*, **15**, 336–8.
Tanabe, T., Konishi, M., Mizuta, T., Noma, T. and Honjo, T. (1987) Molecular cloning and structure of the human interleukin-5 gene. *J. Biol. Chem.*, **262**, 16580–4.
Tracey, K. J., Lowry, S. F. and Cerami, A. (1988) Cachectin: A hormone that triggers acute shock and chronic cachexia. *J. Infect. Dis.*, **157**, 413–20.
Weisbart, R. H. and Golde, D. W. (1989) Physiology of granulocyte and macrophage colony-stimulating factors in host defense. *Hematol./Oncol. Clin. N. Amer.*, **3**, 401–9.
Wolpe, S. D. and Cerami, A. (1989) Macrophage inflammatory proteins 1 and 2: members of a novel superfamily of cytokines. *FASEB J.*, **3**, 2565–73.
Zatz, M. M. and Goldstein, A. L. (1985) Thymosins, lymphokines, and the immunology of aging. *Gerontology*, **31**, 263–77.

10 The complement system

BORDET, when working in PASTEUR's laboratory in Paris in 1896, observed that serum of animals immunized with bacteria (*Vibrio cholerae*), could agglutinate and lyse those bacteria *in vitro* (either in a test tube or on a slide). If, however, the serum was left at room temperature for one week, or briefly heated at 60 °C, the agglutinating activity remained unchanged but the ability to lyse bacteria was lost. The lytic activity could be restored by adding fresh serum from a non-immunized animal to the bacteria-heated serum mixture. On the basis of this experiment, BORDET assumed that lysis of bacteria required the presence of two factors; specific, agglutinating antibodies (discovered by BEHRING in 1889) and a non-specific component that was activated secondarily and was responsible for bacterial lysis and killing. BORDET called the latter component *"alexin"*. Soon after this discovery, particularly when further data were obtained by EHRLICH, the non-specific serum component was renamed **"complement"**.

It soon became obvious that complement was not a single substance but a complex system, containing about 30 various glycoproteins. These are present in the blood serum in an inactive form and are activated by immune complexes (the *"classical" pathway*) or by other substances, mainly of bacterial origin (the *"alternative" pathway*). The glycoproteins which participate in the classical activation are called the *components*, whereas those of the alternative pathway are called *factors* (MÜLLER-EBERHARD, 1988).

Both the components and factors possess effector and regulatory functions. They participate not only in the defence, but also in immunopathological reactions of the organism. The symbol for complement is C and individual components are designated C1, C2 ... up to C9. The C1 component has three subunits: C1q, C1r and C1s. During the classical activation pathway, individual components react gradually and chronologically. The only exception is the C4 component that reacts immediately after C1 activation. The sequence of the classical activation pathway is therefore: C1–C4–C2–C3–C5–C6–C7–C8–C9. During the alternative activation pathway, factors B and D are activated which then, together with C3, activate C5 and further components up to C9, without involving C1, C4 and C2. Individual components are activated step-wise by a cascade mechanism: during activation of a specific component a proteolytic enzyme is generated that splits the next component into two fragments (limited proteolysis), one of them being an enzyme which cleaves the next component in the sequence. Certain fragments do not have enzymatic activity but act as cofactors or bioregulatory substances.

Components C1 to C5 and factor B are cleaved into fragments, whereas components C6 to C9 gradually bind to the surface-bound fragment C5b and do not undergo fragmentation. The generated fragments are designated by small letters added to the symbol of the parent component, *e.g.* C2a, C3b, C5a *etc.* The letters "a" and "b" refer to the small and large fragments respectively, generated during initial degradation of the native component or factor. An exception are fragments C2a and C2b, where the larger fragment is C2a. The enzymatic activity of the component, group of components or their fragments, is denoted by a horizontal line above the numbers or letters, *e.g.* $C\overline{1}$, $C\overline{4b2a}$ *etc.* The component that lost the enzymatic activity is denoted by the letter "i" after the symbol, *e.g.* C2ai. If, however, the fragment lost certain activity due to further degradation (cleavage) of the polypeptide chain, it is also denoted by "i", but before the symbol, *e.g.* iC3b. The letters "c", "d" *etc.* characterize fragments that are usually generated from the primary fragment "b" by further proteolytic processes (C3c, C3g *etc.*).

In addition to effector components and factors (*Table 10.1*), the complement system also contains regulatory glycoproteins that control the activity of effector components (*Table 10.2*). Regulators of the alternative pathway are denoted by the letters H, I and P, whereas other regulatory (glyco)proteins are marked according to their function, *e.g.* C1-inhibitor — C1-INH, C4-binding protein — C4-bp *etc.* In addition, receptors for various fragments of complement components that are present on the surface of various cell types, participate in the regulation of complement system activity.

Complement significantly influences various defence mechanisms: opsonization and phagocytosis, activation and mobilization of immunologically active cells towards the site of inflammation, processing and removal of immune complexes and direct lysis of numerous target cells such as Gram-negative bacteria and eukaryotic cells recognized as foreign by the organism, but also the virus envelope. In addition, complement may initiate various immunopathological processes and its fragments act as regulators of various immune functions and mechanisms.

10.1 Complement activation

10.1.1 The classical complement pathway

The classical and alternative complement activation pathways are shown in *Fig. 10.1.*

C1 is the first component of the classical pathway. C1 is a macromolecular complex composed of three different subunits: C1q, C1r and C1s. **C1q** acts as a *recognition unit,* whereas the effector function is executed by the tetramolecular complex (C1r, C1s)$_2$. All five subunits may be joined by coordination bonds in the presence of Ca^{2+}. C1q is a glycoprotein with a specific amino acid composition. It contains hydroxylysine, hydroxyproline and a large amount of glycine, and thus resembles collagen. C1q contains three types of polypeptide chains — A, B and C; one molecule contains six

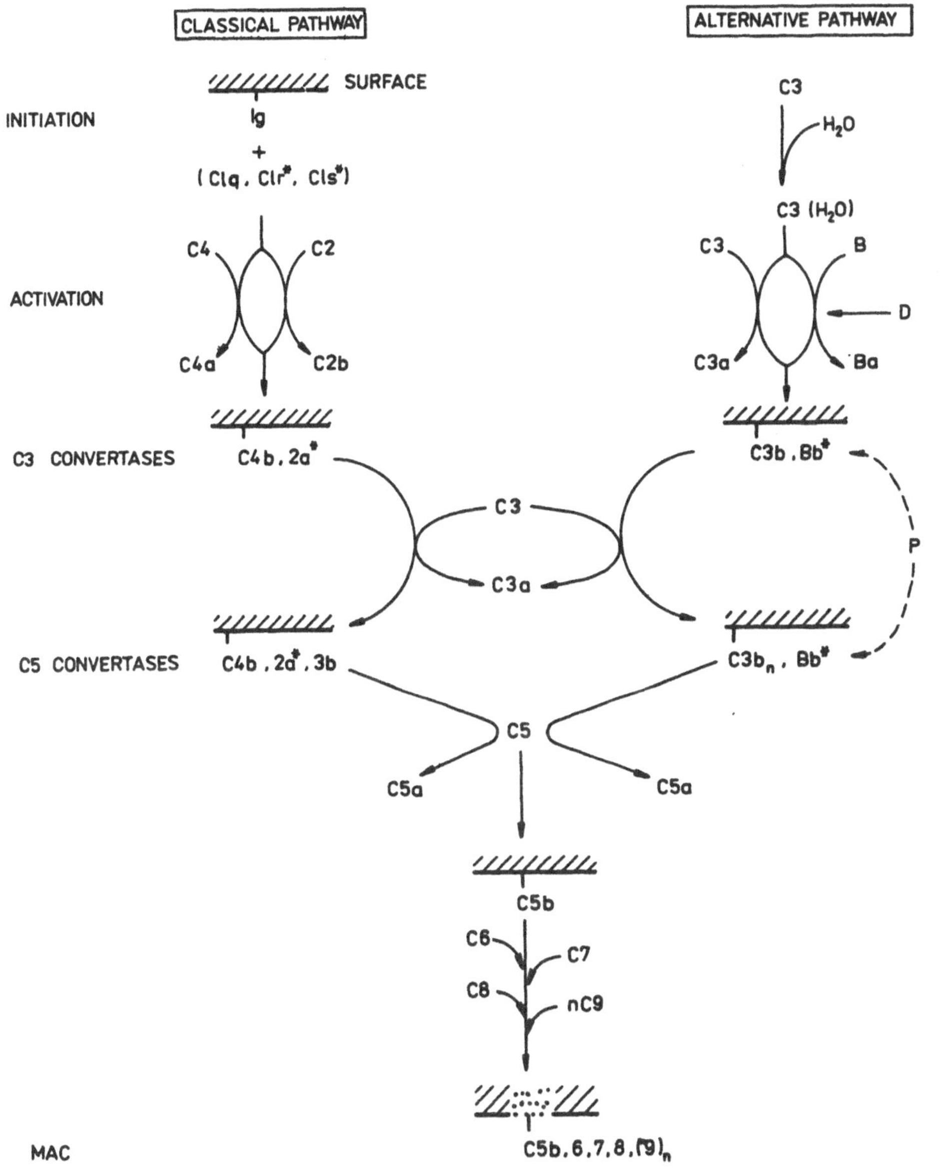

Fig. 10.1. Pathways of classical and alternative complement activation.
The asterisks indicate those fragments with enzymatic activity. MAC, membrane attack complex; P, properdin (stabilizes both convertases of the alternative pathway).

Table 10.1 Properties of human complement components and factors

No.	Complement component	Mol. wt. $\times 10^3$	% saccharides	Number of polypeptide chains	Fragments and their mol. wt. $\times 10^3$	Approx. serum concn. (mg/l)	Number of amino acid residues
Classical activation pathway							
	Recognition unit						
1	Clq	410	8.3	18 (6A + 6B + 6C)	—	150	3 120
2	Clr	166	9.4	2 (A + B)	—	50	242 (*B*-chain only)
3	Cls	83	7.1	1	—	50	?
	Activation unit						
4	C4	204	7.0	3 (α, β, γ)	C4a-6	400	1 722 (single-chain precursor)
					C4b-198		
					C4c-148		
					C4d-35		
5	C2	115	15.9	1	C2a-76	15	732
					C2b-39		
6	C3	180	1.7	2 (α, β)	C3a-9	1 200	1 637
					C3b-171		
					iC3b-168		
					C3c-140		
					C3d-30		
					C3e-12		
					C3f-2		
					C3g-5		
Alternative pathway							
	C3						
7	B	93	8.6	1	Ba-33	200	739
					Bb-60		
8	$\bar{D}$	24	0	1	—	1	222

Table 10.1 (continued)

No.	*Complement component*	*Mol. wt. $\times 10^3$*	*% saccharides*	*Number of polypeptide chains*	*Fragments and their mol. wt. $\times 10^3$*	*Approx. serum concn. (mg/l)*	*Number of amino acid residues*
Membrane attack complex							
9	C5	190	15.0	2 (α, β)	C5a-11 C5b-170 C5d-30	80	434
10	C6	128	7.1	1	—	70	913
11	C7	110	6.4	1	—	55	821
12	C8	163	5.8	3 (α, β, γ)	—	60	α-553 β-537
13	C9	66	7.8	1	—	60	537

Table 10.2 Regulators of human complement

No.	*Complement component*	*Mol. wt. × 10^3*	*% of saccharides*	*Number of polypeptide chains*	*Fragment binding (specific ligand)*	*Approx. serum concentration (mg/l) or presence on cells*	*Number of amino acid residues*
Regulators of convertase activities							
1	C4-bp	570	?	7 (identical)	C4b	180	549 (×7)
2	H	155	?	1	C3b	500	1216
3	P	100–200	?	2–4	C3b, Bb	20	?
4	I	88	28	2	C4b, C3b	30	565 (single chain precursor)
5	DAF (CD55)	75	45	1	C4b2a, C3bBb	Er, PMN, Ly, Eo, Tr	381 or 440
6	CR1 (CD35)	160–250	10–16	1	C3b, C4b, iC3b	Ne, Eo, Mo, Ma, B and other	1 400–1 800
7	MCP (CD46)	45–70	various	1	C3b, iC3b	Ne, Mo, Ly, Tr	?
Other regulators							
8	C1-INH	104	49	1	$\overline{C1r}$, $\overline{C1s}$	165	478
9	Anaphylatoxin inactivator (SCPN)	300	?	1 (2)	C3a, C4a, C5a	50	?
10	Protein S (vitronectin)	80	?	?	C5b67, C9	350	485
11	CFI	160	?	?	C5a	50	?
12	HRF	65	?	1	C8, C9	Er, Tr	?
13	CR2 (CD21)	140	?	1	C3d, C3dg, iC3b, C3b	B and other	950–1 200
14	CR3 (CD11b/CD18)	260	?	2 (α, β)	iC3b, C3d	Ne, Mo, Ma, NK	?
15	CR4 (CD11c/CD18)	240	?	2 (α, β)	iC3b	Ne, Mo, Ma	?
16	CR5	90	?	1	C3dg, C3d	Ne, Tr	?
17	C3aR	?	?	?	C3a, C4a	MC, Eo, Ba, Mo, Ne, Tr	?
18	C5aR	52	?	1	C5a, $C5a_{desArg}$	MC, Ne, Ba, Mo, Ma	?
19	C1qR	65	?	1	C1q	B, Ne, Mo	?
20	Conglutinin	66	?	1	iC3b, C3c	?	?
21	HR	150	?	1	H	B, Mo, Ma, Ne	?
22	C1-INHR	?	?	?	C1-INH	?	?
23	BaR	?	?	?	Ba	B	?

C4-bp, C4b-binding protein; I, C3b-inactivator; DAF, decay-accelerating factor; MCP, membrane cofactor protein; C1-INH, C1-inhibitor; SCPN, serum carboxypeptidase N; CFI, chemotactic factor inactivator; HRF, homologous restriction factor; HR, factor H receptor; Er, eryhtrocytes; Ly, lymphocytes; Tr, thrombocytes; Ne, neutrophils; Eo, eosinophils; Ba, basophils; Mo, monocytes; B, *B*-lymphocytes; MC, mast cells. CR1 and DAF are also in a soluble form.

copies of each type (*i.e.* 6A + 6B + 6C), and has a molecular weight of 410 000. A single chain has a molecular weight of approximately 23 000 and contains 183–194 amino acid residues. The A + B + C chain triplet forms one subunit of the C1q molecule. Their *N*-ends are mutually twisted in the form of a triple-stranded α-helix, that represents a stalk of a structure resembling a tulip. Its head is represented by a globular formation formed by the *C*-ends of all three chains. Six subunits form a complete C1q molecule resembling a bunch of tulips (*Fig. 10.2*).

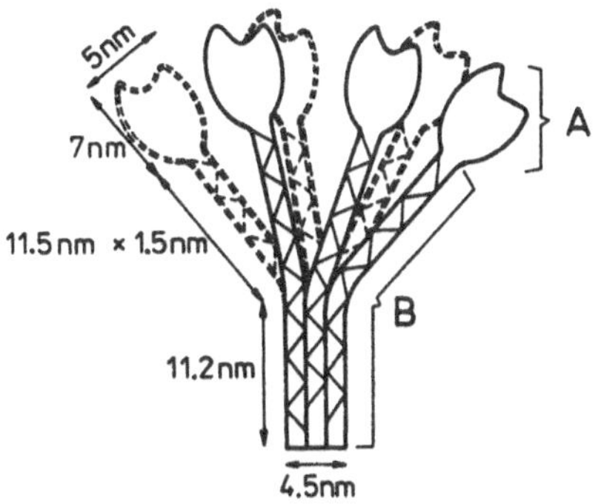

Fig. 10.2. The "tulip" model of the human C1q subunit (Reid and Porter, 1976).
To illustrate the three-dimensional character of this model the parts drawn in solid lines extend towards the viewer whereas those drawn in broken lines extend away from the viewer (A) The *C*-terminus of the polypeptide chains form globular structures always containing one combining site for the IgG C_H2-domain or IgM C_H4-domain (B) The N-halves of the polypeptide chains form the "tulip stalks" containing a collagen-like triple-stranded α-helix

Each globular head possesses one combining site that can bind to the IgG C_H2-domain or IgM C_H4-domain. Such binding provides a signal for the initiation of the classical complement activation pathway. If the immunoglobulins are soluble and in a monomeric form, the binding affinity is low and the bond does not occur. However, if the immunoglobulin is combined with the corresponding antigen and an immune complex is generated, or if immunoglobulin molecules form aggregates, the affinity of the combining site for C1q increases sharply which allows it to bind and initiate the complement cascade. C1q binds most strongly to IgM, then to IgG3, IgG1 and IgG2. It exhibits almost no binding to IgG4 and to other immunoglobulins. A firm bond is accomplished when at least two C1q heads bind to one IgM molecule or to two IgG molecules.

Originally, it was thought that C is activated *via* the classical pathway only in the presence of antibodies that had formed immune complexes with corresponding antigens; it was therefore thought that antibodies had the ability to label the cells to be destroyed or lysed by complement. It was subsequently shown, however, that only IgM and IgG (except IgG4) possessed such an ability even in the absence of specific antibody activity under conditions that form aggregates, or if they were immobilized by adsorption to various surfaces (by physicochemical conditions, PEG, staphylococci *etc.*).

The ability to bind C1q and initiate the classical activation pathway is also possessed by various polycations and polyanions including RNA and DNA, LPS of Gram-negative bacteria (particularly the R-forms, p. 44), certain surface proteins of oncogenic viruses (retroviruses) and **C-reactive protein (CRP)**. Activation of the classical pathway initiated by CRP is of biological importance. Under physiological conditions, only trace amounts of CRP are present in the blood. However, its amount increases more than 100-fold within a few hours of tissue damage. CRP binds to polysaccharides of the surface of bacteria, fungi and other parasites, to phospholipids of animal cells and is as active in C1q binding as IgM or IgG. Unlike these immunoglobulins, however, it does not require interaction with specific antigen.

C1q binding causes an increase in the association strength of the $(C1r-C1s)_2$ complex and generation of conformational changes that induce catalytic cleavage of the C1r proenzyme into the active enzyme $C\overline{1r}$. Its substrate is C1s, which is then changed into the active serine proteinase $C\overline{1s}$. Spontaneous autoactivation of the $(C1r-C1s)_2$ complex is limited and regulated by a non-competitive **inhibitor (C1–INH)** that forms stoichiometric complexes with C1r and C1s (thus blocking their activity) and facilitates dissociation of the $(C1r-C1s)_2$ tetramer from the complete C1 component.

The activated $C\overline{1s}$ subunit may cleave C4, the second glycoprotein of the

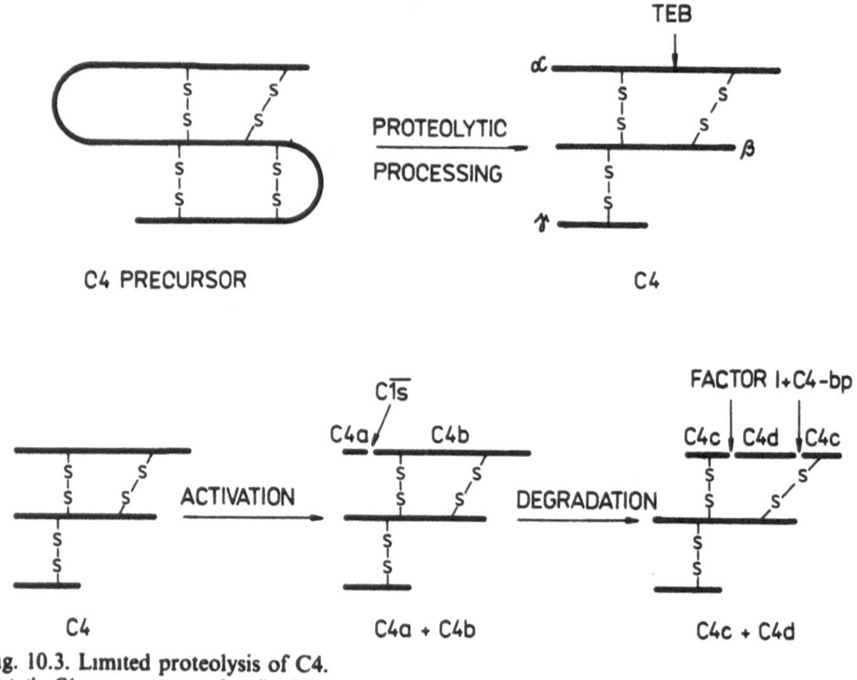

Fig. 10.3. Limited proteolysis of C4.

First, the C4 precursor is proteolytically cleaved into the C4 component, present in the serum. Native C4 contains three polypeptide chains (α, β, γ). The α-chain contains a thiolester reactive bond (TEB) that is uncovered after activation through $C\overline{1s}$, at the same time, fragments C4a and C4b are generated. C4b is further degraded by factor I, in cooperation with cofactor C4-bp, into fragments C4c and C4d. The thiolester bond remains after degradation in fragment C4d that has already lost its activity classical pathway, and its natural substrate. **C4** is split into a small fragment C4a, which possesses an anaphylatoxin activity, and a large fragment C4b, which binds to the target surface (a surface having the previously bound C1q that initiated the classical activation pathway) near the activated C$\overline{1}$. The bound C4b acts as a binding site for the further glycoproteins of the classical pathway, *i.e.* **C2**, which is then split by C$\overline{1s}$, localized in its vicinity. The C4b activity is immediately regulated by two proteins — the **C4-binding protein (C4-bp)** and **factor I.** C4-bp competitively inhibits C2 binding to C4b and serves as a cofactor for factor I, which is a C4b-degrading proteinase (*Fig. 10.3*). The C$\overline{14b}$ enzyme splits C2 into a large fragment C2a, and a small fragment C2b. The C2a fragments are re-bound to the target surface and form another proteolytic enzyme, C$\overline{4b2a}$ (sometimes also known as C$\overline{14b2a}$). This is known as the **C3-convertase of the classical pathway** since its substrate is C3. Degradation and thus also inactivation of the C4b2a convertase is accelerated by several regulatory proteins, the most important being the cell receptor for C3b (CR1), decay accelerating factor (DAF) and C4-bp. C3-convertase splits C3 into a small fragment C3a with anaphylatoxin activity, and into a large fragment C3b that may, similarly to C4b, bind to the target surface through an internal thiolester bond (*Fig. 10.4*). The bound C3b then becomes a binding site for the next glycoprotein C5 and a component of the **C5-convertase of the classical pathway** that initiates activation of terminal components of the complement system.

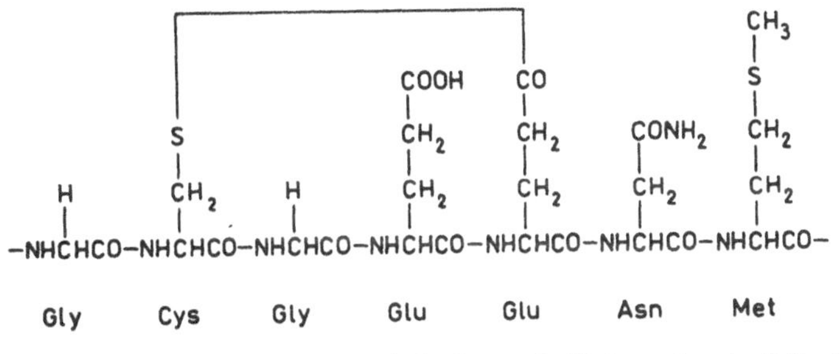

Fig. 10.4. The amino acid sequence around the internal thiolester bond of the human C3b α-chain.
A similar sequence also exists around the thiolester bond of the human C4b α-chain, although Asn is replaced by Thr. Native C3 and C4 molecules have a hidden thiolester bond. Following fragmentation, the bond appears on the surface of C3b and C4b which results in its rapid degradation; in addition, it permits covalent binding of C3b or C4b fragments (through transesterification) to the surface of a particle or other suitable structure in the immediate vicinity.

The **C4 component** is synthesized (mainly in macrophages) as a single polypeptide chain containing 1 722 amino acid residues with a known sequence. This precursor undergoes intensive post-translation processing (*Fig. 10.3*) which consists of glycosylation, limited proteolysis and amino acid modification. This processing results in formation of the serum molecule C4 with a molecular weight of 204 000, containing three polypeptide chains: α (mol. wt. = 95 000), β (mol. wt. = 70 000) and γ (mol. wt. = 33 000). The α-chain contains a functionally important *thiolester bond.* Oligosaccharide units are present in all three polypeptide chains although their function is unknown. If macrophages are incubated in the presence of tunicamycin that inhibits glycosylation, a pure C4 protein possessing similar haemolytic activity to normal C4 (containing a polysaccharide component), is obtained.

Two molecular forms of C4, known as *C4A* and *C4B*, are present in the plasma. These are coded by two different genes that are present in the HLA-complex. Therefore, many workers originally consider C4, together with C2 and B, to be the class III of HLA-antigens (p. 56). The genes coding C4A and C4B are almost identical, but are highly polymorphic, *i.e.* they may contain over 35 alleles, including the null alleles. Products of individual alleles differ only in 15 amino acid residues.

The **C2 molecule** (mol. wt. about 115 000) consists of a single polypeptide chain, to which oligosaccharide units, representing 15.9% of the total weight, are bound. The molecule contains 732 amino acid residues. $\overline{C1s}$, an active proteinase of the $\overline{C14b}$ complex, splits one bond between Arg and Lys of the native C2 component which results in the generation of the catalytic C2a fragment, composed of 509 amino acid residues, and the fragment C2b, containing 223 amino acid residues. C2a is a *serine proteinase* and its molecule is composed of two globular domains joined by a short segment.

The gene for C2 is also present in the major histocompatibility complex. Biosynthetic studies have revealed that three primary translation products with various lengths of the polypeptide chain are generated from this gene. After synthesis, one product (mol. wt. = 84 000) is immediately secreted outside the cells (macrophages), while the other two products (mol. wts. = = 79 000 and 70 000) remain inside the cells for several hours.

10.1.2 Alternative complement pathway

The basis of the alternative complement activation pathway is activation of C3 without the participation of components C1, C4 and C2 (*Fig. 10.1*). To understand the alternative activation pathway, the **internal thiolester bond** in the α-chain of the native C3 component is very important. It has a much higher stability (t/2 of several days) than that of a simple thiolester in aqueous solution (t/2 = 30 min). Such a high stability is due to the fact that the C3 thiolester bond is enclosed by the native chain and is therefore not easily accessible for water molecules that might cause hydrolysis. Following activation and splitting of C3a or C4b, the degradation half-time is markedly shortened to 0.1–1.0 ms.

Nevertheless, the thiolester bond in the native C3 is hydrolysed in serum by surrounding water molecules, albeit slowly. The product of hydrolysis, $C3(H_2O)$, possesses certain properties of the C3b molecule, including the ability to bind **factor B** — the second glycoprotein of the alternative pathway. As soon as factor B has bound, it may be split by $\overline{D}$ (the only factor circulating in an active form) into fragments Bb and Ba. Thus, a new enzyme complex $\overline{C3(H_2O)Bb}$ is generated. This complex can cleave further C3 into a true C3b that can be covalently bound to molecules in the immediate vicinity of this fragment. Binding occurs between the acyl group of thiolester, activated during C3 fragmentation, and the corresponding reactive group (—OH, —NH_2) on the surface of a target structure. In case the target structure is the surface of a molecule, capable of complement activation *via* the alternative pathway, then the randomly bound C3b fragment reacts further with factor B, with which (following cleavage by circulating $\overline{D}$), it forms the proteolytic enzyme $\overline{C3bBb}$, *i.e.* the **C3-convertase of the alternative pathway**. The C3-convertase repeatedly cleaves further C3 and the generated C3 becomes a component of the proteinase $\overline{C3bBb3b}$ (designated also $\overline{C3b_nBb}$). The substrate for this proteinase is C5, *i.e.* the first of the terminal complement components, forming the membrane attack complex (MAC). The $\overline{C3b_nBb}$ complex represents the **C5-convertase of the alternative pathway**. C3b is both the initiating and main end-product of this pathway which permits positive amplification and regulation of the whole process by a feed-back mechanism.

Among those molecules possessing an ability to activate complement *via* the alternative pathway, are aggregated IgG4, IgA1, IgA2, IgE, various polysaccharides such as inulin, agar, agarose, dextran, zymosan, bacterial polysaccharides and the cobra venom factor (CVF). Activators of the alternative pathway are widely distributed among microorganisms. Besides lipopolysaccharides of Gram-negative bacteria and teichoic acids of Gram-positive bacteria, the alternative pathway may also be activated by various viruses, fungi and parasites (helminths and protozoa). The alternative pathway is further activated by heterologous erythrocytes and lymphocytes of certain animal species and by numerous tumour cells. All these microorganisms and cells may be subsequently killed by activated complement. Activation of the alternative pathway is also initiated during contact of blood plasma with plastics which may result in adverse effects during haemodialysis.

The alternative pathway is thought to be particularly important at the initial stage of infection when specific antibodies are still absent. It seems that a properly functional alternative pathway may prevent development of many infections.

As far as these mechanisms of complement activation are concerned, it is clear that the alternative pathway is closely regulated, primarily by serum factors H and I. **Factor H** binds to C3b, competitively inhibiting interaction with B and rendering C3b susceptible to permanent inactivation by factor I. **Factor I** splits C3b thereby generating the inactive degradation product **iC3b**. The relative affinity of factors B and H, competing for C3b, is regulated by the surface to which C3b is bound. On surfaces that do not activate the alternative pathway, the affinity of H is higher than the affinity of B, resulting in the inhibition of active C3-convertase formation. Conversely, C3b immobilized on activating surfaces has a decreased affinity for factor H which permits binding of B and its subsequent activation by factor $\overline{D}$. Degradation of the enzymatically active convertases $\overline{C3bBb}$ and $\overline{C3bBb3b}$ into inactive components is accelerated by factor H, CR1 and DAF. CR1 and DAF are glycoproteins found in the cytoplasmic membrane of numerous cell types. On the other hand, properdin (P) exerts a stabilizing effect on both convertases.

Factor B is a functional analogue of the C2 component of the classical pathway. Its molecule is composed of 739 amino acid residues of known sequence. The factor B molecule is cleaved during activation by factor $\overline{D}$ at the site of one –Arg-Lys– bond generating fragments Ba (234 amino acid residues) and Bb (505 amino acid residues — a serine proteinase). The primary structures of B and C2 are 39% homologous. Fragment Bb, like C2a, consists of two globular domains joined by a short segment. The gene coding for factor B is localized in the MHC region and contains 18 exons.

Factor D substitutes for the C1 component of the classical pathway in the alternative pathway. Factor D is a *serine proteinase* and presumably is only present in the plasma in an active form ($\overline{D}$). It consists of a single polypeptide chain containing 222 amino acids with known sequence (mol. wt. = 23 748) and possesses a relatively low homology with other complement serine proteinases (C1r, Bb, I). Factor D has the highest homology with plasmin and has somewhat lower homology with pancreatic elastase, trypsin, chymotrypsin and kallikrein.

10.2 Glycoproteins regulating complement convertase activity

C3-convertases may be generated in the fluid phase or on the cell surfaces (solid phase). Although C3-convertase is generated during the alternative pathway to a low degree only, it may nevertheless permanently activate C3 and the resultant C3b may be released into the fluid phase or deposited on the surface of autologous normal cells, eventually resulting in their damage. Therefore, a regulatory system must exist to maintain the homeostasis of spontaneous C3 activation and this therefore represents a recognition mechanism capable of distinguishing self and non-self structures. The principle of this control mechanism is the regulation of C3b and C4b fragment activities which are the basic components of both convertases:

C3-convertases — C$\overline{4b2a}$ (classical), C$\overline{3bBb}$ (alternative)
C5-convertases — C$\overline{4b2a3b}$ (classical), C$\overline{3bBb3b}$ (alternative)

Seven glycoproteins are involved in the regulation of convertase activities. Four of them are soluble plasma factors — factors H and I, C4b-binding protein (C4-bp) and properdin; three occur in the form of various cell surface receptors — C3b and C4b (CR1) receptors, decay-accelerating factor (DAF) and gp45-70 (MCP).

Factor I (*C3b-inactivator*) is a serine proteinase of arginyl specificity, the specific substrate of which is the α-chain of C3b or C4b. It consists of two subunits, one large (mol. wt. 48 000) and one small (mol. wt. 38 000), joined by a disulphide bridge. It is synthesized in the liver and by monocytes and is encoded as a single-chain precursor of 565 amino acids with an additional 18-residue leader sequence. The single-chain form undergoes post-translation glycosylation and limited proteolysis which results in the formation of functional glycoprotein I. Factor I cleaves C3b and C4b regardless of whether they are in the fluid phase or covalently bound to soluble immune complexes or cell surfaces. Such degradative cleavage however, takes place only in the presence of cofactors. It proceeds in two steps: during the first step, haemolytically inactive fragments iC3b (*Fig. 10.5*) and iC4b (*Fig. 10.3*) are formed under the influence of factor I and cofactors H, C4-bp and CR1. After the

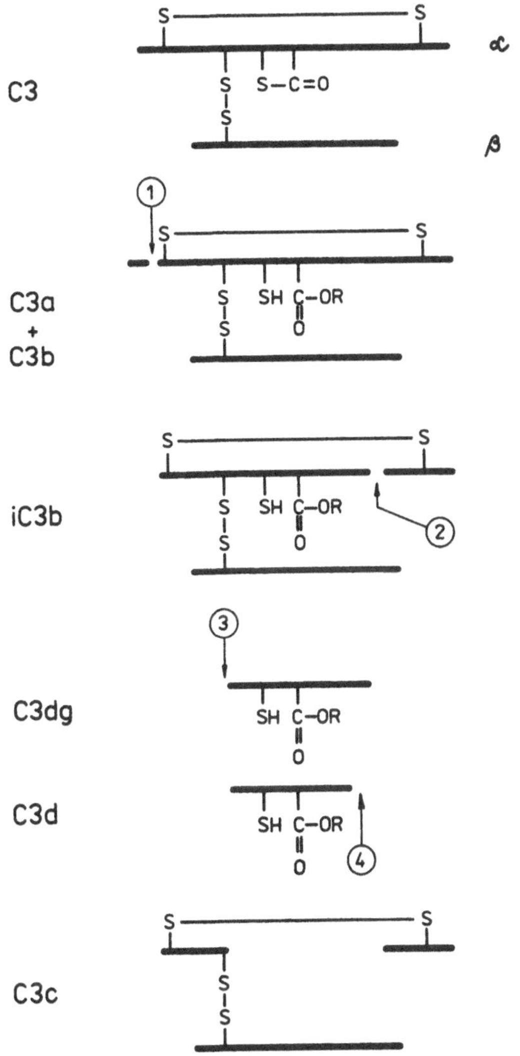

Fig. 10.5. C3 component and its fragments showing the disulphide bridges and thiolester bond. 1, C3-convertase that splits off C3a-fragment; 2, factor I, that splits off a small fragment C3f, the remainder (fragment iC3b) remains unchanged; 3, 4, further proteolytic degradation of iC3b into various fragments.

first proteolysis, a new site on fragments iC3b or iC4b becomes available, allowing factor I to cleave a polypeptide chain, thereby generating C3c and C3dg, and/or C4c and C4d. The role of cofactors during the second cleavage is played by CR1 and C4-bp, but not by H.

C3-convertases are highly labile complexes — they decay spontaneously and release C2a or Bb. The decay is accelerated by DAF, CR1, MCP (formerly called gp45-70), H and C4-bp; therefore they may control local activation and deposition of complement components. Factor H only accelerates the decay of $\overline{\text{C3bBb}}$-convertase, while C4-bp only potentiates the decomposition of $\overline{\text{C4b2a}}$-convertase. The remaining three glycoproteins cause the decay of both convertases.

Factor H (original name β1H), the most abundant cofactor, is a single-chain glycoprotein of molecular weight 155 000. It forms five allotypic variants (polymorphic forms) — H1, H2, H3, H4 and H5 — that are segregated codominantly. The variants of H1, H2 and H3 occur in the human population at a frequency of 0.691, 0.302 and 0.006 respectively.

C4-bp is a glycoprotein with a molecular weight of about 570 000. It consists of seven identical polypeptide chains joined by disulphide bonds. Each chain contains 549 amino acid residues and possesses eight internal homologous segments, each being composed of 60 amino acids with a framework of four conserved cysteine residues. These segments are highly homologous with similar segments present in factor H, DAF, MCP, CR1, CR2, factor B, C2, C1r and C1s molecules as well as in non-complement proteins, *i.e.* the IL-2 receptor, β_2-glycoprotein and haptoglobin. This suggests that all these glycoproteins constitute a single family which originated from a common ancestor and are known as the **regulators of complement activation (RCA)** (Hourcade *et al.*, 1989). The C4-bp molecule has four combining sites for C4b. Two phenotypic variants — C4-bp1 and C4-bp2 — have been described; both are coded by two genes from the autosomal locus with codominant expression. The frequency of C4-bp1 and C4-bp2 in the human population is 0.981 and 0.018 respectively.

Properdin is a glycoprotein that stabilizes the C3- and C5-convertases of the alternative pathway; it therefore has an opposite effect to H, CR1 and DAF. It consists of several non-covalently bound identical subunits, each having a molecular weight of about 50 000. Originally it was presumed that the properdin molecule had four subunits. However, electron microscopic studies revealed that it consists of a mixture of dimers, trimers and a small amount of higher polymers.

CR1 is an integral glycoprotein present in the membrane of erythrocytes, PMN leukocytes, monocytes, all *B*-lymphocytes and some *T*-lymphocytes, mast cells and glomerular podocytes. This receptor has its highest affinity for C3b, but it also binds iC3b, C4b and C3 containing a hydrolysed thiolester bond. It acts as a factor I cofactor during C3b and C4b degradation, plays a key role in the binding and removal of circulating immune complexes (particularly on erythrocytes), facilitates phagocytosis of C3b-opsonized particles and participates in the secretory functions and metabolic activity of macrophages. CR1 expression on erythrocytes is regulated by two codominant alleles — one determines a high number of receptors (650–950), the second determines a low number of receptors (150–350) on the surface of a single erythrocyte. Approximately 34% of the human population are homozygous for the first allele, 12% are homozygous for the second allele and the rest (54%) are heterozygous with 340–620 receptor sites on each erythrocyte. An increased frequency of immunopathological conditions, caused by inadequate removal of circulating immune complexes, may be seen in individuals with few CR1 on erythrocytes. The number of CR1 on phagocytes is not constant. Circulating neutrophils and monocytes have about 5 000 surface receptors; however, following stimulation with chemotactic factors for example, their number may increase ten-fold.

CR1 occurs as four codominantly expressed allotypic variants of various molecular weights: CR1-A, mol. wt. 190 000 (frequency in the human population 0.83), CR1-B, mol. wt. 220 000 (0.16), CR1-C, mol. wt. 160 000 (0.01) and CR1-D, mol. wt. 250 000 (<0.01). Each variant contains a saccharide component (mol. wt. about 15 000). CR1-A contains at least 33 homologous segments, each being composed of 60 amino acids, followed by a 25-residue hydrophobic (transmembrane) segment and a non-homologous *C*-terminal region (the cytoplasmic chain). The sequence of CR1 is much more repetitive than that of H or C4-bp.

DAF (decay-accelerating factor) regulates the activity of both convertases of the classical and alternative pathways by preventing fusion of their individual components. At the same time, it accelerates their dissociation (decay) by binding to corresponding enzymes or their precursors (C2a, Bb). DAF is present as an integral glycoprotein in the cytoplasmic membrane of erythrocytes, neutrophils, monocytes, lymphocytes, platelets, epithelial and endothelial cells. It is synthesized as a precursor (mol. wt. 43 000) that contains an oligosaccharide, mainly composed of mannose units; it binds to polypeptide chain through an *N*-glycosidic bond. It is modified in the Golgi complex by post-translation processing into a form with a molecular weight of 46 000 which is changed by further glycosylation (joining of oligosaccharides through an O-glycosidic bond) into a form with a molecular weight of 70 000 (erythrocytes) or 80 000 (leukocytes).

In addition, there is a soluble form of DAF which is present at low concentrations in plasma, tears, saliva and urine. Isolation and nucleotide sequencing of overlapping cDNA clones for DAF demonstrated the presence of two groups of mRNA coding for two distinct polypeptide chains of unequal length. The longer abundant message codes for a 440-residue protein, while the shorter message codes for a 381-residue polypeptide chain. The product of the longer mRNA is thought to correspond to the soluble form of DAF. The *N*-terminal region of DAF molecule contains four homologous 60-amino acid units.

The membraneous DAF is anchored to lipid bilayers through the 1,2-diacylglycerol moiety of a phosphatidylinositol molecule. The covalent attachment involves ethanolamine and glucosamine moieties which link an oligosaccharide to the COOH-terminus of the protein and the inositol ring of membrane phosphatidylinositol, respectively. It has been suggested that the phospholipid anchor might be sterically more favourable than a membrane-spanning hydrophobic peptide. With respect to DAF, this mode of attachment might have the advantage of an enhanced rate of lateral mobility within the membrane.

Glycoprotein 45–70 (gp 45–70, recently called **MCP — membrane cofactor protein**) is relatively widely distributed; it is present on the surface of human peripheral blood leukocytes and platelets, but not in erythrocytes. It regulates complement activation on cell surfaces in a similar way to DAF; in contrast to DAF, however, MCP possesses a higher affinity for C3b and for C3b-containing enzymes. Its molecular weight varies between 45 000 and 75 000, probably due to differences in glycosylation.

10.3 The biochemistry of C3

The C3 complement component plays a key role in activating both complement pathways and its fragments possess significant biological properties; it therefore belongs to those complement glycoproteins with the best defined structure (Hostetter and Gordon, 1987; Lambris, 1988). Human C3 is a two-chain glycoprotein; its α-chain (mol. wt. 115 000) is joined to the β-chain (mol. wt. 75 000) by one disulphide bond. The complete C3 molecule contains 1.7 % of saccharides. It is synthesized, like C4 and C5, as a single-chain precursor. Human pre-pro-C3 contains 1 663 amino acid residues; the first 22 residues represent a signal peptide followed by 645 residues of the β-chain, arginine tetrapeptide (absent in native C3), and 992 residues of the α-chain. The primary structure of C3 is very similar to that of C4 and α_2-macroglobulin, suggesting their common evolutionary origin. Activation of the C3 component by the corresponding convertase of the classical or alternative pathway results in proteolytic cleavage of C3a from the *N*-terminus of the α-chain. The peptide bond is cleaved between the arginine and serine residues at position 77. The α-chain residue (designated α') and intact β-chain represent the C3b fragment (*Fig. 10.5*).

The α-chain contains an **internal thiolester** (*Fig. 10.4*) which plays an important role in various functional manifestations of C3. The thiolester bond is located between the γ-carboxylic group of the glutamine residue and the sulphydryl group of the nearest cysteine residue in the sequence -Cys-Gly-Glu-Gln-. It is only present in the non-hydrolysed form in the native C3 molecule; it is hydrolysed during activation and cleavage of C3. The reactive glutamine residue provides its carbonyl group for transacylation reactions which results in the origin of an oxy-ester bond with alcohol acceptors, or in an amide bond with amine acceptors (*Fig. 10.6*). If these acceptors are not close to the C3 activation site the thiolester is hydrolysed by water, and, consequently, the glutamyl carbonyl forms a carboxylic group, which in turn prevents further covalent reactions of this glycoprotein. In addition to these interactions, the thiolester bond may also be inactivated by binding to small molecules, *e.g*. methylamine.

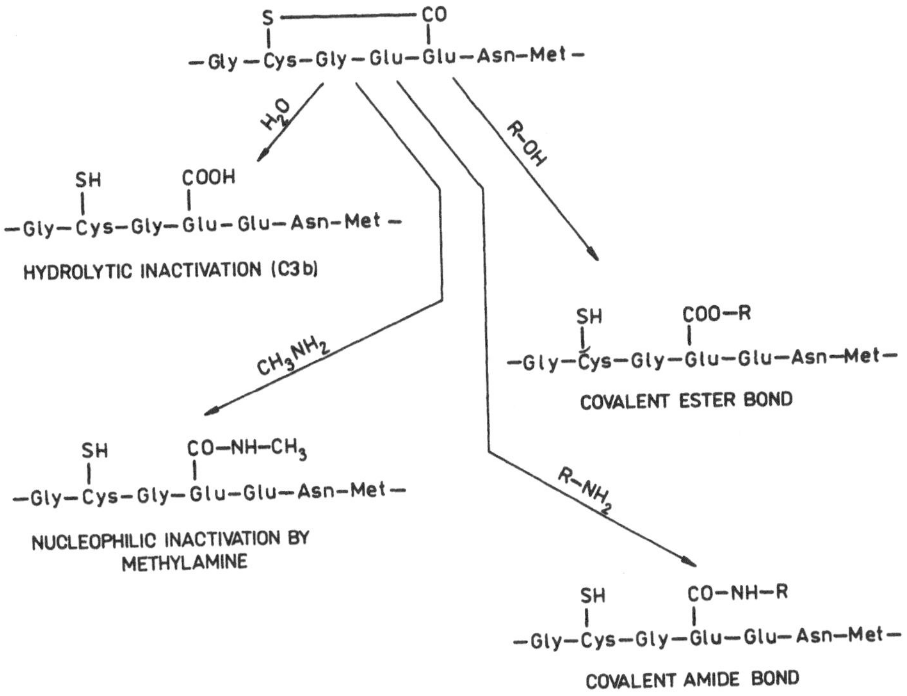

Fig. 10.6. Reaction of the thiolester bond in the α-chain of native C3 component.

The transacylation reactions that yield glutamyl thiolester bonds play an important role in *the opsonization* of microorganisms and other particles. During these reactions, C3b may bind to the particle through one site and to the corresponding receptor (CR1) on the phagocyte surface through the other. This greatly facilitates the contact between the phagocyte and particle. C3b, therefore, is one of the most effective opsonins. Following C3 activation by a convertase of the classical or alternative pathways, a transacylating reaction with an alcohol or amine group occurs on the surface of the bacterial cell within 60–120 μs. Afterwards, the thiolester becomes unreactive through hydrolysis. It thus follows that the C3 activation must occur in the immediate vicinity of bacterial cell acceptors. The bond between the surface acceptor and C3b is covalent, but non-specific, because it may be mediated by any hydroxyl or amine group on the bacteria surface. The bond is resistant to extreme temperature and pH and is irreversible. These properties guarantee its stability even during episodes of fever when the microenvironment is acidified by bacterial cell metabolites. Irreversibility of the opsonic bond means that once C3b has bound to the particle, it cannot be displaced by any other complement glycoprotein. However, even covalently bound C3b may be gradually degraded by proteinases to iC3b, C3dg and C3d that also have important biological functions. They may bind to various types of complement receptors (CRs), present on different cells (*Table 10.2*). Occupation of these receptors provides a signal to the cell to begin phagocytosis of opsonized particle or other reactions.

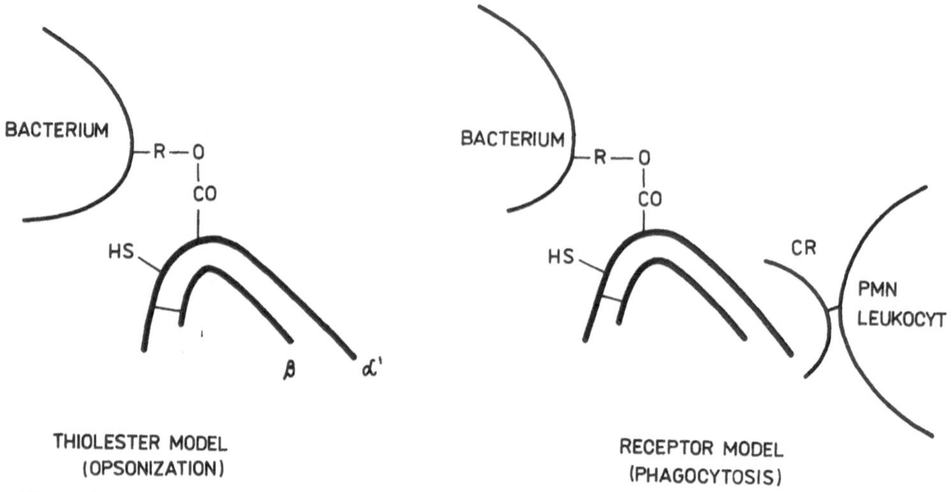

Fig. 10.7. Role of C3 thiolester in mediating opsonin and receptor bond (Hostetter and Gordon, 1987).
Interaction with the bacterial surface may be mediated by both the ester and amide covalent bond. The bond to the complement receptor (CR) is non-covalent and requires conformational changes of the C3 molecule, generated during cleavage of the thiolester bond. α′, β-, polypeptide C3 chains (α′—α-chain after splitting of C3a).

A different site on the C3b molecule to the glutamyl carbonyl of the thiolester bond (*Fig. 10.7*) is involved in binding to the leukocyte surface receptor. Of primary importance is the fact that native C3 with an intact thiolester bond is inactive and cannot act as a ligand for any complement receptor. Only the conformational changes generated during cleavage of the internal thiolester allow recognition of C3 fragments by corresponding receptors. The bond between certain C3 fragments and the leukocyte surface receptor may be achieved either with a free fragment or with a fragment previously bound by an opsonic bond to the bacterial cell or other particle. The following differences can be observed between the interaction with receptor and the opsonin bond:

1. The bond with receptor is not covalent but non-covalent and hydrophobic, hydrostatic and Van der Waals forces are mainly involved.
2. The binding site serves the C3c subdomain and not C3d as for the opsonin bond (*Fig. 10.5*).
3. The interaction with the receptor is specific. Only those fragments which have the appropriate conformation of the binding site can participate in this bond. It is largely limited by the affinity and number of receptors, as

well as by the recycling of the receptors between their cytosol store and the cell surface.

4. When the ligand is in excess, the bond with the receptor is reversible.

Nucleophilic modification of the thiolester during incubation of C3 with nucleophilic low-molecular-weight substances such as methylamine (*Fig. 10.6*) or hydrazine, results in generation of haemolytically inactive molecules (cannot participate in the haemolytic reaction during the classical complement activation pathway). The biological consequences of nucleophilic modification have only recently been elucidated. It appears that modified C3 acquires the ability to bind factor B and form a convertase of the alternative pathway. Under *in vivo* conditions, ammonia acts as a nucleophilic substance. It therefore seems, that, for initiation of the alternative complement activation pathway, the amide C3(NH_3) is even more important than hydrolysed C3(H_2O). In addition, amide C3 may be bound to CR1 on the surface of PMN leukocytes and stimulate release of lysosomal enzymes, even without phagocytosis. Thus, complement may also initiate the inflammatory reaction and damage autologous tissues.

Besides C3 and C4, the internal thiolester is also present in the *α_2-macroglobulin* molecule and in the *pregnancy-zone protein* (PZP). α_2-Macroglobulin contains four thiolester bonds that are identical to those in the C3 molecule. Their main function lies in binding proteolytic enzymes. Therefore both α_2-macroglobulin and PZP are important proteolytic inhibitors.

10.4 Terminal complement components: the membrane attack complex

The C3b fragment, bound in a C5-convertase complex in the classical or alternative pathway, becomes a binding site for C5 — the first of the terminal complement components. Thus, effective convertase enzymes (C2a or Bb) gain the ability to cleave C5 to C5a and C5b. C5b binds to the target surface and C5a is released into the fluid phase. Formation of C5b in the presence of a further glycoprotein, C6, forms the stable complex C5b6. C6, like subsequent complement components, is not degraded into fragments. The C5b6 complex does not, therefore, have enzymatic activity, but can rapidly combine with C7 which yields the hydrophobic complex C5b67. The hydrophobic nature of this complex permits its anchorage (insertion) into the lipid bilayer of the target cell membrane. Insertion of the C5b67 complex into the membrane phospholipid bilayer creates a binding site for C8. The C5b678 complex may create a hole in the target cell membrane and thus initiate its lysis. Its lytic effectiveness is significantly enhanced after C9 binding when a complete **membrane attack complex (MAC)** — C5b6789 — has formed (BHAKDI and TRANUM-JENSEN, 1983; 1984; MÜLLER-EBERHARD, 1984).

Interaction of individual MAC components is of a non-enzymatic nature and is primarily achieved through hydrophobic bonds. MAC contains one molecule of C5b, C6, C7 and C8, but usually between 6 and 19 C9 molecules; its molecular weight is between 1.3×10^6 and 1.6×10^6. With respect to C9, the membrane complex C5b678 fulfils three different functions: (1) it acts as a receptor for C9 and thus directs the lytic effect of C9 to the target cell; (2) it facilitates insertion of C9 into the carbohydrate core of the cell membrane; (3) it catalyses C9 polymerization (Di Scipio and Hugli, 1985).

The cytolytic activity of terminal complement components is regulated by several proteins that are collectively termed **C5b67 inhibitors**. These inhibitors bind the nascent C5b67 and thereby prevent its insertion into the cell membrane. To date, three proteins possessing such inhibitory activity have been identified. (1) Lipoproteins that probably have a similar structure to the cell membrane, allowing the irreversible binding of the C5b67 complex to them. (2) C8 is itself inhibitory if it is bound to C5b67 in the fluid phase, because C5b678 in this phase cannot be incorporated into the cell membrane. (3) **Protein S** that binds to the target membrane at the same time as C5b67. The SC5b67 complex enables binding of C8 and C9; on the other hand, however, it inhibits C9 polymerization. This results in MAC generation that only contains between two and nine C9 molecules and that is not cytolytic for cells other than heterologous erythrocytes. Recently it has been shown that protein S is identical to *plasma vitronectin*, a member of the family of substrate adhesion molecules which include fibronectin and laminin (Podack and Tschopp, 1984; Preissner, 1989).

It has long been known that complement is inefficient in the lysis of homologous erythrocytes. This phenomenon, known as homologous species restriction, is associated with reduced insertion of C9 into the lipid bilayer. Recently, a human erythrocyte membrane protein, responsible for this inefficiency, has been identified. This protein, termed **homologous restriction factor (HRF)**, has a molecular weight of 65 000, a binding affinity for human C8 and C9 and inhibits channel assembly. The importance of HRF in protecting erythrocytes from the lytic action of autologous complement can be seen in its deficiency state. It has been shown that type III *paroxysmal nocturnal haemoglobinuria* (PNH) erythrocytes, which are abnormally sensitive to lysis by homologous and autologous complement, are deficient in HRF. PNH erythrocytes are also deficient in DAF, and thus they lack efficient control of complement convertases. The combined deficiencies of DAF and HRF may explain the intermittent haemolytic anaemia associated with PNH.

The **C5 component** is synthesized as a single-chain precursor protein that splits — prior to secretion into the plasma — into two chains joined by a disulphide bridge. The α- and β-chains contain 262 and 172 amino acid residues respectively. The molecular weight of the whole C5 glycoprotein is 190 000 and its molecule contains 15 % saccharides.

The polypeptide chain of the **C6 component** contains 913 amino acids. It is homologous with the other terminal components of complement C7–C9 (Di Scipio and Hugli, 1989). Specifically, C6 has 29 % of its residues identical with C7. The C6 polypeptide chain is cross-linked by 32 disulphide bonds and has two oligosaccharide groups attached to asparagines located near the amino and the carboxyl termini of the molecule. The organization of secondary structural elements in C6 was elucidated using circular dichroism spectroscopy and an empirical method based on sequence analysis. C6 has an estimated 12 % α-helix, but is comparatively richer in β-sheet (29 %) and β-turns (21 %).

The **C7 component** is formed by a single chain with 821 amino acid residues and mol. wt. 110 kD.

The **C8 component** is composed of three polypeptide chains, α (553 amino acids), β (537 amino acids) and γ with mol. wt. 163 kD.

The **C9 component** (mol. wt. 66 000) has one polypeptide chain with 537 amino acid residues and contains 7.8 % saccharides. Human C9 is a metalloprotein that binds 1 mol of Ca^{2+}/mol of C9 with a dissociation constant of 3 μmol. It is becoming clear that the components of the membrane attack complex, C6, C7, C8 and C9 are structurally, antigenically and functionally related.

The main site of biosynthesis of all complement components are the liver parenchymal cells. Most components are also synthesized in mononuclear phagocytes.

The main function of terminal complement components is to bind to the cytoplasmic membrane of target cells and mediate their damage and lysis. Although the activation of early complement components also takes place on the membrane, only the C5b67 complex is able to penetrate the hydrophobic interior of the membrane. This complex acquires amphiphilic properties and thus also a high affinity for protein–protein interactions. They provoke the conformational changes and exposure of hydrophobic domains, which allow insertion of complement complex components into the hydrophobic interior of the cell membrane. Complexes, containing all terminal C components, can penetrate across the entire cytoplasmic membrane of target cells. The presence of C9 is not required to form channels in the red cell membrane. The C5b678 complex forms channels in the erythrocyte membrane with an inner diameter of 1 nm, whereas in the presence of C9, such membrane channels have an inner diameter of 10 nm and an outer ring-shaped structure with a diameter of 20 nm, as revealed by electron microscopy (HUMPHREY and DOURMASHKIN, 1969) (*Fig. 10.8*).

Based on these findings and known properties of low-molecular-weight ionophores, MAYER (1972) proposed a theory that *stable holes* are formed by the C5b9 complex in the cell membrane (*Fig. 10.9*). The inner walls of the holes are hydrophilic whereas the outer walls are hydrophobic. Hydrophilicity allows the ions and low-molecular-weight substances to flow through the holes.

Under normal conditions, the cells have a lower osmotic pressure than the surrounding medium. This is largely maintained by the sodium pump (Na^+/K^+-ATPase) present in the membrane, which continually pumps Na^+ ions from the intracellular environment against the concentration gradient. In addition, the high protein content in the cell cytoplasm results in only a low osmotic pressure. As soon as the influx of Na^+ and other ions becomes higher than the capacity of the sodium pump (due to the holes in the cell membrane), a phenomenon known as the Donnan effect occurs. The basis of this effect is the increase of the internal osmotic pressure by colloid swelling (namely intracellular proteins), caused by water molecules and low-molecular-weight ions flowing through the holes into the cells. If such osmotic swelling affects a cell with a high intracellular protein content, such as

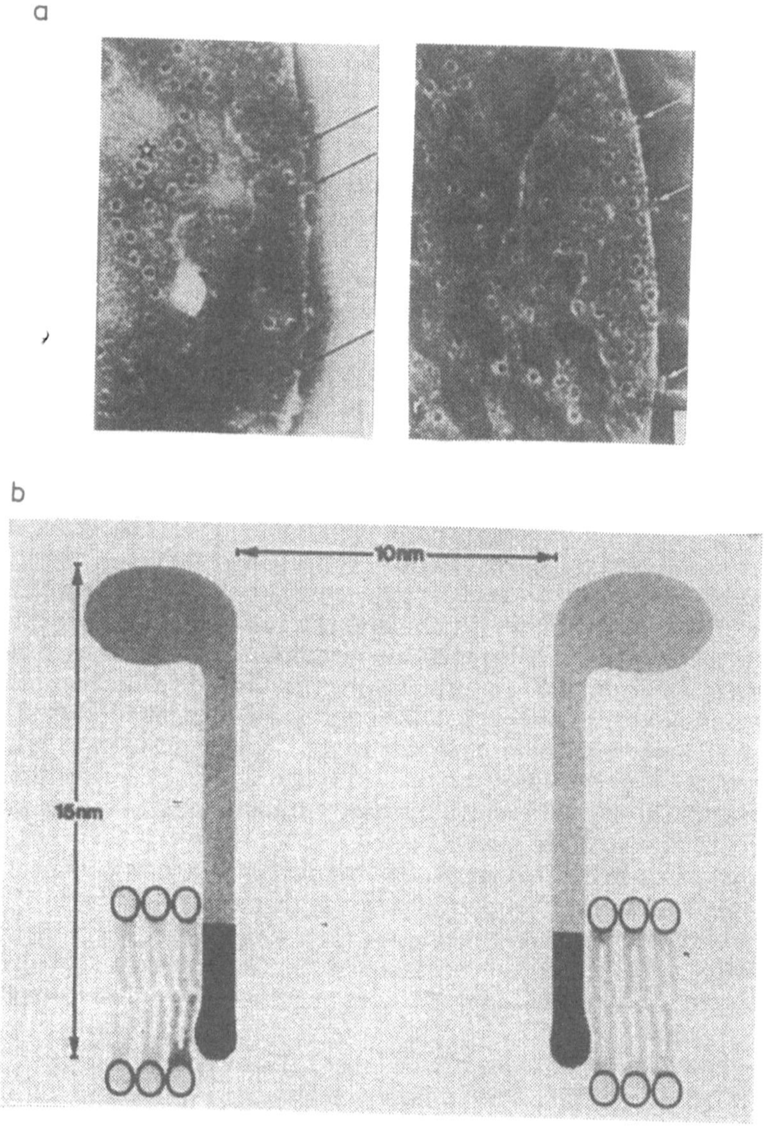

Fig. 10.8. Incorporation of C5b9 complexes into the erythrocyte membrane.
(a) Electron microscopic visualization of complement-mediated channels (arrows mark the profiles of the outer parts of MAC complexes) (b) Insertion of complex into the membrane phospholipid bilayer

erythrocytes, the cell membrane will disrupt and cell lysis occurs. The contents leak into the surrounding environment, and in the case of erythrocytes, results in clearing of their suspension (haemolysis). If anti-erythrocyte antibodies and complement are involved, the phenomenon is called **immune haemolysis** (*Fig. 10.10*). The principle of immune haemolysis is used to determine the total haemolytic complement level.

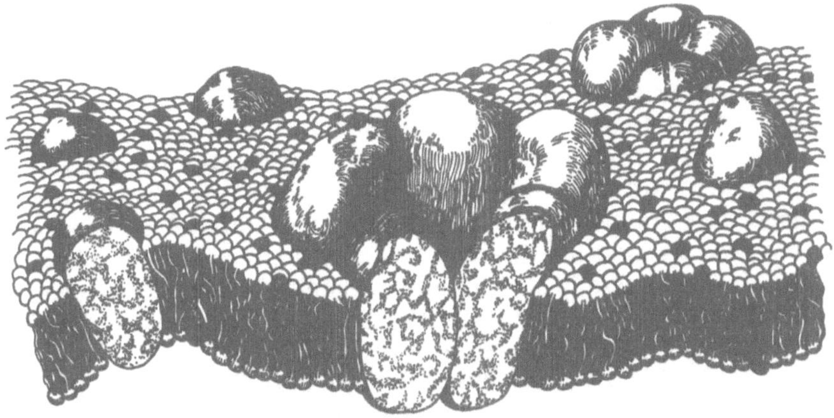

Fig. 10.9. A conception of the mechanism of hole generation in the cell membrane phospholipid bilayer by five terminal complement components.

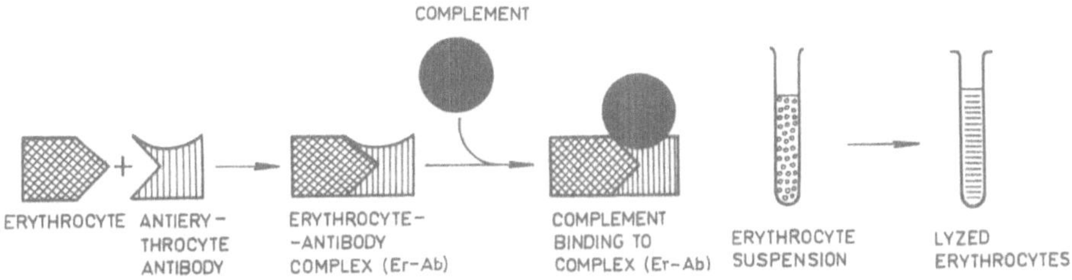

Fig. 10.10. Mechanism of immune haemolysis.

The Mayer theory of permanent holes is not the only theory that attempts to explain the immune haemolysis phenomenon. LACHMANN and THOMPSON (1970) advanced another theory according to which cell lysis is achieved by a *detergent-like effect* of terminal complement components which dissolve the target membrane and subsequently cover it with a "perforated patch". Experimental data, however, including studies on the properties of isolated C5b9 immune complexes, are not in full agreement with any of these theories. BASHFORD *et al.* (1986) suggested that the holes which are observed following MAC insertion into the membrane, are not "true" holes, because they are actually covered with a phospholipid bilayer. Thus, an uncontrolled transport of molecules between the intracellular and extracellular space occurs only at the internal rim of the ring formed by MAC, where the phospholipid bilayer covering the hole does not fit tightly.

Better understanding of the exact mechanism of cell membrane damage is further hampered by two additional facts: first, the ability of even pure poly-C9 to induce cytolysis and, second, the high diversity of cell surfaces which may be damaged by complement.

The **C9-monomer** is a typical *amphipathic protein* — the amino-terminal half of its chain mainly contains hydrophilic amino acid residues, whereas the carboxy-terminal half contains primarily hydrophobic amino acids (DI SCIPIO and HUGHLI, 1985). Under appropriate conditions, even in the absence of the C5b-8 complex, the C9-monomer spontaneously polymerizes. The C9 polymer contains 12 to 18 monomeric units, and its molecular weight is between 700 000 and 1 100 000. Sterically, it forms hollow tubes measuring 2.1 × 1.6 nm. Individual monomers are oriented in the tubes with their hydrophobic ends outwards and their hydrophilic ends inwards. Such an orientation allows the C9-polymer to be inserted into the phospholipid membrane: the membrane components that had been originally present at these sites must therefore be laterally suppressed or completely released. Insertion of large numbers of poly-C9 tubes into the membrane drastically expands its surface, forming ruptures that in turn cause cytolysis.

Besides erythrocytes and other animal cells, complement may also damage the cells of microorganisms; indeed the effect on bacteria is the best understood. Because of the substantially more complex structure of bacterial surfaces (p. 42), the bactericidal effect of complement has certain specific features. Complement cannot directly kill Gram-positive bacteria as their thick cell wall acts as a mechanical inhibitor. However, fragments C3b and iC3b can opsonize these bacteria and thereby significantly facilitate their phagocytosis.

Gram-negative bacteria may be directly killed by complement, although the exact molecular form of MAC required for optimal killing is unknown. To date, it is clear that the cytolytic complex must contain all five terminal components (C5b, C6, C7, C8 and C9). In addition, two membranes — an outer and cytoplasmic membrane — must be damaged. For Gram-negative bacteria, the partial or complete dissolution of the outer membrane and cell wall by complement prior to lysis, is essential. Only then does the cytoplasmic membrane become accessible to MAC. Dissolution of the outer membrane probably results from a detergent-like mechanism and not by osmotic lysis. It thus follows that the terminal complement complexes that lyse erythrocytes and bacteria need not be structurally identical. It appears that the key factor in bacterial killing is the number of MAC-bound C9 molecules, because these molecules are decisive for the size of channels formed in the membrane. For example, JOINER *et al.* (1985) have shown that if the C5b9 complexes contained less than 3.3 C9 molecules, they were unable to kill *E. coli* cells. If, however, MAC contained more than 11 C9 molecules, the *E. coli* cells were killed at a rate directly proportional to the mean total number of C9 molecules bound per bacterium.

10.4.1 Membrane damage by other channel-forming proteins

Formation of channels (or pores) in the cytoplasmic membrane of target cells is not only induced by terminal complement components. A similar capability is also possessed by certain viruses, bacterial and animal toxins, and endogenous and synthetic substances (*Table 10.3*). The term *"channel"* or *"pore"* is not used literally in this context. These terms refer to any structural change in the cytoplasmic membrane of the target cell which results in enhanced membrane permeability.

Table 10.3 Cytolytic substances producing channels and damage to transport functions of cell membranes (Pasternak, 1987)

Group	*Example*
Viruses	Paramyxoviruses, *e.g. Sendai* virus, other viruses with surface envelope at low pH, *e.g.* influenza virus
Toxins — Microbial	*α*- and *δ*- toxin of *Staphylococcus aureus*, streptolysin O, toxin of *Entamoeba histolytica*
— Animal	melitin of bees and wasps
Endogenous substances	MAC (C5b-9), cytolysins of eosinophils, *NK*-cells and cytotoxic *T*-lymphocytes
Synthetic substances	Polylysine and other cationic proteins, detergents (at low concentrations), other "membrane active" compounds

The cytoplasmic membrane of certain cells, *e.g.* erythrocytes, is not folded and these cells are therefore unable significantly to increase their volume. In such cells, complement and other channel-forming proteins produce "true" holes that result in osmotic lysis. Most animal cells, however, possess a folded cytoplasmic membrane and may therefore enlarge their volume by a certain degree; these cells may be demaged by complement without significant cytolysis. The basis of such damage is the abnormal regulation of ion and low-molecular-weight substance transport because of the changes in transmembrane potential induced by poorly fitting "patches" around the channel-forming membrane proteins. An increase in extracellular divalent ion concentration ($Zn^{2+} > Ca^{2+} > Mg^{2+}$) may prevent leakage of substances (*e.g.* phosphorylated metabolites) from cells damaged by the second mechanism. The protective effect of Ca^{2+} and Zn^{2+} for animal cells is manifested at concentrations approaching physiological values. Conversely, however, erythrocytes require extremely high Ca^{2+} concentrations (> 10 mmol/l), which is another cause of their high haemolytic potential.

Nucleated cells can actively defend themselves against the effect of complement or other **channel-forming proteins**. The damaged membrane area, where MAC or other cytolytic substances have formed a channel, can be either excluded from the cell by exocytosis, or, conversely, ingested by endocytosis; exocytosis is usually the predominating mechanism. For example, leukocytes, using such a mechanism, may eliminate MAC-mediated channels from 2% of their surface. This mechanism clearly represents an important defence against random MAC incorporation into the membranes of autologous cells (MORGAN, 1989).

As well as the non-enzymatic (physicochemical) mechanisms mentioned above, the phospholipid bilayer of cell membranes may also be damaged enzymatically, *e.g.* by snake and bacterial toxins which possess phospholipase activity.

From the immunological point of view, it is also important that cytotoxic *T*-lymphocytes (CTL), and probably also *NK*-cells, produce channel-forming proteins (**perforins**) that may penetrate the phospholipid bilayer of the target cell membrane and induce damage visible by electron microscopy. The perforins are localized in cytoplasmic granules from CTL and *NK*-cells and include perforin, seven members of serine proteases termed granzyme A (mol. wt. = 60 000), granzyme B (mol. wt. = 29 000) up to granzyme G, leukolexin and other cytolysins that also participate in the cytolytic process (TSCHOPP and NABHOLZ, 1990). The perforin molecules polymerize in the target cell membrane to form tubular lesions that resemble those produced by the complement component C9. Indeed, there are structural and functional similarities between perforin and C9, C8α, C8β and C7.

Of the various bacterial toxins which are capable of channel formation in cell membranes, the staphylococcal α-toxin and the streptococcal streptolysin-O are the best defined. Most pathogenic strains of *S. aureus* produce **α-toxin** which is assumed to be one of the main pathogenic factors of staphylococci. It is secreted as a monomer with a molecular weight of 34 000. Following contact with an appropriate target membrane, it polymerizes to hexamers that form tubes with an inner diameter of 2–3 nm, and an outer diameter of 8.5–10 nm. These hexamers may be inserted into the cell membranes thereby producing the holes (*Fig. 10.11*). Hexamers of the α-toxin are also formed after contact of α-toxin monomers with serum low-density lipoproteins (LDL). In such cases, however, LDL causes hexamer inactivation, which represents a significant non-immune defence mechanism against the effect of α-toxins.

Streptolysin O (SLO) is secreted by group A β-haemolytic streptococci. It comprises two haemolytically active forms; the first form (mol. wt. 69 000), generating the second (mol. wt. 57 000) by proteolytic degradation. SLO represents a prototype form of bacterial cytolysins with activated —SH groups that are reversibly inactivated by atmospheric oxygen. Following binding to cholesterol in the cell membrane, these cytolysins polymerize, forming twisted rods that drill holes into the phospholipid bilayer (*Fig. 10.11*). SLO oligomers are highly heterogenous and form large channels (diameter 30–35 nm) in the membrane.

10.5 The biological effects of complement

The complement system influences the activity of numerous cells and physiological mechanisms of the organism. These effects may involve either the

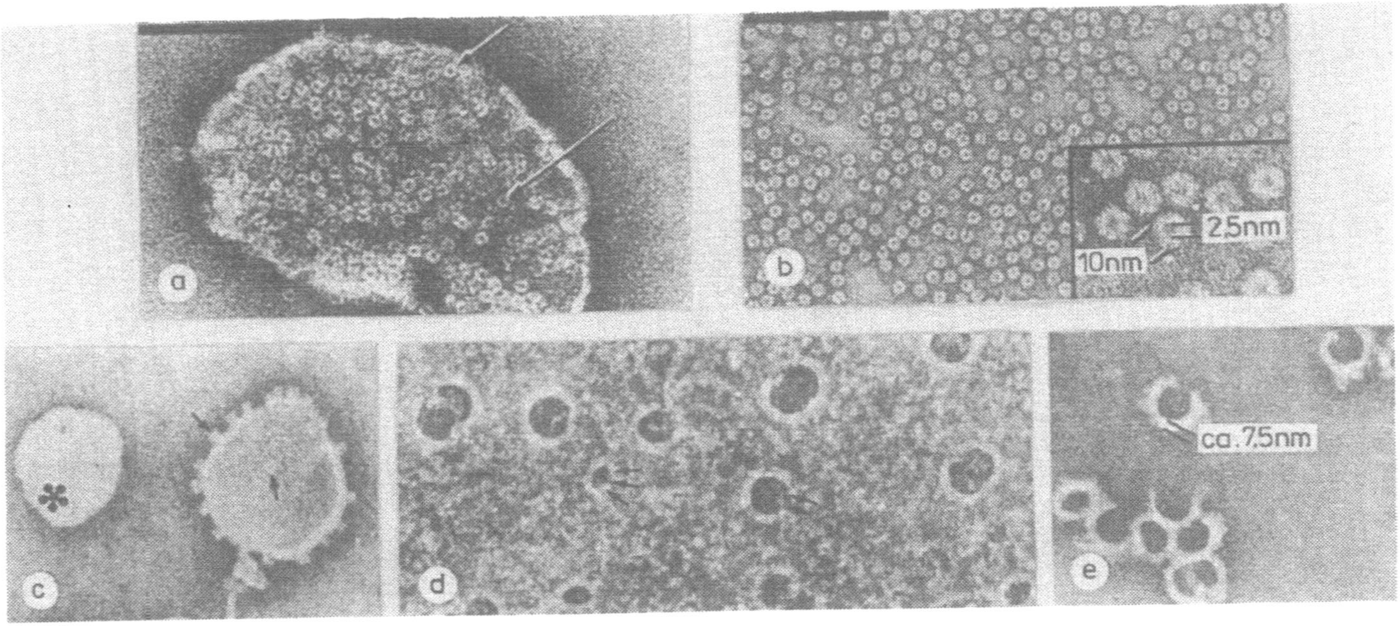

Fig. 10.11. Effect of bacterial toxins on animal cell membrane.

(a) Electronmicrograph of a rabbit erythrocyte lysed by *Staphylococcus aureus* α-toxin (scale bar represent 100 nm) Numerous holes (diameter 10 nm) produced by the α-toxin hexamers are visible in the erythrocyte membrane (b) Isolated α-toxin hexamers (c) Lecithin liposome with α-toxin hexamers incorporated into the membrane (right) The asterisk shows a liposome undisturbed by α-toxin (d) Holes in the erythrocyte membrane produced by streptolysin O (e) Isolated streptolysin O oligomers (Bhakdi and Tranum-Jensen, 1984) whole complement, or only individual components or fragments. Activation of the whole complement system, with the formation of the effector MAC unit, results in cytotoxic and cytolytic reactions. Cytolysis, induced by MAC generated after complement activation *via* the classical pathway in the presence of antibodies, is termed **immune cytolysis**. Target cells for MAC action may be erythrocytes, nucleated animal cells (autologous or foreign), bacteria (Gram-negative, susceptible to serum), microscopic fungi, viruses with a surface envelope and virus-infected cells. The result of cytotoxic complement reaction may be beneficial for the organism (elimination of the infectious agent or damaged cells) or harmful (damage to autologous normal cells by immunopathological reactions).

Different fragments, released from individual components during complement activation, operate a non-cytolytic mechanism through specific receptors present on various cell types. The direction and intensity of the biological response depends on the state of the receptors (affinity and density) and on the function of cells bearing these receptors. From the functional standpoint, complement receptors can be divided into two types: the adherent type and the other receptors. **Adherent receptors** (sometimes also called C3/C4 receptors) mediate adherence of cells and other particles with bound C3 or C4 fragments. These receptors are known as CR1 to CR5 (*Table 10.2*). Adherence reactions mediated through the CR receptors on phagocytes lead to stimulation of phagocytosis, activation of metabolism and secretory functions and movement of phagocytes into the inflammatory site. These receptors, present on other cells of the immune system, are involved in various immunoregulatory reactions. CR1 on erythrocytes may bind circulating immune complexes (that had activated complement) and transport them to the liver where the immune complexes are partially degraded and thus become more soluble.

As far as the adherence receptors are concerned, most data have accumulated about CR1 (p. 266). **CR2 (CD21)**, expressed on human *B*-lymphocytes, is an integral membrane glycoprotein (mol. wt. 140 000) having a close sequence similarity with CR1. It consists of a single polypeptide chain. Its function is unknown, but presumably it acts as a receptor for the Epstein–Barr virus.

CR3 (CD11b/CD18) is a glycoprotein (mol. wt. 249 000) belonging to the trio of adhesive proteins known as the *"LFA-1 family"*. All three glycoproteins (LFA-1, Mo-1 and Leu-M5) are heterodimers composed of two non-covalently linked polypeptide chains α and β; α-chains are different (mol. wt. 155 000 for CR3), β-chains (mol. wt. 94 000) are identical. The CR3 glycoprotein can be detected using the monoclonal antibody Mo-1. It is a lectin-type receptor, similar to bovine conglutinin. Besides iC3b, it may also bind unopsonized yeasts (*Saccharomyces cerevisiae*) and *zymosan* (a polysaccharide in yeast cell walls based on polyglucose). CR3 is distinguished from other C3 receptors by its divalent cation-dependent affinity for iC3b.

CR4 (CD11c/CD18) is the glycoprotein p150,95 that can be detected on cells using the Leu-M5 monoclonal antibody. CR4 has a similar specificity to

CR2; it is present, however, on different cell types (*Table 10.2*) (Ross, 1989; Becherer *et al.*, 1989).

Conglutinin is a glycoprotein present in mammalian sera belonging to the family *Bovidae*. Conglutinin causes aggregation of cells or other material (conglutination). Conglutination may be of immune or non-immune character. Conglutinin is responsible for non-immune conglutination, whereas immune conglutination is induced by immunoconglutinins. **Immunoconglutinins** are actually antibodies against activated complement components and are synthesized at higher rates during various, particularly infectious, diseases.

CR1 and CR3 on neutrophils, monocytes and macrophages are of basic importance for phagocytosis of opsonized particles and for phagocyte metabolism stimulation. Resting non-activated neutrophils and monocytes are unable immediately to begin phagocytosis after contact with C3b- or iC3b-opsonized particles; further stimulus or activation is required. Monocytic phagocytosis is activated by fibronectin or laminin. Monocyte CR1 may also

Table 10.4 Occurrence and biological function of some complement receptors

Receptor	*Ligand*	*Cells*	*Biological function*
Clq	Clq region with collagen structure	PMN, monocytes, lymphocytes	Oxygen metabolism stimulation, stimulation of ADCC
C3a	C3a, C4a	neutrophils	Stimulation of aggregation, secretion of lysosomal enzymes and oxygen metabolism
		monocytes, macrophages	Activation, stimulation of spreading, IL-1 secretion
		mast cells	Induction of release of histamine, leukotrienes and other mediators of anaphylaxis
		T-lymphocytes	Suppression of antibody response
C5a	C5a, $C5a_{desArg}$	neutrophils	Stimulation of chemotaxis, aggregation, lysosomal enzyme secretion, superoxide production
		eosinophils	Stimulation of chemotaxis and granule content release
		basophils	Similar to eosinophils + histamine secretion
		mast cells	Similar to C3a receptor
		monocytes	Stimulation of chemotaxis and PGE_2 production, IL-1 secretion
		macrophages	Activation, stimulation of spreading, lysosomal enzyme secretion, oxygen metabolism, induction of IL-1 production
		T-lymphocytes	Stimulation of antibody response
		platelets	Stimulation of aggregation, secretion of serotonin and lysosomal enzymes
Ba	Ba	monocytes	Stimulation of chemotaxis
Bb	Bb	monocytes	Stimulation of spreading
		lymphocytes	Stimulation of DNA synthesis
H	H	PMN	Stimulation of factor I secretion
		monocytes	Stimulation of oxygen metabolism
		B-lymphocytes	Stimulation of blastogenesis

be selectively activated by a specific lymphokine produced by *T*-cells. Activation of neutrophil CR1 and CR3 for phagocytosis often requires the combined effect of a chemotactic peptide (*N*-formyl-Met-Leu-Phe, C5a) and fibronectin or other activator. Neutrophil CR3 receptors are also specifically activated by zymosan or its active component — β-1,3-glucan.

The second group of receptors reacts with small complement fragments (C4a, C3a, C5a) as well as with C1q, Ba, Bb and H. Stimulation of these receptors results in various biological effects (chemotaxis, secretion of vasoactive amines, mediators of the inflammatory and anaphylactic reactions, *etc.*, *Table 10.4*). Anaphylatoxins are among the most well-defined fragments.

10.5.1 Structure and function of anaphylatoxins

Anaphylatoxins are proteolytic products of the serine proteinases of the complement system: C3a, C4a and C5a. C3-convertase cleaves C3a from C3, C4a is generated by the effect of Cls on C4, and C5a by the effect of C5-convertase on C5. Anaphylatoxins of various animal species contain approximately 75 amino acid residues (human C3a, 77, mol. wt. 9 038; C4a, 77, mol. wt. 8 740; C5a, 74, mol. wt. 11 200). The amino acid sequence has been determined for most of them. All three anaphylatoxins possess considerable homology in their primary structure which confirms their common evolutionary origin. Some of them, *e.g.* human and rat C5a and rat C4a possess a saccharide component. The oligosaccharide chain that binds to human C5a at position 64 has a molecular weight of 2 300. Conformation analyses revealed that 40–50 % of the anaphylatoxin molecule has the α-helix structure (*Fig. 10.12*). The *C*-terminal arginine is of fundamental importance

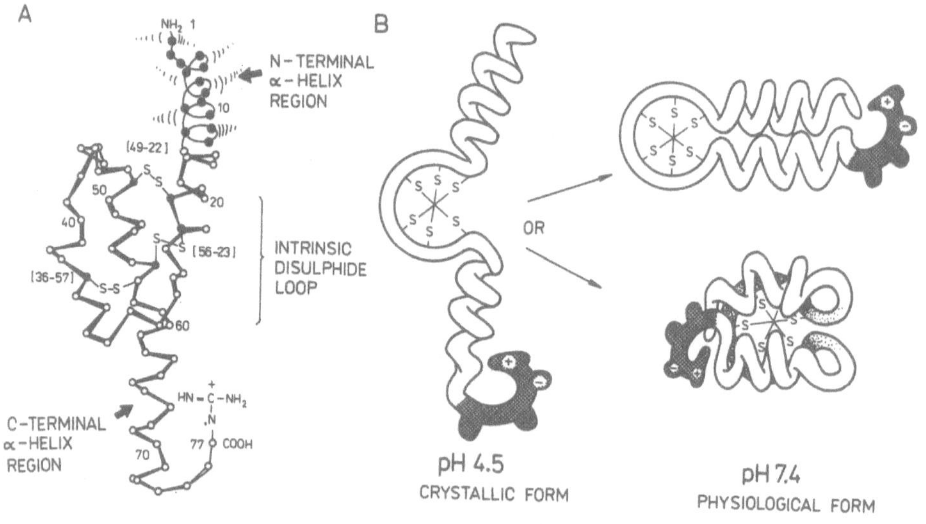

Fig. 10.12. (A) Tertiary structure of human crystalline C3a. (B) Models of human C3a molecules in crystalline form under physiological conditions.
The active site (-Leu-Gly-Leu-Ala-Arg) is cross-hatched (Hugli, 1984).

for the biological activity of C3a. As soon as arginine is removed, the biological activity disappears completely. The peptide, comprising the last 21

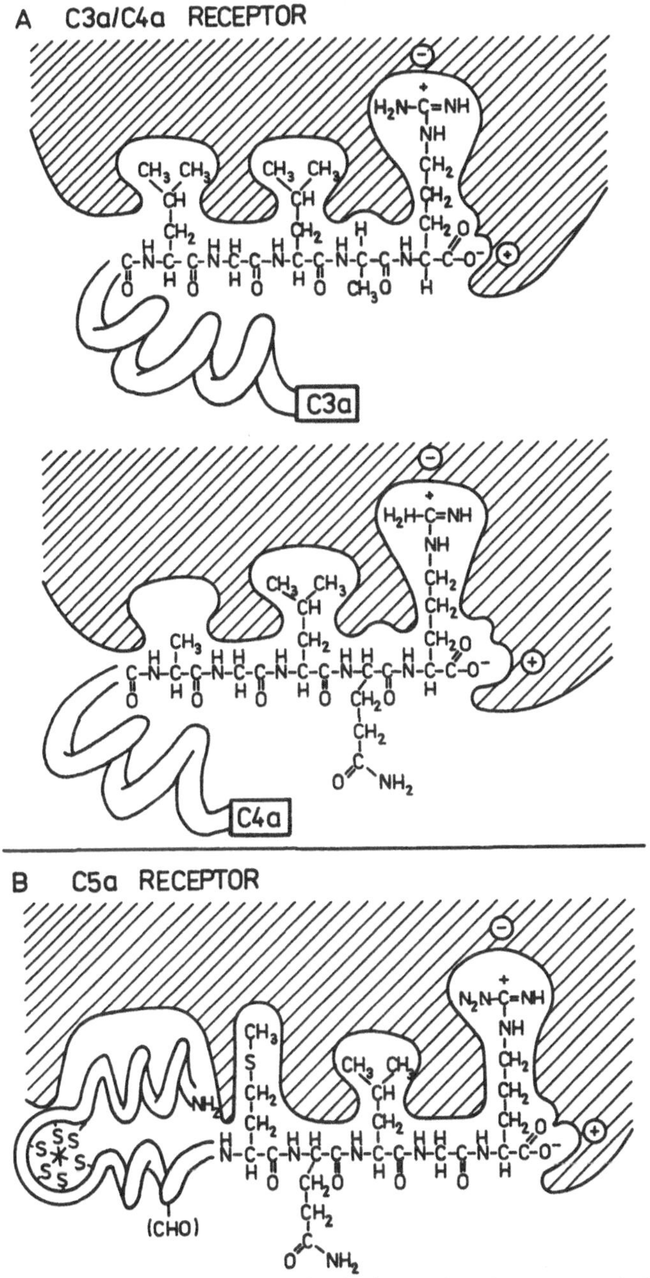

Fig. 10.13. Schematic representation of the molecular interactions between anaphylatoxin active sites and corresponding receptors on the cell surface (cross-hatched) (Hugli, 1984).
(A) C3a/C4a receptor, (B) C5a receptor.

amino acid residues of the *C*-terminus of the C3a molecule, also possesses the biological activity of C3a. For C5a, the situation is more complicated. Removal of *C*-terminal arginine only decreases the biological activity. It seems that C5a binds to a specific receptor near its *C*-terminus, as well as *via* another site in the molecule (*Fig. 10.13*).

Anaphylatoxins meet all the criteria which characterize local hormones. Thus, they possess hormone-like properties, are produced in the blood or other body fluids at the required time and site, are rapidly degraded and influence various tissues and cells *via* specific receptors. The biological response of cells to an anaphylactic signal may be divided into four categories: chemotaxis, morphological and adherence changes, release of granule contents and metabolic activation. **C5a** stimulates *chemotaxis* of neutrophils, eosinophils, basophils and monocytes. It was originally thought that **C3a** also exerts a chemotactic effect on neutrophils; according to contemporary knowledge, however, the C3a chemotactic activity is quite low. One neutrophil possesses an average of $1–3 \times 10^5$ combining sites for C5a. Under the influence of C5a, PMN leukocytes change their shape, increase their adherence to foreign surface and tend to aggregate. Even macrophages change shape and their spreading ability increases. Stimulation of neutrophils, macrophages and other cells with C5a results in numerous metabolic changes such as activation of oxygen metabolism and superoxide production, enhancement of aerobic glycolysis and increased hexose monophosphate shunt activity. Finally, after contact with neutrophils, C5a induces release of lysosomal enzymes. In other cell types, such as mast cells, basophils and eosinophils, C5a induces release of histamine and other vasoactive amines that are contained in the intracellular granules.

Anaphylatoxins also influence other cells and tissues (*Table 10.5*). Among the well-known effects of these substances are the induction of *smooth muscle contraction* and enhancement of *vascular permeability*. These activities decrease in the following order: C5a > $C5a_{desArg}$ > C3a > histamine/bradykinin > C4a. The vasoactive effect, *i.e.* the increase of vascular permeability, is usually secondary due to histamine and prostaglandin (PGE_2) release. The smooth

Table 10.5 Biological activities of anaphylatoxins

Activity	*C3a*	*C4a*	*C5a*
Lethal effect in experimental animals	–	–	+
Lung damage	+	–	+
Increased vascular permeability	+	+	+
Contraction of smooth muscles	+	+	+
Granulocyte chemotaxis stimulation	–	–	+
Induction of neutrophil aggregation	+	–	+
Induction of release from leukocytes:			
lysosomal enzymes	±	±	+
vasoactive amines	+	+	+
reactive oxygen forms	±	–	+
arachidonic acid metabolites	+	±	+

+, activity present; –, activity absent.

muscle contraction in the lungs is primarily mediated by liberated leukotrienes (*slow-reacting substance of anaphylaxis* — **SRS-A**). In this case the activity decreases in the order: C5a > histamine > acetylcholine > C3a ≫ C4a.

Under normal physiological conditions the biological activities of anaphylatoxins are observed during the inflammatory reaction. Most of these activities, however, are components of anaphylactic or allergic (hypersensitivity) reactions that may produce serious damage or even death. Anaphylatoxins are rapidly degraded; anaphylatoxins released into the blood are degraded by *serum carboxypeptidase N* (**SCPN**), whereas receptor-bound anaphylatoxins are internalized (engulfed by the cell) and subsequently degraded within the cell (*Fig. 10.14*).

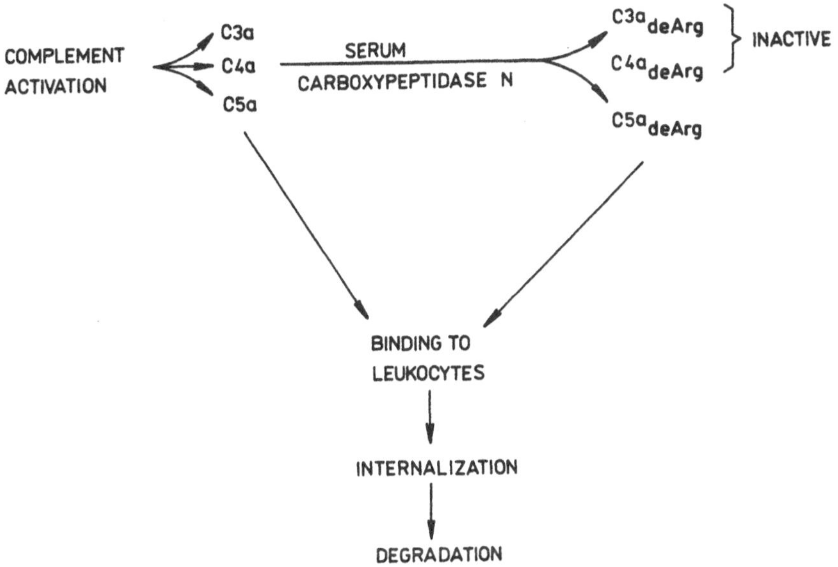

Fig. 10.14. Main mechanisms of anaphylatoxin activity regulation *in vivo*.
Serum carboxypeptidase N rapidly inactivates C3a and C4a by splitting off the *C*-terminal arginine residue. However, human $C5a_{desArg}$ still possesses significant chemotactic activity and its intracellular inactivation occurs only after binding to the corresponding receptor and ingestion (internalization) of this complex

C5a and C3a also possess significant immunoregulatory activities. Thus, C5a enhances antigen-specific and polyclonal antibody response as well as the proliferative response of *T*-cells. In contrast, C3a suppresses these responses. In addition, C5a and to a lesser degree, $C5a_{desArg}$, stimulate monocytes to produce IL-1 and tumour necrosis factor.

10.6 Genetics of complement glycoproteins

The synthesis and function of all complement components and factors is genetically controlled; for most components the localization and arrangement of their structural genes is also known (CAMPBELL *et al.*, 1988). Products of these genes may be structurally and functionally characterized by biochemical and immunological techniques. Studies on these products revealed that the genes for complement components may express two main forms of genetic variability due to mutation changes. The first type is due to the absence of the correct gene, or the existence of null alleles in the corresponding locus. The **null allele** encodes either an afunctional or a missing glycoprotein which results in a specific deficiency of the complement system which, in turn, is manifested by various disease states. Deficiency of complement glycoproteins is transferred to descendants, usually as an autosomal recessive marker. The second form of genetic variability is the existence of **complement component polymorphism**. Occurrence of polymorphic forms (allotypes) has been found in C2, C4, C6, C7, C8, B, D, H, C4-bp, C1-INH, CR1 and others. Most of these allotypic variants have autosomal codominant heredity. Individual variants, except C1-INH, generally do not possess different functional activities; there are, however, differences in the primary structure of their molecules and thus also in charge.

Certain genes which encode complement glycoproteins occur on their own, while others occur in clusters. One such cluster is formed by the human genes for the regulatory components CR1, CR2, C4-bp, DAF, factor H and others on chromosome 1 and is known as the regulator of complement activation (RCA) gene cluster. Since the **RCA gene cluster** encodes the glycoproteins involved in the control of C3-convertases, it acts as the regulatory counterpart of the **class III gene cluster** of the MHC (*Fig. 10.15*) that encodes the structural components of the C3-convertases C2, B and C4 (PORTER, 1984). In spite of their non-syntonic chromosomal location, the functionally related MHC–class III and RCA gene clusters may share a common evolutionary history. The encoded glycoproteins that bind to C3b and/or C4b and share a particular structural organization of repeats of 60 amino acids are characterized by a framework of highly conserved residues.

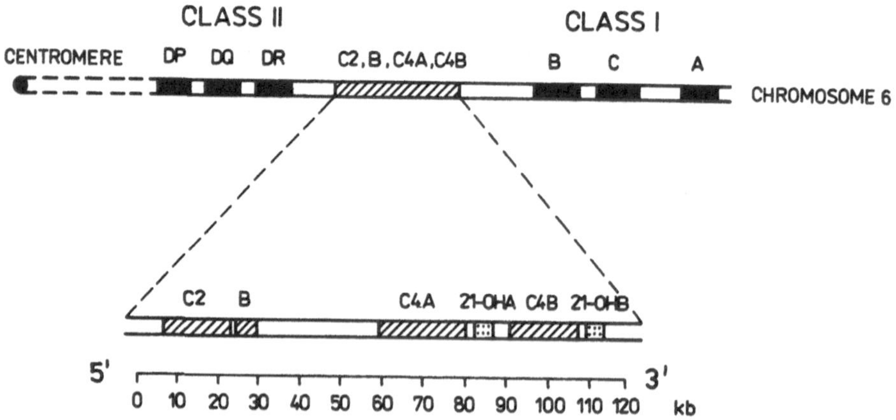

Fig. 10.15. Schematic view of complement genes localized in the HLA region (class III histocompatibility antigens).

21-OHA, 21-OHB, genes for steroid 21-hydroxylases; kb, kilobases in the DNA molecule chain. The direction of gene transcription is from the 5′-end towards the 3′-end.

The loci for C2 and factor B are very close to each other, the C2 locus being nearer to locus DR, which encodes the HLA-DR products (p. 55). C4 synthesis is regulated by two loci, C4A and C4B. Between the gene loci for factor B and C4A in the DNA molecule there is a segment of approximately 30 000 nucleotides. A segment of about 2 000 nucleotides separates the genes for steroid 21-hydroxylase from the 3′-end of C4A and C4B. All four complement genes are found in close proximity which is supported by the fact that no mutual recombinations have been observed. Studies on genetic polymorphism have so far revealed five allotypes of C2, 18 allotypes of factor B, 13 allotypes of C4A and 22 allotypes of C4B.

Phenotypic expression, *i.e.* products of these complement genes, in a given individual is termed a **complotype** (inherited together). There is a characteristic association between complotypes and HLA-haplotypes. This means that some HLA-haplotypes are always associated with certain complotypes which enables better definition of HLA-haplotypes by means of known complotypes.

The similarity in structure, function, biosynthesis and post-translation modification proves that the C2 and B genes have developed from a common locus. Their mutual proximity suggests that the separation arose from tandem DNA duplication. Detailed mapping and sequence analysis have shown that the C2 gene occupies a segment of about 18 000 nucleotides in the DNA molecule, while the factor B gene comprises only 6 500 nucleotides. The difference is caused by longer introns in the C2 gene. The factor B gene consists of 18 exons separated by introns of various length (*Fig. 10.16*).

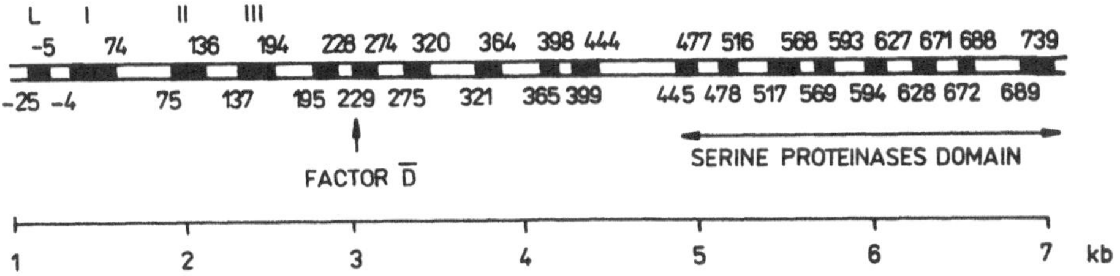

Fig. 10.16. Structure of factor B gene.
Black rectangles represent exons, white rectangles introns L is the leading sequence Exons I, II and III code homologous regions of the Ba fragments The numbers indicate the order of amino acid residues, coded by individual exons Factor D splits the polypeptide chain into Ba and Bb at the site between amino acid residues 234 and 235 The last eight exons code a domain that is responsible for the serine proteinase activity of Bb

Functionally, the products of C4A and C4B are not fully equivalent. Glycoproteins encoded from the C4A locus have significantly lower haemolytic activity than products of C4B genes. This is due to the fact that C4A glycoproteins bind more effectively to protein antigen-containing immune complexes, whereas C4B glycoproteins bind better to polysaccharide antigen-containing immune complexes. Sequence analysis shows that isotypes C4A and C4B differ in only six amino acid residues. Differences in the primary structure among individual allotypic variants within the C4A or C4B isotype might be even smaller. Quite often the difference is based on the exchange of a single amino acid residue. That is enough, however, for the resultant change in the molecular electric charge to allow electrophoretic separation of these two slightly different allotypes. In the human population, the most frequently occurring allele is C4A3 (frequency 0.695), followed by C4A2 (0.080), C4A4 and C4A6 (0.055) *etc.* Even the occurrence of the null allele, C4AQ0, is relatively frequent (0.130). Among C4B products, the most frequent is a product of allele C4B1 (0.760), followed by the null allele C4BQ0 (0.165), C4B2 (0.105) *etc.*

Steroid 21-hydroxylase (21-OH) is a component of the cytochrome P-450 system, present in the microsomal cell fraction; 21-OH genes are localized in close proximity to the C4A and C4B genes. 21-OH participates in steroid biogenesis. Presumably, only the product of the 21-OHA gene (not 21-OHB product) participates in biosynthesis of steroids in the human suprarenal gland (*Fig. 10.15*).

Studies on complement allotypes have contributed to better understanding of the pathogenetic mechanism of some diseases caused by complement glycoprotein deficiency, as well as to the elucidation of the association of some diseases with certain HLA haplotypes.

10.7 Complement deficiencies

Complement disorders may be either primary (genetic, inborn, congenital) or secondary (acquired). The **primary deficiency** is due to an abnormal or absent gene for certain complement glycoproteins. **Secondary disorders** are the result of unfavourable chemical, physical or biological factors. In contrast to the genetic disorders, acquired disorders may be reversible after removal of the harmful factor. Inborn defects of almost all complement glycoproteins have been recorded (*Table 10.6*). These defects, however, are mostly very rare because they are inherited in an autosomal recessive manner and therefore occur only in homozygous individuals. An exception is the **C1-inhibitor deficiency** that is inherited as an autosomal dominant and therefore also affects heterozygotes. Among other defects, the most frequent is C2 and C9 deficiency (several tens of cases). The most severe clinical course is seen in C3, CR3 and C1-INH deficiencies (FRIES *et al.*, 1986; STARŠIA, 1987).

The functional defect of C1-inhibitor may be caused either by partial deficiency (decreased level), afunction of the molecule, or by the presence of autoantibodies against C1-INH. As soon as the C1-INH level has decreased below 30% of the normal level (several hundred recorded cases), the deficiency is manifested as a *hereditary angiooedema*. The disease is characterized by recurrent oedema of the skin and mucous membranes of the respiratory, gastrointestinal and urogenitary tract. Tracheal oedema may cause suffocation. The oedema results from uncontrolled complement activation with subsequent formation of anaphylatoxins and kinins (primarily C4a and C2b) that enhance vascular permeability. Oedema occurs either spontaneously or as a response to a relatively small injury. C1-inhibitor is encoded by two genes (one maternal, one paternal). If one gene is normal and the second is mute (no transcription), a decreased C1-INH level is present in about 85% of patients. About 15% of those suffering from hereditary angiooedema have normal or even increased C1-INH blood levels; however, these molecules are structurally abnormal and afunctional. In these cases one gene is normal while the second is abnormal. C1-INH deficiency may also originate secondarily during certain lymphoproliferative diseases. Despite the uncontrolled activation of early complement components, patients with hereditary angiooedema are not more susceptible to infections, although an increased occurrence of autoimmune diseases is observed.

Table 10.6 Inborn deficiencies or abnormalities of complement glycoproteins

Glycoprotein deficiency	*Disease*	*Heredity*
C1q	SLE, glomerulonephritis, CHPI	AR
C1r	SLE, CHPI	AR
C1s	SLE	?
C4	SLE, Sjögren's syndrome, CHPI, autoimmune diseases, insulin-dependent diabetes mellitus	AR
C2	SLE, juvenile rheumatoid arthritis, glomerulonephritis	AR, HLA-association
C3	CHPI, glomerulonephritis, SLE, rheumatoid arthritis, atherosclerotic vascular disease	AR
C5	Chronic *Neisseria* infections, SLE	AR
C6	Chronic *Neisseria* infections	AR
C7	Chronic *Neisseria* infections	AR
C8-β	*Neisseria* infections, polyarthritis	AR
C8-α, γ	*Neisseria* infections, SLE	AR?
C9	Sclerodermia, purpura	AR
B	Optic neuritis, coeliac disease, ankylosing spondylitis	AR?
D	CHPI	?
P	CHPI	XR
C1-INH	Hereditary angioedema	AD
H	?	AR
I	CHPI	AR
CR1	SLE	AR
CR3	CHPI, immunoadherence defect, leukocytosis	AR

SLE, systemic lupus erythematosus; CHPI, chronic pyogenic infection. Heredity: AR, autosomal recessive; AD, autosomal dominant; XR, X-linked, recessive.

Deficiency of the first complement components and CR1 is usually associated with *immune complex diseases* in which there is a decreased ability to take up and solubilize immune complexes. C3 deficiency is among the most severe conditions. Such patients suffer from frequent infections usually caused by *Streptococcus pneumoniae* and *Neisseria meningitidis*. Some patients also suffer from glomerulonephritis and systemic lupus erythematosus. Even patients with deficiency of the terminal components C5, C6, C7 and C8 suffer from chronic infections caused by *Neisseria meningitidis* and *Neisseria gonorrhoea*. As the C8 molecule is encoded by two genes (one for the β-chain, the second for the α- and γ-chains), two types of deficiency can be observed: C8β and C8α-γ deficiency. C8β deficiency occurs mostly in the Caucasian population, whereas C8α-γ deficiency occurs mainly in the Negro population. About 80% of patients suffering from C9 or factor B deficiency have no infectious complications.

Most complement glycoproteins including regulatory components display extensive polymorphism and individual alleles may be associated with different disorders. For example, C4AQ0B1 allele is associated with systemic lupus erythematosus (SLE), Sjogren's syndrome and insulin-dependent diabetes mellitus; C4A3B3 with rheumatoid arthritis and insulin-dependent diabetes mellitus; C4A4B2 with congenital C2 deficiency; and C4A6B1 with psoriasis (Kay and Papadimitriou, 1990). The electrophoretically fast form of C3, C3F, has been shown to be associated with disorders such as rheumatoid arthritis, atherosclerotic vascular disease, a juvenile onset form of SLE and membrane-proliferative glomerulonephritis. By contrast, SLE is associated with a null allele (C4AQ0) at the C4A locus.

References

Bashford, C. L., Alder, G. M., Menestrina, G., Micklem, K. J., Murphy, J. J. and Pasternak, C. A. (1986) Membrane damage by haemolytic viruses, toxins, complement and other cytotoxic agents: a common mechanism blocked by divalent cations. *J. Biol. Chem.*, **261**, 9300–8.

Becherer, J. D., Alsenz, J., Servis, C., Myones, B. L. and Lambris, J. D. (1989) Cell surface proteins reacting with activated complement components. *Complement Inflamm.*, **6**, 152–65.

Bhakdi, S. and Tranum-Jensen, J. (1983) Membrane damage by complement. *Biochim. Biophys. Acta*, **737**, 343–72.

Bhakdi, S. and Tranum-Jensen, J. (1984) Mechanism of complement cytoplysis and the concept of channel-forming proteins. *Phil. Trans. R. Soc. Lond.*, **B306**, 311–24.

Campbell, R. D., Law, S. K. A., Reid, K. B. M. and Sim, R. B. (1988) Structure, organization, and regulation of the complement genes. *Annu. Rev. Immunol.*, **6**, 161–96.

Di Scipio, R. G. and Hugli, T. E. (1985) The architecture of complement component C9 and poly(C9). *J. Biol. Chem.*, **260**, 14802–9.

Di Scipio, R. G. and Hugli, T. E. (1989) The molecular architecture of human complement component C6. *J. Biol. Chem.*, **264**, 16197–206.

Fries, L. F., O'Shea, J. J. and Frank, M. M. (1986) Inherited deficiencies of complement and complement-related proteins. *Clin. Immunol. Immunopathol.*, **40**, 37–49.

Hostetter, M. K. and Gordon, D. L. (1987) Biochemistry of C3 and related thiolester proteins in infection and inflammation. *Rev. Infect. Dis.*, **9**, 97–109.

Hourcade, D. E., Holers, V. M. and Atkinson, J. P. (1989) The regulators of complement activation (RCA) gene cluster. *Adv. Immunol.*, **45**, 381–416.

Hugli, T. E. (1984) Structure and function of the anaphylatoxins. *Springer Semin. Immunopathol.*. **7**, 193–219.

Humphrey, J. H. and Dourmashkin, R. R. (1969) The lesions in cell membranes caused by complement. *Adv. Immunol.*, **11**, 75–115.

Joiner, K. A., Schmetz, M. A., Sanders, M. E., Murray, T. G., Hammer, C. H., Dourmashkin, R. and Frank, M. M. (1985) Multimeric complement component C9 is necessary for killing of *Escherichia coli* J5 by terminal attack complex C5b-9. *Proc. Natl. Acad. Sci. USA*, **82**, 4802–12.

Kay, P. H. and Papadimitriou, J. M. (1990) What's new in the role of complement in diseases? *Pathol. Res. Pract.*, **186**, 410–4.
Lachmann, P. J. and Thompson, R. A. (1970) Reactive lysis: the complement-mediated lysis of unsensitized cells. II. The characterization of activation reactor as C6b and the participation of C8 and C9. *J. Exp. Med.*, **131**, 644–57.
Lambris, J. D. (1988) The multifunctional role of C3, the third component of complement. *Immunol. Today*, **9**, 387–92.
Mayer, M. M. (1972) Mechanism of cytolysis by complement. *Proc. Natl. Acad. Sci. USA*, **69**, 2954–9.
Morgan, B. P. (1989) Complement membrane attack on nucleated cells: resistance, recovery and non-lethal effects. *Biochem. J.*, **264**, 1–14.
Müller-Eberhard, H. J. (1984) The membrane attack complex. *Springer Semin. Immunopathol.*, **7**, 93–141.
Müller-Eberhard, H. J. (1988) Molecular organization and function of the complement system. *Annu. Rev. Biochem.*, **57**, 321–347.
Pasternak, C. A. (1987) Virus, toxin, complement: common actions and their prevention by Ca^{2+} or Zn^{2+}. *BioEssays*, **6**, 14–19.
Podack, E. R. and Tschopp, J. (1984) Membrane attack by complement. *Mol. Immunol.*, **21**, 589–603.
Porter, R. R. (1984) The complement components of the major histocompatibility locus. *CRC Crit. Rev. Biochem.*, **16**, 1–19.
Preissner, K. T. (1989) The role of vitronectin as multifunctional regulator in the hemostatic and immune systems. *Blut*, **59**, 419–31.
Reid, K. B. M. and Porter, R. R. (1976) Subunit composition and structure of subcomponent C1q of the first component of human complement. *Biochem. J.*, **155**, 19–23.
Ross, G. D. (1989) Complement and complement receptors. *Current Opin. Immunol.*, **2**, 50–62.
Staršia, Z. (1987) *Clinical importance of complement*. Prague, Avicenum, pp. 53–95 (*in Slovak*).
Tschopp, J. and Nabholz, M. (1990) Perforin-mediated target cell lysis by cytolytic *T* lymphocytes. *Annu. Rev. Immunol.*, **8**, 279–302.

11 Antigen–antibody reactions *in vitro*

Antigen can react with antibodies *in vivo* or *in vitro*. The *in vivo* reaction can be beneficial for the organism (immunity), harmful (immunopathological reactions) or indifferent (immune system tolerates, rather than responds to the antigen). The *in vitro* reactions are the basis for immunochemical methods which depend on biospecific binding between binding sites of the antibody and determinant groups of the antigen resulting in formation of antibody–antigen complexes (**immune complexes**).

According to Zikán (1986a, b), the antigen–antibody reaction involves non-covalent interactions similar to those occurring in other biospecific reactions (*e.g.* substrate–enzyme, hormone–hormone receptor, *etc.*). However, compared to enzymes and some hormones, antibodies do not change the structure of the antigen irreversibly.

11.1 Molecular forces responsible for the antigen binding

The forces involved in stabilizing the spatial structure of proteins and other macromolecules are also responsible for the stability of immune complexes. They are based on weak non-covalent interactions between different functional groups of the antigen and antibody. The interactions include hydrogen bonds, non-polar hydrophobic interactions, Coulomb, Van der Waals and London dispersion attractive forces and sterical repulsive forces. True chemical (covalent) bonds are not involved at all.

Hydrogen bonds result from interaction of the electron-deficient hydrogen atom (proton) with two atoms of high electron density (electronegative atoms). During this interaction, bridges between such atoms are formed. In proteins, hydrogen bonds are formed between hydrophilic groups (—OH, —NH_2, —COOH)

—O...H...O—
—O...H...N—
—N...H...N—

According to the type of reactive group, hydrogen bonds have a different energy. Water decreases the hydrogen bond energy between two electronnegative atoms of the polypeptide chain because its molecules can themselves be involved in hydrogen bond formation.

Hydrophobic reactions take place when two hydrophobic surfaces are sufficiently close. In protein molecules, the amino acids leucine, isoleucine, valine and phenylalanine are usually involved. They do not form hydrogen bridges with water molecules and, therefore, prefer mutual interactions in aqueous solutions, during which water molecules surrounding hydrophobic groups are removed. In aqueous environments, protein molecules therefore assume such a conformation, in which their hydrophobic groups come as close as possible and their contact with water molecules is either excluded or limited. Molecules arranged in this way assume an energetically more advantageous state acquiring entropy. In antigen–antibody reactions, hydrophobic interactions are the most important.

Coulomb forces are generated by the mutual attraction of electrically oppositely charged functional groups or molecules. Terminal amino acid residues, as well as lysine, arginine, histidine, aspartic acid and glutamic acid are the main charged groups in protein molecules. Coulomb forces are often involved in antigen–antibody interactions, although their effect is not as important as that of the hydrophobic forces. For instance, antibodies with a net negative charge are produced against positively charged haptens and *vice versa*.

A charge on one molecule or atom group can induce dipole formation on another molecule. Such interactions belong to the so-called **Van der Waals forces**. They are based on the mutual interaction of electron clouds of two polar atom groups. The mutual interaction of two electron envelopes leads to formation of oscillating dipoles on both groups. The dipoles generated in this way (one part of the atom groups has a partial positive charge and the other has a partial negative electric charge) attract each other at sites carrying opposite charges. Van der Waals forces play a major role in stabilization of immune complexes.

Similar attractive forces are also generated during interaction of electron clouds of two non-polar atom groups. They are called **London dispersion forces**. Dispersion forces are determined by fluctuation of electrons and, therefore, do not require permanent dipoles. They can be explained by laws of quantum mechanics.

Bond energy significantly increases with decreasing distance between two reacting groups. In electrostatic Coulomb interactions, the attractive force (F) is indirectly proportional to the square of the distance (d) between charges: $F = 1/d^2$. However, in dispersion forces it is indirectly proportional to the seventh power of the distance ($F = 1/d^7$). This indicates that Coulomb forces act at a greater distance than dispersion forces.

For all forces involved in antigen–antibody interaction both reacting atom groups must be as close as possible. Only thus can attractive forces play a significant role. This is the basis of biospecific antigen–antibody binding. In order to involve attractive forces, steric repulsive forces must be overcome.

Steric (spatial) **repulsive forces** are generated between two atoms which are not bound by chemical bonds, on the basis of a mutual penetration of their electron clouds. The higher the complementarity of the two clouds, the smaller the repulsive forces between them. These forces are the basic factor by which the antibody "selects" a suitable antigen for mutual interaction. Non-complementary (non-specific) antigen determinants do not have binding sites on the antibody molecule and, therefore, repulsive forces between them are high to prevent or minimize binding (cross-reactions). When the electron clouds of the determinant and the binding site are complementary, the repulsive forces are very small and the attractive forces involved in immunospecific binding can predominate. In this case the antibody has a high affinity for the antigen.

11.2 Equilibrium constants

During the reaction of the antigen determinant (it actually acts as a monovalent hapten, H) and the antibody binding site (Ab) an equilibrium is established. It can be expressed as:

$$[\mathrm{Ab}] + [\mathrm{H}] \underset{k_d}{\overset{k_a}{\rightleftharpoons}} [\mathrm{AbH}]$$

Square brackets indicate the equilibrium molar concentrations of the components, k_a is the association constant and k_d is the dissociation constant.

According to the **Guldberg–Waage law** of active mass action, the rate of complex formation is proportional to the concentrations of both reacting components. The rate of association then equals k_a [Ab] [H] and the rate of dissociation equals k_d [AbH]. At equilibrium, the association rate equals the dissociation rate

$$k_a\,[\mathrm{Ab}]\,[\mathrm{H}] = k_d\,[\mathrm{AbH}]$$

or

$$\frac{k_a}{k_d} = K = \frac{[\mathrm{AbH}]}{[\mathrm{Ab}]\,[\mathrm{H}]}$$

where K is the **equilibrium association constant**. This characterizes bond efficiency and its values in antigen–antibody reactions vary between 10^5 and 10^{11} mol/l.

The thermodynamic determination of the equilibrium association constant in the interaction of a polyvalent antigen is highly complicated, and is much simpler for a monovalent hapten than a complex antigen and for a monoclonal antibody or conventional antibody with equivalent (homogeneous) binding sites. The equilibrium association constant can be determined experimentally when solutions of the hapten and the antibody are mixed and, after the equilibrium has been established, concentrations of the free hapten and the hapten bound with the antibody are determined. The results are substituted in the *Scatchard* or *Langmuir equations*, from which K is calculated. Both equations can be derived from the Guldberg–Waag law:

$$\frac{[\text{AbH}]}{[\text{Ab}]} = r = \frac{nK\,[\text{H}]}{1 + K\,[\text{H}]}$$

where r represents moles of hapten bound by one mole of antibody, [H] is the molar concentration of free hapten and n is the antibody valency. It follows from this relationship that:

$$\frac{r}{[\text{H}]} = nK - rK$$

The Scatchard equation can be graphically illustrated as the relationship between $r/[\text{H}]$ and r. A line is obtained, from which both K and n can be easily calculated (*Fig. 11.1*).

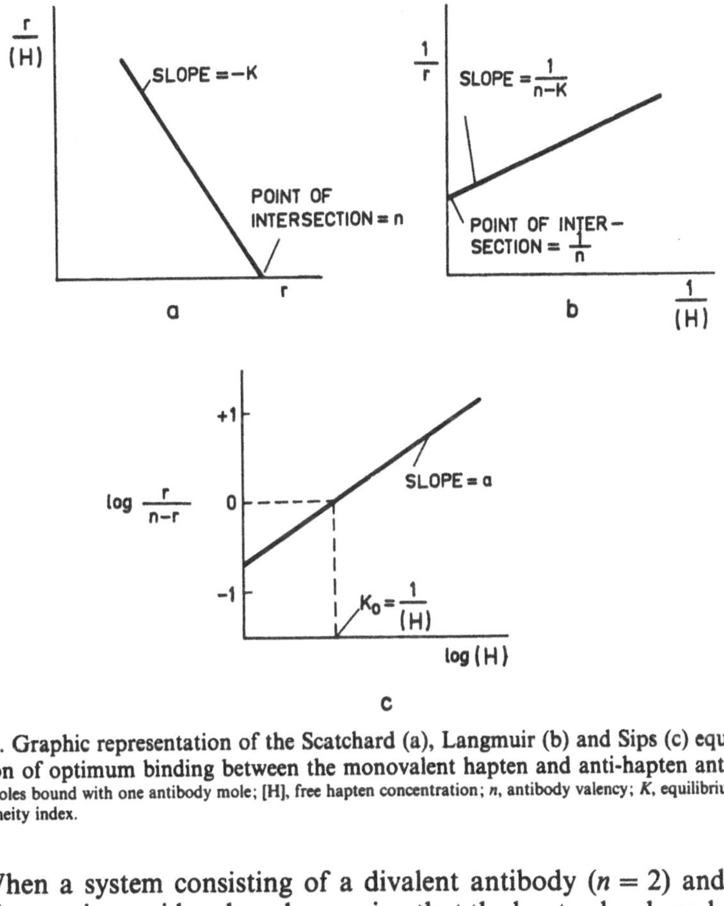

Fig. 11.1. Graphic representation of the Scatchard (a), Langmuir (b) and Sips (c) equation for expression of optimum binding between the monovalent hapten and anti-hapten antibody.
r, hapten moles bound with one antibody mole; [H], free hapten concentration; n, antibody valency; K, equilibrium constant; a, heterogeneity index.

When a system consisting of a divalent antibody ($n = 2$) and monovalent hapten is considered, and assuming that the hapten has been bound to exactly one half of the antibody binding sites ($r = 1$), the preceding equation can be rewritten as

$$\frac{1}{[\mathrm{H}]} = 2K - K = K_0$$

The constant K_0 is known as the **internal association constant**. It equals the reciprocal value of the free hapten concentration when the molar ratio of the antibody–hapten complex is 1 : 1. Both K_0 and K are expressed in l/mol.

By rearranging the equation $r/[\mathrm{H}] = nK - rK$, the relationship:

$$\frac{1}{r} = \frac{1}{n}\frac{1}{[\mathrm{H}]}\frac{1}{K} + \frac{1}{n}$$

is obtained which can also be illustrated graphically, *i.e.* as the relationship of the reciprocal values of r and [H] (*Langmuir adsorption isotherm*).

When a polyvalent antigen reacts instead of the monovalent antigen, the Scatchard plot assumes the shape:

$$\frac{r}{[\mathrm{H}]} = nK - srK$$

where s is antigen valency. In this case, the equilibrium association constant can be determined only if each antibody binding site reacts with a single determinant, which is always on another antigen molecule.

Scatchard or Langmuir equations characterize the affinity of the hapten for the specific antibody and, in addition, the affinity of any ligand for its specific binding site, *e.g.* the binding of a low-molecular-weight substrate to an enzyme binding site. The enzyme–substrate interaction can be expressed by a straight line. During interactions of conventional antibodies with haptens, deviations from this linear relationship can be seen. These are caused by heterogeneity of antibodies and resulting heterogeneity of their binding sites.

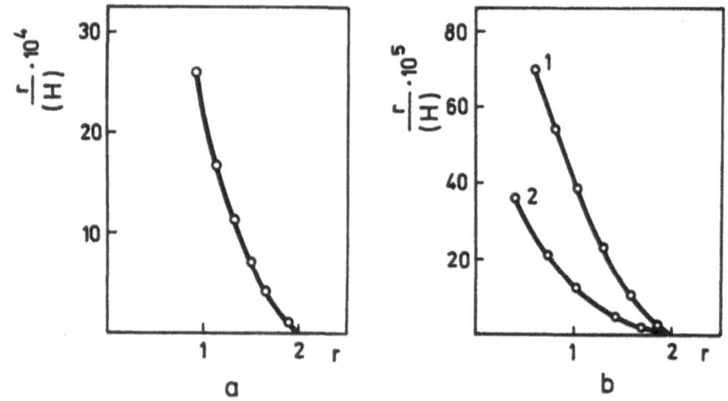

Fig. 11.2. Illustration of binding between certain haptens and anti-hapten antibody according to the Scatchard equation.

(a) Hapten, arsenylic acid, (b) Hapten, dinitrophenol, 1, 2, two anti-dinitrophenyl antibodies with different affinity In these examples the antibodies are divalent

Heterogeneity of antibodies (generated in response to a single antigen) accounts for the fact that, in a certain population, all molecules of antibodies do not have an identical affinity for the antigen. The resultant curve, rather than a straight line, makes it impossible to determine K (*Fig. 11.2*).

Several mathematical solutions have been proposed to solve this problem. One of these is based on expression of distribution of antibodies of the same population but with different affinities using Gauss or Sips distribution functions.

Sips equation:

$$\frac{r}{n} = \frac{(K[\mathrm{H}])^{\mathrm{a}}}{1 + (K[\mathrm{H}])^{2}}$$

can be expressed logarithmically:

$$\log \frac{r}{n - r} = a \log K + a \log [\mathrm{H}]$$

where a is the so-called *heterogeneity index* and the other symbols have the same meaning as in the previous equations. The heterogeneity index determines the extent to which the equilibrium constants deviate from the mean constant and has a function similar to that of mean standard error in statistical analysis. The latter equation can be illustrated graphically as the relationship of $r/(n - r)$ and log [H]. A straight line is thus constructed and its slope determines the heterogeneity index a (*Fig. 11.1*). When

$$\log \left(\frac{r}{(n - r)}\right) = 0, \; K_0 = \frac{1}{[\mathrm{H}]}$$

the Sips equation makes it possible to determine the internal association constant K_0, as well as the heterogeneity index. Values a vary between 1 and 0. When $a = 1$, all antibody binding sites are homogeneous (equivalent). In this case, both the Scatchard and Langmuir equations can be expressed as a straight line. The lower the a value, the higher the heterogeneity of the antibody.

11.3 Affinity

The intensity of interaction between the antibody combining site and the antigen determinant (hapten) characterizes the affinity of antibodies to a particular antigen (hapten). **Affinity** is then a thermodynamic expression of the mean binding energy of the antibody for one antigen determinant. In thermodynamics it is known as the "standard chemical affinity" ($A°$), but only as "affinity" in the immunological literature. At equilibrium it is equal to the negative change of standard Gibbs energy:

$$A° = -\Delta G°$$

and is related to the association constant according to:

$$-\Delta G^\circ = RT \ln K$$

where R is the gas constant and T is the temperature in degrees Kelvin. It follows from this relationship that the higher the antibody affinity, the higher the G° value. Gibbs energy includes two components — enthalpy (H) and entropy (S), so that:

$$\Delta G = \Delta H - T\Delta S$$

where the enthalpy change (ΔH) is the heat absorbed or released during association. Negative ΔH values characterize exothermic reactions (heat release) whereas positive ΔH values indicate endothermic reactions (heat absorption). Standard enthalpy change can be calculated from the temperature dependence of the association constant:

$$\frac{\mathrm{d}\ln K}{\mathrm{d}K} = \frac{\Delta H^\circ}{RT^2}$$

or it can be determined calorimetrically. Entropy, the second component of Gibbs energy, expresses the degree of ordering of the system. The higher the entropy (degree of disorder), the higher its probability.

Hydrogen bond formation is usually associated with negative ΔH° and ΔS° values. Hydrophobic interactions usually have weakly positive ΔH° and highly positive ΔS°. In Coulomb interactions ΔS° is also positive, due to disarrangement of water molecules. Non-polar compounds increase the intensity of Coulombian interactions and decrease that of hydrophobic interactions. Conversely, increasing the ionic strength of the environment decreases the intensity of Coulombian, and increases the intensity of hydrophobic interactions.

Antigens usually include several determinants and, therefore, the term "**avidity**" was introduced to characterize the binding energy between a complex antigen and its antibody. Avidity depends on affinity, although the valency of the antigen and antibody, as well as non-specific factors which affect the strength of the antigen–antibody bond must also be considered. The number of sites on the antibody molecule able to react with determinants of a specific antigen is termed its valency, whereas antigen valency is characterized by the number of determinant groups which can bind to the antibody. Non-specific factors refer to the mutual interactions of parts of molecules other than the antibody binding sites and antigen determinant.

During the reaction of a complete antigen with its antibody, multivalent interactions take place in addition to the specific interaction of the binding site with the determinant group. It therefore follows that the terms "affinity" and "avidity" are not synonymous and that their meanings are different. KARUSH (1962) characterizes multivalent interactions by **functional affinity**, as compared with univalent interactions exhibiting **internal affinity**. Multivalency increases internal affinity by multiple binding between molecules of the antigen and molecules of the antibody.

For instance, polyvalent IgM antibodies, whose binding sites have the same affinity to the antigen as bivalent IgG antibodies, have a higher avidity to the multivalent antigen than IgG antibodies. IgM antibodies are more "avid", including more binding sites and having larger molecules and, therefore, have a greater ability to bind the antigen.

11.3.1 How is antibody affinity measured?

The equilibrium constant of the antigen–antibody reaction is usually a measure of affinity. In practice, the amounts of free antigen and antibody-bound antigen are determined at several antigen concentrations, and after equilibrium has been reached. Analogously, it is also possible to determine the amount of free and bound antibody at several antibody concentrations. For this purpose several methods may be used to determine either the internal affinity or the functional affinity. Secondary phenomena such as precipitation, agglutination and non-specific interactions of immunoglobulins with Fc-receptors and complement components complicate the determination (ZIKÁN, 1986a).

Equilibrium dialysis, membrane filtration, spectrofluorometric (fluorescence quenching) and separation methods (gel filtration, affinity chromatography, electrophoresis and ammonium sulphate precipitation) are the most frequently used procedures.

Equilibrium dialysis is based on the impermeability of a dialysis membrane to the antibody and its permeability to the low-molecular-weight hapten. It is performed in a vessel divided into two parts by the dialysis membrane. One compartment contains a solution of a pure antibody, the other contains a hapten labelled with, for example, a radioactive isotope. The membrane allows the free diffusion of hapten molecules, but antibody molecules cannot penetrate the membrane. The hapten diffusing to the antibody compartment can react with the antibody and thus form immune complexes. Equilibrium is reached when the hapten has an identical concentration in both vessel compartments. Radioactivity in one compartment then reflects the amount of free hapten plus the amount of hapten bound in the immune complex, whereas the radioactivity in the other compartment reflects only the concentration of the free hapten. This process is repeated several times with a different hapten concentration but always with the same antibody concentration. The values obtained are substituted in Langmuir or Sips equations and affinity may then be calculated. Equilibrium dialysis is rather time-consuming since the time required to attain equilibrium is relatively long (16–48 h).

Membrane filtration also involves separation of the bound and free component by a semipermeable membrane. However, it is faster than equilibrium dialysis and can be used for both the system antibody–low-molecular-weight hapten and the system antibody–high-molecular-weight antigen. However, the relative molecular weights of the antigen and antibody must differ significantly. Two methods can be used. In the first procedure, a hapten solution is mixed with a specific antibody solution volume in a vessel separated by a semipermeable membrane (filter) until the concentrations of the inflowing and outflowing hapten are identical. The total hapten concentration in the filtration vessel is then assayed and the amount of hapten bound in the immune complex is thus determined. In the second procedure, a known antibody–hapten mixture is placed in the filtration vessel and hapten concentrations are determined in small filtrate fractions corresponding to the free hapten concentration at a given time interval. The measured concentrations and the known total hapten concentration in the vessel then serve to calculate the bound hapten concentration.

As the antibody-bound hapten has absorption properties different from those of the free hapten, the two hapten forms can be determined **spectrophotometrically** by measuring absorbance at two wavelengths.

The method using **fluorescence quenching** is based on the fluorescence of proteins. From the physical point of view, a molecule fluoresces when it absorbs light with a certain wavelength and then emits the absorbed energy in the form of light with a higher wavelength. Proteins fluoresce when irradiated with ultraviolet light. Phenylalanine, tyrosine and tryptophan residues in protein molecules are mainly responsible for their fluorescence. When pure proteins are irradiated with light with a wavelength of 280–295 nm, they emit light at 330–350 nm. In the case, when the excitation energy is transferred from proteins to non-fluorescing molecules, protein fluorescence decreases. Such a situation occurs when a non-fluorescing hapten is bound to an antibody and the resultant complex is irradiated with ultraviolet light. The light emitted by the antibody is absorbed by the hapten molecules proportionally to the amount of the bound hapten, which appears as a decrease of fluorescence ("quenching") compared to the antibody without the bound hapten (*Fig. 11.3*). It is thus possible to determine the

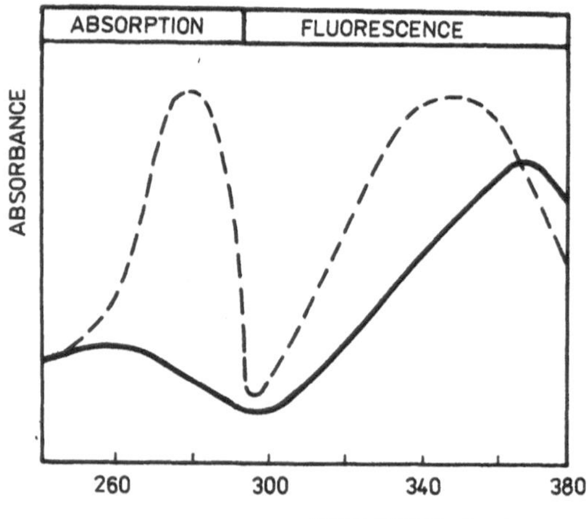

Fig. 11.3. Absorption and emission spectra of rabbit anti-dinitrophenyl antibody and its immune complex with dinitrophenyl lysine as hapten.
Broken line, pure antibody; solid line, hapten–antibody complex.

relative amounts of bound and free hapten. The fluorescence quenching method is advantageous since it only requires small amounts of antibodies, although the antibodies must be quite pure and have certain spectral properties.

The fluoescence method can also be used when the hapten itself exhibits fluorescence. Two possibilities may then occur — fluorescence of the hapten after binding to the antibody decreases (fluorescence quenching is measured) or increases (increase of fluorescence is measured). As well as fluorescence measurements, **polarization of fluorescence** can be measured. In this approach, the antigen is usually labelled with fluorescein isothiocyanate or with 1-dimethyl-aminonaphthalene-5-sulphonyl chloride.

Ammonium sulphate precipitation is the oldest separation method. It is rapid and an immune serum can be used instead of a pure antibody. It is based on the fact that antibodies and their immune complexes with the hapten are precipitated at a 50% saturation with ammonium sulphate. In this method, the hapten is also usually labelled with a radioactive isotope. Radioactivity in the precipitate is proportional to the amount of bound hapten, whereas radioactivity in the supernatant indicates the amount of free hapten. The method can only be used if the hapten remains soluble in ammonium sulphate at 50% saturation.

When the antigen and the antibody have sufficiently different relative molecular weights and are eluted in different peaks during gel chromatography, this technique can also be used to determine the amount of free and antibody-bound antigen.

In **affinity chromatography**, the antigen or the antibody is bound to a suitable carrier and acts as a ligand (p. 209). The second reacting component of this pair can then specifically bind to the ligand. When, for instance, the antibody has been bound to an insoluble carrier, the immunoadsorbent is packed in a column. A solution of the antigen is applied to the column and the antigen is immobilized on the ligand (antibody). The immobilized antigen is then displaced from the column by the free antibody. When the interactions are univalent, displacement zonal elution can be used to calculate the equilibrium association constant directly from the chromatographic experiment using the equation:

$$V - V_0 = K'c'(V_0 - V_z) + \frac{K'c'(V_0 - V_z)}{Kc}$$

where K is the equilibrium association constant for the interaction of the free component with the ligand bound to the chromatographic bed, K' is the equilibrium association constant for the interaction of the free component with the free ligand, V is the elution volume of the free component, V_0 is the void volume of the column (elution volume of a non-reacting protein), V_z is the volume of gel pores, c is the molar concentration of the free displacement ligand, and c' is the molar concentration of the bound ligand. By plotting $V - V_0$ versus $1/c$ a straight line is obtained which enables the determination of K and K' values.

Using binding parameters obtained by affinity chromatography, equilibrium constants can be determined and, in addition, the results of attempts to isolate antigens or antibodies can be predicted.

11.3.2 Biological significance of antibody affinity

BURNET's clonal-selection theory is based on the assumption that antibody synthesizing *B*-cells have immunoglobulin receptors on their surface that are specific for a certain antigen or its determinant group. Cells producing high affinity antibodies should also carry high affinity receptors and, in contrast, cells producing low affinity antibodies should have low affinity receptors. Heterogeneity of antibodies therefore reflects heterogeneity in the affinities of lymphocyte immunoglobulin receptors.

It was found that avidity (functional affinity) of antibodies against numerous complete antigens, as well as affinity of anti-hapten antibodies, usually increases with the time elapsed after antigen administration (immunization), in agreement with SISKIND and BENACERRAF's (1969) hypothesis on the selection of *B*-cells by the antigen. This phenomenon is also referred to as the maturation of antibody response. It usually involves IgG antibodies, whereas IgM antibodies generally have a low internal affinity.

In the presence of high amounts of the antigen, cells whose receptors exhibit a wide range of affinities are stimulated. In other words, cells with both high and low affinity receptors are stimulated. Such a situation occurs immediately after immunization or after administration of high antigen doses. As the administered antigen is gradually degraded in the organism, competition among cells capable of antibody response arises. At a low antigen concentration, it is likely that only cells with receptors exhibiting a high affinity towards a given antigen will be stimulated. When the antigen concentration decreases below a certain level, *B*-cells are no longer stimulated.

However, the antigen can not only stimulate the *B*-cell towards antibody production but also towards suppression (tolerance). The response thus depends on numerous factors, although the interaction between the immunoglobulin receptors and the antigen is particularly important. The amount and size of immune complexes generated on the *B*-cell surface depend on the amount of antigen, the number of reacting *B*-cells, and on the number, localization and mobility of immunoglobulin and Fc-receptors. Redistribution and aggregation of receptors induced by the antigen binding play an important role in stabilizing complexes of polyvalent antigens with the *B*-cell. In general, *B*-cells carrying high affinity receptors are easily tolerable (do not respond at a high antigen concentration). When the antibody response of the cells carrying high affinity receptors is suppressed, the cells with low affinity receptors begin to operate.

Antibodies with high affinity are more efficient than low affinity antibodies in most immune and immunochemical reactions. Low affinity antibodies can produce small immune complexes which sediment in the walls of blood vessels, or in certain tissues, and induce inflammatory immune complex diseases. For example, formation of antibodies with a low functional affinity is associated with glomerulonephritis induced in rabbits by long-term administration of an antigen.

Low affinity antibodies are also characterized by so-called **cross-reactivity**, *i.e.* the ability of one antibody to react with several similar antigens. This is due to the fact that low affinity antibodies are more likely to remain free and react with other antigens. Low affinity antibodies are usually produced after administration of a high dose of antigen which stimulates many cell clones and also increases the probability of a cross-reaction. In immunopathology, cross-reaction with a host antigen leading to autoimmune diseases is quite important. It therefore follows that production of low affinity antibodies is not advantageous for the producing individual since these antibodies may induce serious diseases.

Both low affinity and high affinity antibodies can be involved in *in vitro* reactions (immunochemical methods). Low affinity antibodies are usually used in immunoaffinity chromatography or electrophoresis. On the other hand, many methods (RIA, EIA *etc.*), which require washing of the resultant insoluble immune complexes, preferably employ antibodies with the highest possible functional affinity.

11.4 Secondary phenomena

Studies of the kinetics of antigen–antibody reactions are complicated by many other interactions known as *secondary phenomena.* These include precipitation of immune complexes, agglutination of particles, binding of complement components, binding of cytophilic antibodies and immune complexes to cells and non-specific interactions of antibodies and their complexes with various compounds.

11.4.1 Immunoprecipitation

Precipitation refers to a *precipitate* which is formed after the reaction of a soluble antigen with a soluble antibody. The precipitate is produced by the origination of a three-dimensional structure during interaction of a polyvalent antigen with a polyvalent antibody. According to Marrack (1938) and Heidelberger (1939) the **precipitation reaction** has two phases. Soluble immune complexes are formed during the first, rapid phase. They can only be detected by special methods. The second, much slower phase, results in aggregation of soluble complexes produced during the first phase. Aggregation gives rise to visible (insoluble) complexes. Because of this, the antigen–antibody mixture is allowed to react for several hours or even days for quantitative precipitation. The amount of the precipitate may then be determined. Antigens reacting during the precipitation reaction are called **precipitogens** and corresponding antibodies are called **precipitins**.

Not all antibodies can precipitate antigens. According to MARRACK'S (1938) original *lattice theory* of formation of large antigen–antibody complexes it was assumed that monovalent or univalent antibodies, *i.e.* conventional antibodies having only a single binding site, are *non-precipitating antibodies*. However, the existence of such antibodies could not be demonstrated. Only Fab fragments have a single binding site, compared to complete immunoglobulin molecules. The lack of precipitation ability of such antibodies is apparently due to their low affinity. For non-precipitating antibodies a large excess of antigen must be added, so that the antigen can react with at least some binding sites. The excess antigen then eliminates the second phase of the precipitation reaction, *i.e.* aggregation of small complexes is avoided.

Monoclonal antibodies also cannot usually precipitate antigens. However, their inability to produce large agregates is not associated with low affinity (their affinity is usually high) but rather with a high specificity, *i.e.* the ability to react only with a single determinant (epitope) on the complex antigen molecule (p. 21).

The precipitation reaction is also affected by the antibody isotype. IgM antibodies usually precipitate poorly or not at all. In contrast IgG antibodies are good precipitins.

MARRACK and HEIDELBERGER assumed that interaction between the antigen determinant and the binding site is particularly important for both phases of the precipitation reaction. However, according to present data, the formation of large complexes (aggregates) cannot be explained solely on the basis of interaction between the epitope and paratope. During aggregation of soluble complexes to large insoluble complexes (the second phase of the immunoprecipitation reaction) non-specific, particularly hydrophobic, bonds involving other parts of the antibody and antigen molecules, are involved. Participation of the Fc region of the antibody molecule in production of secondary (insoluble) immune complexes is also confirmed by the fact that $F(ab)_2$ fragments, compared to entire molecules with the same affinity, cannot usually precipitate a multivalent antigen. In addition, due to hydrophobic changes originating in the antibody molecule after binding with the antigen, its solubility, and therefore also the solubility of the whole complex, decreases. The rate of precipitation is directly proportional to temperature, increasing between 0 and 56 °C, but the amount of the precipitate decreases with increasing temperature. The pH optimum is 6.5–8.2.

The ratio of the amount of antigen and antibody, and their concentrations, are important factors influencing the amount of the precipitate produced. The amount of the precipitate can be determined by liquid (serological) or gel (immunodiffusion) methods. In a liquid medium it is possible to measure the amount of the precipitate (weighing, protein determination) and, in addition, changes in the optical properties of the reaction mixture can be measured (turbidimetry, nephelometry, spectrofluorimetry) and, occasionally, the quantity of the antigen in the precipitate or supernatant can be determined. When measuring the amount of free antigen and of antigen bound in the immune complex, the antigen is usually labelled with a radioactive isotope. When the precipitation occurs in a gel, only the amount of the precipitate is determined, usually according to its area or other parameters characterizing its shape.

Quantitative precipitation is performed in a series of test tubes. Each test tube contains an equal quantity of the antibody (antiserum) and increasing concentrations of the antigen. Equal volumes of the two components are added. According to the original macromethod (HEIDELBERGER and KENDALL, 1937) 0.5 ml of antiserum is always used. Micromodifications require special microtest tubes and microcentrifuges and enable as little as 1–10 μl of the antiserum to be used. The micromethods can be performed in wells of microtitration plates. The mixture is stirred, allowed to react for 1–2 hours at 37 °C and then for 2–10 days at 4 °C in order to achieve maximum precipitation of immune complexes. Precipitation time in the cold can be reduced to 1–2 hours when the reaction proceeds in 2% polyethyleneglycol with mean molecular weight of 6000. In this case, the antigen should be dissolved in a mixture of 4% polyethyleneglycol in phosphate-buffered saline (pH 7.2, PBS). After completion of the precipitation, the precipitate is isolated by centrifugation and washed three times with PBS or, occasionally, with 2% polyethyleneglycol in PBS. The washed precipitate is allowed to dry in the test tubes and then dissolved in 0.2 mol/l NaOH. The protein content of the solutions is determined by measuring absorbance at 280 nm or according to LOWRY *et al.* (1951).

When the amount of the protein precipitate is plotted graphically against the concentration of the added antigen a **precipitation curve** is obtained (*Fig. 11.4*). From this curve it is apparent that in test tubes containing little antigen there is also relatively little precipitate. This is the region of excess antibody. As the antigen concentration increases, the amount of the precipitate also increases. Its maximum quantity is produced in the *equivalence zone*. In this zone, free molecules of the antigen and antibody do not occur in the supernatant after centrifugation. With increasing concentration of the

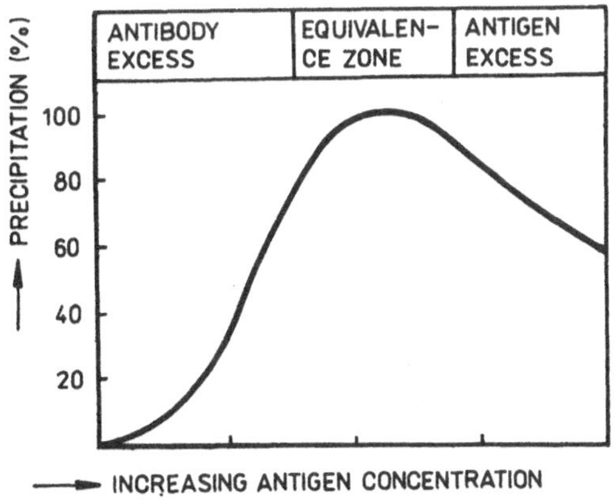

Fig. 11.4. Precipitation curve.

antigen the equivalence zone is shifted to the region of antigen excess. When the excess of antigen is small, formation of the precipitate remains maximal. With extreme antigen excess the amount of the precipitate may decrease due to formation of soluble complexes. The shape of the precipitation curve depends on the type of antigen. Protein antigens with a relative molecular weight of 40 000–160 000 usually yield precipitation curves with sharp peaks in the equivalence zone (*Fig. 11.5*). Denatured proteins and polysaccharides yield curves with broad peaks. Certain antigens and antibodies do not precipitate readily and the method of quantitative precipitation cannot therefore be applied. By using precipitation curves it is possible to determine binding properties of the antibody or antiserum with a given antigen. It is expressed as the amount of antigen which can, in the equivalence zone, bind 1 mg of the antibody or 1 ml of the antiserum.

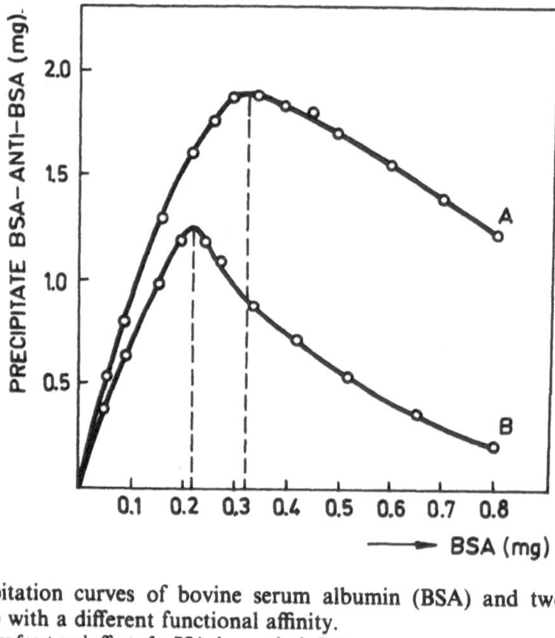

Fig. 11.5. Precipitation curves of bovine serum albumin (BSA) and two rabbit anti-BSA antibodies (A, B) with a different functional affinity.
Antibody A has a higher functional affinity for BSA than antibody B Increasing amounts of the antigen (BSA) were added to 0 1 ml of the antiserum It follows from the curves that, in the equivalent zone, 0 1 ml of antibody A binds 0 325 mg of the antigen and, therefore, 1 ml of antibody A would bind 3 25 mg of BSA Similarly, 1 ml of antibody B binds 2 20 mg of BSA

The presumed structure of immune complexes in the presence of excess antibody, in the equivalence zone and in the presence of excess antigen is illustrated in *Fig. 11.6*.

Certain ionic detergents, *e.g.* sodium dodecylsulphate at a low concentration, inhibit immunoprecipitation reactions (more than 90% inhibition at a 0.2% concentration). Conversely, non-ionic detergents (Triton X-100, Nonidet P 40, Tween 80, Lubrol WX) do not significantly affect the precipitation reaction.

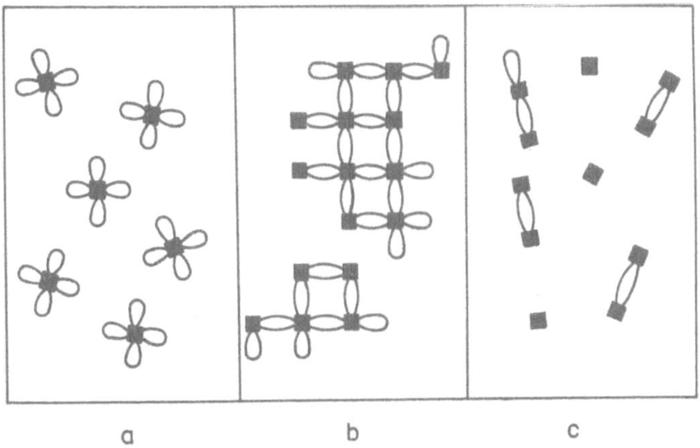

Fig. 11.6. Schematic illustration of possible antigen–antibody complexes in various stages of the precipitation reaction.
(a) Excess of antibody, (b) equivalence zone (lattice structure), (c) excess of antigen Squares represent molecules of tetravalent antigen and ellipses depict molecules of bivalent antibody

11.4.2 Immunoagglutination

When an antibody reacts with antigen determinants on the surface of erythrocytes, bacteria or other cells, *i.e.* with an insoluble (corpuscular) antigen, clustering of such particles (immunoagglutination) occurs. In this case, the insoluble antigen is known as an **agglutinogen**, agglutination-inducing antibodies are called **agglutinins** and the complex of agglutinated particles is an **agglutinate**. When the agglutinogen is an integral part of the particle surface, the term *"active"* or *"direct" agglutination* is used. Inert particles (latex) and

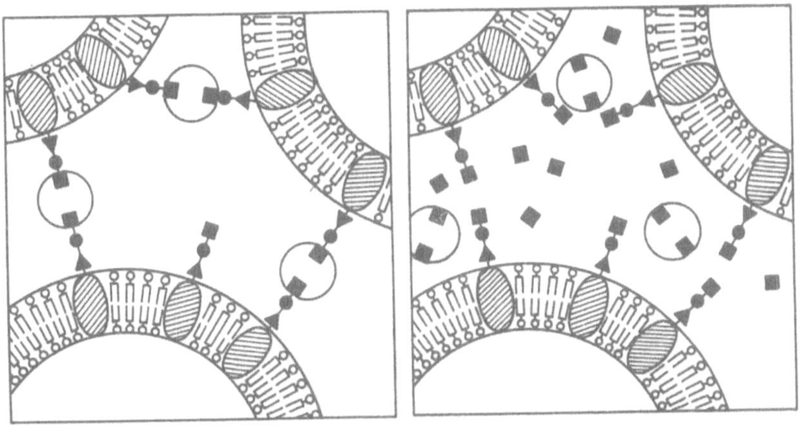

Fig. 11.7. Mechanism of agglutination (left) and inhibition of agglutination (right).
The large ring sections represent cells In their membranes, consisting of a phospholipid bilayer, there are protein molecules (ellipses) having antigen determinants or haptens on their surface (black squares, circles and triangles) Specific antibody (small ring) combines with haptens bound to the cells forming bridges which agglutinate several cells When a free hapten is present in the mixture, antibodies bind preferentially to its molecules (having a higher affinity to them), which therefore inhibits agglutination of cells (or other particles)

erythrocytes, and occasionally also other cells, can be passively coated by a certain antigen. When an antibody directed against this antigen is then added, the particles begin to agglutinate. This is the basis of *passive* or *indirect agglutination*. Passive agglutination thus depends on an added antigen coating a particle, and not the antigen proper on the particle surface.

Immunoagglutination is thus based on cross-linking of the polyvalent antigen by antibody molecules that must at least be bivalent. Univalent fragments of antibodies, *e.g.* Fab, do not induce immunoagglutination although they can inhibit it. Conversely, $F(ab)_2$ fragment is just as efficient in agglutination as the complete IgG antibody. Compared to immunoprecipitation, immunoagglutination does not therefore involve non-specific interactions of Fc fragments. Agglutination reactions can be classified like precipitation reactions. Agglutination can be inhibited by both excess antibody and excess cellular antigen (*Fig. 11.7*). The valence of antibodies and antigens is one of the most important factors. Particles with a small number of determinants agglutinate less readily than particles with numerous determinants. Due to the number of binding sites and the additional effect of multivalence, IgM antibodies are more effective in agglutination than IgG antibodies.

References

Heidelberger, M. (1939) Quantitative absolute methods in the study of antigen-antibody reactions. *Bacteriol. Rev.*, **3**, 49–95.

Heidelberger, M. and Kendall, F. E. (1937) A quantitative theory of the precipitin reaction. IV. The reaction of pneumococcus specific polysaccharides with homologous rabbit antisera. *J. Exp. Med.*, **65**, 647–60.

Karush, F. (1962) Immunologic specificity and molecular structure. *Adv. Immunol.*, **2**, 1–40.

Lowry, O. H., Rosebrough, N. J., Farr, A. L. and Randall, R. J. (1951) Protein measurement with the folin phenol reagent. *J. Biol. Chem.*, **193**, 265–75.

Marrack, J. R. (1938) *The chemistry of antigens and antibodies*. London, His Majesty's Stationery Office, 194 pp.

Siskind, G. W. and Benacerraf, B. (1969) Cell selection by antigen in the immune response. *Adv. Immunol.*, **10**, 1–50.

Zikán, J. (1986a) Antibody-antigen interaction I. Character of interacting components, thermodynamic and kinetic expression of this interaction. *Chem. Listy*, **80**, 935–53. (*in Czech*).

Zikán, J. (1986b) Antibody-antigen interaction. II. Methods of measurement of parameters of this interaction and their importance. *Chem. Listy*, **80**, 1048–70. (*in Czech*).

12 Immunochemical methods

All immunochemical methods are based on a biospecific linkage between an antigen (hapten) and an antibody, or occasionally, on the non-specific interactions of free or bound antibodies with certain compounds (complement, protein A) or cell receptors (Fc-receptors). Immunochemical methods are used to demonstrate the presence of an antigen or to determine its quantity (when a specific antibody is available) or the quantity of an antibody (when a specific antigen or hapten is available). The course of the antigen–antibody reaction is measured by a decrease of one of the pair of reacting components or, more frequently, by the formation of a reaction product, the antigen–antibody complex. Qualitative methods to measure immune complex formation include measurement of turbidity, precipitate, agglutinate or complement fixation. Quantitative methods make it possible to determine the amount of immune complexes produced.

In the formation of the antigen–antibody complex four basic factors should be considered — type of antigen, type of antibody, environment in which the reaction takes place and the method of identifying immune complexes.

An **antigen** can be complete or incomplete (hapten). A complete antigen may have a different number of determinants, different solubility (colloidal, soluble or cellular, insoluble) and form (free or bound). The bound form is an antigen bound to a carrier (cell, inert particle, immunoadsorbent, surface of a microtitre plate well *etc.*) and also the antigen–antibody complex. As well as immunogenicity and specificity, the antigen can also exhibit biological activities (*e.g.* enzymatic, hormonal or receptor activities) which can be used directly or indirectly to detect its complex with the antibody.

The **antibody** can either be conventional (binding sites on all molecules are not equivalent) or monoclonal. It is always an immunoglobulin of a certain class or its fragment. The internal affinity is important for the reaction with the hapten, whereas the functional affinity (avidity) and the number of binding sites on its molecule are important for its reaction with the antigen. The antibody can also be free or bound. Bound (insoluble) forms of the antibody include immunoglobulin receptors on the surface of *B*-lymphocytes, immunoglobulins bound to Fc-receptors of different cells, antibodies adsorbed non-specifically to the surface of various particles or wells of microtitration plates, in the form of ligands in immunoadsorbents and immune complexes. The antibodies bound in immune complexes can be resolubilized (in the presence of antigen or antibody excess).

The **medium** in which the *in vitro* antigen–antibody reaction takes place can be liquid, semisolid (gel) or combined (one component is in the liquid, the other in the solid phase).

The immune complex produced can be detected by the naked eye (precipitate, agglutinate), after staining with common histochemical stains (usually reacting with proteins), by different optical methods (turbidimetry, conventional and laser nephelometry, spectrofluorimetry) or after labelling one of the reacting components (antigen, hapten or antibody) with a radioactive isotope, or, occasionally, with another compound functioning as a label. The reacting compound thus labelled becomes a component of the immune complex, whose presence and quantity can thereby be determined. The antigen or antibody must be labelled before their mutual interaction. Radioactive isotopes such as ^{14}C, ^{3}H, ^{131}I and ^{75}Se (radioimmunoassay — RIA), enzymes (enzyme-immunoassay — EIA), fluorochromes (fluorescence immunoassay — FIA), luminophores (chemiluminescence immunoassay — CIA), particles (latex, glycomethacrylate, erythrocytes, liposomes), bacteriophages (viroimmunoassay — VIA), electron-dense compounds (immunoelectronoptic techniques) *etc.* may be used as labels.

The above four basic factors and their varied properties make it possible to design various combinations of procedures that can be used under laboratory conditions to measure the antigen–antibody reaction. Such combinations are actually individual immunochemical methods and their various modifications.

Some authors distinguish several **generations of immunochemical methods**. The oldest methods for determining the presence and concentration of antibodies and complete antigens in an aqueous environment belong to the first generation. These are, in fact, the classical serological methods. The second generation methods make it possible to analyse even complex mixtures of antigens and to quantitatively determine antigen and antibody concentrations. Immunodiffusion and immunoelectrophoretic methods are among these methods. The third generation includes highly sensitive methods to quantitate antigens, antibodies and haptens. Most of these methods are readily automated and are thus labour-saving. This group of methods includes various types of immunoassay (RIA, EIA, CIA *etc.*), immunonephelometry and immunofluorimetric and other techniques. Some of these can be combined, often resulting in a higher sensitivity, specificity and, occasionally, other advantages. The fourth generation methods should make it possible to measure continuously concentrations of an antigen, hapten or antibody. These methods include, for example, the use of analytical antibody electrodes, which might also be used to regulate biotechnological or physiological processes in artificial organs, *e.g.* in the artificial kidney.

However, not all immunochemical methods can be precisely classified in a particular category.

12.1 Serological methods

Serological methods are performed in a liquid (aqueous) medium. Using such techniques it is possible to determine the presence and concentrations of antibodies or antigens, and sometimes, haptens as well.

When a known antigen is available, it can be used to detect antibodies in serum or other solutions. For particle-bound antigens, agglutination reactions are used, whereas precipitation reactions are employed for soluble antigens. For example, it is possible to determine the serum antibody level in humans and animals with suspected infectious diseases. The concentration of serum antibodies determined after addition of a known antigen is expressed as a **titre**, *i.e.* as the highest serum dilution which can react with the antigen. The serum is diluted with physiological saline (1 : 2, 1 : 4, 1 : 8 *etc.*) or if the antibody concentration is high 1 : 20, 1 : 40, 1 : 80.

Similarly, it is possible to detect an unknown antigen with a known antibody. Certain microbial antigens can be determined in this way and serotyping is used to identify bacteria.

12.1.1 Precipitation methods

A serological precipitation reaction can be performed on a microscope slide (a drop of serum is added to a drop of antigen solution or *vice versa* and precipitate formation observed), in a glass tube (diameter 2–4 mm) sealed at the bottom with plasticine (ring precipitation — formation of a precipitation ring at the interphase of antigen and antibody) or in a test tube. For qualitative antigen or antibody determinations the precipitate is measured visually (*e.g.* + +, + or −), whereas quantitative determinations require an accurate measurement of the amount of precipitate (quantitative precipitation).

Many precipitating antibodies are not strictly specific for a certain antigen but can also precipitate other similar antigens, albeit to a lesser extent. The reaction with such related antigens is known as cross-reaction.

When a conjugate of hapten with a protein carrier is used to prepare the antibody, quantitative precipitation can also be used to determine the concentration of the free hapten. Two series of test tubes are used. The antibody and the hapten–carrier conjugate are added to both sets, while the free hapten is added to one of them. Due mainly to spatial phenomena (lower steric repulsive forces) the antibodies have a higher affinity for the free hapten than for the hapten—carrier conjugate. The free hapten reacts first, occupying the antibody binding sites and producing small soluble hapten–antibody complexes. Formation of large insoluble hapten–carrier–antibody complexes does not occur since the antibody has no free binding site. The presence of a free hapten is therefore indicated by **inhibition of precipitation**. The exact hapten concentration may then be determined from a calibration curve constructed from data obtained with known hapten concentrations.

12.1.2 Immunonephelometry and immunoturbidimetry

Modern (third generation) nephelometric and turbidimetric methods are also based on the serological precipitation reaction. However, this differs from classical quantitative precipitation by detection of the immune complex produced, increased sensitivity (nanogram quantities of the antigen–antibody complex can be determined) and by being readily automated (several hundred samples per day can be analysed). Turbidity formed after mixing the antigen and antibody solutions is measured turbidimetrically (standard spectrophotometers can be used) or nephelometrically or fluoronephelometrically, using purpose-built instruments, usually with laser light sources. The volume of the antigen solution (*e.g.* serum) required for a single analysis is small (10 µl or less) and it is therefore possible to analyse 20 different antigens in a total volume of 0.2 ml in several minutes (when a *laser immunonephelometer* is used). However, it is necessary to use appropriate dilutions of high-quality monospecific antibodies. It can be seen from the precipitation curve that the concentration of the antigen tested is directly proportional to the amount of the immunoprecipitate, and thus to turbidity, only within the region of excess antibody (a linear relationship between the amounts of antigen and precipitate). Therefore, a suitable dilution must be determined experimentally for each antibody and each antigen (if not stated by the manufacturer of the antiserum or diagnostic kit). In practice, the immunonephelometric determination of a particular antigen also requires the use of standard (known) concentrations (in addition to test samples) and construction of a calibration curve. The lower limit of protein antigen concentration which can be determined turbidimetrically is 5–20 mg/l. Laser immunonephelometry is 5–10 times more sensitive. For immunonephelometric or immunoturbidimetric determinations of antigen (particularly in blood serum) it is also possible to use high-speed centrifugal analysers.

12.1.3 Agglutination methods

Agglutination techniques are analogous to precipitation methods, only differing in the type of antigen used. Agglutination methods enable the identification of various bacteria and other microorganisms and animal cells and different types of erythrocytes with a known antiserum. Conversely, known agglutinogens enable the simple demonstration of specific antibodies.

When testing for serum antibodies, constant amounts of the cellular antigen are placed in a series of test tubes or wells of microtitre plates and gradual geometric dilutions of the test (antibody-containing) serum are added. The contents are mixed and allowed to stand at an appropriate temperature for a fixed time. Agglutination is formed in positive test tubes below a clear solution. A control test tube containing only the antigen must also be included. In this tube, agglutination should not occur.

The agglutination reaction may be performed in test tubes, on microtitre plates or microscope slides. With the latter, the results can be observed with the naked eye directly or under the microscope (microscopic agglutination). Microtitre plates are usually made of plastic, in which wells are drilled or moulded. This technique is advantageous because only small amounts of the antigen and antibody are required and because sample handling can be automated. The results obtained can be determined easily and accurately.

In an agglutination reaction, agglutination is sometimes positive at higher serum dilutions but negative at lower dilutions. This so-called **"prozone" phenomenon** is caused by the presence of non-agglutinating (blocking) antibodies. *Blocking antibodies* bind the antigen more rapidly than agglutinating antibodies and are in fact analogous to non-precipitating antibodies (they also have a very low affinity). When their concentration in the test serum is too high, they block surface determinants of the antigen, so that normal agglutinins cannot bind. When the serum is diluted the blocking antibodies are also diluted. At a certain dilution, the concentration of the blocking antibodies decreases to such an extent that the agglutinins predominate and agglutination therefore occurs (*Table 12.1*).

Table 12.1 Agglutination pattern in the presence of blocking antibodies (Prozone phenomenon)

Tube number	*1*	*2*	*3*	*4*	*5*	*6*	*7*	*8*	*9*
Antiserum dilution	1:20	1:40	1:80	1:160	1:320	1:640	1:1260	1:2560	1:5120
Normal agglutination	+++	+++	+++	+++	++	++	+	–	–
Prozone phenomenon	–	–	–	+++	+++	++	+	–	–

Agglutination reactions can be divided into direct and indirect (passive) reactions. **Direct agglutination reactions** are those, in which integral, cell-surface agglutinogens react with the antibody, whereas **indirect agglutination reactions** involve soluble antigens or haptens that have been passively adsorbed on to the surface of a cell or inert particle. Agglutination reactions are much more sensitive than precipitation reactions. However, their standard versions cannot be used for the accurate, quantitative determination of antibodies. Concentrations of antigens or antibodies can only be determined semi-quantitatively.

Agglutinogens isolated from the surface of cells can be used to inhibit direct agglutination and appropriate antigens or pure haptens can inhibit passive agglutination. Both methods are analogous to inhibition of precipitation and can be used for the quantitative determination of antigen or hapten concentrations.

When erythrocytes are used as agglutinogen, the term **haemagglutination** is used. The red cell surface contains several antigens which react with both isologous and heterologous antibodies. Suspensions of erythrocytes are not stable and sediment after a certain period. However, agglutination is readily distinguished from sedimentation.

The use of haemagglutination methods became quite common after the demonstration that different soluble antigens and haptens can be passively adsorbed onto erythrocytes. When erythrocytes with a passively bound antigen are treated with antiserum specific for this antigen, they agglutinate. In these **passive haemagglutination** techniques the test antiserum (when it contains heterophilic antibodies) can non-specifically agglutinate even normal erythrocytes, in addition to those with a passively conjugated antigen. Such non-specific agglutination of normal erythrocytes must be removed. This is performed by mixing the antiserum with washed normal erythrocytes to adsorb the cross-reacting antibodies. The mixture is centrifuged and the saturated antiserum no longer reacts with normal erythrocytes. It can therefore be used to determine an antigen or hapten by passive haemagglutination.

Some antigens adsorb to the erythrocyte surface spontaneously, whereas others require chemical pretreatment of the red cell. Spontaneously adsorbed antigens include viral and some bacterial antigens, egg albumin, bovine serum albumin, desoxyribonucleic acids, many haptens, including most antibiotics, and other compounds. Other antigens bind to the surface of erythrocytes after treatment of the cells with tannin, bisdiazotized benzidine, 1,3-difluoro-4,6-dinitrobenzene, chromic chloride, tolylene-2,4-diisocyanate, glutaraldehyde or water-soluble carbodiimides. Erythrocytes with a passively bound soluble antigen must not agglutinate spontaneously, but only in the presence of specific antibodies.

When the corresponding soluble antigen is added to a mixture of erythrocytes coated by that antigen and its antibody, their agglutination is inhibited. The **passive haemagglutination inhibition** technique is very sensitive and can be used to demonstrate very low amounts of a soluble antigen or hapten.

In haemagglutination tests, IgM antibodies are usually used (because of their multivalence and molecular size). The participation of IgM and IgG antibodies in haemagglutination can be distinguished with 2-mercaptoethanol. When this compound is added to an immune serum, the IgM pentamer is cleaved and loses its ability to agglutinate erythrocytes. Thus, in the absence of 2-mercaptoethanol, haemagglutination may be due to either IgM or IgG antibodies, whereas in the presence of 2-mercaptoethanol, it is due only to IgG antibodies.

As well as erythrocytes, *latex* or other particles made from synthetic polymers are used in passive agglutination, or passive agglutination inhibition methods.

12.1.4 Methods utilizing complement fixation

The antigen–antibody complex, and thus also the presence of the antigen or the antibody, can also be determined on the basis of complement fixation or consumption. This is a relatively sensitive method (less than 1 μg of the antigen or antibody can be determined) because even relatively low amounts of the resultant immune complex can activate complement *via* the classical pathway and thus fix or consume large amounts of complement.

The method used to determine antigens and antibodies involves two steps. In the first step, the antigen reacts with the antibody in the presence of a known amount of the complement which is being consumed (binding to the immune complexes produced). Fresh guinea-pig serum usually serves as the source of complement. The second step aims to detect the first phase since the mutual binding of the antigen, antibody and complement is usually not accompanied by any visible changes. In this step, the residual haemolytic activity of the complement that has not been bound to the immune complexes is measured. This value is subtracted from the total amount of complement added (analogy of reverse titration) and the amount of bound complement, which is proportional to the amount of antibody or antigen, is determined. In practice, the second step is performed so that sheep erythrocytes coated by the *amboceptor* (rabbit antibodies against sheep erythrocytes) are added to the components that have reacted in the first step. If the total amount of complement added in the first step has been consumed, haemolysis in the second step cannot occur. This is a positive **complement-fixation test** (CFT). If complement has not been consumed, haemolysis occurs because the complement is free to bind to the erythrocyte–anti-erythrocyte antibody complex. This is a negative complement-fixation test.

Complement-fixing reactions may be qualitative or quantitative. The qualitative version can only demonstrate the presence or absence of the antibody or antigen in a given system, whereas the quantitative version can also measure their concentrations. Quantitative determinations are made in a series of test tubes (or microtitrate plate wells) to which constant amounts of complement and one component, and serial dilutions of the second component, are added. The results are expressed as the highest antiserum dilution (antibody determination) or the lowest antigen concentration (antigen determination) which still yields a positive CFT.

Complement fixation also preferentially reflects IgM antibodies since these have a higher haemolytic activity than IgG antibodies. It is a disadvantage of CFT methods that it only enables the relative rather than the absolute amounts of antibody or antigen to be determined and that only antibodies of certain classes (human IgG1, IgG3, IgM and possibly also IgG2) can be determined.

12.2 Immunodiffusion methods

Immunodiffusion methods are based on immunoprecipitation. However, they are performed in gels rather than in liquid media. Several natural and synthetic materials have been tested as gel media but agar and agarose gels are most suitable for immunodiffusion tests. Commercial agar used to make bacteriological media is not suitable for such purposes and has to be specially treated.

Agar is a mixture of polysaccharides extracted from cell walls or the intracellular matrix of red sea algae (*Rhodophyta*). Most of these polysaccharides have the disaccharide *agarobiose* — consisting of *β*-D-galactose bound with 3,6-anhydro-*α*-L-galactose by *β*-1,4-glycosidic bond as their basic unit. Agarobiose subunits are interconnected with *α*-1,3-glycosidic bonds. Other groups including sulphate, pyruvate, methoxyl and glucuronate are linked with the basic chain at some sites. The number of side groups differs in different agar batches and is responsible for their slightly different properties. In addition, agar gels rather efficiently absorb certain compounds (*β*-lipoproteins, thyreoglobulins, fibrinogen, lysozyme *etc.*) rendering their diffusion or migration during electrophoresis impossible. For these reasons, natural agar is further fractionated into agarose and agaropectin. **Agarose** (*Fig. 12.1*) has hardly any side anionic groups and therefore lacks the ion-exchange properties and molecular-sieving features predominate. It has a much more standard composition than agar and significantly lower non-specific adsorption and electroendoosmosis. In contrast, the *agaropectin* fraction is rich in anionic groups, and has even less advantageous properties than non-fractionated agar.

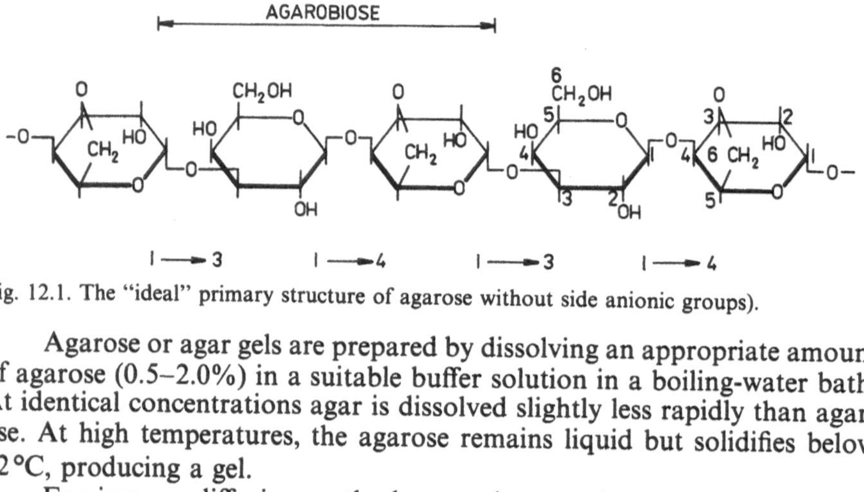

Fig. 12.1. The "ideal" primary structure of agarose without side anionic groups).

Agarose or agar gels are prepared by dissolving an appropriate amount of agarose (0.5–2.0%) in a suitable buffer solution in a boiling-water bath. At identical concentrations agar is dissolved slightly less rapidly than agarose. At high temperatures, the agarose remains liquid but solidifies below 42 °C, producing a gel.

For immunodiffusion methods, agar is poured on to an appropriate support to which antibodies and antigens are applied in various ways to migrate (diffuse) in the gel. Depending on the type of application, only one (**single immunodiffusion**) or both components can diffuse (**double immunodiffusion**). Both single and double immunodiffusion may proceed in one direction (one-dimensional immunodiffusion) or in two directions (two-dimensional).

The precipitation reaction occurs at the site where the diffusing components meet. This is manifested by formation of precipitation lines, arcs or circles. They can be observed with the naked eye (*in situ*), after staining (usually with a protein stain), by autoradiography in which one component is labelled with a radioactive isotope, or with a secondary antibody labelled with an enzyme or other label. In the latter case, secondary antibody reacts specifically with the first antibody. When for example a specific rabbit IgG antibody (first, *primary antibody*) reacts with the antigen, a pig anti-rabbit IgG antibody (second or *secondary antibody*) can be used to detect the primary antibody and thus also the immune complex produced.

Until they meet, both the antigen and antibody migrate almost independently in the gel, *i.e.* by free diffusion. Diffusion results from a thermodynamic gradient of free energy determined by temperature, entropy, composition of the system and other factors. It is an irreversible process in the sense of the second law of thermodynamics. Diffusion rate can be characterized on the basis of the *second Fick's law* as:

$$\frac{dx}{dt} = D\frac{d^2c}{dt^2}$$

where c is the molar concentration of the diffusing component (antigen or antibody), t is the diffusion time, x is the diffusion path per time t and D is the diffusion constant.

Diffusion proceeds in the direction of decreasing concentration of the diffusing component. The diffusion constant D is a characteristic property of any compound. Its value depends on temperature, size and shape of diffusing molecules and, for polyelectrolytes such as proteins, also on pH and ionic strength. It follows from this law that the diffusion rate is directly proportional to the diffusion constant and concentration of the diffusing compound and indirectly proportional with time.

Diffusion of the antigen and antibody proceeds according to Fick's law only until they meet. At the site of their interaction an immunoprecipitate is produced which is impermeable to molecules of both components. Any antigen molecule that has diffused to this site encounters a corresponding antibody molecule and the resultant immune complex is incorporated into the precipitate. In double immunodiffusion methods (when both antigen and antibody diffuse) only molecules of a single component may reach the site of mutual encounter after a certain time, particularly when the concentrations of the two components are not equivalent. For example, when the antigen is in excess, all antibody molecules may have already been bound in the precipitate and can no longer bind with the remaining diffusing antigen molecules. Under such conditions, the antigen molecules will continue to diffuse, even across the precipitation line, which will itself broaden by diffusion.

When the system includes one antigen and one antibody, a single precipitation line is formed at the site of their interaction. When two or more antigens and antibodies are present, they behave independently of each other and each pair produces an independent precipitation line.

12.2.1 Single immunodiffusion

In techniques based on simple immunodiffusion, one reacting component (antigen or antibody) is added to the molten agarose or agar and the other reacting component is subsequently applied to the formed gel. The component that has been evenly mixed in the gel cannot diffuse, since such a diffusion cannot conform to the thermodynamic laws described above. Only the second reacting component diffuses into the gel, where it encounters molecules of the first reacting component. The OUDIN method, and single radial immunodiffusion in particular, are the most important immunodiffusion methods.

The **Oudin method** (*single, one-dimensional diffusion*) uses test tubes of 3–4 mm in diameter, into which is placed 0.3–0.5 ml of 1% agarose in a suitable buffer (*e.g.* veronal-acetate buffer, pH 8.6, and ionic strength, 0.05) and containing an antibody. After melting, the agarose solution is maintained at 50–55 °C. The antibody (antiserum) solution is also heated to this temperature and appropriate volumes of the two solutions are mixed before transferring to the test tubes. After cooling and solidification the gel is overlaid with 0.05–0.1 ml of the antigen or antigens. A drop of paraffin oil is sometimes placed on the surface of the antigen solution to prevent evaporation. Under these conditions only antigen molecules are free to diffuse into the gel (*Fig. 12.2*). In the gel they encounter antibody molecules that were mixed in the agar which, after their interaction, form a ring-like precipitate appearing as a narrow band from the side view.

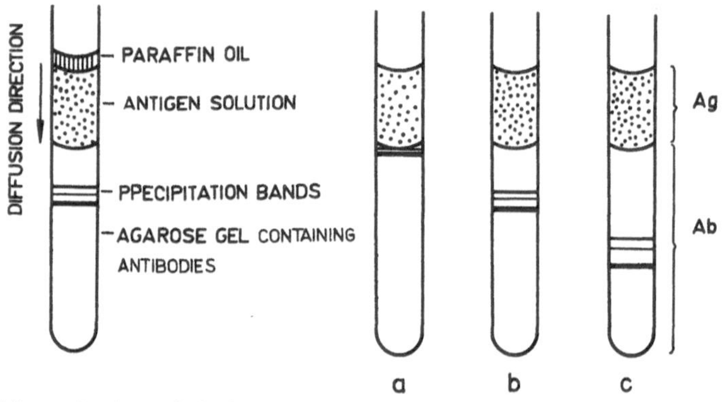

Fig. 12.2. Schematic view of single immunodiffusion in one dimension according to Oudin (1956).
(a) Concentration of antigens (Ag) and antibodies (Ab) are equivalent; (b) antigen is in excess; (c) antigenis in greater excess than in (b).

When additional antigens and antibodies are present in the system, several precipitation lines are formed. Their localization in the agarose gel depends primarily on the ratio of antigen and antibody concentrations. When these concentrations are equivalent, the precipitation bands are formed at the beginning of the agarose layer and the distances between them are minimal. It is therefore difficult to determine the number of precipitation lines. The higher the antigen concentration (with respect to the antibody concentration), the further are the precipitation lines from the gel–antigen solution boundary and the greater the distances between them.

The Oudin technique can be used to demonstrate the number of antigens present in the test mixture and also to determine them quantitatively. This method uses a monospecific antibody mixed with the gel and a solution containing an excess of the test antigen is used to overlay the gel. The antigen forms a single precipitation line with the antibody whose width is proportional to the antigen concentration. Precipitation rings can be accentuated by 7.5% acetic acid.

Single radial immunodiffusion (RID) is performed on glass plates, Petri dishes or plastic foils onto which hot (50 °C) 1.5% agarose solution containing a suitable concentration of the specific antibody is poured. The agarose solution is usually prepared in veronal buffer containing merthiolate (0.01–0.001%) as a bacteriostatic agent. After cooling and solidification (the gel layer is 1–2 mm deep) circular wells are cut in the gel into which the antigen solution is pipetted. Some of the holes are filled with solutions containing known antigen concentrations (standards). The plate is then placed exactly level in a humidified chamber to prevent gel drying. The antigen is the only component that can diffuse in this system. The antigen diffuses radially, and produces a precipitation ring after its interaction with the antibody (*Fig. 12.3*). The higher the antigen concentration, the larger the diameter of the precipitation ring. Diameters of precipitation rings around the wells containing standard antigen solutions are measured first and plotted against corresponding antigen concentrations. The resultant calibration curve makes it possible to determine the antigen concentrations of the test samples from the diameters of their precipitation rings.

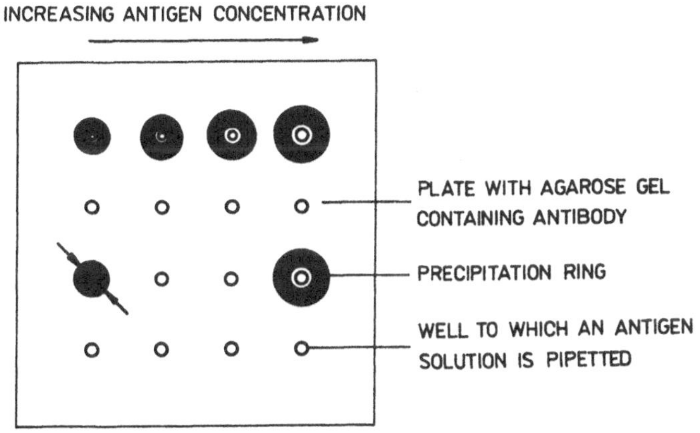

Fig. 12.3. Principle of single radial immunodiffusion.

Two methods may be used. According to Mancini *et al.* (1965) the diameters of precipitation rings should only be measured after immunodiffusion is complete (48–72 h with most antigens). In this case, the ring diameter, or its square, is directly proportional to the antigen concentration and the calibration curve is a straight line on linear graph paper. When the diameters are measured before completion of immunodiffusion, a curve rather than a straight line is obtained.

In the second procedure (FAHEY and MCKELVEY, 1965) the ring diameters are measured after only 6–24 h, *i.e.* before completion of immunodiffusion. They are then directly proportional to the logarithm of concentration:

$$\log c = \frac{d - d_0}{K}$$

where c is the antigen concentration, d the precipitation ring diameter, d_0 the diameter of the well containing the antigen solution and K the slope of the straight line constructed using the above equation. The calibration curve is plotted on semilogarithmic paper (*Fig. 12.4*). The second method is much faster. However, it is less accurate and the relationship between the ring diameter and the logarithm of the antigen concentration is only linear within a certain range of antigen–antibody ratios and, therefore, erroneous results in some samples may be obtained. Sensitivity of the simple two-dimensional immunodiffusion is usually around 0.02 mg/ml.

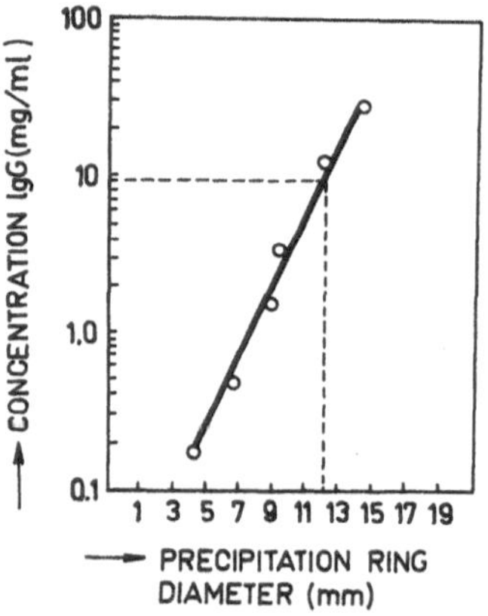

Fig. 12.4. Calibration curve for determination of human IgG by single radial immunodiffusion according to Fahey and McKelvey (1965).
The test sample yielded a precipitation ring 12.5 mm in diameter, which was equivalent to 9.5 mg IgG/ml (dotted line).

The single radial immunodiffusion is a basic method for the quantitative determination of antigens. In clinical laboratories it is used to determine the levels of IgG, IgA, IgM, IgD, various complement components, albumin, α_1-antitrypsin, ceruloplasmin, α_2-macroglobulin, transferrin and other plasma proteins. Such methods require high-quality antisera and antigen solutions with standard antigen concentrations. Agarose gel plates can be prepared in the laboratory, or ready-made plates in various diagnostic kits are available commercially.

Such kits consist of plates with a ready-made gel containing an appropriate antibody. Wells are pre-cut in the gel and the kit also contains a standard antigen solution. The standard is placed in some of the wells and the unknown test samples placed into the remaining ones. After a certain diffusion time, the diameters of all precipitation rings are measured precisely and concentrations of the antigen in the test samples are determined from the calibration curve.

Both modifications of the standard radial immunodiffusion method are fairly reproducible; their coefficient of variation is usually less than 10%. The MANCINI modification has a lower coefficient of variation — usually less than 2%. With high-quality antisera, the single radial immunodiffusion can usually detect as little as 1–3 μg/ml of IgG or albumin.

Precipitation rings are usually measured with the naked eye or with the aid of a magnifying glass. They can also be stained with protein-specific dyes (Amido Black 10 B, Bromophenol Blue, Coomassie Blue *etc.*). A chromogenic or other substrate can act as the detector if the antigen is an enzyme which does not lose its catalytic activity after reacting with the antibody. When the antigen is labelled with a radioactive isotope, precipitation rings can be detected by autoradiography.

The principle of radial diffusion can be used to detect antigens using corresponding antibodies as well as to detect other compounds with special activities, *e.g.* enzymes. In the latter case, an enzyme substrate is added to the agarose gel instead of the antiserum. The enzyme solution is placed in the wells and then diffuses radially and degrades the substrate. The substrate degradation is indicated by a lighter ring or, alternatively, it can be detected using a compound which reacts differently with the degraded and native substrate. This technique is particularly useful for the microdetermination of lysozyme, proteolytic enzymes, phosphatases and other enzymes in tissue extract and biological fluids. When the agarose gel contains sheep erythrocytes coated with the amboceptor (haemolytic system) instead of the antibody, rings of haemolysis are formed around the wells containing complement solution. The diameters of haemolytic rings are directly proportional to the amount of complement.

12.2.2 Double immunodiffusion

In double immunodiffusion, both reacting components can diffuse in the agarose gel (*i.e.* antigen and antibody). Compared to single immunodiffusion, it is necessary to have almost equivalent antigen and antibody concentrations. The double immunodiffusion methods can be used particularly for the qualitative determination of antigens and to determine whether or not they are related. They are not suitable for quantitative determinations.

The double immunodiffusion is performed in test tubes, microscope slides and, occasionally, in Petri dishes. When using test tubes, they contain three layers — bottom, middle and upper (*Fig. 12.5*). The bottom gel contains the antibody and the upper gel contains the antigen (or *vice versa*). The middle gel does not contain either the antigen or the antibody. The antigen and the antibody diffuse towards each other, eventually interacting in the middle (free) gel to produce precipitation bands. In this case, both antigen and antibody molecules diffuse, although only in a single direction. Therefore, the technique is correctly termed **one-dimensional double immunodiffusion**.

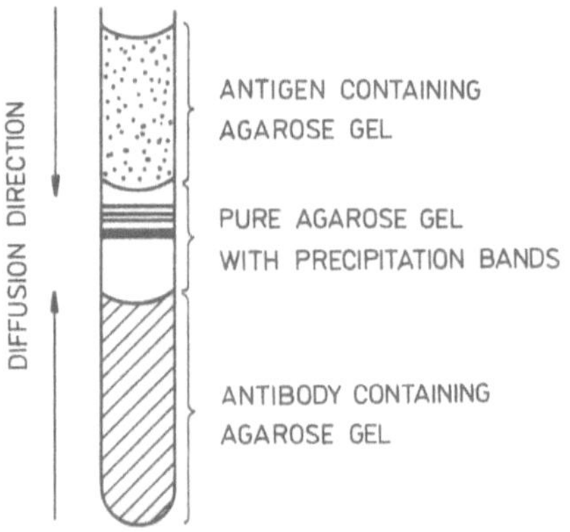

Fig. 12.5. Double immunodiffusion in one dimension.

Double immunodiffusion in an agarose gel prepared on a glass plate is most frequently performed in the modification of OUCHTERLONY (1948, 1953). In this modification the gel contains neither the antigen nor the antibody. After solidification, circular, rectangular or square wells are cut and the antigen or antibody is placed in them. Circular holes, 3–6 mm in diameter, are punched 5–8 mm apart. The antigen and the antibody diffuse radially from the holes (*Fig. 12.6*) and, therefore, the technique is also known as double radial immunodiffusion or **two-dimensional double immunodiffusion.** When the antigen and the antibody are in two neighbouring wells cut in the gel, their molecules interact after a certain period and form a characteristically shaped precipitate–precipitation arc. This is formed after several hours or days of incubation of the plate in a wet chamber (usually at room temperature). The number of antigens and antibodies can be determined from the number of lines between the two wells.

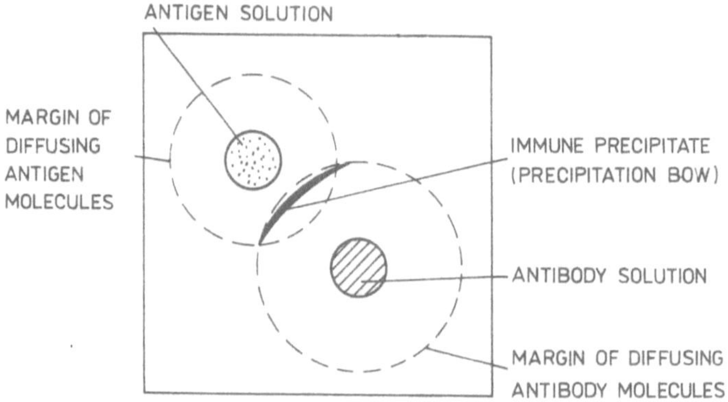

Fig. 12.6. Two-dimensional double immunodiffusion according to Ouchterlony.

When several wells are punched around the antibody-containing gel and different antigens are placed in them, mutually connected or differently crossed arcs are formed. This method can determine whether individual antigens are identical or non-identical and, occasionally, whether their antigenic determinants are related (*Fig. 12.7*).

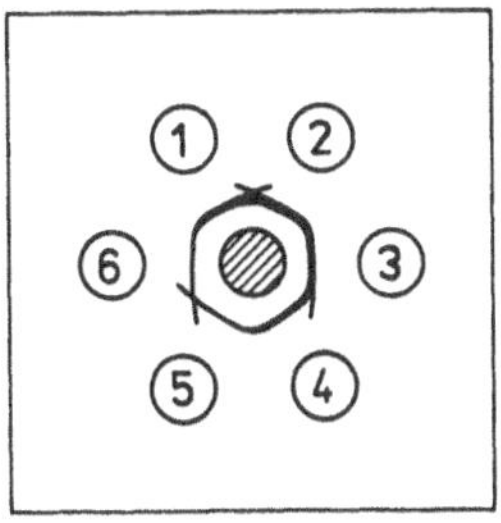

Fig. 12.7. Different precipitation patterns in two-dimensional double immunodiffusion.
The central well contains antibodies while neighbouring wells contain antigen Identity phenomenon may be observed among antigens 1, 2, 3, 4, 5 and antigens 6 and 1 Antigens 5 and 6 are non-identical Non-identity phenomenon is also observed among some determinant groups of antigens 1 and 2 Bilateral identity cannot be observed between antigens 3 and 4 Antigen 3 contains additional determinants

The **Ouchterlony method** is not suitable for quantitative determinations. It is only used for the qualitative demonstration of antigens when specific antisera are available or for demonstration of antibodies using pure antigens. A visible precipitate is formed when the ratio of the antigen and antibody concentrations is optimal (in the equivalence zone). In practice it is necessary

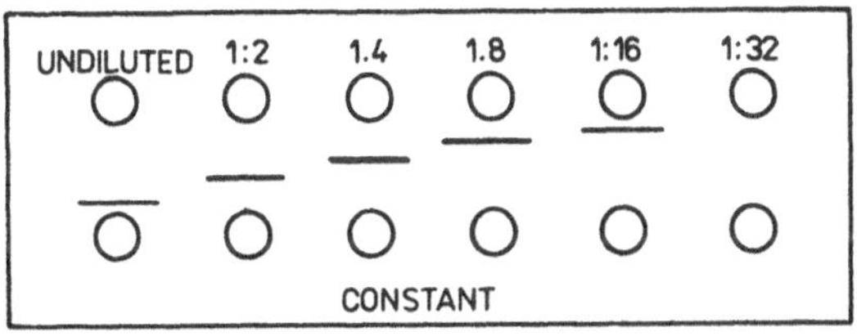

Fig. 12.8. Effect of antigen–antibody ratios on the position of the precipitation line in Ouchterlony two-dimensional double immunodiffusion.
Precipitate disappears in the presence of excess antibody (antigen dilution 1 32)

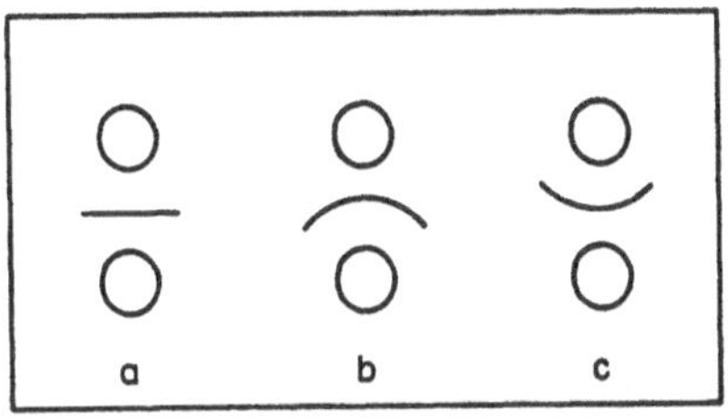

Fig. 12.9. Relationship between relative molecular weight of the antigen and shape of precipitation line in Ouchterlony two-dimensional double immunodiffusion.
(a) Mol wt of antigen and antibody are similar (*e g* IgG and anti-IgG), (b) mol wt of antigen is less than antibody (*e g* serum albumin and IgG antibody), (c) mol wt of antigen is greater than antibody (*e g* ferritin and specific IgG)

to test dilutions of a certain antigen or, alternately, dilutions of the antiserum, to establish the optimum ratio for precipitate formation. Localization of the precipitation line between two wells in the gel also depends on the antigen : antibody ratio (*Fig. 12.8*). The relative molecular weights of the antigen and antibody determine the shape of the precipitation line (*Fig. 12.9*).

12.3 Immunoelectrophoretic methods

As their name suggests, these techniques are a combination of electrophoretic and immunodiffusion methods. They are usually performed in agarose or agar gels prepared on glass or plastic supports of various sizes. 1–2% Agarose with suitable properties (high purity, low electroendoosmosis and low non-specific adsorption) is usually used to prepare the gel. Agarose gel is prepared in veronal, borate, Tris-barbital or Tris-Tricine buffer, of pH 8–9 and ionic strength 0.025, containing 0.02% sodium azide or 0.01% merthiolate as bacteriostatic agents.

Immunoelectrophoretic methods can be used to determine both the presence and amount (concentration) of antigens and can be performed much faster than immunodiffusion methods.

On a methodological basis immunoelectrophoretic methods can be divided into five main groups:

1. Classical immunoelectrophoresis according to GRABAR and WILLIAMS (1953).
2. Rocket immunoelectrophoresis.
3. Counter-immunoelectrophoresis.
4. Two-dimensional (cross) immunoelectrophoresis.
5. Immunofixation.

Classical and counter-immunoelectrophoresis yield qualitative data only, rocket immunoelectrophoresis provides only quantitative results and two-dimensional immunoelectrophoresis yields both qualitative and quantitative data (*Table 12.2*).

12.3.1 Immunoelectrophoresis according to Grabar and Williams

This technique, developed by GRABAR and WILLIAMS (1953), is a combination of zonal electrophoresis and OUCHTERLONY two-dimensional double immunodiffusion. It includes two steps: in the first wells are cut in the agarose gel and solutions of the test antigens are added. The plate is placed in an electrophoresis chamber and the antigens are separated under the influence of direct current (*Fig. 12.10*). Human serum serves as the antigen and can be separated into several fractions at a voltage of 5 V/cm for about 60 min. In the second step, longitudinal troughs (about 2 mm wide) are cut in the direction of the separated antigens and the antiserum containing corresponding antibodies is placed in them. The troughs should be cut with the aid of a template. The plate is then placed level in a wet chamber, during which time

Table 12.2 Lowest limits of detection of various immunochemical methods and their potential for qualitative or quantitative use

Method	*Antigen or antibody concentration (mg/l)*	*Application*[a]	
		qualitative	*quantitative*
Precipitation in liquid medium	10–20	++	++
Precipitation in agarose gel			
Single radial immunodiffusion	10–20	–	++
Double immunodiffusion according to Ouchterlony	10–30	++	–
Immunoelectrophoresis			
According to Grabar and Williams	10–20	++	–
Rocket	0.2–1.0	–	++
Counter-immunoelectrophoresis	0.5–5.0	++	–
Two-dimensional	0.1–1.0	++	+
Radioimmunoelectrophoresis	0.001–0.01	+	+
Agglutination	0.02–0.2	++	±
Passive haemagglutination	0.005–0.02	++	+
Complement fixation	0.1–0.5	++	±
Immunoadsorbent binding	5–10	+	+
Immunoradioisotope methods	0.0001–0.01	–	++
Immunoenzyme methods	0.001–0.1	–	++
Viroimmunoassays	0.0001–0.01	–	++

[a] Very suitable (++), suitable (+), unsuitable (–) for qualitative or quantitative use

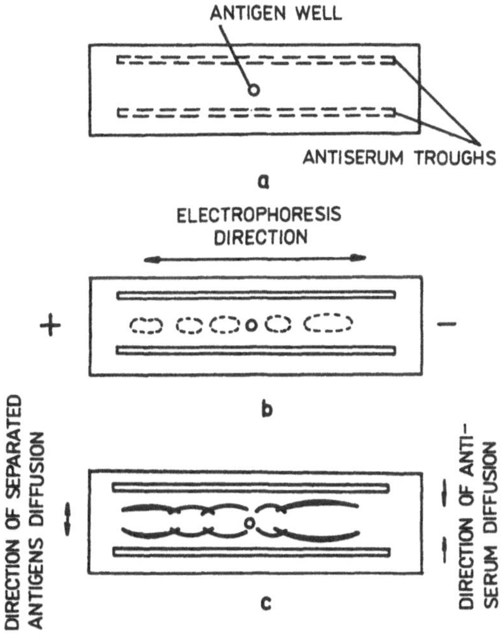

Fig. 12.10. Immunoelectrophoresis.
(a) Wells are cut in an agar plate and filled with a mixture of antigens to be separated by electrophoresis (b) After electrophoresis, antigens are separated but their zones are usually not visible, longitudinal troughs are cut and antiserum added (c) Antiserum diffuses from troughs at right angles to the direction of antigen separation forming precipitation lines (arcs) The plate is finally dried and stained to accentuate the precipitation arcs

the antigen and the antibody diffuse towards each other (24–48 h) and produce characteristic precipitation arcs at the trough margin. After completion of immunodiffusion the precipitation arcs may be visualized directly (with the naked eye) or the excess antiserum is washed from the plate and the plate dried and stained with a suitable dye (Amido Black 10 B, Coomassie Brilliant Blue *etc.*) to accentuate the arcs.

Immunoelectrophoresis according to GRABAR and WILLIAMS has been widely used in clinical laboratories and has contributed substantially to the study of serum proteins and their alterations in various diseases. During zonal electrophoresis in agarose, plasma proteins are separated into 5–6 fractions, whereas immunoelectrophoresis and a high-quality polyspecific antiserum can detect up to 35 precipitation lines. Each precipitation arc indicates one protein. The precipitation lines of normal proteins have characteristic shapes and positions on the immunoelectrophoretogram (*Figs. 12.11* and *12.12*). For example, the prealbumin arc is the first anodal line, whereas IgG appears close to the cathodal end. When for example the myeloma monoclonal immunoglobulin is detected in the analysed serum, the precipita-

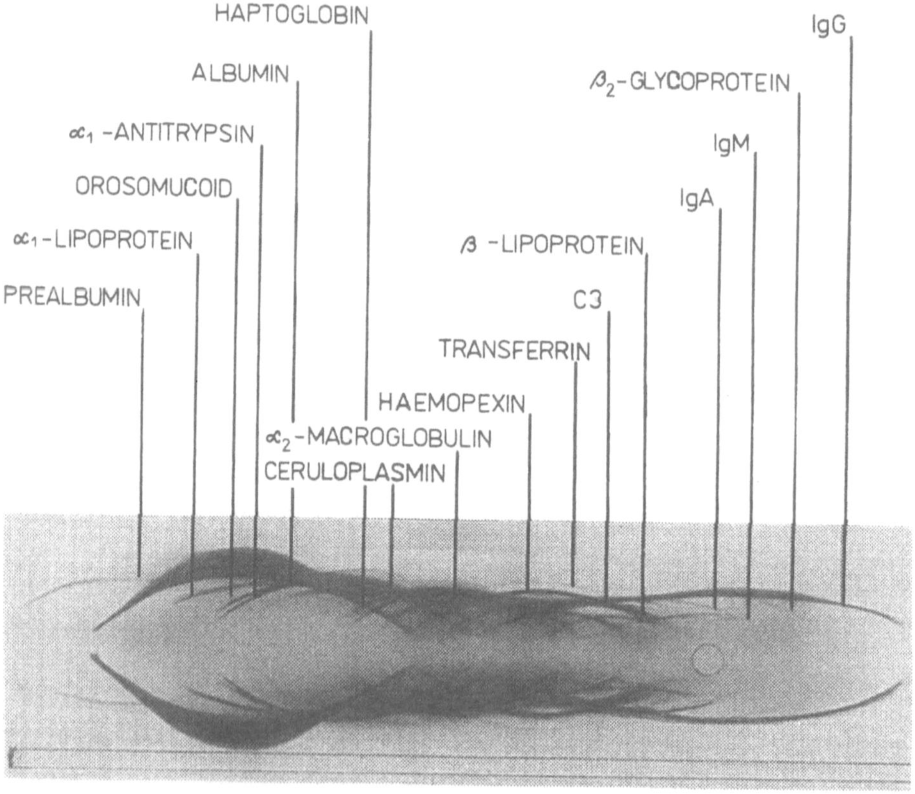

Fig. 12.11. Immunoelectrophoretogram of normal human serum I.
The longitudinal basins contain rabbit antiserum against human serum proteins

tion line of the corresponding immunoglobulin fraction is usually a different shape. This can be used for the detection of myeloma immunoglobulins. The serum of a patient with a suspected gammopathy (p. 140) or other disorder

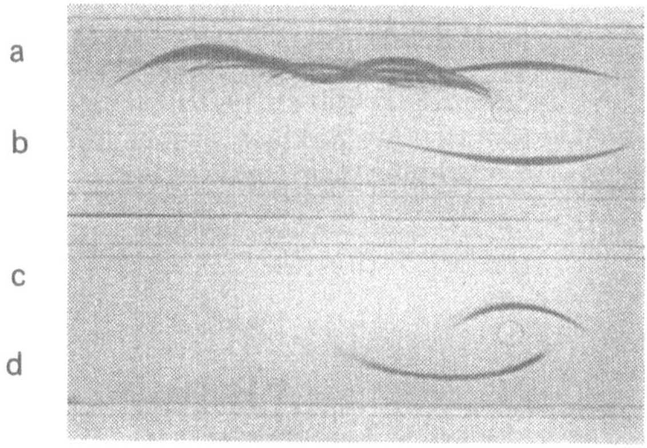

Fig. 12.12. Immunoelectrophoretic pattern of normal human serum II.
(a) Polyspecific antiserum against human serum proteins, (b), monospecific antiserum against IgG, (c), antiserum against IgM, (d), antiserum against IgA

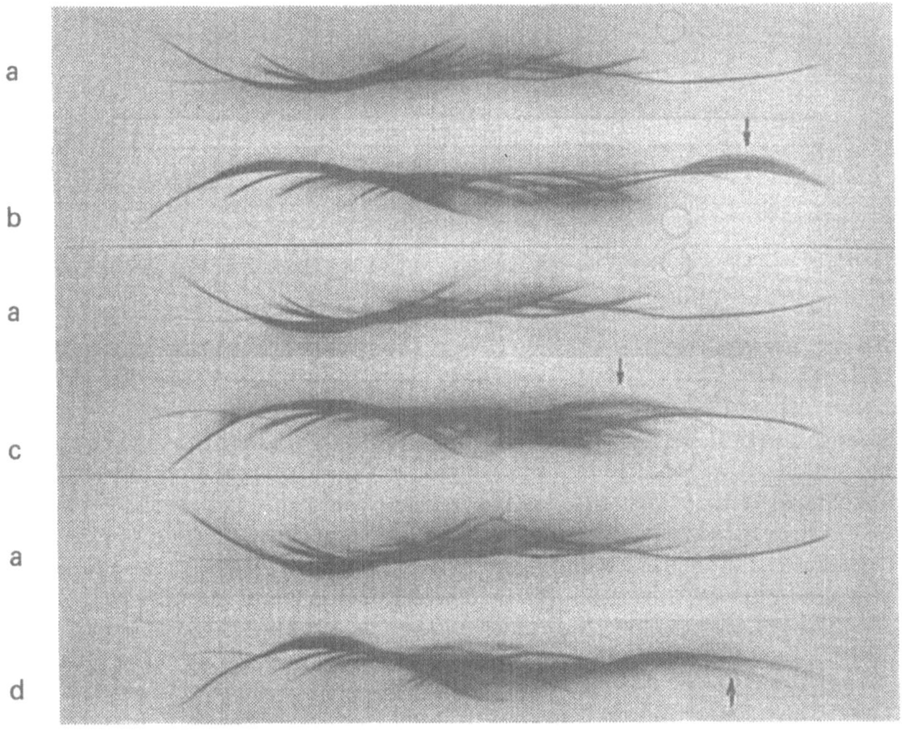

Fig. 12.13. Immunoelectrophoretic analysis of normal human sera (a) and sera with myeloma IgG (b), myeloma IgA (c) and Waldenstöm's macroglobulin IgM (d).
Longitudinal basines contained polyspecific antiserum against human serum proteins Monoclonal (pathological) immunoglob-lins are indicated by the arrows

of immunoglobulin biosynthesis is subjected to immunoelectrophoresis. An antiserum against all plasma proteins and specific antisera against immunoglobulins or their chains are used. On the basis of the shape of the precipitation line and its position, it is possible to decide whether a normal heterogeneous mixture of immunoglobulins of a certain class is involved, or whether a pathological, monoclonal immunoglobulin is present (*Fig. 12.13*).

During immunoelectrophoretic analysis of a complex antigen mixture individual precipitation arcs can produce various phenomena (*Fig. 12.14*), which can be used to determine their characteristics.

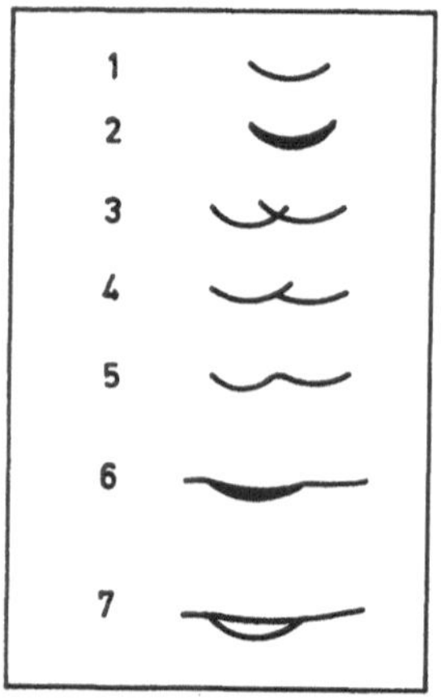

Fig 12 14. Precipitation phenomena in immunoelectrophoresis.
1, One antigen, 2, excess of antigen, 3, two immunochemically non-identical (different) antigens, 4, two partially identical antigens, 5, two identical antigens, 6, excess of certain determinants in a heterogeneous antigen, 7 the phenomenon of antigen "splitting" which, for immunoglobulins, may indicate pathological monoclonal immunoglobulin

12.3.2 Rocket immunoelectrophoresis

Rocket immunoelectrophoresis (RIE), also called **electroimmunodiffusion** (GILL *et al.*, 1971), is based on simple immunodiffusion, but the antigen molecules do not migrate by free diffusion but are accelerated by direct electric current. The method was introduced by LAURELL in 1966. It is performed in gels containing regularly dispersed monospecific, precipitating antibodies. It may therefore be viewed as *affinity electrophoresis,* in which only the antigen which can bind biospecifically with the antibody in the gel, is detected in the separated mixture. Although agarose is used most frequently as a carrier it is also possible to use cellulose acetate membranes.

The gel must be the same depth all over (*e.g.* 8.5 × 8.5 cm). Circular wells are punched in the gel at one end (diameter 2 mm, and about 5 mm apart for a 1.5 mm thick gel). Standard antigen solutions are placed into at least four wells and the test antigen solutions are placed into the remaining wells. The volumes of the test solutions must be identical. The plate is placed in an electrophoresis chamber with the starting wells at the cathodal end. Veronal buffers of slightly alkaline pH are used as electrolytes and under these conditions, most protein antigens migrate towards the anode. When the power is switched on, the antigen molecules migrate into the gel, where they encounter the antibody molecules. Rocket-shaped precipitates (hence "rocket" immunoelectrophoresis) are formed after an equivalent ratio of their concentrations has been reached. The height of the rocket is directly proportional to the antigen concentration. When electrophoresis proceeds without cooling, a voltage of 10–15 V/cm can be applied (depending on the ionic strength of the buffer) and electrophoresis is usually performed for 30–180 min. The rocket heights are measured directly, after staining, or after intensification in molybdophosphoric acid for 5–10 min.

Rocket heights and their corresponding standard antigen concentrations are plotted on linear graph paper. A calibration curve is thus obtained (*Fig. 12.15*) which can be used to determine the concentrations of unknown samples.

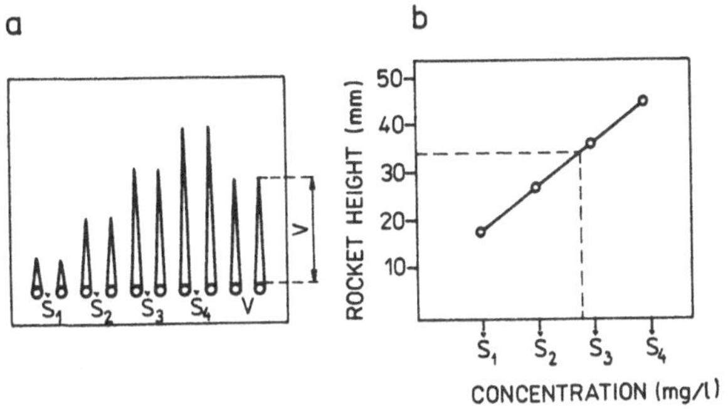

Fig. 12.15. Schematic representation of rocket immunoelectrophoresis (a) and corresponding calibration curve (b) to determine antigen concentrations in test samples.

At present, RIE is one of the most useful ways to quantitate macromolecules against which monospecific antibodies can be prepared and which readily migrate in the agarose gel. Its sensitivity is about 50-fold higher than RID. It is thus possible to detect protein concentrations as low as 0.2–0.1 mg/l with an accuracy of $\pm 5\%$ and usually within 2 h. With more slowly migrating antigens (with mol. wt. usually greater than 150 000) RIE must proceed for a longer time or a mixture of hydroxyethylcellulose with agarose must be used instead of pure agarose. The mobility of very slowly migrating antigens (*e.g.* IgM and α_2-macroglobulin) can be increased by carbamylation (reaction with KCNO).

RIE can also be performed in its reverse form to measure antibody concentrations in different sera. In this case the gel contains the antigen instead of antibodies.

Fused rocket immunoelectrophoresis is a modification of RIE which makes it possible to quantitate proteins in individual fractions separated by various chromatographic techniques (*Fig. 12.16*). In this technique, two types of gel are poured on the plate. The bottom, thin gel is made from pure

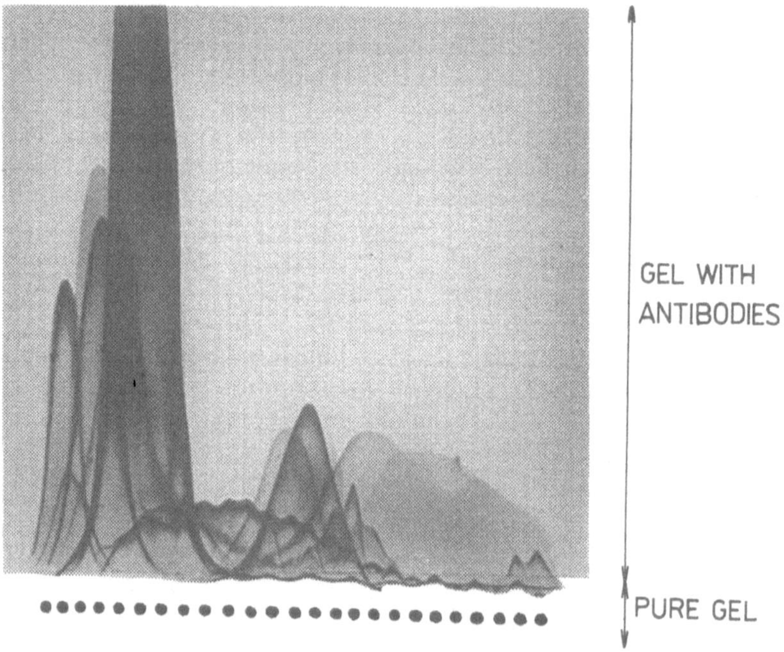

Fig. 12.16. Fused rocket immunoelectrophoresis of fractions (each well contains one fraction) after gel filtration of human serum on a Sephadex G-200 column.

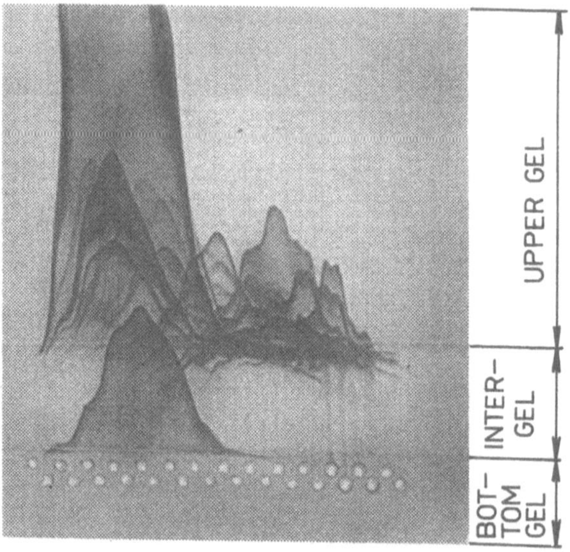

Fig. 12.17. Fused rocket immunoelectrophoresis with an intergel for localization of transferrin. Normal human serum was fractionated on a Sephadex G-200 column and samples of the individual fractions (3 µl) were pipetted into wells of the lower agarose gel which did not contain antibody The intergel contained monospecific anti-transferrin serum and the upper gel contained polyspecific antiserum against serum proteins

agarose. Two rows of wells are punched in the gel and samples of individual fractions are gradually introduced into the wells. The upper, thick gel contains polyspecific antibodies against the separated proteins. After electrophoresis and optional staining, fused precipitation lines appearing as peaks of various shapes are obtained instead of individual rockets. Each peak belongs to a single protein and its concentration in a given fraction is proportional to the height of the peak above the well containing the fraction sample.

When it is necessary to demonstrate that a certain peak belongs to the protein under study, a third agarose layer containing a specific antibody against the test protein is placed between the pure agarose gel and the agarose containing polyspecific antibodies. The test protein then forms a peak in this interlayer, whereas other proteins freely diffuse through it and form precipitation peaks in the third gel layer containing polyspecific antibodies (*Fig. 12.17*).

12.3.3 Counter-immunoelectrophoresis

Counter-immunoelectrophoresis (CULLIFORD, 1964; NAKAMURA, 1966) is a modification of one-dimensional double immunodiffusion. However, migration of antigen molecules and antibodies is accelerated by direct current, so that the results may be obtained after only 30 min. The technique is performed on narrow glass plates (*e.g.* microscope slides) in a pure (1–1.5%) agarose gel, usually prepared in veronal buffer (pH 8.6). Two parallel rows of wells are cut in the gel (*Fig. 12.18*). The wells are 2–3 mm in diameter, 4–5 mm apart and the two rows are 6–10 mm distant. Antigen solutions are placed in one row of wells and an antibody solution in the other. The slide is then subjected to electrophoresis and both components migrate against each other and ultimately form a precipitation arc in the equivalence zone which can be stained or is visible with the naked eye. When, under these conditions, the antibodies migrate towards the cathode, this method can only be used to demonstrate antigens with an anodal mobility. Therefore, the antigen is always placed in the wells on the cathodal side in the electrophoretic chamber.

Counter-immunoelectrophoresis is usually used for the rapid, qualitative demonstration of antigens, providing that specific antibodies are available, or, conversely, to demonstrate an antibody with a pure, standard antigen.

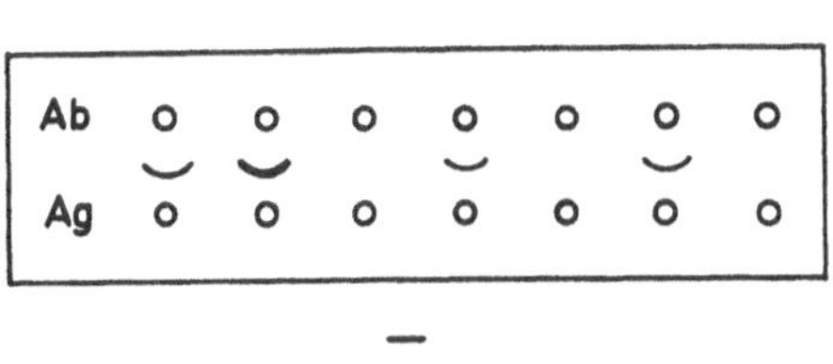

Fig. 12.18. Counter-immunoelectrophoresis.
Ab, wells with antibody; Ag, wells with antigens. Samples in wells 1, 2, 4 and 6 contain the test antigen.

It can also be used for the semiquantitative determination of numerous diagnostically important antigens, *e.g.* HBs antigen in serum hepatitis B, carcinoembryonal antigen (CEA), α-fetoprotein (AFP) *etc.*

12.3.4 Two-dimensional (crossed) immunoelectrophoresis

This technique is actually a combination of zonal electrophoresis and rocket immunoelectrophoresis. **Two-dimensional immunoelectrophoresis** was developed in the 1960s (RESSLER, 1960; LAURELL, 1965; CLARKE and FREEMAN, 1966) and is also known as **crossed immunoelectrophoresis**.

Electrophoresis is performed in two directions. The antigen mixture is first separated by zonal electrophoresis in agarose or polyacrylamide gel. Isoelectric focusing or isotachophoresis can be performed instead of electrophoresis. After electrophoresis has been completed a narrow gel strip containing the separated antigens (*e.g.* 2 × 8.5 cm) is cut and transferred to one side of a clean glass plate (8.5 × 8.5 cm) (*Fig. 12.19*). The remaining area of the plate is filled with agarose gel containing polyspecific antibodies against these antigens and the second electrophoresis is then performed at right angles (cross-wise — hence *"crossed"* immunoelectrophoresis) to the first dimension. The antigens migrate in the antibody-containing gel and form broad precipitation rockets or peaks in the equivalence zones (*Fig. 12.20*). Their position on the immunoelectrophoretogram characterizes the type of antigen. The area of the precipitation peak and, after a sufficient period of electrophoresis, its height, are proportional to the antigen concentration.

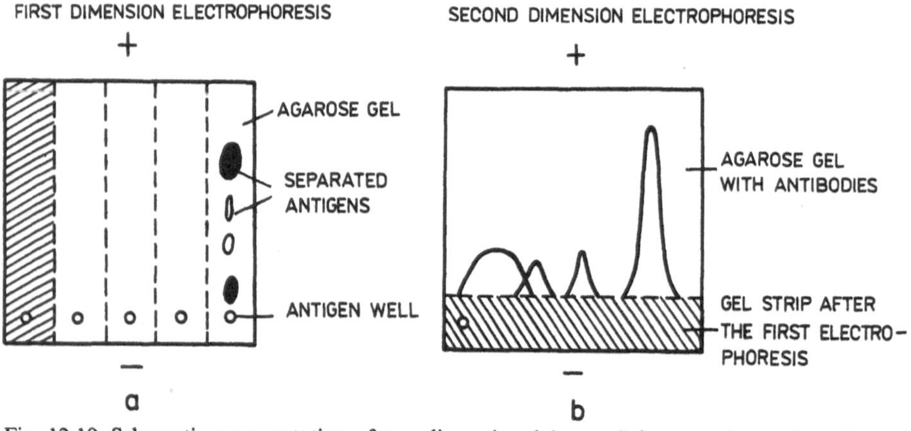

Fig. 12.19. Schematic representation of two-dimensional (crossed) immunoelectrophoresis.

Veronal buffer is usually used as the electrolyte. When human serum is used as the antigen, electrophoresis lasts for 45–60 min in the first direction, and 2–2.5 h in the second direction (voltage 5–8 V/cm, temperature 20 °C). The duration of electrophoresis depends on the voltage applied (potential gradient), temperature, ionic strength of the buffer, concentration of separated antigens and the antibody concentration. Two-dimensional immunoelet-rophoresis has so far been the most useful immunochemical method for the analysis of mixtures of numerous antigens. This method can detect about 50

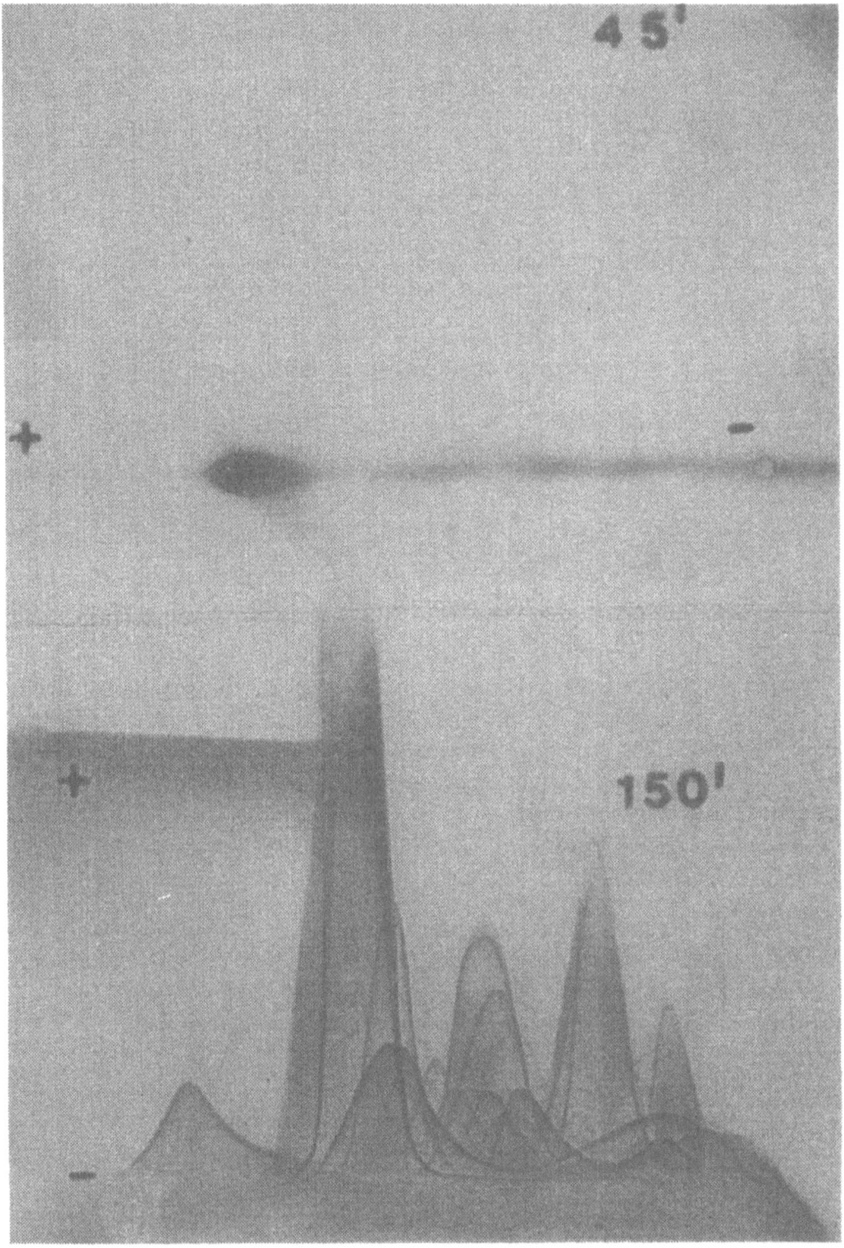

Fig. 12.20. Crossed immunoelectrophoresis of normal human serum in agarose gel (bottom). In the upper half, zonal electrophoresis of the same serum (also in agarose gel) is shown for comparison. Numbers indicate electrophoresis time in minutes.

different proteins in human serum. When isoelectric focusing is used in the first direction, the number of detected antigens is even higher. The sensitivity of the method for proteins is 0.1–1.0 mg/l.

In addition to agarose gels, two-dimensional immunoelectrophoresis can also be performed on a cellulose acetate film (Cellogel, *Fig. 12.21*).

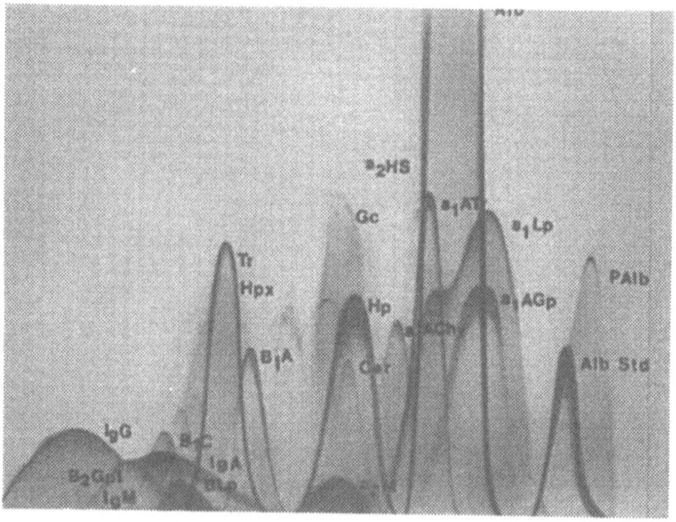

Fig. 12.21. Two-dimensional immunoelectrophoresis of normal human serum on cellulose acetate (Cellogel) film.

Precipitation peaks produced during crossed immunoelectrophoresis can overlap or fuse and can indicate whether particular antigens are related or not (*Fig. 12.22*). Individual peaks (and corresponding antigens) can be identified by comparing their positions and shapes with those of a standard antigen subjected to two-dimensional immunoelectrophoresis under identical conditions, by using monospecific antisera, by using specific staining proce-

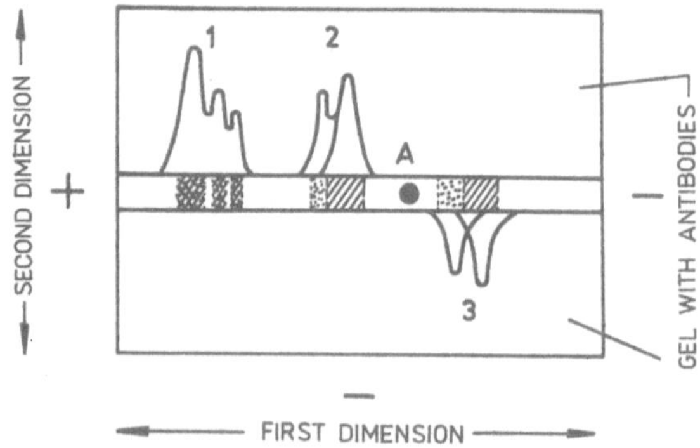

Fig. 12.22. Precipitation patterns in two-dimensional immunoelectrophoresis.
A, site where the antigen mixture was applied, 1, identical antigens migrating towards anode, 2, partially identical antigens migrating towards anode; 3, non-identical antigens migrating towards cathode

dures or by using pure antigens as internal standards. The pure antigen can also be applied to the second well occurring before or after the starting well containing the test antigen mixture. In fact, this yields a **tandem two-dimensional immunoelectrophoresis** in which an identical antigen contained in both wells forms a tandem (pair) of precipitation peaks with identity markers (*Fig. 12.23*). A peak can also be identified using a gel interlayer containing a monospecific antibody (*Fig. 12.17*), the **two-dimensional line immunoelectrophoresis**, or occasionally other modifications.

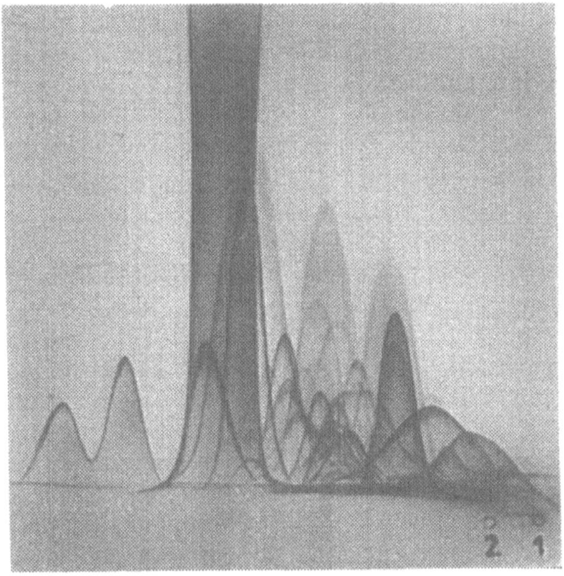

Fig. 12.23. Tandem two-dimensional immunoelectrophoresis of human serum.
The first well contained human serum while the second well contained pure prealbumin The upper gel contained rabbit antibodies against human serum proteins A double peak of prealbumin with the identity marker appears on the left-hand side

12.3.5 Immunofixation

In addition to the four basic methods described above, related immunoelectrophoretic methods exist, among which **immunofixation** is particularly important. In the older scientific literature it is also known as *"direct immunoelectrophoresis"* (WILSON, 1964) or *"quantitative immunoelectrophoresis"* (ALFONSO, 1964). In principle, zonal electrophoresis, isoelectric focusing or isotachophoresis is used to separate a mixture of antigens in agarose, polyacrylamide or acetylcellulose gels, while immunofixation is used to detect some of the separated antigens. After electrophoresis, specific antibodies are applied to the gel to produce immune complexes with the corresponding antigen at the site where it migrated during electrophoresis. The gel is then washed with a suitable buffer to remove non-precipitated antigens and excess antibody. The insoluble immune complexes are stained with protein-reacting dyes on dried gels or visualized with a second antibody labelled with an enzyme or radioactive isotope.

Immunofixation requires antisera with a high titre of monospecific antibodies. It is only usually applied to the part of the gel where the presence of the test antigen is expected. One or two drops per cm^2 of gel are applied and the antiserum is carefully spread with a glass rod (taking care not to damage the gel surface) or a piece of acetyl cellulose is wetted with the antiserum, dried and placed on the gel surface. If several acetylcellulose strips with different antisera are applied to a single gel, their mutual contact must be avoided. It should be noted that small amounts of a concentrated antiserum yield better immunoprecipitates than larger amounts of the diluted antiserum. For immunofixation, 0.02–0.2 ml of high-quality antiserum is usually used for a single sample. The correct antigen to antibody concentration ratio is essential for the formation of the immunoprecipitate and must always be determined experimentally for any antigen. When blood serum is the antigen, it must usually be diluted before electrophoresis, so that the test protein has a concentration of 0.1–0.5 g/l.

After electrophoresis and application of the antiserum, the gel is incubated in a wet chamber at room temperature for 30–120 min. The incubation time depends on the gel type and thickness: 1–1.5 mm agarose gels are incubated for 30–60 min, whereas comparable polyacrylamide gels require up to 2 h. When acetylcellulose strips are used to apply the antiserum, replicas of antigens from the gel surface can appear on them after washing and staining. The gel itself is then washed, dried and stained using procedures similar to those of other types of immunoelectrophoresis. Washing the gel to remove nonspecific proteins is a very important step. Physiological saline is usually used for washing and agarose gels and polyacrylamide gels are washed for 16–24 and 48–72 h, respectively.

Immunofixation of antigens after their separation by zonal electrophoresis or isoelectric focusing is used particularly to study polymorphism of proteins, particularly of those occurring at low concentrations and which are not easily detected by their specific characteristics, *e.g.* some complement components (C2, C3, C4, B), proteinase inhibitors (α_1-antitrypsin) and some other serum glycoproteins.

12.4 Immunoblotting

Blotting is a biochemical analytical method with which it is possible to transfer molecules from a semi-solid phase (*e.g.* a polyacrylamide or agarose gel) to a solid phase (*e.g.* a nitrocellulose membrane). The transfer can be performed by diffusion or by electroblotting (using direct electric current). Molecules transferred to the solid phase are visualized using different chromogenic reactions, by autoradiography or after reaction with specific antibodies, the latter method being termed **immunoblotting**. Blotting methods are used particularly for transfer of DNA, RNA and protein and glycoprotein molecules after separation by electrophoresis or isoelectric focusing on gel carriers.

The blotting technique was originally designed for the needs of molecular biology, namely for the analysis of DNA fragments. In 1975, SOUTHERN described a method to transfer DNA fragments from agarose gels to nitrocellulose membranes by capillary action. DNA fragments transferred to nitrocellulose were then hybridized with radioactive RNA making it possible to detect homologous DNA regions. Homologous DNA fragments were detected as distinct bands on autoradiograms. In 1977, ALVINE *et al.* improved this method by using diazobenzyloxymethylated (DBM) paper instead of nitrocellulose. Even small DNA fragments and RNA molecules binding only weakly to nitrocellulose can bind covalently to the DBM paper. Some authors called this method **Northern blotting** (as compared with previous **Southern blotting**), although neither method has anything to do with geography. RENART *et al.* (1979) and TOWBIN *et al.* (1979) modified the two methods for protein analysis (immunoblotting). In 1981, BURNETTE used specific antibodies and radioactively labelled protein A as the second ligand to detect immobilized antigens. The method has been termed **Western blotting**.

Since then, the immunoblotting techniques have been widely used for the analysis of enzymes, serum proteins, proteohormones, receptors, nucleoproteins, ribosomal proteins, antigens of pathogenic microorganisms and viruses, oncogens, allergens and many other proteins and glycoproteins (BEISIEGEL, 1986). They can be easily performed, they are highly sensitive and specific, and even in very complex mixtures it is possible to detect a single protein in positions with characteristic size and charge. It is possible to obtain an identical copy (even several of them) of the gel surface and it is therefore subsequently possible to use detection and analytical procedures that are not directly applicable to the gel. It is also possible to use monoclonal and conventional antibodies *etc.*

There are two basic steps in immunoblotting methods. During the first step, the test antigen mixture is separated by means of one-dimensional or two-dimensional electrophoresis or isoelectric focusing in agarose or polyacrylamide gel. In the second step, *i.e.* after electrophoresis, the gel is placed in the blotting apparatus, where antigens are transferred from the gel to a suitable matrix. After immobilization of the antigens to the matrix, the matrix is treated with a blocking reagent to saturate (block) non-specific binding sites. Non-specific interactions between the detection probe and matrix are thus avoided. Antibodies, specific binding proteins or other ligands can be used as detection probes. If non-specific interactions were not blocked, the quality of detection of immobilized proteins would decrease because of the high background intensity. Visualization of the detection probe (*e.g.* by autoradiography, colour reaction, fluorescence *etc.*) is the last step.

The transfer of proteins from the gel to the immobilization matrix can be achieved by simple diffusion (relatively ineffective), by capillary pressure (*Fig. 12.24*), negative pressure (negative blotting) or by electrophoretic forces (electroblotting). **Electroblotting** is most effective and is therefore most widely used. It has the advantage that several copies can be obtained and different detection reactions performed. The size of the protein and electroblotting conditions determine which proteins are preferentially detected on individual copies. Low-molecular-weight proteins usually occur on the first copy, whereas larger proteins are found on subsequent copies.

Electroblotting devices may be vertical or horizontal (*Fig. 12.24*), and use platinum or graphite electrodes. Some of them, *e.g.* the LKB Multiphor II Nova Blot System, are of "brick-box" type and are in fact options of electrophoresis systems produced by the company.

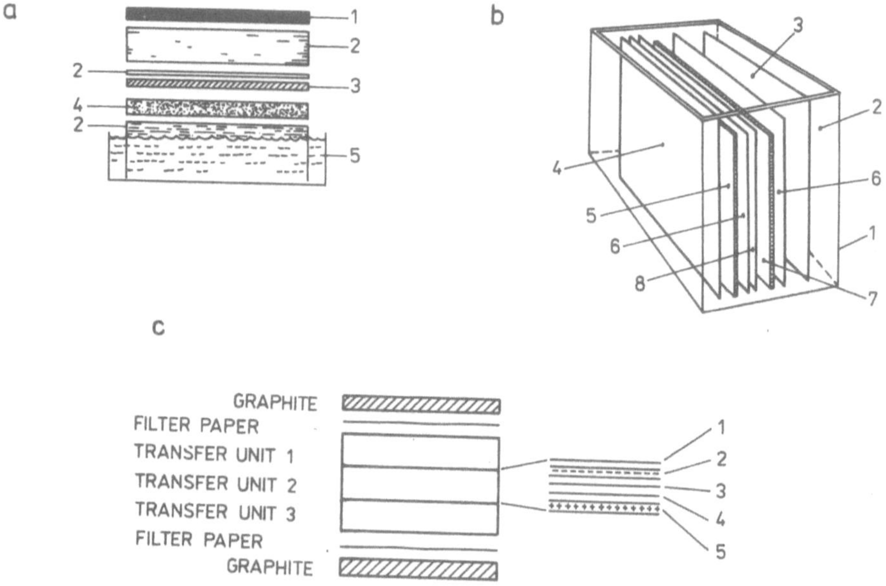

Fig. 12.24. (a) Apparatus for blotting of antigens by capillary pressure.
1, Weight, 2, filter paper, 3, immobilization matrix (e g nitrocellulose), 4, gel with separated antigens, 5, buffer
(b) Apparatus for vertical electroblotting.
1, Chamber, 2, buffer, 3, cathode, 4, anode, 5, foam rubber insert, 6, filter paper, 7, gel, 8, immobilization matrix
(c) Apparatus for horizontal electroblotting (Multiphor II Nova Blot System).
1, Dialysing membrane, 2, filter paper (cathode), 3, gel, 4, nitrocellulose membrane, 5, filter paper (anode)

Two types of immobilization matrices, *i.e.* nitrocellulose and DBM paper, are widely used. Other matrices have either a low binding capacity (acetylcellulose) or it is difficult to detect proteins on them (Zetabind, Gen-Screen). **Nitrocellulose** which binds proteins almost quantitatively (probably by hydrophobic bonds), without any pretreatment, is the most useful. Its binding capacity for proteins is about 80 mg/cm^2. When transferring small polypeptides, the pore size becomes critical. The use of nitrocellulose membrane with a pore size of 0.1–0.2 µm is therefore recommended. Proteins are bound covalently (capacity 10–20 μg/cm^2) to **DBM paper**. A major disadvantage of this immobilization matrix is that it must always be diazotized, *i.e.* commercial benzyloxymethylated paper must be converted to its diazobenzyloxymethylated derivative before use, and this may cause problems in reproducibility of the results.

For "electroblotting" of proteins on nitrocellulose Tris–glycine buffer, pH 8.3, containing 20% methanol is usually used. Transfer takes about 5 h when 60 V and 0.2 A are applied. When transferring to DBM paper, phosphate buffer (pH 6.5) or borate buffer (pH 9.2) are used. The most important factors determining conditions of protein electroblotting are the separation gel layer and the efficiency of their elution from the separation gel layer and the binding capacity of the immobilization matrix for the protein(s) of interest.

After their transfer to the nitrocellulose membrane, the proteins can be stained with protein-specific stains. However, when specific antibodies or another detection probe are to be applied, the nitrocellulose matrix must be preincubated with a **blocking reagent**, *e.g.* bovine serum albumin (3–5%), haemoglobin (5%), ethanolamine (10%), Tween 20 (0.05%) or other compounds. When using DBM paper, reactive diazo-groups that have not reacted with the transferred proteins are usually blocked with Tris–HCl buffer (pH 9) containing 0.25% gelatin and 10% ethanolamine.

Proteins on the nitrocellulose matrix can be detected by one of three basic techniques — direct, non-specific protein staining, direct specific staining (with the first ligand) or indirect specific staining (with the second, or occasionally third ligand). Amido Black 10B, Coomassie Brilliant Blue R-250, Indian ink (all these have a sensitivity of 0.5–1.0 μg of protein), silver stain (sensitivity 4–10 ng) and colloidal gold (sensitivity about 50 pg) can be used for direct non-specific staining. For **direct, specific labelling** of selected proteins, specific antibodies or other ligands (labelled with radioactive isotopes, fluorochromes or enzymes) may be used to visualize the proteins on the nitrocellulose membrane.

Indirect specific labelling is one of the most widely used methods. Two or three ligands may be used. The first ligand (usually an antibody) acts as a specific probe which binds selectively to the target protein and does not react with other proteins. As the first ligand is not labelled, a second ligand — an additional (secondary) antibody carrying a label — must be used for its detection. **Protein A**, which reacts specifically with the Fc-domains of most immunoglobulins is often used as the second ligand. Sometimes, the secondary antibody is also unlabelled and, thus, a labelled third antibody is used for its detection.

From the immunological point of view, electroimmunoblotting is widely used to identify pathogenic bacterial and viral antigens, to detect antibodies against these antigens (for diagnostic purposes), autoantigens (in autoimmune diseases), and pathological monoclonal immunoglobulins (paraproteins) and in immunogenetics, to determine phenotypes of proteins occuring in polymorphous forms (*e.g.* some complement components such as factor B

— *Fig. 12.25.*). Immunoblotting can also be used to identify individual peaks in crossed immunoelectrophoresis (*Fig. 12.26*).

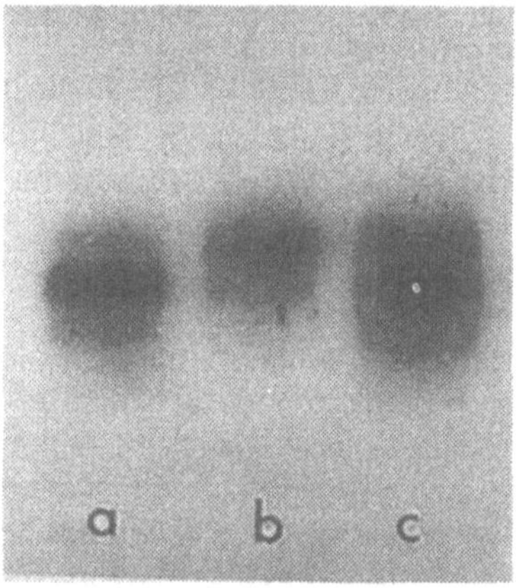

Fig. 12.25. Blot of phenotypes of human factor B on nitrocellulose membrane after separation of human serum by isoelectric focusing in polyacrylamide gel.

Proteins were transferred to nitrocellulose by electroblotting. A specific monoclonal antibody served as a probe (first ligand) to detect isomorphous factor B forms Visualization was by peroxidase-labelled secondary antibody. a, Slow phenotype (BS); b, fast phenotype (BF); c, mixed phenotype (BS/F)

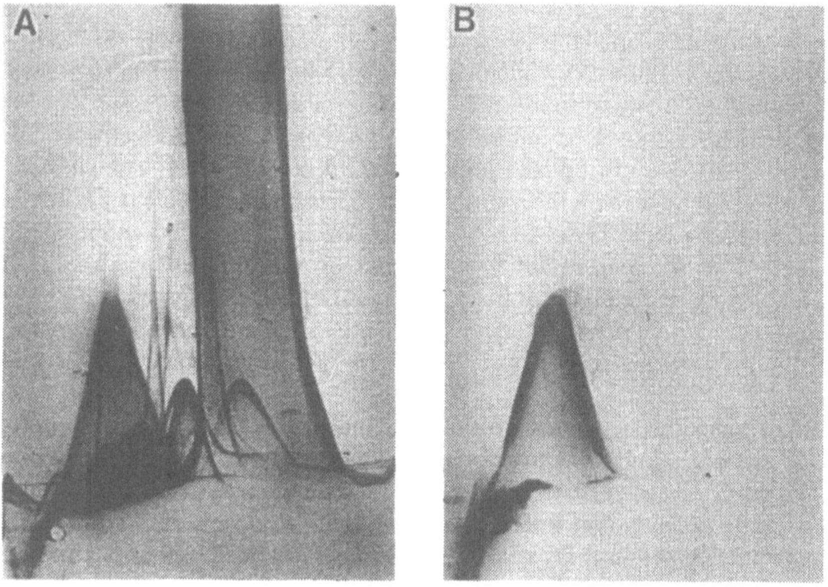

Fig. 12.26. Identification of a peak obtained during crossed immunoelectrophoresis by electro-immunoblotting.

(a) Crossed immunoelectrophoresis of human serum in agarose gel (b) Blot on nitrocellulose membrane after reaction with mouse monoclonal antibody against human IgM and detection with a peroxidase-labelled second antibody (goat anti-mouse antibody)

12.4.1 Dot immunobinding assay

Nitrocellulose membranes are also used in the **dot immunobinding assay (DIBA)**. This technique was introduced by HAWKES *et al.* (1982) as a modification of the immunoblotting technique. In practice, the test sample, usually in several dilutions, and standards containing known amounts of the appropriate antigen, are applied to the surface of nitrocellulose wetted with PBS. Dots with the antigen are washed with PBS in a vacuum. Free reactive groups which could non-specifically react with antibodies are then blocked with a blocking reagent. The nitrocellulose membrane is then incubated with specific antibodies against the tested antigen. The antibodies (usually monoclonal) only react with the membrane where the specific antigen is present. The resultant immune complexes are detected by a secondary antibody labelled with a suitable enzyme. Coloured spots (*dots*) occur after addition of the appropriate substrate, and their colour intensity can be estimated semi-quantitatively with the naked eye (compared to the colour intensity of the standard dots) or evaluated accurately by densitometry in a reflectance spectrophotometer. The secondary antibody is usually conjugated with peroxidase, and hydrogen peroxide in the presence of 3,3′-diamino-benzidine is used as the substrate.

The DIBA makes it possible to rapidly analyse, both semi-quantitatively and quantitatively, the antigen concentration in a single dot. The method is simple and readily reproducible, only small quantities of the sample are required and the sensitivity of the method is high. The lower limit for the quantitative determination of, for example, immunoglobulins (including the IgG subclasses) is 1–5 ng (JOL-VAN DER ZIJDE *et al.*, 1988). The greatest sensitivity was achieved by using the peroxidase–avidin–biotin complex together with 4-chloro-1-naphtol as substrate, which detected 1 pg of immunoglobulin molecules (LAVANCHY *et al.*, 1990).

Small peptides and other low-molecular-weight ligands do not attach to conventional membranes. Thus, chemically reactive membranes should be used for this purpose. Studies conducted on peptides have shown that they can be detected with high sensitivity by immunoblotting and dot immunobinding techniques by chemical crosslinking of the peptides to gelatin-coated nitrocellulose (VAN DER SLUIS *et al.*, 1987), by direct crosslinking to the matrix (KAKITA *et al.*, 1982) or without crosslinking. In the latter case, the nitrocellulose membrane is preactivated in divinyl sulphone, and derivatized by ethylenediamine and glutaraldehyde. The divinyl sulphone-ethylenediamine-glutaraldehyde spacer is introduced in order to minimize steric hindrance of peptide–antibody interaction. The aldehyde groups on the activated nitrocellulose (Nit-CHO) were stable for one month at 4 °C. Peptides were attached to the membrane by reaction of the amino group with the free carbonyl, forming peptide bonds. With this technique the decapeptide angiotensin I and the octapeptide angiotensin II could be detected with a sensitivity of 500 or 20 pg/cm^2 respectively (LAURITZEN *et al.*, 1990).

12.5 Immunoradioisotope methods

In 1959, YALOW and BERSON conducted *in vivo* metabolic studies of ^{131}I-labelled insulin and showed that, in patients suffering from diabetes, previously treated with insulin, antibodies against insulin could be detected. The antibodies formed complexes with the radioactively labelled insulin and addition of non-labelled insulin displaced the labelled insulin from the complex. This observation formed the basis of a method for insulin determination. Numerous such **radioimmunoassays** were later used for the determination of proteohormones and of many other compounds. In 1977, YALOW was awarded the Nobel prize for the discovery of radioimmunoassay methods.

Immunoradioisotope methods are highly sensitive and specific (YALOW, 1978). They can be used to determine any compound, against which it is possible to prepare antibodies, at nanogram (10^{-9} g), to picogram (10^{-12} g) quantities. Due to the high specificity, individual compounds can be determined directly in complex biological fluids, such as blood serum, cerebrospinal fluid, urine *etc.* **Radioimmunoassay (RIA)** is the basic method of this group.

The principle of RIA is as follows. When the concentration of a compound (*e.g.* X) is to be determined, it is first necessary to prepare a specific antiserum against the test compound. The highly purified compound X is used to immunize a suitable animal. When X is only a hapten rather than an immunogen, it must be bound to a macromolecular carrier and the conjugate used for immunization. As well as polyclonal sera against X, monoclonal antibodies can also be prepared, and this is now a commonly used approach. Compound X is then radioactively labelled, usually with ^{125}I for a protein, or with ^{3}H or, less frequently, also with ^{14}C, when low-molecular-weight haptens are involved. The labelled compound X* is thus produced. The antibody (Ab) does not discriminate between X and X*. The equilibrium constants K are identical for both X-Ab and X*-Ab complexes. When different concentrations of X are added to the X*-Ab complex, its molecules displace some of the X* molecules from the complex and the X-Ab complex is formed in addition to X*-Ab. The more X molecules added, the fewer X* molecules complexed with the antibody. The radioligand X* thus competes with the normal, non-labelled ligand X for the antibody binding site. Therefore, radioimmunoassay is sometimes also known as a *competitive protein binding assay* if a protein is the ligand.

The analytical system used in RIA consists of four components: the anti-X antibody, the labelled compound X*, the non-labelled compound X at a known concentration (standard) and compound X at an unknown concentration (the test sample).

When determining X by RIA a series of test tubes are required. Equal amounts of X* and of the anti-X antibody (antiserum) are added. Their mutual ratio should be such that about 50–60% of X* could bind to the antibody. A known concentration of X is then added to some tubes (standard solutions) and the test samples of unknown concentrations are added to the remaining tubes. The mixtures are allowed to react and, after a suitable incubation period, the equilibrium:

$$X^* + X + Ab \rightleftharpoons X^*\text{-}Ab + X$$
$$X^* + X + Ab \rightleftharpoons X\text{-}Ab + X^*$$

is established.

At equilibrium, the mixture contains the free labelled compound X* and unlabelled compound X, as well as complexes of the two forms of the antigen with the antibody (X*-Ab and X-Ab). The more X molecules in a tube, the fewer X* molecules able to bind with the antibody. The amount of the radioactive ligand X* is identical in each tube and, therefore, the number of its molecules bound in the X*-Ab complex depends only on the concentration of X. A set of standard solutions containing known concentrations of X is therefore used. Thus, it is possible to determine the amount of X* bound in the immune complex with the antibody at known concentrations of X, and a calibration curve is constructed on the basis of the data obtained. The curve is then used to calculate concentrations of X in the test samples (*Fig. 12.27*). The curve may have a different shape (sigmoidal, linear, hyperbolic) depending on which parameters are illustrated. The concentration of X is plotted on a logarithmic scale on the abscissa, whereas radioactivity, expressed as dpm or cpm, of the radioactive isotope, or percentage of the compound X* bound in the immune complex are plotted on the ordinate.

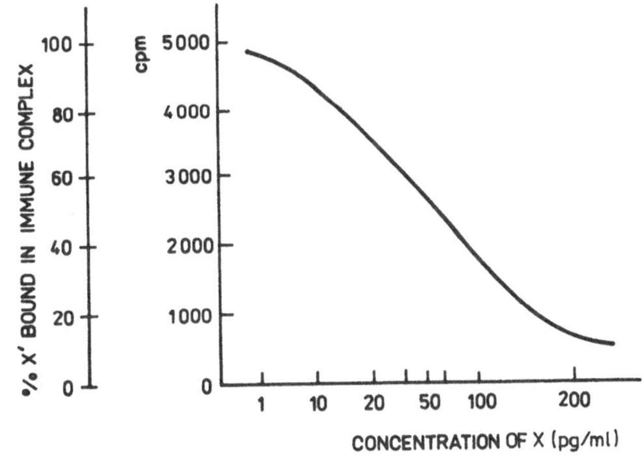

Fig. 12.27. Calibration curve for radioimmunoassay compound X.

In order to determine the amount of the labelled compound X* bound in the complex with the antibody at a certain concentration of X, the X*-Ab complexes must be separated from the free compounds X and X*. The immune complexes may be separated by immunochemical or physicochemical methods.

The immunochemical method is based on the use of a *secondary antibody* which is specific for a primary antibody (anti-isotype immunoglobulin).

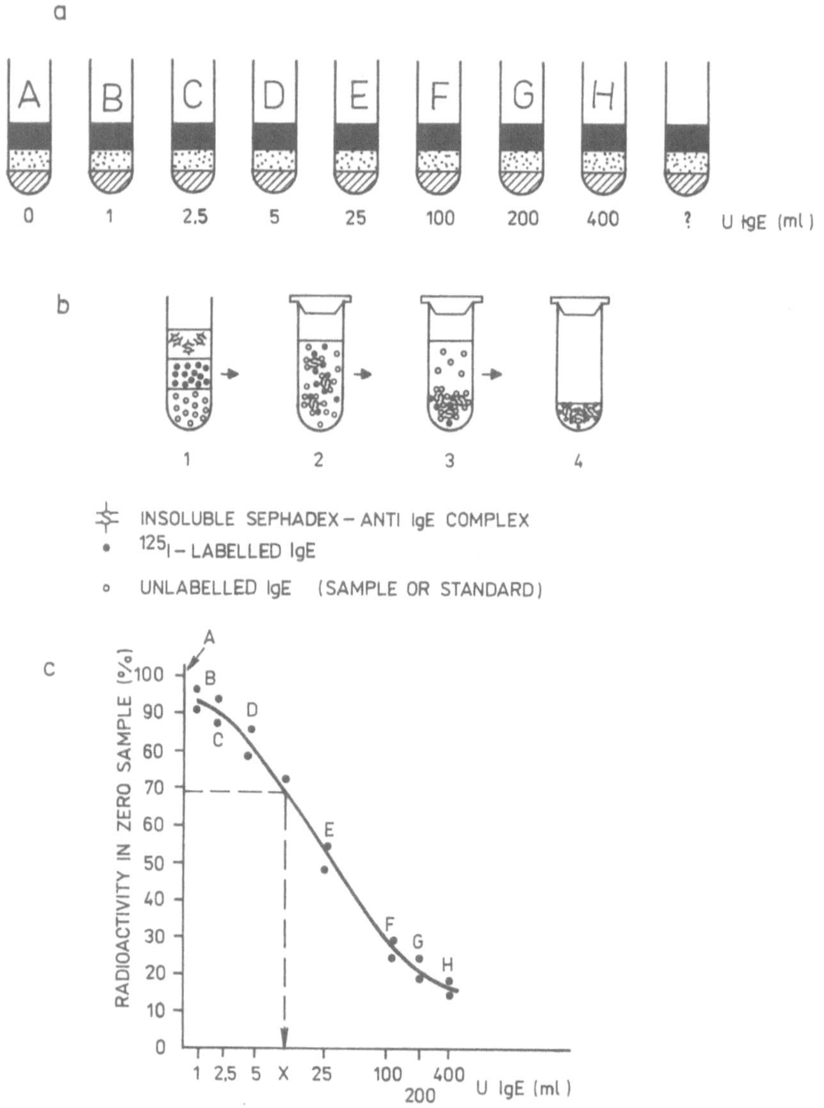

Fig. 12.28. Principle of solid phase radioimmunoassay (Phadebas IgE Test).

(a) Identical amounts of anti-IgE antibody bound to Sephadex particles (hatched) and IgE labelled with ^{125}I (dotted) are pipetted into a set of test tubes The indicated amounts of non-labelled IgE (black) are added to tubes B–H Tube A is a blank without non-labelled IgE, whereas the B–H tubes contain standard and the remaining tubes the test samples (patient sera) (b) Working procedure 1 In each tube Sephadex-anti-IgE particles are mixed with labelled and non-labelled IgE 2 The mixture is incubated overnight at room temperature Equilibrium between immune complexes containing labelled or non-labelled IgE is thus established 3 Solid particles (immune complexes) are separated by centrifugation 4 Radioactivity is assayed and a calibration curve is constructed from the values obtained (c) Typical calibration curve for IgE radioimmunoassay

When such a secondary antibody (Abs) is added to the X*-Ab complex, the complex X*-Ab-Abs is produced which precipitates and can be separated by centrifugation. Similarly it is also possible to use non-precipitating monoclonal antibodies as the primary antibodies. This technique can also be used in a simplified form; the secondary antibody is bound to an insoluble carrier resulting in an immunoadsorbent which reacts with the X*-Ab complex.

Physicochemical methods utilize the different mobility or relative molecular weight of the free ligand X and its complexes with the antibody (ion-exchange chromatography, electrophoresis, gel-filtration chromatography), precipitation of the X*-Ab complexes (*e.g.* after adding neutral salts), adsorption of free X and X* on to activated charcoal, ion exchangers or ligand binding to an immobilized antibody (an immunoadsorbent). The latter procedure is particularly useful. The anti-X antibody is bound to an insoluble carrier and the immunoadsorbent is added to the X* and X mixture. The X*-Ab and X-Ab complexes are insoluble and therefore, no other techniques are required to separate soluble free X and X* ligands. This modification is called **solid phase radioimmunoassay** and an example of this technique is shown in *Fig. 12.28*.

A good quality antiserum is essential for the successful determination of an antigen or hapten by RIA. The antiserum is prepared by multiple immunization (up to six months) of a suitable animal (usually a rabbit or guinea-pig) with a purified antigen. The purity of the antigen is essential for the antiserum specificity. It usually has a high titre and should be diluted for use in the actual assay (sometimes up to $1:10^6$). Such conventional antisera obtained from individual animals may not exhibit an identical specificity, even though a pure antigen was used for their preparation. Thus, they may cross-react with other structurally related antigens or haptens. Monoclonal antibodies are much more specific, recognizing only a single epitope on the antigen molecule. In addition, a highly purified antigen is not required for their preparation and long-term immunization of the experimental animals is not necessary.

Numerous companies produce ready-made kits containing a specific antiserum (antibody), radioactively labelled antigen or hapten and appropriate standards. Using these kits it is possible to determine the concentrations of about 50 different hormones, numerous biologically important antigens, immunoglobulins, cyclic nucleotides, prostaglandins, opiates, barbiturates and other pharmaceuticals (*Table 12.3*).

RIA techniques are highly sensitive and only very small amounts of the individual compounds are required. The entire procedure can be automated albeit with certain difficulties. On the other hand, it is a heterogeneous technique which always requires a separation step (at least centrifugation in the solid phase radioimmunoassay).

In addition to radioimmunoassay, radioassay and immunoradiometric assays are used.

Radioassay is performed in a similar way to RIA. However, the ligand is not bound to a specific antibody but rather to another carrier or transport protein usually also found in the organism. The determination of thyroxine (T4) and tri-iodothyroxine (T3) using thyroxine-binding globulin is an example of this technique.

Table 12.3 Examples of compounds which can be measured in biological fluids by radioimmunoassay (RIA) using commercially available kits

Compound	*Isotope*	*Sensitivity (pg/ml)*
ACTH	^{125}I	1–2
Aldosterone	^{3}H	5
Angiotensin II	^{125}I	50
c-AMP, c-GMP	^{125}I	10–20
Barbiturates	^{3}H	10 000
Digoxin	^{3}H	100
IgD	^{125}I	0.5 U[a]
IgE	^{125}I	1.0 U[a]
Insulin	^{125}I	5
Calcitonin	^{125}I	10
Corticosterone	^{3}H	100
Cortisol	^{3}H	100
LSD	^{125}I	100
Morphine	^{125}I	500
	^{3}H	5 000
Ouabain	^{3}H	200
Progesterone	^{3}H	20
PGE_2	^{3}H	10
PGF_{2d}	^{3}H	5
Testosterone	^{3}H	50
Thyroxine	^{125}I	2 000

[a] Sensitivity is expressed in international units per ml.

Immunoradiometric assays involve labelling of the antibody with a radioactive isotope, whereas the antigen or hapten are unlabelled. An excess of labelled antibody is added to the antigen and reacts with all antigen molecules. It is then necessary to separate the resultant immune complexes from the free antibody rather than from the free antigen, in contrast to RIA.

The reactivity of the labelled antibody may be impaired by binding of the isotope to the binding site. This can be prevented by previously binding the antibodies to the immunoadsorbent with the antigen. This blocks the binding sites, the antibodies may then be labelled and the labelled antibodies eluted.

12.6 Immunoenzyme methods

In spite of their sensitivity and specificity, radioimmunoassays have numerous disadvantages. These include radiation hazards in particular, the short half-life of many of the radioactive isotopes used which give rise to transport problems between the manufacturer and the laboratory, and expensive instruments are required to measure radioactivity. It is therefore not surprising that other assays which use the high specificity, accuracy and sensitivity of the antigen–antibody reaction, but which avoid the use of radioisotopes, have been developed. In practice, labelling a single reacting component with an enzyme (hence the terms immunoenzyme methods and **enzyme-immunoassays — EIA**) is widely used. The methods are similar to RIA, but enzyme activity is assayed instead of radioactivity.

The use of enzymes to label antigens, haptens and antibodies with the aim of their quantitative determination was introduced independently by ENGVALL and PERLMANN, and VAN WEEMEN and SCHUURS in 1971. Actually, as early as in 1966 AVRAMEAS and URIEL used enzyme-labelled antigens or antibodies to visualize weak precipitation lines following immunodiffusion in agarose gels and NAKANE and PIERCE (1966) used enzyme-labelled antibodies to detect antigens in histological sections.

Compared to RIA, EIA avoids the risks of irradiation, no expensive counters are required and the reagents used are relatively stable; EIA is also cheaper and less time-consuming. Most EIA methods are almost as sensitive as RIA and achieve the theoretical potential imposed by the formation of the immune complex, *i.e.* the equilibrium constant K of the antigen–antibody reaction. Under optimal conditions, the association constants of the antigen–antibody complex reach values of about 10^{-12} mol/l, (fmol/ml). Thus for example antigens of molecular weight about 10 000 can be determined at concentrations of approximately 10 pg/ml. EIA methods are suitable for automation and computerized data analysis.

At present, many different types of enzyme immunoassays have been developed (SCHUURS and VAN WEEMEN, 1977; TIJSSEN, 1985; MANČAL, 1987). They can be classified according to the type of compound to be determined (antigen or antibody), according to which component is labelled, according to reaction type (competitive or non-competitive) and according to whether the free labelled component must be separated from the bound component. For any type of EIA, the most advantageous enzyme and most suitable procedure for conjugating it to the antigen or antibody must be chosen. All immunospecifically reacting sites, *i.e.* antigenic determinants or binding sites on the antibody, in the resultant conjugate must be preserved. Damage to immunospecifically reacting sites would reduce reaction specificity. The enzyme activity of the conjugate may remain unchanged or may change. According to these criteria EIA methods can be divided into two main groups:

1. *Homogeneous EIA methods* — the catalytic activity of the enzyme in the conjugate decreases or increases.
2. *Heterogeneous EIA methods* — the catalytic activity of the enzyme in the conjugate does not change.

12.6.1 Homogeneous EIA

Homogeneous EIA relies on modifying the catalytic activity of certain enzymes conjugated with an antigen after its binding to the antibody. In other words, the enzyme activity of the free conjugate is different from that of the conjugate–antibody complex. When a mixture of a known amount of the enzyme-labelled antigen and standard, or occasionally an unknown amount of the non-labelled hapten (antigen), reacts with a limited amount of the antibody, the enzyme activity is directly proportional to the amount of the conjugate bound to the antibody. It follows that it is not necessary to separate the free conjugate from that bound in the immune complex. Thus, the method involves only a single step — it is homogeneous. In the literature, homogeneous EIA is often abbreviated to **EMIT (enzyme multiplied immunoassay technique)**. The enzyme activity of the conjugate after its binding to the antibody may decrease or increase (*Fig. 12.29*).

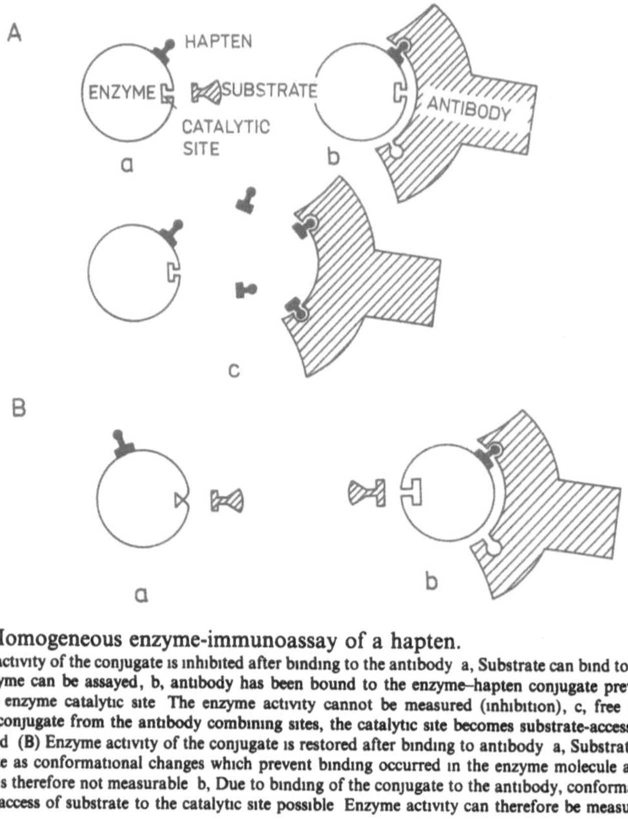

Fig. 12.29. Homogeneous enzyme-immunoassay of a hapten.
(A) The enzyme activity of the conjugate is inhibited after binding to the antibody a, Substrate can bind to the enzyme catalytic site and the enzyme can be assayed, b, antibody has been bound to the enzyme–hapten conjugate preventing access of the substrate to the enzyme catalytic site The enzyme activity cannot be measured (inhibition), c, free hapten displaces the enzyme–hapten conjugate from the antibody combining sites, the catalytic site becomes substrate-accessible and the enzyme activity is restored (B) Enzyme activity of the conjugate is restored after binding to antibody a, Substrate cannot bind to the enzyme active site as conformational changes which prevent binding occurred in the enzyme molecule after hapten binding Enzyme activity is therefore not measurable b, Due to binding of the conjugate to the antibody, conformational changes were induced making access of substrate to the catalytic site possible Enzyme activity can therefore be measured

The decrease in activity is caused by inhibition of the enzyme activity of the conjugate after its reaction with the antibody. The antibody binding sites are complementary to the hapten and not to the enzyme determinants. The free enzyme hapten conjugate (EH) has enzymatic activity and degrades the substrate (S) giving rise to the product (P):

$$EH + S \rightarrow P$$

When the antibody is added to EH, a complex that does not exhibit the enzyme activity is formed:

$$EH + Ab \rightarrow EH\text{-}Ab$$

$$EH\text{-}Ab + S \rightarrow \text{no } P \text{ (inhibition of the activity)}$$

The antibody has a higher affinity for the free hapten and, therefore, H displaces the EH conjugate from its binding sites (competitive reaction)

$$EH\text{-}Ab + H \rightarrow EH + H\text{-}Ab \text{ (activity is restored)}$$

The amount of the added hapten is then proportional to the restored enzyme activity.

Lysozyme (EC 3.2.1.17), malate dehydrogenase (EC 1.1.1.37) and glucose-6-phosphate dehydrogenase (EC 1.1.1.49) are usually used to label haptens in homogeneous EIA methods.

The inhibition of the enzyme activity after binding to the specific antihapten antibody is due to conformational changes induced near the active (catalytic) site of the enzyme, or, in contrast, by suppression of conformational changes essential for substrate binding. In the former case, preventing access of the substrate to the enzyme active site may be involved (*Fig. 12.29A*). Lysozyme is an example. Lysozyme has one of its six lysine residues with a free ε-amino group in position 97. The hapten is covalently bound to this group. The hapten is close to the enzyme catalytic site and the antibody therefore directly affects its spatial arrangement. In malate dehydrogenase and glucose-6-phosphate dehydrogenase the haptens do not bind close to their catalytic sites and, it is therefore assumed that the antibody binding prevents conformational changes induced by the substrate immediately before its binding to the enzyme.

The second type of homogeneous EIA, in which the enzyme activity after binding of the conjugate to the antibody is reactivated (*Fig. 12.29B*), is only rarely used, *e.g.* for thyroxine determination.

The homogeneous EIA method is used particularly for the quantitative determination of low-molecular-weight compounds (haptens), although homogeneous EIA methods for certain high-molecular-weight compounds, *e.g.* immunoglobulins, have also been described. The above techniques may be used to determine concentrations of low-molecular-weight haptens, such as some drugs (antibiotics, opiates, hypnotics, sedatives, cytostatics *etc.*) and hormones (thyroxine, tri-iodothyroxine, cortisol, oestriol *etc.*) in body fluids. The quantitation of these compounds in blood is important for the rational treatment of some individual patients (pharmacological monitoring). In fact, some drugs are absorbed differently in different individuals, and are also differently distributed and excreted from the body. The antibiotic gentamicin and digoxin used in the treatment of cardiac failure are examples of such drugs. However, they have a relatively narrow range of therapeutically efficient levels in blood. When these limits are exceeded, unfavourable (toxic) effects may occur. The physician therefore cannot decide on the basis of clinical findings alone, whether the doses given are sufficient or excessive and potentially harmful. The levels of the drug in the blood must be controlled, and the doses adjusted in individual patients (monitoring). Homogeneous EIA using commercially-available diagnostic kits is one of the methods that can be used to control treatment.

Homogeneous EIA methods have the advantage of simplicity and speed. Usually, a single test only takes a few minutes, whereas heterogeneous EIA methods may take up to several hours. In addition, they are readily automated. However, they are less sensitive than heterogeneous EIA and preparation of reagents is relatively complicated.

12.6.2 Heterogeneous EIA–ELISA techniques

Compared to homogeneous EIA, heterogeneous EIA methods require that the free labelled reacting component is separated from the labelled reacting component bound in complex with the antibody. Therefore, the technique is two-step (non-homogeneous), *i.e.* heterogeneous. Heterogeneous EIA methods use enzymes with identical activity in both free conjugates with the antigen (hapten) and in conjugates bound in immune complexes. Separation of the free and bound conjugates can be performed by classical precipitation of the immune complexes with polyethyleneglycol or a second antibody. In this respect the method is completely analogous to RIA. However, immobilization of one of the pair of reactants by binding to a solid carrier (usually the wall of the test tube or well of the microplate) is used most frequently. Separation of the free and bound components is readily achieved by washing. As the procedure is based on immunoadsorption the technique is known as the **enzyme-linked immunosorbent assay (ELISA).**

Enzymes used in these techniques should have a low relative molecular weight, high stability and enzyme activity, they should bind covalently to antibodies and to various functional groups of antigens, the product of the enzyme reaction should be coloured or otherwise readily detectable and the enzymes must be readily available and inexpensive. Of the enzymes fulfilling these conditions, horseradish peroxidase (EC 1.11.1.7) is most widely used, although alkaline phosphatase (EC 3.1.3.1), β-D-galactosidase (EC 3.2.1.23) and glucose-oxidase (EC 1.1.3.4) are also used.

There are several types of ELISA methods which differ in the analytical procedure (competitive or non-competitive) and the type of labelled reactant (antigen or antibody). Competitive ELISA for antigens and haptens, sandwich ELISA for antigens (non-competitive or competitive), indirect ELISA for antigens, sandwich ELISA for antibodies, ELISA for IgM detection are the most important forms of this technique.

(a) **Competitive ELISA for antigens and haptens**

The antibody specific for the test antigen is bound (immobilized) to the solid phase (*Fig. 12.30*). A known amount of the enzyme-labelled antigen and either a known (standard) or unknown (test) amount of the unlabelled antigen are then added. During incubation, the labelled antigen competes with the standard or test unlabelled antigen. After incubation, the reaction vessel (*e.g.* the microtitrate plate well) is washed with buffer to remove free reactants. The labelled and unlabelled antigens which are bound to the immobilized antibodies remain. The more unlabelled antigen in the reaction mixture, the less labelled antigen complexes with the antibody. A substrate solution is then added, which, following degradation by the enzyme present in the immune complexes, yields a coloured product. The colour intensity (absorbance) of the reaction mixture is assayed spectrophotometrically (usually in a 1 cm cuvette in a standard spectrophotometer). When the enzyme immunoassay is performed in microtitre plates rather than test tubes (which has the advantage of low reagent volumes) a special spectrophotometer designed for ELISA techniques (*"ELISA-reader"*) is used. The absorbance values of wells containing known (standard) amounts of the antigen are used to construct a calibration curve from which concentrations in the test samples may be determined.

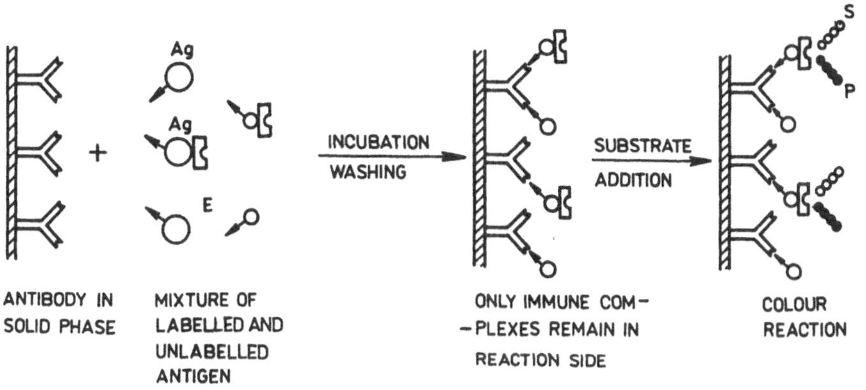

Fig. 12.30. Competitive ELISA of haptens and antigens.
Ag, antigen; E, enzyme; S, substrate; P, product.

This type of ELISA is mainly used to measure low-molecular-weight compounds that are easy to prepare in sufficient amounts and high purity. The principle of competitive ELISA is utilized in many commercial kits used to assay hormones (*e.g.* thyroxine, tri-iodthyroxine, prolactin, insulin, cortisol *etc.*), drugs (*e.g.* theophyllin, digitoxin, digoxin *etc.*) and some proteins (*e.g.* IgE, AFP, CEA).

(b) **Sandwich ELISA for antigens**

In this type of assay, the antibody is also first bound to the solid phase (*Fig. 12.31*). Known (standard) or unknown (test) amounts of the antigen are then bound to it. After washing, a second, enzyme-labelled antibody, which reacts with the remaining free determinants of the antigen bound to the immobilized antibody, is then added. The reaction vessel is washed again and a substrate to detect the enzyme activity is added. A calibration curve is constructed from the standard values and used to calculate concentrations of the test antigen in the samples.

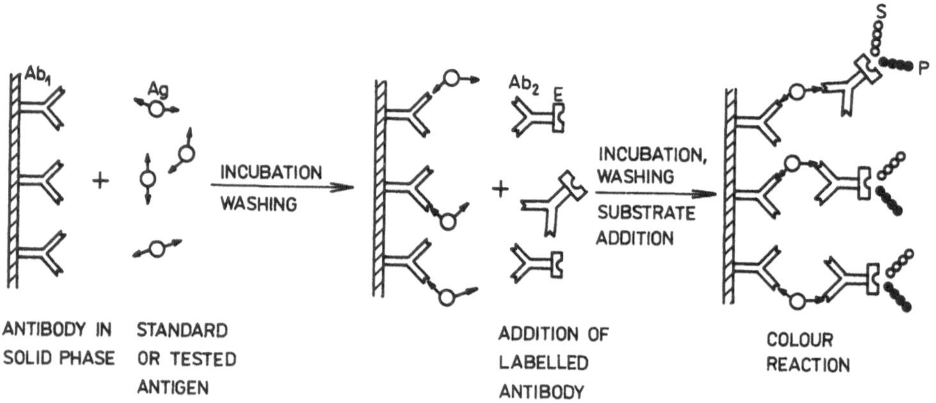

Fig. 12.31. Sandwich ELISA of antigens.

This type of ELISA does not require a labelled pure antigen. A standard antigen, even in a complex mixture with other antigens, *e.g.* human blood serum, is sufficient. However, the concentration of the antigen to be measured in this mixture must be known. The antiserum used must have a high titre of strictly monospecific antibodies. It is recommended that the antibody which is immobilized on the solid phase is of a different isotype than the free labelled antibody. Both antibodies must have an identical specificity. An additional advantage of this technique is that the antibody, irrespective of its specificity, can be labelled with the enzyme *via* the same conjugation reaction. As two molecules of the antibody always react with one antigen molecule, the sandwich ELISA can only be used with antigens having at least two determinants. This criterion is best fulfilled by macromolecular antigens (*e.g.* HBsAg, IgE, AFP, CEA, ferritin) which are difficult to purify.

The sandwich ELISA can also be performed as its double modification (the **double sandwich ELISA**). Instead of the second labelled antibody, an unlabelled antibody is used and a trimolecular complex is thus formed (Ab_1–Ag–Ab_2) which can be detected by a labelled anti-immunoglobulin (anti-isotype antibody against Ab_2). A distinct advantage of this system is that a single commercial labelled anti-immunoglobulin can be used to quantitate different antigens. However, the assay takes longer to perform and non-specific reactions may occur.

(c) **Indirect ELISA for antigens**

The antigen is bound to the solid phase (after its saturation) and the reaction vessel is washed (*Fig. 12.32*). When a small antigen is used, it must first be bound to a larger carrier, *e.g.* bovine serum albumin. The test sample containing the antigen to be measured is then mixed with a specific labelled antibody and the mixture incubated with the immobilized antigen. A control, which does not contain the free antigen, is incubated in parallel. After incubation, the reaction wells are washed, the substrate is added and a colour reaction develops. The difference between the absorbance of the sample containing the free antigen and the absorbance of the control well (without the free antigen) is indirectly proportional to the amount of antigen in the test sample. It is a competitive method in which the immobilized antigen and the antigen in the test sample (free) compete for the antibody binding site. The more antigen in the sample, the less antibody can bind to the immobilized antigen and the less colour develops in the well. The strongest colour reaction occurs in the control well which does not contain the free antigen. Indirect ELISA for antigens can also be modified so that an unlabelled antibody (*e.g.* rabbit IgG) is used instead of a labelled antibody. The immobilized immunocomplexes are detected by a labelled anti-immunoglobulin as the second antibody (*e.g.* pig anti-rabbit IgG). A distinct advantage of this modification is that the commercial labelled antibody can be used.

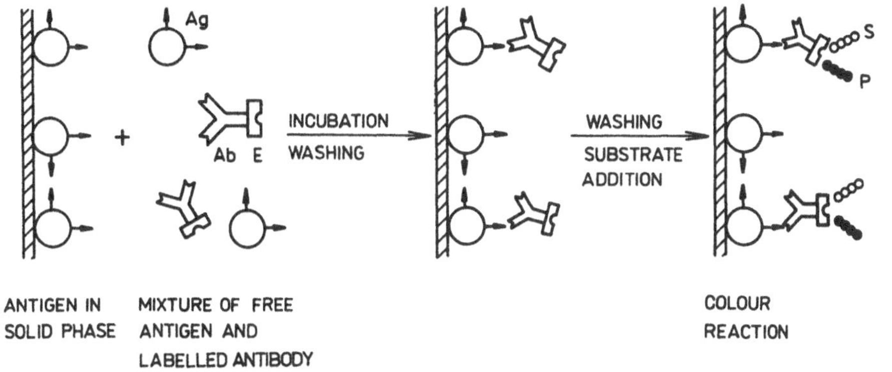

Fig. 12.32. Indirect ELISA of antigens.

(d) **Sandwich ELISA for antibodies**

This modification is also known as indirect ELISA for antibody detection and may be used to demonstrate or titrate antibodies specific for a certain

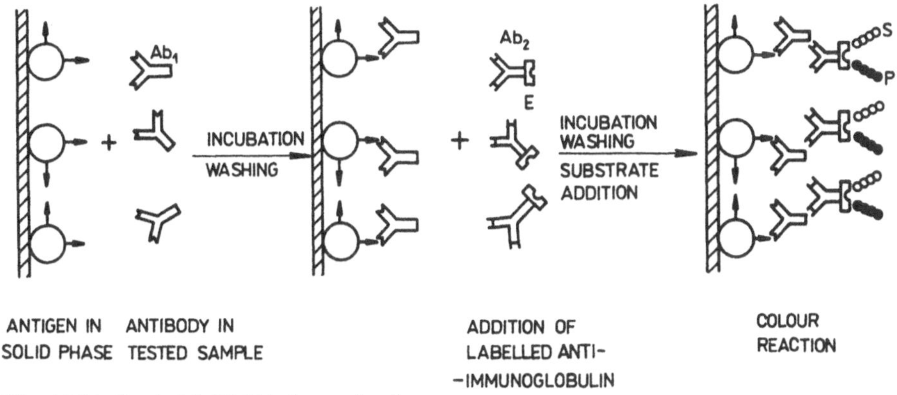

Fig. 12.33. Sandwich ELISA for antibodies.

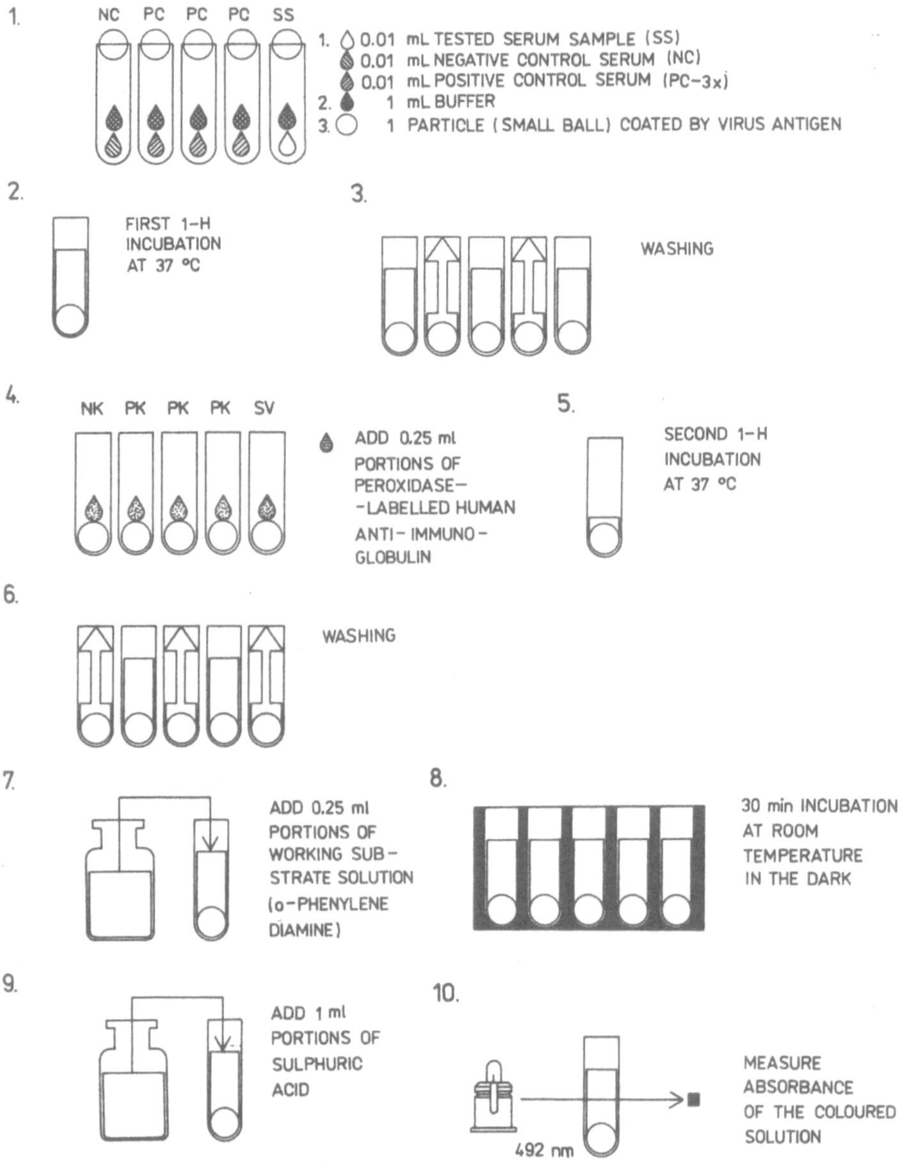

Fig. 12.34. Sandwich ELISA for anti-HIV-1 antibodies (Roche).

Microamounts of sera to be tested for anti-HIV-1 (HIV, virus inducing AIDS) antibodies are pipetted into test tubes. Two tubes contain control sera in which anti-HIV-1 antibodies are present (positive control) or absent (negative control). A buffer and a plastic bead coated with immobilized chimeric viral antigen are then added to each tube. The antigen is actually a pair of antigens, ENV (gp41) and GAG (p24) having 270 amino acid residues and prepared by recombinant DNA technology. These antigens were selected because their antibodies are considered to be the most characteristic and stable markers of HIV-1 infection. After all components have been added, the mixture is incubated, washed and labelled anti-immunoglobulin (against human IgG + IgM) is added. The incubation and washing steps are then repeated. A substrate solution is added to the washed trimolecular complex and, after incubation, sulphuric acid is added to induce a colorimetric reaction. The absorbance of the reaction mixture is determined spectrophotometrically. A negative control, *i.e.* a tube that did not contain antibodies against gp41 + p24, serves as a blank. In this case, human antibodies and therefore the labelled anti-immunoglobulin, could not bind to the bead. Colour intensity is directly proportional to the concentration of antivirus antibodies. The whole procedure is automated so that the laboratory technician does not come into contact with the infectious material (except for serum collection).

antigen. Its principle is shown in *Fig. 12.33*. An antigen is bound to the solid phase and washed; the diluted antiserum, in which the presence of antibodies is to be determined, is then added. The mixture is incubated to allow the antibodies present to react with the immobilized antigen. The wells are washed and the enzyme-labelled second antibody is added. After another washing, the enzyme substrate is added and the intensity of the colour reaction is measured.

The method is simple and rapid, *e.g.* the demonstration of serum antibodies against HIV-1, one of the viruses causing *acquired immunodeficiency syndrome (AIDS)*, only takes about 3 h (*Fig. 12.34*). The labelled anti-immunoglobulins used as the second antibody are available from various manufacturers and only the target antigen is required. This technique is widely used in diagnostic kits to demonstrate the presence of antibodies against infectious agents. The kits contain an appropriate viral, bacterial or parasitic antigen (often immobilized on a solid phase), positive control serum (containing antibodies against the test antigen) and negative control serum (without any antibodies). During the actual assay the test serum is compared with both control sera.

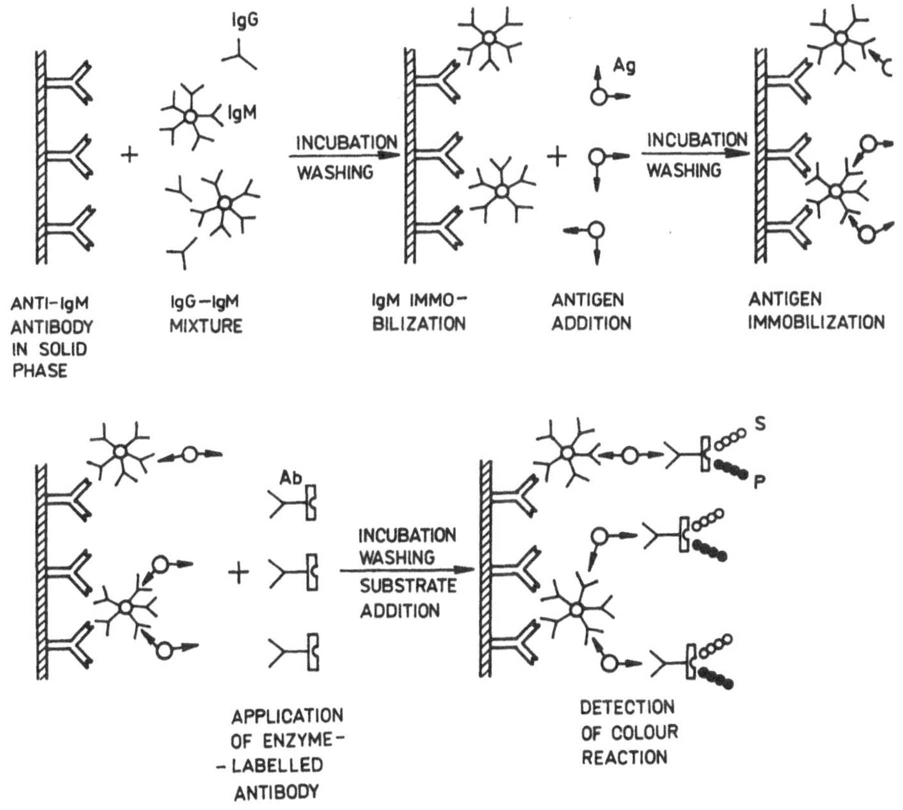

Fig. 12.35. ELISA for the detection of IgM antibodies

This technique can also be used as a double sandwich ELISA (with an identical principle to that used in the double sandwich ELISA for antigens) which is more sensitive than the simple sandwich ELISA.

Both methods are suitable for titration of IgG and IgM antibodies. They differ only in the specificity of the labelled anti-immunoglobulin (anti-IgG or anti-IgM). However, the demonstration of IgM antibodies is sometimes complicated by a high concentration of IgG antibodies with identical specificity competing successfully for the antigen binding. Therefore, when antibodies against IgM are to be demonstrated, the use of a modified technique is recommended.

This ELISA method is called **IgM-capture** (*Fig. 12.35*). The anti-IgM antibody is first bound to the solid phase and the antibody immobilizes all IgM antibodies present in the test serum. After washing, an antigen solution is added to the reaction vessel which then binds to those immobilized IgM molecules having specific binding sites. The reaction vessel is washed again and the specific antibody against the antigen is added. After incubation and washing, the substrate is added and the resultant absorbance determined. The absorbance is proportional to the amount of the antigen-specific IgM immobilized during the second step. The procedure is complex, but often the only one by which the presence of IgM antibodies can be demonstrated. In practice, it is often used to demonstrate antibodies against some viral antigens, *e.g.* antibodies against the hepatitis A virus.

12.6.3 Other immunoenzyme methods

As well as homogeneous and heterogeneous EIA, many other modifications using enzyme-labelling have been introduced into analytical practice. These modifications aim to increase sensitivity and speed of classical ELISA methods or to simplify the assay procedure. For increased sensitivity, it is possible to combine two methods, *e.g.* RIA and ELISA (enzyme-amplified RIA) or two enzymes may be used; the first enzyme increasing the activity of the second (enzyme amplification). The technique of enzyme immunochromatography is one of the most useful modifications.

EPRIA — enzyme potentiated radioimmunoassay — combines RIA and ELISA principles and makes antigen determination highly sensitive. It contains two basic steps. In the first, the antigen reacts with a specific antibody (which is usually immobilized) labelled with an enzyme. In the second step, a radioactive substrate is added to the enzyme immune complex and, after its degradation, radioactivity of the product of the enzyme reaction is assayed. For example:

1. HBsAg + Ab-glutamate decarboxylase
2. HBsAg — Ab-glutamate decarboxylase + L-[^{14}C]glutamic acid
   → $^{14}CO_2$ + γ-aminobutyric acid

The radioactivity of $^{14}CO_2$ is assayed in a liquid scintillation spectrometer. This method for HBsAg determination is about 100 times more sensitive than the standard RIA (FIELDS *et al.*, 1981). Another advantage of EPRIA is that a radioactively labelled antigen is not required, and only a commercially available substrate labelled with a radioactive isotope is used.

The **enzyme-potentiated ELISA** is based on the fact that the reaction product of the enzyme by which the antibody or antigen is primarily labelled, is not detected, but, instead, is used to induce catalytic activity of a secondary enzyme system that was inactive previously. In this way, even a few molecules of the primary product, that cannot be detected by common procedures, can initiate the second enzyme reaction, giving rise to numerous, and readily detectable product molecules.

The effect of such *enzyme "accelerators"* are also found in numerous biochemical systems *in vivo, e.g.* in the complement or blood coagulation systems (proteolytic accelerators are involved), in glycogen mobilization (phosphorylation accelerator) *etc.*

An example of an analytically applicable enzyme potentiator is shown in *Fig. 12.36*. In this case NAD^+ that is produced in various primary enzyme reactions, most readily by dephosphorylation of $NADP^+$, acts as a catalytic switch. Dephosphorylation is catalysed by alkaline phosphatase (EC 3.1.3.1), and the oxidation–reduction potentiator $NAD^+/NADH$ can therefore be used to amplify all ELISA methods which use alkaline phosphatase as a label. When this oxidation–reduction potentiator reacts with a tetrazolium salt (*e.g.* iodonitrotetrazolium — INT), 50 molecules of readily detectable INT-formazan are produced from one molecule of NAD^+ per minute of incubation. Thus, during a 10-min incubation, the secondary accelerator amplifies the primary enzyme reaction approximately 500 times.

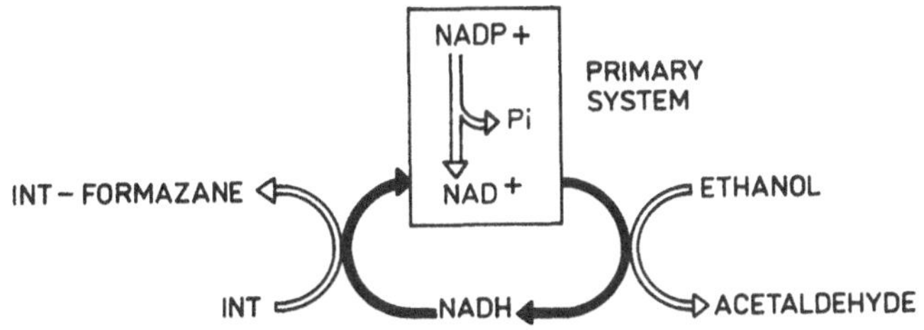

Fig. 12.36. The oxidation–reduction system as a secondary enzyme amplifier.
Alkaline phosphatase, an enzyme of the primary system, produces NAD^+, which is a substrate for the oxidation-reduction cycle driven by alcohol dehydrogenase and lipoamide dehydrogenase. In the presence of excess ethanol and both enzymes, the cycle yields coloured formazan in direct proportion to the amount of NAD^+ produced. INT, 3-(4-iodophenyl)-2-(4-nitrophenyl)-5-phenyltetrazolium chloride, a colourless tetrazolium salt giving rise to red formazan after reduction.

Enzyme immunochromatography is a potentially useful technique as no instruments or special devices are required. So far it has been used to quantify certain drugs and biologically active compounds in biological fluids (ZUK *et al.*, 1985). Due to its rapidity and simplicity it is suitable even for doctors' surgeries and first aid services. In principle, it is a combination of homogeneous EIA with immunochromatography and capillary migration. The assay procedure is schematically illustrated in *Fig. 12.37* and includes the following steps:

1. Specific antibodies (against the target ligand) are first applied to a strip of chromatography paper (Whatman 31 ET) and the paper is dried.
2. The enzyme reagent consisting of glucose oxidase (EC 1.1.3.4) and horseradish peroxidase (EC 1.11.1.7) conjugated to the ligand (hapten or antigen) is prepared. The test sample containing the free ligand (*e.g.* serum from a patient) is added to the enzyme reagent and mixed. One end of the paper strip containing the immobilized antibodies is immersed in this mixture and its components then migrate along the paper strip due to capillary action. During this step the strip absorbs a reproducible volume of the enzyme reagent mixture and the test sample. During migration, the ligand and its enzyme conjugate bind immunospecifically with the immobilized antibody, whereas glucose oxidase is homogeneously distributed all over the paper strip. In fact, the technique is ascending paper chromatography (for glucose oxidase) combined with immunoaffinity chromatography (for the analysed ligand) lasting for 10 min. The complexes produced after reaction of the migrating ligand with the immobilized antibody form a zone on the paper strip whose height is directly proportional to the concentration of the ligand in the analysed sample. It is, in fact, a competitive immunoassay during which the enzyme conjugate is used to determine length of migration of the ligand bound to the immobilized antibody.

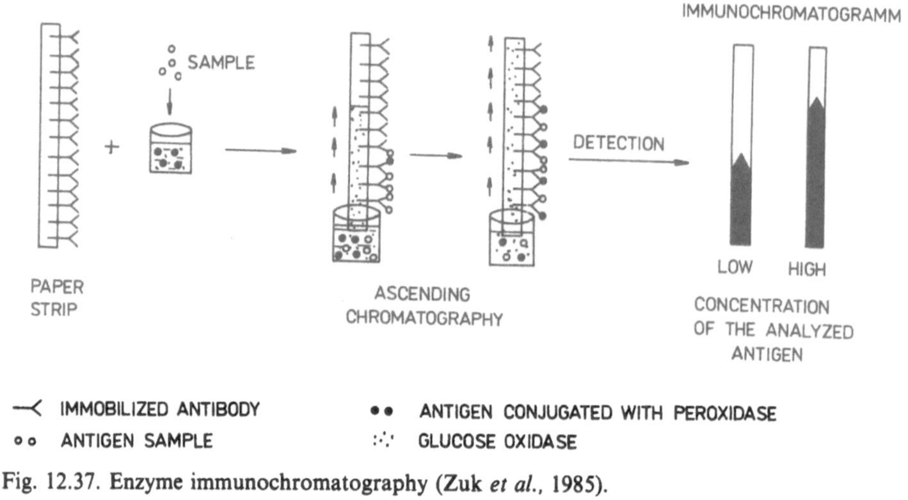

Fig. 12.37. Enzyme immunochromatography (Zuk *et al.*, 1985).

3. Measuring the height of the zone with the immobilized ligand and thus its concentration is performed as follows. The paper strip is immersed in a developing solution containing substrates of both enzymes present in the enzyme reagent (glucose and 4-chloro-1-naphthol). Glucose incubated with glucose oxidase yields hydrogen peroxide which is required for the

peroxidase-catalysed oxidation of the chromogenic substrate 4-chloro-1-naphthol. This reaction gives rise to the final reaction product, a blue-grey compound which adheres firmly to surface of the paper strip. Although peroxide is produced all over the strip, the visible product is formed only in the zone with the immobilized ligand, where its peroxidase conjugate is also present. The entire detection procedure takes 5 min. The zone of the reaction product is clearly delineated, and its height can be easily measured. The method is highly reproducible and it is not therefore necessary to use a standard calibration curve for each batch of assays.

Several companies supply chromatographic kits based on enzyme immunochromatography. These contain paper strips with the immobilized antibody and additional reagents required for the analysis, as well as a table which gives the antigen concentration in appropriate units from the height of the immobilized hapten or antigen. The whole analysis lasts for 15 min at the most. The sensitivity of the method (lower detection limit) for low-molecular-weight haptens is 0.1–1.0 mg/l. Blood, serum and plasma can be analysed.

Analytical systems utilizing paper strips containing all necessary reactants in a dry state are based on a similar principle. They can be used for both homogeneous and heterogeneous enzyme immunoassays. The **apoenzyme reactivation immunoassay system (ARIS)** is an example. Using this procedure it is possible, for example, to determine theophylline or phenobarbital levels of 3–60 mg/l in 80–90 s (THOMPSON and BOGUSLASKI, 1987).

In the ARIS technique, the following reactants are adsorbed on to filter paper strips (Whatman 31 ET) in a two-step procedure: an antibody against the test hapten, apoglucose oxidase, peroxidase, glucose (in aqueous buffer — first step), the hapten–FAD conjugate and 3,3′,5,5′-tetramethylbenzidine (TMB) (in an organic solvent — second step).

ARIS is based on the fact that the free hapten–FAD conjugate regenerates apoglucose oxidase (it does not contain the FAD coenzyme) to a fully active glucose oxidase, whereas the hapten–FAD conjugate bound to the specific anti-hapten antibody does not exhibit this effect. As a competitive analysis is involved, the amount of the free hapten–FAD conjugate depends on the concentration of free hapten in the sample. The activated glucose oxidase catalyses conversion of glucose to H_2O_2 which, together with peroxidase, oxidizes the colourless TMB to a blue product. The analysis is performed in such a way that a drop of the test sample (*e.g.* serum or urine) is applied to the paper containing the dried reactants and the blue colour of the spot is measured in a reflectance spectrophotometer. The concentration of the hapten in the test sample is determined with a calibration curve constructed from standard hapten solutions. The method is particularly suitable for monitoring drugs and their metabolites in body fluids during therapy.

Genetic engineering of *β-galactosidase* (EC 3.2.1.23), an enzyme commonly used in enzyme-linked immunochemistry systems, has led to the development of a homogeneous immunoassay system **CEDIA (combined enzyme donor immunoassay).** The *lac* operon from *E. coli* consists of an operator, a promotor, and the Z, Y and A genes (*Fig. 12.38*). The Z gene encodes a 1 021 amino acid polypeptide of β-galactosidase. This polypeptide is enzymatically inactive, but several identical polypeptide chains spontaneously aggregate into a tetrameric form which is the active β-galactosidase enzyme.

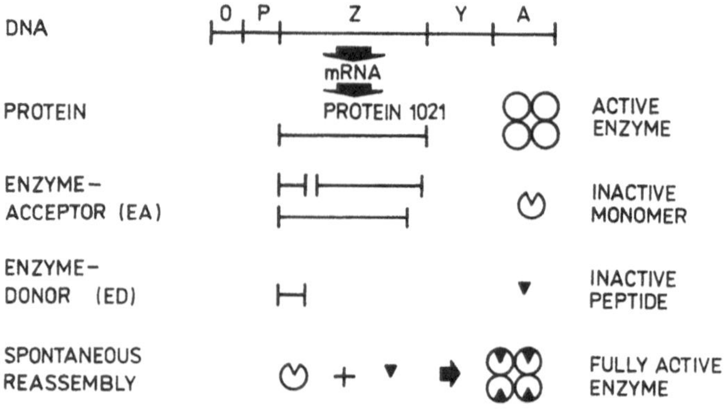

Fig. 12.38. Lac operon of *E. coli* (Henderson *et al.*, 1986).
O, operator; P, promotor; Z, Y, A, genes.
The Z gene encodes β-galactosidase; mutant Z genes encode enzyme-acceptors and enzyme-donors.

Using recombinant DNA techniques, HENDERSON *et al.* (1986) have constructed enzyme acceptors (EAs) and enzyme donors (EDs). The EAs are large polypeptides having small deletions, or missing sequences, in encoded protein. In solution, these large polypeptides are free (unbound) inactive monomer chains. The EDs are small polypeptides containing some of the sequences that have been omitted from the EAs. They are also enzymatically inactive. The unique property of the CEDIA technology is that EA combines in solution with ED to form tetrameric enzyme spontaneously that is enzymatically as fully active as if it were natural β-galactosidase.

The CEDIA homogeneous assay system operates by controlling the spontaneous assembly of the EDs and EAs through an antigen—antibody reaction. A hapten can be covalently attached to the ED in such a way that there is no interference in forming active β-galactosidase enzymes when the ED conjugate is mixed with EA. Adding to the system an antibody to the hapten will inhibit the spontaneous assembly of enzyme. Placing this system in competition for free analyte from a patient serum will create active enzyme in direct proportion to the amount of free unknown analyte or hapten. The amount of enzyme created is monitored through the hydrolysis of an appropriate enzyme substrate such as o-nitrophenyl-β-D-galactopyranoside or chlorophenol red-β-D-galactopyranoside.

Typical CEDIA assays are developed by formulating two reagents. The first reagent consists of EA and anti-hapten antibody. The second reagent contains ED–hapten conjugate and enzyme substrate. Mixing of the test sample with the first reagent results in prebinding of the sample hapten to anti-hapten antibody; EA does not participate in any reaction at this stage. Upon addition of the second reagent, the remaining anti-hapten antibody binds to the ED–hapten conjugate and thereby reduces the amount of enzyme formed. After the addition of both reagents, the role of change in absorbance is measured. CEDIA kits contain two or three calibrators. The enzyme rates observed with these calibrators can be used to construct a linear standard curve and the concentration of analyte in the test sample is determined by comparison to observed enzyme rates from a calibration curve.

CEDIA reagents are applicable not only for haptens but for high-molecular-weight analytes as well using a wide variety of instruments, from those found in the physician's office to the larger high–throughput discrete analysers used for health profiles. This technology is highly specific, sensitive and time saving. For example, CEDIA assay for digoxin requires only about 10 min with the sensitivity below 1 ng/ml (HENDERSON *et al.*, 1986).

The procedure termed **ELISPOT** or **ELISA plaque assay** has been employed as an alternative to conventional *plaque-forming cell assays* to enumerate specific as well as total immunoglobulin-secreting cells and also to detect a variety of cells secreting antigenic substances. The original techniques employed either alkaline phosphatase (SEDGWICK and HOLT, 1983) or horseradish peroxidase (CZERKINSKY *et al.*, 1983) labelled antibodies and corresponding chromogen substrates as indicator systems for the detection of antibody-producing cells. A combination of both indicator systems can be used for the concurrent detection of distinct types of antibody-secreting cells. In the procedure, zones of solid-phase-bound IgG and IgA antibodies secreted from distinct cells can be visualized as blue or red spots corresponding to either of these isotypes (CZERKINSKY *et al.*, 1988).

The assay consists of five stages: first, a solid-phase immunoadsorbent is prepared; the second stage, incubation of the cell suspension, is followed by the addition of a mixture of alkaline phosphatase- or horseradish peroxidase-conjugated antibodies. The fourth stage comprises the stepwise addition of alkaline phosphatase and horseradish peroxidase chromogen substrates which will yield insoluble blue and red spots, respectively. These spots are enumerated in the fifth and final stage. Their diameter ranges from 0.05 mm to 0.2 mm. A stereomicroscope equipped with a vertical white light source under low magnification ($\times 40$ to $\times 60$) is ideal for this enumeration.

The **enzyme-linked immunospot (ELISPOT)** assay can also be used for the detection of various cytokines; for example, for the detection of either IFN-γ or IL-5 secreted by individual *T*-cells (TAGUCHI *et al.*, 1990). The production of these two cytokines is important for the determination of helper *T*-cell subsets. In mice, cloned CD4$^+$ T_H-cells can be divided into T_{H1}-cells secreting IFN-γ, IL-2 and TNF-β and T_{H2}-cells which fail to produce these cytokines and instead secrete IL-4, IL-5 and IL-6 (MOSMANN and COFFMAN, 1989).

To detect IL-5-producing cells, 96-well nitrocellulose-based plates are coated with monoclonal antibody anti-IL-5. Control wells are coated with PBS. All wells are then blocked with culture medium. *T*-cell mitogen (ConA or PHA) stimulated or nonstimulated cells resuspended in medium are added to individual wells and incubated for 20 h at 37 °C in humidified atmosphere of 10% CO_2 in air. The plates are thoroughly washed with PBS containing 0.05% Tween and then incubated overnight at 4 °C with biotinylated monoclonal antibody anti-IL-5. The plates are then washed with PBS-Tween and incubated with avidin-peroxidase. Spots representing single IL-5-secreting cells are developed with the substrate 3-amino-9-ethylcarbazole. The number

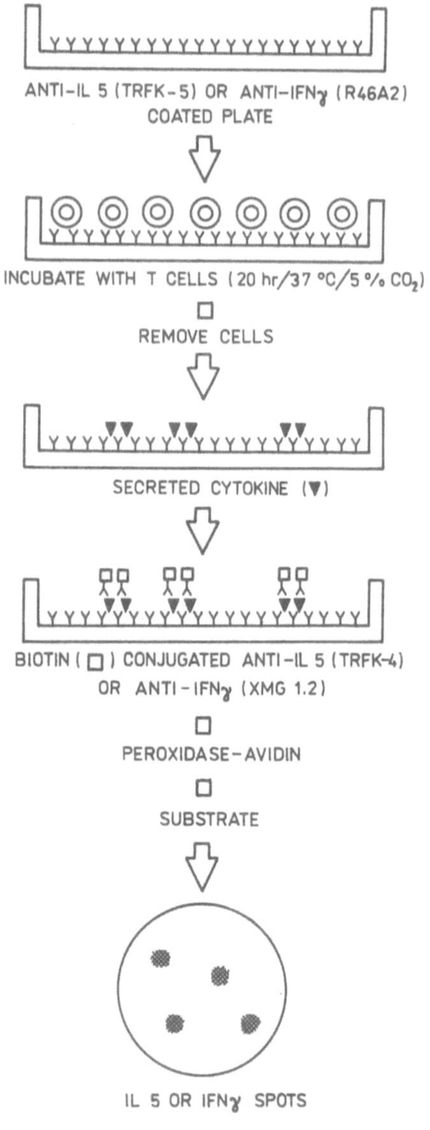

Fig. 12.39. Protocol for the cytokine IL-5 and IFN-γ-specific enzyme-linked immunospot (ELISPOT) assay (Taguchi *et al.*, 1990).

of spots is enumerated with the aid of a dissecting microscope. In order to determine the number of IFN-γ-specific spot-forming cells, a similar assay is performed (*Fig. 12.39*).

12.7 Fluorescence immunoassay (FIA)

This technique is based on principles identical to those of RIA and EIA except that a fluorescence probe (usually fluorescein-isothiocyanate (FITC) or tetramethylrhodamin-isothiocyanate (TMRITC)) is used to label the antibody or antigen molecules. Fluorescence is detected in fluorimeters or photon counters. The use of fluorescing molecules in immunoassay was introduced in 1961 by DANDLIKER and FEIGEN.

The ability of a molecule to absorb energy (*via* a chemical reaction or by direct absorption of different types of radiation) and re-emit the energy in the form of photons, is called **luminescence**. The absorbed energy excites the electron field of the luminophore molecule from the ground singlet state to an unstable state with a higher energy. Such an excited molecule can irradiate its energy in the form of light or in another form (*e.g.* as heat) and return to its ground state. When the molecule has been excited by light, its return to the ground state may proceed directly, when the process is known as **fluorescence**, or *via* a semi-stable triplet state in a process termed **phosphorescence**. Luminiscence phenomena have only a low energy loss (the difference between excitation energy and emission energy). As far as light is concerned, the energy loss is manifested by a shift in wavelength between the excitation and emitted light. In fluorescence, the wavelength of the emitted light is 30–50 nm higher than that of the excitation light; in phosphorescence this difference is over 200 nm. The basic characteristics of fluorescence molecules include the wavelength of the absorption and emission maxima, the molar absorption coefficient, the quantitative yield (ratio of absorbed and emitted photons) and the fluorescence time (time of fluorescence after excitation). The properties of the most widely used fluorescence probes (fluorophores) are listed in *Table 12.4*.

In fluorescence immunoassays, fluorescence probes are usually excited by laser light sources. In fact, FIA is a highly sensitive technique. However, in practice its sensitivity significantly decreases because many proteins and components of biological materials may also fluoresce and this increases the background fluorescence level. Because of this, the sensitivity of FIA varies between 10^{-9} and 10^{-12} mol/l.

Molecules used as **fluorescence probes** must have a high fluorescence intensity, the fluorescence signal must be readily discriminated from the background and binding of the probe to an antibody or antigen must not alter its properties. In addition to traditional probes (*Table 12.4* and *Fig. 12.40*), fluorescence probes of rare earths and their chelates have recently been used, due particularly to their high sensitivity. Antibodies and antigens labelled with these probes can even be detected at 10^{-14} mol/l.

Table 12.4 Properties of the most important fluorescence probes used in FIA (Hemmilä, 1985)

Probe	*Maximum wavelength (nm)*		ε *(l/mol)*	*Quantitative yield*	*Fluorescence time (ns)*
	Absorption	*Emission*			
Fluorescein isothiocyanate (FITC)	492	520	7×10^4	0.85	4.5
Rhodamin B isothiocyanate (RBITC)	550	585	1.2×10^4	0.70	3.0
Tetramethylrhodamin isothiocyanate (TMRITC)	550	580	5×10^4	0.60	1.0
Umbelliferones	380	450	2×10^4		
Fluorescamin	394	475	6.3×10^3	0.10	7.0
2-Methoxy-2,4-diphenyl-3(2H)-furanone (MDFP)	390	480	6.4×10^3	0.10	
Porphyrins	400–410	619–633			
Chlorophylls	430–453	648–669			
Phycobiliproteins	550–620	580–660	7×10^5 up to 2.4×10^6	0.50–0.98	

ε, molar absorption coefficient.

FIA methods can be classified according to the criteria used for other immunoassays. Their division into heterogeneous and homogeneous techniques has been widely used.

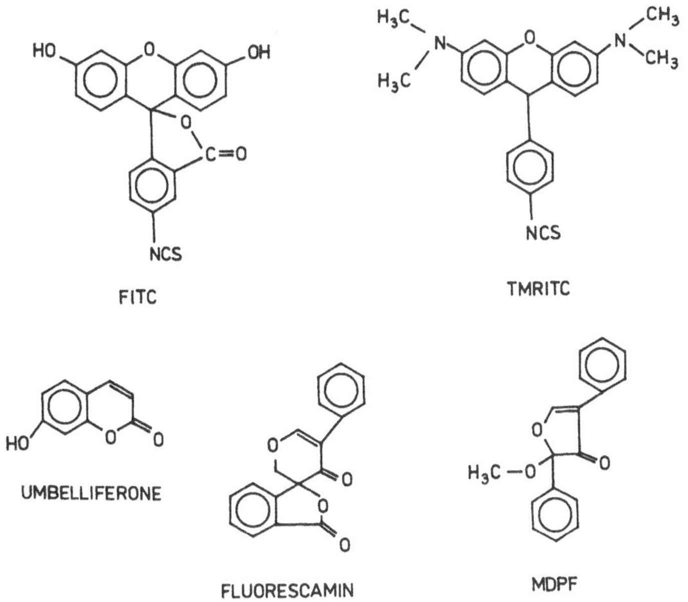

Fig. 12.40. Structure of some fluorochromes.
FITC, fluorescein isothiocyanate; TMRITC, tetramethylrhodamin isothiocyanate; MDPF, 2-methoxy-2,4-diphenyl-3(2H)furanone.

12.7.1 Heterogeneous FIA methods

When using heterogeneous FIA methods, the free and bound labelled antigens or antibodies must be separated before measurement. Individual molecular species can be separated by precipitation of the immune complexes or using a labelled reactant (antigen or antibody) bound to the solid phase. This technique is called separation FIA. However, solid phase sandwich techniques are used much more frequently and are known as immunofluorometric assays (IFMA).

Separation fluoroimmunoassay (Sep-FIA) is based on competition between a fluorochrome-labelled and an unlabelled antigen for a limited amount of antibody. The fluorescence probe can be used to label the antibody. However, in the latter case, the labelled antibodies compete for free antigen in the test and standard samples bound to the solid phase. Separation of immune complexes with a second antibody, polyethyleneglycol or ammonium sulphate precipitation, commonly used in RIA, are only rarely used since these procedures increase the background fluorescence. Immobilizing one of the reacting components to the solid phase is usually used to achieve separation. Polysaccharide, polyacrylamide, polystyrene microbeads or magnetic beads are most commonly used as the solid phase. These particles are also included in commercial kits to measure various drugs and hormones by FIA. After the reaction has been completed, particles carrying the labelled immune complexes are separated by centrifuging and their fluorescence measured. Polyglutaraldehyde microbeads with a ferrofluid (Fe_3O_4) dispersion and FITC admixture or nickel particles coated with the antigen are magnetic. Their separation is readily achieved in a gravitational or magnetic field.

Immunofluorometric assays (IFMA) developed from immunofluorescence microscope techniques used to detect antigens, antibodies, microorganisms or receptors in tissue sections as early as in 1941 (Coons *et al.*). IFMA are analogous with IRMA. In IFMA, excess fluorochrome-labelled antibodies are also used. The primary antibody which binds a specific antigen (microorganisms, virus, protein, proteohormone) is immobilized on the solid phase. The amount of the bound antigen is quantitated using a second antibody (sandwich technique). When IFMA is used to determine antibodies, an antigen which binds specific antibodies in the sample is bound to the solid phase and the antibodies are then measured with anti-isotype-labelled antibodies (*e.g.* antibodies against human IgG *etc.*). In IFMA methods, polyacrylamide beads, polymethacrylate and, occasionally, acetate-nitrate cellulose discs serve as the solid phase.

12.7.2 Homogeneous FIA methods

In homogeneous FIA methods, separation of the free and immune complex-bound antigen before the fluorescence measurement is not required. Therefore, the assay procedures are simple and rapid. However, the sensitivity of homogeneous FIA is often limited by interference from certain compounds present in the sample, particularly in blood, and small fluorescence changes. Almost all homogeneous FIA methods are based on competition and are used mainly to determine relatively high antigen concentrations (mg/l). Haptens can be determined at about 100 times lower concentrations. The antibody binding induces changes in the fluorescence properties of the fluorochrome-labelled antigen. In homogeneous FIA methods, fluorescence polarization, quenching or amplification of fluorescence, excitation fluorescence transfer and fluorescence-labelled substrates are used.

The **fluorescence polarization immunoassay (FPIA)** uses irradiation of a fluorophor by polarized light. As the molecule returns to its non-excited state the absorbed energy is released as polarized photons. This secondary polarization is indirectly proportional to fluorescence time and rotational movement of the fluorophor conjugate. When the fluorophor is bound to the small hapten molecule, its random rotation decreases the polarization signal. After the conjugate binds to a specific antibody, its rotation decelerates and the polarization signal is amplified. The FPIA method is simple, rapid and accurate but due to small changes in polarization it can only be used to quantify compounds with a molecular weight below 20 000 (Dandliker and De Saussure, 1970). Automatic fluorometric instruments and assay kits are available for determining the levels of various drugs and hormones. The apparatus makes it possible to perform up to 20 assays in 10 min. Digoxin for example, can be determined with extreme sensitivity (down to 0.2 µg/l). Other compounds can be determined with a lower sensitivity.

Immunoassays based on a gradual increase of fluorescence measure the increase of fluorescence after antigen binding to a specific antibody. In this case dansyl derivatives or aniline naphthalene sulphonic acid are used as fluorescence probes. These compounds are usually not used in FIA.

Fluorescence quenching is used much more frequently than gradual increase. Such methods are called **fluorescence quenching immunoassays**. Fluorescein derivatives are mainly used as fluorophores. Their fluorescence decreases (is quenched) when they non-specifically conjugate with proteins, bind to specific (anti-fluorescein) antibodies or when an FITC-labelled antigen reacts with a corresponding antibody. The technique has two modifications, *i.e.* direct or indirect quenching. The direct method in mainly used to measure haptens (opiates, antibiotics, hormones) as well as some complete antigens (immunoglobulins, albumin). FITC-labelled hapten or antigen react with an antibody directed against the hapten or the antigen epitope. FIA with an indirect quenching is based on a similar principle to that of the direct quenching FIA. However, an additional antibody directed against the fluorescence probe is used. In this case two antibodies compete for the labelled antigen — one specific for the test antigen and the second — for the fluorophore. When the sample contains only bound antigen, the lowest fluorescence quenching is observed since steric hindrance prevents the contact with the anti-FITC antibody necessary for the highest quenching effect. At high free antigen concentrations, the anti-FITC antibodies can react directly with the fluorophore and maximal fluorescence quenching is observed. The indirect quenching FIA is generally used to determine large antigens.

The **fluorescence excitation transfer immunoassay (FETIA)** uses two fluorescence labels with different fluorescence wavelengths (Ullman *et al.*, 1976). One functions as a donor and the other as an acceptor. As a donor, FITC is usually used to label the antigen, and TMRITC (acceptor) to label the antibody. The wavelength of FITC emission is close to that of TMRITC excitation (*Table 12.4*). Therefore, FITC fluorescence in immune complexes is absorbed by TMRITC which then quenches the FITC fluorescence. In this case a self-calibration immunoassay is in fact involved. The TMRITC fluorescence is proportional to the antibody concentration and the antigen concentration is determined from the FITC/TMRITC ratio. FETIA is used to quantify proteins and even low-molecular-weight haptens.

Substrate-labelled fluoroimmunoassay (SLFIA) is a competitive FIA, which uses a substrate labelled with a fluorescent probe (fluorogen). The fluorogen has specific chemical features in that it consists of two components: an enzyme substrate and a fluorophore precursor. The fluorogen itself does not exhibit fluorescence. However, when it is cleaved by an appropriate enzyme in the reaction mixture, the precursor is converted into a fluorescent molecule (fluorophore). When a fluorogen-labelled antigen or hapten reacts with an antibody, steric hindrance prevents access of the enzyme and thus also catalytic release of the fluorophore and onset of fluorescence. Fluorescence inhibition can be removed by adding an unlabelled antigen which releases a proportional amount of the labelled antigen from the immune complex. Galactosylumbelliferone is usually used as the fluorogen (the enzyme is β-galactosidase), although other glycosidic or ester derivatives of umbelliferone may also be used. Several commercial kits for determination of low-molecular-weight drugs and some proteins are based on SLFIA. The technique can be readily automated using continuous flow and centrifugal analysers.

When chelates of lanthanides (usually europium chelates) are used instead of traditional fluorescence probes, fluorescence intensity may be increased by a factor of 10^6 times. This increase is not direct but occurs only after reaction of the primary chelate with a special activator. Eu^{3+} then dissociates from the chelate and a new chelate enclosed in a micelle surrounded by water molecules is produced. After irradiation with light at 320–360 nm, strong fluoresence with a sharp peak at 613 nm is emitted. This principle is used in **dissociation-enhanced lanthanide fluoroimmunoassays.** Using the fluorimeter and assay kits supplied by LKB it is possible to measure various hormones, drugs and proteins with a sensitivity equal to that of RIA.

12.8 Chemiluminescence immunoassay

Chemiluminescence is the generation of visible light in some energy-releasing chemical reactions. In such reactions intermediates or end-products in an electron-excited state are produced. When the excited molecules return to the ground state, they emit a light quantum (photon). Compared to fluorescence this phenomenon does not require light excitation. When chemiluminescence occurs in living organisms it is called bioluminescence. **Bioluminescence** reactions are usually catalysed by the enzyme luciferase degrading its specific substrate, luciferin (Carrico *et al.*, 1976). Chemiluminescence and bioluminescence are in fact identical phenomena and, therefore, in the medical and scientific literature they are often termed **luminescence**, even though this term is actually wider and includes fluorescence, photoluminescence and other phenomena. A compound which is capable of chemiluminescence is known as a *chemiluminophore*. The amount of chemiluminescence or bioluminescence is measured by a *luminometer*. As a relatively large amount of energy is required to generate a single photon (about 200 kJ depending on the wavelength), most chemiluminescence reactions are of the oxidation type.

Chemiluminophores used in **chemiluminescence immunoassays (CIA)** are either consumed during the assay reaction (luminol, isoluminol, acridine esters) or not consumed (peroxidase, luciferase). Chemiluminophores of the first group are usually oxidized, generating light. *Luminol oxidation* (*Fig. 12.41*) mainly proceeds *via* hydrogen peroxide in the presence of a catalyst. Aminophthalate dianion which has strong luminescence in an alkaline environment (pH 10–11) is thus produced. All aromatic hydrazides react in this way although only luminol and isoluminol that bind readily to a ligand (antigen or antibody) *via* the primary amino group are used for CIA. *Isoluminol*, whose conjugate with the ligand generates significant luminescence, is more advantageous.

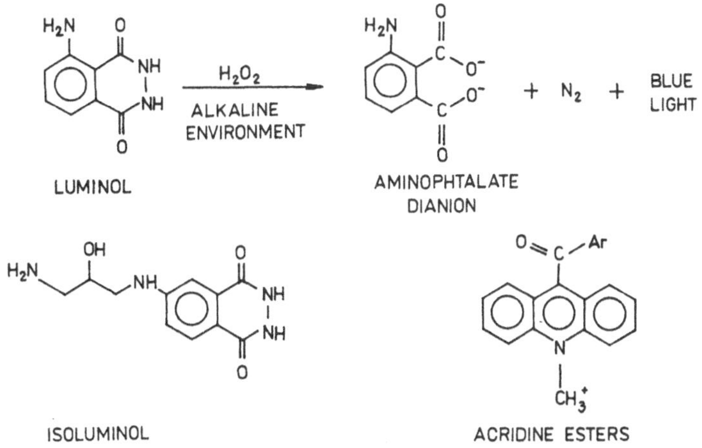

Fig. 12.41. Oxidation of luminol and structure of isoluminol and of acridine dyes.
Luminol, 3-aminophthalalhydrazide; isoluminol, 4-(3-amino-2-hydroxypropylamino)-phthalhydrazide; Ar, aryl.

Chemiluminescence immunoassay (sometimes simply termed luminescence immunoassay, (LIA) can, like other immunoassays, also be performed in several modifications (competitive, non-competitive, homogeneous, heterogeneous). Antigen (hapten) or antibody can be labelled with the chemiluminophore (SCHROEDER *et al.*, 1976; SEITZ, 1984). When an antigen is labelled and the heterogeneous modification is used, the competitive reaction is allowed to proceed (labelled antigen in the presence of standard or unknown amounts of the antigen), the labelled immune complexes are separated and allowed to react with an oxidation agent. The generated light is then measured in a luminometer and the unknown values derived from a calibration curve. This method is used particularly to measure low-molecular-weight haptens (hormones) and some proteins (HBsAg, immunoglobulins). Its sensitivity is comparable to that of RIA. For example, the assay range for thyroxine determination is 15–150 μg/l and for cortisol determination 20–400 μg/l. Homogeneous CIA with the isoluminol labelled antigen uses the fact, that, after binding to a specific antibody, chemiluminescence increases by about four times. When the antibody is labelled with the chemiluminophore, the antigen is usually immobilized on the solid phase (polystyrene beads), and after immune complexes have been formed, luminescence is generated using an oxidant and a catalyst. The method is therefore known as the **solid-phase**

antigen luminescence technique (SPALT). Instead of a specific antibody, a secondary antibody can be labelled with the chemiluminophore (sandwich arrangement; *Fig. 12.42*).

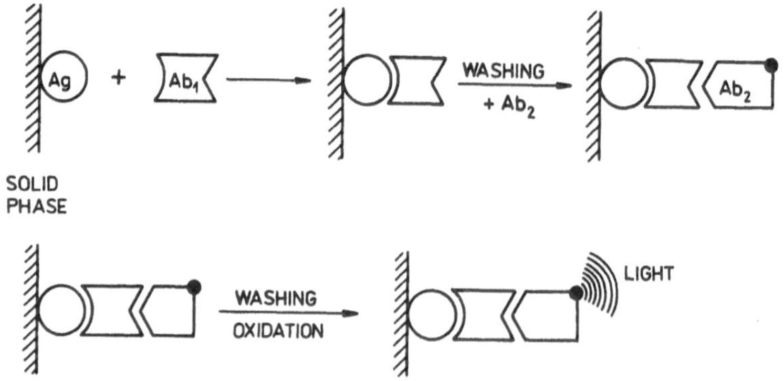

Fig. 12.42. Sandwich chemiluminescence immunoanalysis with a labelled secondary antibody. The solid circle represents chemiluminophore.

Peroxidase and luciferase are among those labels that do not change during the chemiluminescence reaction. In fact, peroxidase substitutes for the catalyst accelerating oxidation of luminol by hydrogen peroxide, yielding aminophthalate dianion. In this case reactant (antigen or antibody) is labelled with peroxidase but, unlike EIA its activity is not assayed. The amount of the peroxidase-labelled immune complex is detected by adding an excess of hydrogen peroxide and luminol, and measuring the intensity of the generated chemiluminescence in a luminometer.

Luciferase is the enzyme which catalyses the bioluminescence of fireflies. It is used less frequently in CIA, but is widely used for a highly sensitive determination of ATP based on the reaction:

$$\text{ATP} + \text{luciferin} + \text{Mg}^{2+} + \text{O}_2 \xrightarrow{\text{luciferase}} \text{oxyluciferin} + \text{light} + \text{other products}$$

This bioluminescence method can be used to measure ATP directly as well as to determine the viability of various cells. When the cell is metabolically active (viable) it contains ATP. After cell death, ATP is rapidly degraded so that ATP content becomes a good measure of cell viability. For example, using the bioluminescence technique it is possible to detect even a single vital leukocyte (containing ATP) in a suspension of 10^7 leukocytes.

Chemiluminescence generated by luminol or isoluminol oxidation can be used to measure antigens or antibodies in CIA and, in addition, for the sensitive assay of oxidants, particularly of hydrogen peroxide and other reactive forms of oxygen (p. 430).

In addition to classical CIA, methods combining CIA with FIA (**excitation transfer CIA**) have recently been used. These reactions use both chemiluminophore and fluorophore. Photons released from the chemiluminophore are not measured but rather used to excite the fluorophore before fluorescence measurement. The antigen is labelled with the chemiluminophore (aminobutyl-ethylisoluminol), whereas the antibody is labelled with fluorescein isothiocyanate. Free unlabelled antigen emits light at 460 nm, whereas the immune complex emits light at 520 nm, and the method can therefore be used in the homogeneous modification. Its sensitivity is comparable with that of RIA. Excitation transfer CIA is particularly used to measure various hormones, cAMP, cGMP as well as immunoglobulins and other protein antigens.

12.9 The biotin–avidin and biotin–streptavidin reactions as universal amplification systems in immunoassays

Immunoassays such as RIA, ELISA, or protein blotting all require a sensitive, specific, flexible and convenient detection system. For this purpose, the biotin–streptavidin system offers considerable advantages.

Both egg-white avidin and bacterial streptavidin are able to bind biotin with high specificity. This non-covalent interaction is one of the strongest interactions known in biology (*Table 12.5*). Only the intact ureido ring of

Table 12.5 Dissociation constants of some binding proteins and their ligands

Binding protein	*Ligand*	*K_d (mol/l)*
Lectins	Glycoproteins	10^{-3}—10^{-5}
Enzymes	Substrates	10^{-3}—10^{-6}
Antibodies	Antigens	10^{-7}—10^{-11}
Receptors	Hormones	10^{-9}—10^{-12}
Avidin	Biotin	10^{-15}
Streptavidin	Biotin	10^{-15}

biotin (*Fig. 12.43*) is required for this reaction. Consequently, the carboxyl group of the valeric acid side chain can easily be modified chemically for the design of reactive biotinyl derivatives (Wilchek and Bayer, 1984). Thus, biotin, bound to a macromolecule (*e.g.* antibody or enzyme), is still available for interaction with the active **avidin** (mol. wt. 67 000). Because of the tetrameric structure of avidin, the biotinyl residues of two different biotinylated proteins (antibody and enzyme for example) can be accommodated simultaneously. These properties of the avidin–biotin complex have prompted the

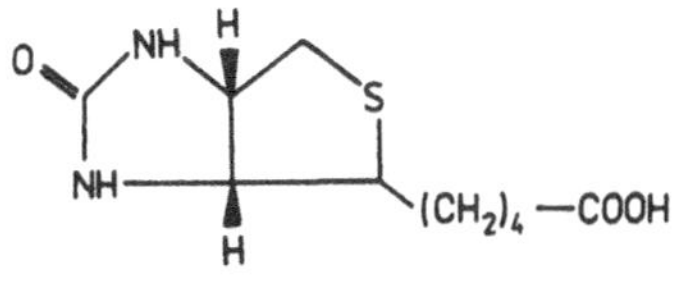

Fig. 12.43. The structure of biotin.

widespread exploitation of this system for a variety of purposes, which include the detection, localization, and purification of proteins, polysaccharides and nucleic acids, as well as immunoassay procedures and the mapping of epitopes in hybridoma technology.

The general applications of **avidin–biotin technology** in various analytical systems are shown in *Table 12.6*. Use of this system, however, may be restricted because of the high basicity (pI = 10.5) and the presence of carbohydrate moieties on the avidin molecule which cause high level of non-specific binding. Therefore, the use of streptavidin instead of avidin is more advantageous. **Streptavidin** (mol. wt. 60 000) is produced by the bacterium *Streptomyces avidinii*. Like avidin, the bacterial protein forms a very strong and specific non-covalent complex with biotin, and consists of four identical subunits, each of which contain a single biotin-binding site. However, streptavidin differs from avidin in that the former is not glycosylated and exhibits a neutral pI.

Table 12.6 Streptavidin conjugated probes for use in immunoassay systems (Strasburger *et al.*, 1988)

Probe	*End-point*
Enzymes	
Penicillinase	Disappearance of colour
β-Galactosidase	Colorimetry, fluorometry
Alkaline phosphatase	Colorimetry
Horseradish peroxidase	Colorimetry, chemiluminescence
Glucose-6-phosphate dehydrogenase	Bioluminescence
Chemiluminescent labels	
Isoluminol	Chemiluminescence
Acridinium esters	Chemiluminiscence
Fluorescent labels	
FITC	Fluorometry
Texas red	Fluorometry
Phycobiliproteins	Fluorometry
Metals	
Europium	Time-resolved fluorescence
Colloidal Gold	Colorimetry
Radioactive markers	
^{125}I	Counting of radioactivity

In principle, assay systems based on the interaction between biotin and streptavidin can simply employ an antibody onto which biotin has been conjugated while the signal is provided by a streptavidin-bound label (*e.g.* enzyme or fluorophor). This approach is very flexible in use since the antibody and label are not directly joined but coupled *via* biotin–streptavidin (*Fig. 12.44*).

In spite of the very high affinity interaction between biotin and streptavidin, the binding ability may be hindered when the biotin has been conjugated to a macromolecule, such as IgG or peroxidase, due to steric constraints. This results in a marked drop in the strength of interaction and reduces the inherent advantages of using a biotin–streptavidin-based system. This effect can be alleviated by the use of an appropriate "spacer arm" between the biotin and the macromolecule, allowing the binding between biotin and streptavidin to proceed unhindered. Such a spacer arm between biotin and proteins may be formed using long-chain *N*-hydroxysuccinimide ester derivatives of biotin.

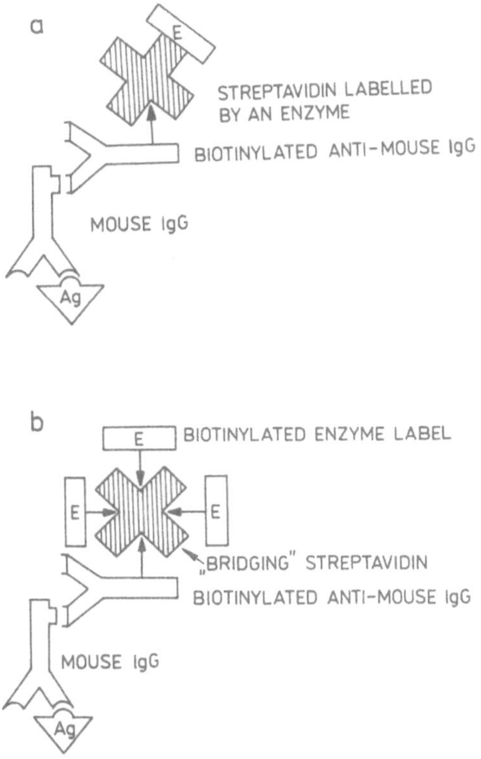

Fig. 12.44. Use of labelled streptavidin to detect an antigen (a) or as a bridge to obtain a stronger signal (b).

Labelled streptavidin may be used as a universal marker in the competitive-type immunoassay procedures for haptens. In this approach, the solid-phase antigen (consisting of a hapten–macromolecule conjugate in which the macromolecule is different from that used for immunization) is reacted with the sample and biotinylated specific antibody. After the immunological reactions, and a subsequent washing step, labelled streptavidin is added. Depending on the marker used, the end-point is determined by colorimetry, fluorometry, or luminometry (*Fig. 12.45*).

The advantages of the streptavidin—biotin interaction as a mediator in two-site sandwich-type and competitive immunoassays are summarized as follows (STRASBURGER *et al.*, 1988):

1. Unifies immunoassay methodology (competitive and two-site, haptens and large molecules).
2. Biotinylation of antibodies or enzymes can be achieved under mild conditions.
3. Biospecific activities of conjugates are retained.
4. The end-point in the assay is versatile (calorimetry, luminometry, fluorescence, *etc.*).
5. Mediation *via* the streptavidin–biotin interaction serves to amplify the sensitivity of the assay.
6. Commercial availability of biotinylated secondary antibodies, enzymes and enzyme-labelled avidin and streptavidin conjugates.

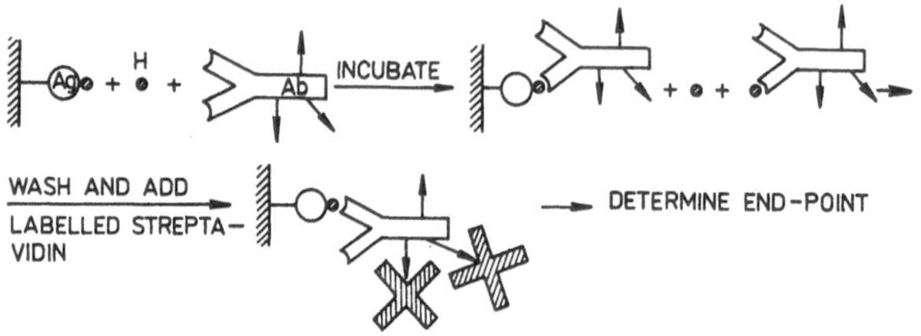

Fig. 12.45. Solid-phase antigen assay for hapten (H) using a streptavidin–biotin amplification.

12.10 Immunoassay using particles

One of the requirements for an ideal immunoassay label is that visual detection methods (with the naked eye) can be used. Various types of particles can fulfil this requirement. One of the immunochemically reacting components (antigen or antibody) is bound to the particle and the second reactant induces particle agglutination, agglutination inhibition, lysis of the particles with subsequent release of their contents or changes in their properties which facilitates the detection of the reaction in question; for example, measuring light reflection (nephelometry), turbidity (turbidimetry), non-agglutinated particles, and fluorescence or colour intensity of the released content (*e.g.* haemoglobin after lysis of erythrocytes) *etc.* Haemagglutination is one of the oldest methods used. However, in its original form it could only be used for the semiquantitative determination of antigens and antibodies. For this purpose, various colloidal particles, polystyrene latex particles and liposomes have been used in addition to erythrocytes.

The use of colloidal particles, consisting of a metal or its insoluble salt, as a label for the immunoassay was introduced by LEUVERING *et al.* (1980). As the dispersion of colloidal particles in a liquid is called a *sol*, the method has been termed **sol particle immunoassay (SPIA)**. Colloidal gold had been used as an electron-dense label in immunoelectron microscopy in 1971 (FAULK and TAYLOR). In addition to gold, silver, silver iodide, barium sulphate and occasionally other metals and their compounds can be dispersed and used as colloids. *Colloidal gold* particles are used most frequently, because of their optical properties, absence from body fluids (avoiding interference), and easy preparation. Immunochemically reactive particles are obtained by adsorbing antibody molecules on to the surface of gold particles with a diameter of 50 nm (Au_{50}–Ab). The antibodies labelled in this way can be used in heterogeneous or homogeneous immunoassay systems.

Of the heterogenous techniques, sandwich SPIA and inhibition of sandwich SPIA have been most widely used. The antigen is detected by the red colour of Au_{50}–Ab dispersion (visually or photometrically) or by atomic absorption spectrometry as the particles contain a metal. For example, the lower detection levels of human chorionic gonadotropin (the placental hormone excreted in the urine of pregnant women and therefore used as a pregnancy test) are 170 pmol/l (visual detection), 5 pmol/l (colorimetric determination) and 0.02 pmol/l (atomic absorption spectrometry), which is as sensitive as RIA.

All homogeneous SPIA are based on agglutination or on inhibition of agglutination of gold particles coated by the antibody. The principles are analogous to those of haemagglutination and haemagglutination inhibition assays. Also in these assays the antigen bridges the distance between individual Au_{50}–Ab particles inducing their fusion and aggregation. During the formation of larger aggregates (agglutination), the red colour of the original Au_{50}–Ab dispersion decreases and finally disappears completely (depending on the degree of aggregation). These colour changes can easily be detected with the naked eye. The principle of direct agglutination can be used to determine larger molecules having at least two determinant groups. However, haptens are immunochemically monovalent and, therefore, cannot be analysed by direct agglutination. By conjugating the hapten to a carrier molecule (*e.g.* BSA) an immunochemically polyvalent hapten complex able to agglutinate Au_{50}–Ab particles is formed. However, when in addition to the hapten –carrier conjugate and antibody-coated gold particles, the free hapten is present in the assay system (in the test or standard samples), agglutination is inhibited, since the free hapten reacts preferentially with the antibody on gold particles and thus prevents binding of agglutinating conjugates. The degree of agglutination is then indirectly proportional to the concentration of the free hapten (determined with a calibration curve constructed from a series of known hapten concentrations).

Both homogeneous SPIA can be readily automated with centrifugal analysers, in which the analysis time is reduced to a few minutes.

A distinct advantage of SPIA is that more detection methods can be used, non-radioactive labels can be employed, the reagents are relatively stable (more stable than those used in EIA) and the system is simple and requires no additional chemical reactions. Visual detection makes it possible to measure specific compounds with commercial kits, even in a doctor's surgery or at home (*e.g.* human chorionic gonadotropin in pregnancy tests).

For **immunoassays using particle counting** latex particles 0.8 μm in diameter are usually used. *Latex particles* had already been used in various semiquantitative agglutination tests but have since become the basis of sensitive quantitative methods known as **PACIA (particle-counting immunoassay)** or **IMPACT (immunoassay by particle counting)** (CAMBIASO *et al.*, 1977). They are based on direct agglutination or on its inhibition.

In *direct agglutination* assays, latex particles are coated by antibodies against the test antigen. When the antigen is present in the sample, the particles agglutinate and the number of non-agglutinated particles present in the mixture before addition of the antigen decreases. The antigen concentration in the test sample is then proportional to the decrease of non-agglutinated particles. Compared to classical agglutination techniques, the number of non-agglutinated particles rather than the amount of the agglutinate is measured. Special cell counters capable of detecting only particles of 0.6–1.0 µm diameter are used, to avoid counting aggregated particles.

A disadvantage of the direct agglutination assay is that latex particles can be agglutinated non-specifically by the rheumatoid factor (an IgM antibody against IgG) that may be present in the sample. To exclude this interference, only $F(ab)_2$ fragments of specific antibodies are used to coat the latex particles, as the rheumatoid factor binds to the Fc-domains of the antibody.

This technique is mainly used to measure various protein antigens. In this case, the latex particles are coated by the antigen rather than the antibody.

Inhibition of agglutination is used for the immunoassay of haptens. Its principle is identical to agglutination inhibition of colloidal gold particles, although the detection method is different — using a particle counter.

Irrespective of whether latex particles are used as a marker to determine antigens, antibodies or haptens, the assay system is always homogeneous and does not require any separation step. The detection procedure is always identical — counting of non-agglutinated particles. For this purpose commercially available automatic devices (*e.g.* Acade Diagnostic Systems) are used. Reagents are stable — latex particles retain their properties for 18 months — and the detection range for typical proteinaceous antigens is 3–400 µg/l.

Latex particles may also be used in particle agglutination immunoassays in which the agglutination is measured quantitatively by spectrophotometers or nephelometers. The direct agglutination assay has also been called **PETIA (particle enhanced turbidimetric immunoassay)**; the indirect assay has been called **PETINIA (particle enhanced turbidimetric inhibition immunoassay)**.

The intensity of light scattered by particles dispersed in water varies with the number of particles, the diameter of the particles, the wavelength of the incident light, and the angle of the detector to the incident light. Particles which scatter light best have diameters approximately equal to the wavelength of light being scattered. Therefore, for visible light (wavelength 390–760 nm) the best scattering particles have diameters of 0.39–0.76 µm. Particles outside this range will not scatter as well. In practice, particles

<0.1 μm are poor scatters; however, agglutinated particles quickly grow to a size where they scatter light much better.

Theoretical sensitivity of latex agglutination tests for an antigen with mol. wt. 100 000 is 10 ng when using particles with a diameter of 0.1 μm; if particles are 1.0 μm diameter, sensitivity will be 10 pg; if 10 μm diameter particles are used, sensitivity improves to 10 fg (BANGS, 1990).

For quantifying particle immunoassays instruments such as coulter counter, nephelometers or spectrophotometers can be used. Agglutinated particles may also be trapped on a filter while single particles pass through. In this case (Kodak Surecell), red-dyed particles coated with antibody are mixed with a sample and poured on a filter. Single particles pass and no colour appears on the surface. If the sample contains the appropriate antigen, the particles agglutinate and the agglutinated clumps are caught upon the filter resulting in red (or pink) positive colour test for the antigen. This principle could easily be applied to an assay procedure where the reflected colour intensity would correlate with antigen content in the sample.

In **particle capture ELISA** tests, antibody is adsorbed onto latex particles, the particles are caught on a filter and dried. The procedure is as follows: first, a sample is passed through the filter and specific antigen is caught by the antibody. Next, a second antibody–enzyme conjugate is put through the filter; the second antibody is caught by the Ag–Ab complex on the particles to complete the sandwich. Third, enzyme substrate is passed through the filter and the enzyme reacts with it to create a colour which is proportional to the amount of antigen caught.

First described by CLEVELAND and RICHMAN (1987), **particle immunofiltration assays** were originally designed to capture infected cells or other organisms on the surface of a microporous membrane. Conjugated antibodies were then allowed to react with these organisms and, after washing away the unbound conjugate, a chromogen/substrate solution was added to the membrane. Early versions of this methodology require microscopic examination of the membrane to detect the presence of coloured organisms or cells.

Today, these procedures usually employ a binding step in which microspheres coated with capture reagent (antibody) are allowed to react with the patient specimen. The particles are themselves captured on a membrane sieve and washed. An appropriate conjugate (*e.g.* secondary antibody—enzyme) is then added, the beads are again washed to remove any unbound conjugate, and finally, a chromogen/substrate solution is added. If the desired analyte was present in the clinical specimen, the membrane will contain a coloured precipitate, whereas the absence of the desired analyte yields no colour formation.

Particle immunofiltration assay are generally faster than microtitre plate EIA methods because the particles have a much greater surface area than the microtitre plate well and mixing the particles with the specimen increases the likelihood that the desired analyte will react with the capture reagent. The sensitivity and specificity of these assays compare favourably with microtitre plate EIA methods.

Erythrocytes can be used instead of latex particles. At the end of the assay they are haemolysed and the released haemoglobin is measured instead of counting the non-agglutinated particles. This method is known as **passive haemolysis**. Erythrocytes are lysed by complement which is activated by immune complexes produced on the surface of the erythrocyte. The immune complexes are formed in such a way that the antigen or antibody passively bound on the cell surface react with the second member of this pair present in the test sample. **Passive haemolysis inhibition (PHI)** can also be used for determination of antigens and haptens. Erythrocytes coated by the antigen, free antigen and antibody are then present in the assay system. The degree of haemolysis inhibition (after addition of complement) is directly proportional to the amount of the free antigen (hapten) present. When the antigen contains amino groups or carboxyl groups, it can be covalently bound to the erythrocyte surface by carbodi–imide. In this case PHI is equivalent to RIA with respect to its sensitivity and accuracy. When the antigen is bound non-covalently (*e.g.* by chromic chloride) the method is less reliable and less sensitive.

Liposomes are artificial particles with an envelope formed by one or more phospholipid bilayers and which can contain various compounds. Liposomes surrounded by one membrane resemble lysosomes. Liposomes can be used as a label. A lipid is first bound to the antigen and the conjugate is incorporated into the liposome phospholipid bilayer. When a specific antibody reacts with the labelled particle an immune complex is produced which activates the added complement and results in disintegration of the liposome. The label is then released from their interior and can be detected. For example, a fluorophore or stable free radical can be used as labelling compounds. In the latter case, the method is known as **spin membrane immunoassay (SMIA)** and its sensitivity approaches that of RIA. When multilayered liposomes are used to label the antigen (they contain several concentric phospholipid bilayers) they disintegrate after formation of the immune complex, even in the absence of complement.

12.11 Other non-isotopic immunoassay methods

As well as EIA, ELISA, FIA, CIA and particle immunoassay, other non-isotope immunoassay methods have been developed, although they are less commonly used in practice. This group of methods includes the use of spin immunoassay, metalloimmunoassay, viroimmunoassay, electrodes with immobilized antigen and antibody electrodes.

Spin immunoassay (SIA) is also known as **free radical assay technique (FRAT)** since a stable free radical (usually dinitrophenyl) serves as a label (STRYER and GRIFFITH, 1965; LEUTE *et al.*, 1972).

The term *"spin"* designates the mechanical moment of mobility of electrons determining their quantum character. In principle, spin is based on the constant rotation of the electron around its axis. This spin can be "to the right" or "to the left" (+1/2 or −1/2). Electrons in atoms and various compounds tend to combine in pairs to neutralize their opposite spins. When some atoms or compounds have a spin unpaired electron (or electrons), they become **free radicals**. When the free radical is placed in a magnetic field with a suitable intensity and is excited by radiowaves of suitable frequency, the unpaired electrons absorb and reflect (scatter) energy on the basis of resonance. Resonance absorption of unpaired electrons can be modified by changes in their close chemical proximity, *e.g.* by binding of antibodies to antigen determinants in their proximity. These changes can be detected by electron-spin resonance spectroscopy. The detection is based on the fact that the free antigen labelled with an unpaired electron yields a resonance spectrum with three sharp peaks in a spin resonance spectrometer. However, when a specific antibody is bound to such an antigen, the spectrum changes — only a weak signal of electron paramagnetic resonance without sharp peaks is observed.

The principle of SIA is identical to that of other homogeneous immunoassays. SIA is mainly used to determine different haptens, *e.g.* hormones, barbiturates, opiates and their metabolites. The modification with a free radical bound to the antibody can also be used. The sensitivity of the method is comparable to RIA, although the instruments required are very expensive.

Metalloimmunoassay (MIA) uses soluble metal compounds (compounds of iron, mercury, chromium, copper, platinum, manganese *etc.*) as labels. The metal can bind directly to a carbon atom (organometallic compounds) or the binding may be mediated *via* a heteroatom (coordination complexes). MIA is used to quantitate both antigens and antibodies using procedures analogous to those of other immunoassays. After immunoreaction and separation of labelled free and bound reactants, the content of the metal used is determined (usually by flameless atomic absorption spectrometry). Sensitivity (20–500 μg of antigen/l) is comparable to that of RIA. For detection of the metal bound in the immune complex other analytical techniques, such as amperometric titration, differential pulse anode dissolving voltametry or differential pulse polarography may be used.

Viroimmunoassays (VIA) are based on similar principles to other homogeneous immunoassay methods. However, bacteriophages are used to label haptens or antigens. **Bacteriophages** (or simply phages) are viruses which infect bacteria, replicate in them and which subsequently induce their lysis and death. Conjugates of haptens or antigens with a suitable phage also exhibit this property, although in complexes with an antihapten antibody, the phage lytic activity is lost. When bacteria are inoculated on an agar nutrient medium they produce a continuous layer on its surface. A bacterial culture infected with a phage (*e.g. Escherichia coli* infected with bacteriophage T4) does not grow as a confluent layer but shows lighter spots without bacteria that have been lysed by the phage. These "holes" in the bacterial cultures are called *plaques*. When an antihapten antibody is added to the test system containing the bacteriophage-labelled hapten and the bacterial culture, the number of plaques decreases. Because of competition for the antibody binding site, the addition of free unlabelled hapten removes inhibition of the phage lytic activity and increases the number of plaques. The degree of the restored phage activity is thus proportional to the concentration of the free hapten in the test sample.

Using VIA it is possible to quantitate various steroids, prostaglandins and other low-molecular-weight and high-molecular-weight compounds with an identical or even higher sensitivity to that of RIA. However, since VIA techniques require a special laboratory and are time-consuming, they have not yet been widely used.

In addition to plaque counting, detection of the intracellular content of bacteria can be used. When an increased production of an enzyme (*e.g.* β-D-galactosidase) is induced in *E. coli*, the enzyme is released, after bacteriophage lysis, into the culture medium where its activity can be determined spectrophotometrically with a suitable substrate (*e.g.* *o*-nitrophenyl-β-D-galactopyranoside).

Immunoaffinity electrodes belong to the group known as biosensors. **Biosensors** are based on a membrane-bound biologically active compound which is in direct contact with a sensitive analytical sensor. Biospecific interactions of this compound are transferred to a sensor, whose signal is amplified and converted into a suitable output (recorder, scale, digital output, *etc.*). According to the character of the biologically active compound, the biosensors are divided into enzyme sensors, affinity sensors, cell, organ (tissue) and mixed electrodes, and occasionally, transistors or thermistors. In addition to the laboratory analysis of some compounds, bioelectrodes may also be used *in vivo,* for the continuous monitoring of physiologically important compounds in the body, as well as in modern biotechnologies for the operational control of various bioprocesses.

Affinity electrodes are based on the specific interaction between the ligand and the binding glycoprotein. The immunoaffinity electrodes utilize the interaction between the antigen (hapten) and antibody. One of this pair is covalently bound to a metal detecting electrode. When the antibody is immobilized on a thin layer of a hydrophobic polymer (usually polyvinyl chloride) covering the metal conductor, an **antibody electrode** is formed (Boitieux *et al.*, 1979; Solsky and Rechnitz, 1981). In the interface between the solution and the electrode surface there is a charge which depends on the charge of the immobilized antibody. After binding of the antigen (hapten) present in the solution this charge changes. This change can be measured potentiometrically in comparison with a reference electrode immersed in the same solution. The potential difference is proportional to the antigen concentration in the test sample. Similarly antigen immobilization results in an **antigen electrode** which can then be used to measure antibodies.

Various low-molecular-weight haptens and protein antigens can be measured with antibody electrodes. The lower detection level of prostaglandins, for example, is 0.1–0.5 μg/l, and of HBsAg about 1 μg/l (Connell *et al.*, 1983).

12.12 General immunoassay methodology

Quantitative analytical methods with antibodies or antigens as primary reagents are now integral to many clinical, immunological, biochemical, pharmaceutical, and basic scientific investigations (GOSLING, 1990). For numerous clinically important analytes that are proteins or peptides (including hormones, cytokines, lipoproteins, oncoproteins, pathogen antigens, and specific antibodies), there are, as yet, no viable alternatives. Even where an analyte can feasibly be determined by chromatographic, colorimetric, or other standard procedures, quantitative immunochemical methods (*i.e.* immunoassays) are often used because of their speed, simplicity, and relatively low cost. This trend has been greatly encouraged by the availability of convenient, reliable commercial kits.

The nomenclature of immunoassay systems may be confusing to anyone wishing to understand the similarities between different assays as well as the diversity of their designs. In general, all assay names contain "immuno", the combining form of the adjective "immune", and another combining word indicating the type of label used, along with the word "assay": *e.g.* radioimmunoassay.

To distinguish reagent excess from competitive assays, the common practice is to reverse the order of the combining forms, as in immunoradiometric assay (IRMA) or immunofluorometric assay (IFMA). However, similar names are sometimes used for competitive assays with labelled antibody. The acronym ELISA is generally used for reagent excess assays of specific antibodies or antigens, but sometimes also interchangeably with enzyme immunoassay and immunoenzymometric assay.

Much of the variety in immunoassay names comes from the numerous labelling substances that have been exploited in the development of immunoassays. These substances include radioisotopes (radio), enzymes (enzymo or enzyme), fluorescent (fluoro), chemi-, and bio-luminiscent compounds (lumino) *etc.*

GOSLING (1990) sorted the relevant immunoassay systems into six groups (*Table 12.7*):

1. The first group includes immunoassays of antigens or haptens in which labelled "analyte" is used; these are basically equivalent to classical RIA. They involve the use of a limited concentration of antibody and, despite being a misnomer, the term "competitive" is almost universally used to describe them.
2. Labelled-antibody reagent-limited ("competitive") assays for antigen or hapten. They have the advantage that labelled antigens or haptens with undesirable properties (*e.g.* low solubility in aqueous media) may be avoided.
3. Assays in this group include precipitation, nephelometric, and turbidimetric immunoassays as well as particle agglutination and particle-counting immunoassays. In general, their end-points involve the direct detection of

immune complexes, and some are characterized by the lack of any labelled reagent.

Table 12.7 Some important immunoassays arranged in groups according to Gosling (1990)

Group	*Assay name*	*Acronym*	*Labels used*	*Analytes*
1.	RadioIA	RIA	Radioisotopes	Unrestricted
	ChemiluminoIA	CIA	Chemiluminophores	Unrestricted
	EnzymoIA	EIA	Enzymes	Unrestricted
	FluoroIA	FIA	Fluorophores	Unrestricted
2.	Chemiluminescent labeled-antibody IA		Chemiluminophores	Unrestricted
	"Enzyme-linked immunosorbent assay"	"ELISA"	Enzymes	Unrestricted
3.	Immunonephelometric assay		None	Large antigens
	Immunoturbidimetric assay		None	Large antigens
	Latex agglutination test	LAT	Coloured latex particles	Large antigens
	Particle counting IA	PACIA	Estapor K150 particles	Large antigens and haptens
	Particle-enhanced turbidimetric inhibition	PETINIA	Latex particles	Large antigens
4.	Enzyme-linked immunosorbent assay	ELISA	Enzymes	Large antigens
	Immunochemiluminometric assay	ICLMA	Chemiluminescent compounds	Large antigens
	Immunofluorometric assay	IFMA	Fluorescent compounds	Large antigens
	Immunoradiometric assay	IRMA	^{125}I	Large antigens
5.	Enzyme-linked immunosorbent assay	ELISA	Enzymes	Specific antibodies
	Radioallergosorbent test	RAST	^{125}I	Allergen-specific IgE
6.	Combined enzyme donor IA	CEDIA	β-Galactosidase fragment	Digoxin
	Enzyme-monitored IA technique	EMIT	G6PDH, MDH, lysozyme	Drugs and other haptens
	Homogeneous luminescence IA		Isoluminol	Haptens
	Fluorescence energy transfer IA	FETIA	2-Fluorescein derivatives	Immunoglobulins
	Fluorescence polarization IA	PFIA	Fluorescein	Drugs and other haptens
	Prosthetic-group-label IA	PGLIA	FAD	Drugs
	Substrate-labelled fluorescence IA	SLFIA	Umbelliferone	Theophylline

IA, immunoassay; G6PDH, glucose-6-phosphate dehydrogenase; MDH, malate dehydrogenase.

4. This group contains assays involving labels in which all the principal reagents are used in excess. They include two-site "sandwich" assays such as immunochemiluminometric assays, IRMA, IFMA, and most ELISAs,

all of which have the fundamental advantage that their performance characteristics are not as dependent on antibody affinity as the competitive assays are. In effect, very low detection limits are attainable by maximizing the signal/noise ratio of the label.

5. Immunoassays for quantifying specific antibodies are included in this group. In principle they resemble equivalent assays for receptors or specific binding proteins. Most often, diluted test serum is added to excess antigen immobilized on a solid phase. The amount of specific antibody that binds (or is "captured") may then be quantified by the use of labelled antibodies that specifically bind to the constant region of the immunoglobulin class of interest (secondary antibodies).
6. The assays in this group involve use of labelled reagents and have in common features that result in modulation of signal from the label by the binding reaction, thereby allowing binding to be monitored without the necessity for a separation step. Consequently they are referred to as separation-free or homogeneous immunoassays. Characterized by simplicity and speed, they are widely used to monitor concentrations of therapeutic drugs and drugs of abuse in blood and urine when low detection limits ($< 10\,\mu$mol/l) are not required.

References

Alfonso, E. (1964) Quantitative immunoelectrophoresis of serum proteins. *Clin. Chim. Acta*, **10**, 114–22.

Alvine, J. C., Kemp, D. J. and Stark, G. R. (1977) Method for detection of specific RNAs in agarose gels by transfer to diazobenzyloxymethyl-paper and hybridization with DNA-probe. *Proc. Natl. Acad. Sci. USA*, **74**, 5350–9.

Avrameas, S. and Uriel, J. (1966) Methode de marquege des antigens et des anticorpes avec des enzymes et son applications en immunodifusion. *C. R. Acad. Sci., Ser. D* **262**, 2543–5.

Bangs, L. B. (1990) New developments in particle-based tests and immunoassays. *J. Int. Fed. Clin. Chem.*, **2**, 188–93.

Beisiegel, U. (1986) Protein blotting. *Electrophoresis*, **7**, 1–18.

Boitieux, J. L., Desmet, G. and Thomas, D. (1979) An "antibody electrode", preliminary report on a new approach in enzyme immunoassay. *Clin. Chem.*, **25**, 318–21.

Burnette, W. N. (1981) "Western blotting". electrophoretic transfer of proteins from sodium dodecyl sulfate–polyacrylamide gels to unmodified nitrocellulose and radiographic detection with antibody and radiolabelled protein A. *Anal. Biochem.*, **112**, 195–204.

Cambiaso, C. L., Leek, A. E., de Steenwinkel, F., Billen, J. and Masson, P. L. (1977) Particle counting immunoassay (PACIA). I. A general method for the determination of antibodies, antigens and haptens. *J. Immunol. Methods*, **18**, 33–44.

Carrico, R. J., Yeung, K. K., Schroeder, H. R., Boguslaski, R. C., Buckler, R. T. and Christner, J. E. (1976) Specific protein-binding reactions monitored with ligand–ATP conjugates and firefly luciferase. *Anal. Biochem.*, **76**, 95–110.

Clarke, H. G. M. and Freeman, T. (1966) A quantitative immuno-electrophoresis method (Laurell electrophoresis). *Protides Biol. Fluids Proc. Colloq.*, **14**, 503–9.

Cleveland, P. H. and Richman, D. D. (1987) Enzyme immunofiltration staining assay for immediate diagnosis of herpes simplex virus and varicella-zoster virus directly from clinical specimens. *J. Clin. Microbiol.*, **25**, 416–20.

Connell, G. R., Sanders, K. M. and Williams, R. L. (1983) Electroimmunoassay. A new competitive protein-binding assay using antibody-sensitive electrodes. *Biophys. J.*, **44**, 123–6.

Coons, A. H., Creech, H. J. and Jones, R. N. (1941) Immunological properties of an antibody containing a fluorescent group. *Proc. Soc. Exp. Biol. Med.*, **47**, 200–2.

Culliford, B. J. (1964) Precipitin reactions in forensic problems. *Nature*, **201**, 1092–4.

Czerkinsky, C., Moldoveanu, Z., Mestecky, J., Nilsson, L.-Å. and Ouchterlony, Ö. (1988) A novel two colour ELISPOT assay. *J. Immunol. Methods*, **115**, 31.–7.

Czerkinsky, C., Nilsson, L.-Å., Nygren, H., Ouchterlony, Ö. and Tarkowski, A. (1983) A solid-phase enzyme-linked immunospot (ELISPOT) assay for enumeration of specific antibody-secreting cells. *J. Immunol. Methods*, **65**, 109–21.

Dandliker, W. B. and Feigen, G. (1961) Quantification of the antigen-antibody reaction by the polarization of fluorescence. *Biochem. Biophys. Res. Commun.*, **5**, 299–304.

Dandliker, W. B. and De Saussure, V. A. (1970) Fluorescence polarization in immunochemistry. *Immunochemistry*, **7**, 799–828.

Engvall, E. and Perlmann, P. (1971) Enzyme-linked immunosorbent assay (ELISA). Quantitative assay of immunoglobulin G. *Immunochemistry*, **8**, 871–4.

Fahey, J. L. and McKelvey, E. M. (1965) Quantitative determination of serum immunoglobulins in antibody-agar plates. *J. Immunol.*, **94**, 84–91.

Faulk, W. P. and Taylor, G. M. (1971) An immunocolloid method for the electron microscope. *Immunochemistry*, **8**, 1081–3.

Fields, H. A., Davis, C. L., Dreesman, G. R., Bradley, D. W. and Maynard, J. E. (1981) Enzyme potentiated radioimmunoassay (EPRIA): a sensitive third-generation test for the detection of hepatitis B surface antigen. *J. Immunol. Methods*, **47**, 145–59.

Gill, C. W., Fischer, C. L. and Holleman, C. L. (1971) Rapid method for protein quantitation by electroimmunodiffusion. *Clin. Chem.*, **17**, 501–12.

Gosling, J. P. (1990) A decade of development in immunoassay methodology. *Clin. Chem.*, **36**, 1408–27.

Grabar, P. and Williams, C. A. (1953) A method permitting the combined study of the electrophoretic and immunochemical properties of a mixture of proteins: application to blood serum. *Biochem. Biophys. Acta*, **10**, 193–4.

Hawkes, R., Niday, E. and Gordon, J. (1982) Dot-immunobinding assay for monoclonal and other antibodies. *Anal. Biochem.*, **119**, 142–7.

Hemmilä, I. (1985) Fluoroimmunoassays and immunofluorometric assays. *Clin. Chem.*, **31**, 359–70.

Henderson, D. R., Friedman, S. B., Harris, J. D., Manning, W. B. and Zaceoli, M. A. (1986) CEDIA, a new homogeneous immunoassay system. *Clin. Chem.*, **32**, 1637–41.

Jol-Van der Zijde, C. M., Labadie, J., Vlug, A., Radl, J., Vossen, J. M. and Van Tol, M. J. D. (1988) Dot-immunobinding assay as an accurate and versatile technique for the quantification of human IgG subclasses. *J. Immunol. Methods*, **108**, 195–203.

Kakita, K., O'Connell, N. and Permutt, A. (1982) Immunodetection of insulin after transfer from gels to nitrocellulose filters. *Diabetes*, **31**, 648–55.

Laurell, C. B. (1965) Antigen–antibody crossed electrophoresis. *Anal. Biochem.*, **10**, 358–72.

Laurell, C. B. (1966) Quantitative estimation of proteins by electrophoresis in agarose gel containing antibodies. *Anal. Biochem.*, **15**, 45–52.

Lauritzen, E., Masson, M., Rubin, I. and Holm, A. (1990) Dot immunobinding and immunoblotting of picogram and nanogram quantities of small peptides on activated nitrocellulose. *J. Immunol. Methods*, **131**, 257–67.

Lavanchy, D., Stroun, J. and Frei, Ph. C. (1990) Modified dot assay with increased sensitivity: detection of small amounts of immunoglobulin molecules and the importance of different detection systems. *J. Clin. Lab. Anal.*, **4**, 251–5.

Leute, R., Ullman, E. F. and Goldstein, A. (1972) Spin immunoassay of opiate narcotics in urine and saliva. *J. Am. Med. Assoc.*, **221**, 1231–4.

Leuvering, J. H. W., Thal, P. J. H. M., Van der Waart, M. and Schuurs, A. H. W. M. (1980) Sol particle immunoassay. *J. Immunoassay*, **1**, 77–91.

Mancini, G., Carbonara, A. O. and Heremans, J. F. (1965) Immunochemical quantitation of antigens by simple radial immunodiffusion. *Immunochemistry*, **2**, 235–54.

Mančal, P. (1987) *Methods in Enzyme Immunoassays*. Prague, Institute of Sera and Vaccines, 83 pp.

Mosmann, T. R. and Coffman, R. L. (1989) Th1 and Th2 cells: different patterns of lymphokine secretion lead to different functional properties. *Ann. Rev. Immunol.*, **7**, 145.

Nakamura, S. (1966) *Cross electrophoresis: Its principles and applications.* New York, Elsevier.

Nakane, P. K. and Pierce, G. B. (1966) Enzyme-labelled antibodies: preparation and application for the localization of antigens. *J. Histochem. Cytochem.*, **14**, 929–31.

Ouchterlony, Ö. (1948) *In vitro* method for testing the toxin-producing capacity of diphteria bacteria. *Acta Pathol. Microbiol. Scand.*, **25**, 186–91.

Ouchterlony, Ö. (1953) Antigen–antibody reactions in gels. IV. Types of reactions in coordinated systems of diffusion. *Acta Pathol. Microbiol. Scand.*, **32**, 231–40.

Oudin, J. (1956) Réaction de précipitation spécifique des sérums d'animaux de même espèce. *C. R. Acad. Sci.*, **242**, 2489–90.

Renart, J., Reiser, J. and Stark, G. R. (1979) Transfer of proteins from gels to diazobenzyloxymethyl-paper and detection with antisera: a method for studying antibody specificity and antigen structure. *Proc. Natl. Acad. Sci. USA*, **76**, 3116–20.

Ressler, N. (1960) Two-dimensional electrophoresis of serum protein antigens in an antibody containing buffer. *Clin. Chim. Acta*, **5**, 795–803.

Schroeder, H. R., Vogelhut, P. O., Carrico, R. J., Boguslaski, R. C. and Buckler, R. T. (1976) Competitive protein binding assay for biotin monitored by chemiluminescence. *Anal. Chem.*, **48**, 1933–7.

Schuurs, A. H. W. M. and Van Weemen, B. K. (1977) Enzyme-immunoassay. *Clin. Chim. Acta*, **81**, 1–40.

Sedgwick, J. D. and Holt, P. G. (1983) A solid-phase immunoenzymatic technique for the enumeration of specific antibody-secreting cells. *J. Immunol. Methods*, **57**, 301–9.

Seitz, W. R. (1984) Immunoassay labels based on chemiluminescence and bioluminescence. *Clin. Biochem.*, **17**, 120–5.

Solsky, R. L. and Rechnitz, G. A. (1981) Preparation and properties of an antibody-selective membrane electrode. *Anal. Chim. Acta*, **123**, 135–41.

Southern, E. M. (1975) Detection of specific sequences among DNA fragments separated by gel electrophoresis. *J. Mol. Biol.*, **98**, 503–17.

Strasburger, C. J., Amir-Zaltsman, Y. and Kohen, F. (1988) The avidin–biotin reaction as a universal amplification system in immunoassays. In: *Non-Radiometric Assays: Technology and Application in Polypeptide and Steroid Hormone Detection.* Alan R. Liss, pp. 79–100.

Stryer, L. and Griffith, O. H. (1965) A spin-labelled hapten. *Proc. Natl. Acad. Sci. USA*, **54**, 1785–91.

Taguchi, T., Mc Ghee, J. R., Coffman, R. L., Beagley, K. W., Eldridge, J. H., Takatsu, K. and Kipono, H. (1990). Detection of individual mouse splenic *T*-cells producing IFN-γ and IL-5 using the enzyme-linked immunospot (ELISPOT) assay. *J. Immunol. Methods*, **128**, 65–73.

Thompson, S. G. and Boguslaski, R. C. (1987) Homogeneous dry reagent immunoassay strips for the determination of therapeutic drugs in human serum or plasma. *J. Clin. Lab. Anal.*, **1**, 293–9.

Tijssen, P. (1985) *Practice and Theory of Enzyme Immunoassays.* Amsterdam, Elsevier Sci. Publ., 549 pp.

Towbin, H., Staehelin, T. and Gordon, J. (1979) Electrophoretic transfer of proteins from polyacrylamide gels to nitrocellulose sheets: procedure and some applications. *Proc. Natl. Acad. Sci. USA*, **76**, 4350–4.

Ullman, E. F., Schwarzberg, M. and Rubenstein, K. E. (1976) Fluorescence excitation transfer immunoassay. A general method for determination of antigens. *J. Biol. Chem.*, **251**, 4172–8.

Van der Sluis, P. J., Pool, C. W. and Sluiter, A. A. (1987) Press-blotting on gelatin-coated nitrocellulose membranes. A method for sensitive quantitative immunodetection of peptides after gel isoelectric focusing. *J. Immunol. Methods*, **104**, 65–74.

Van Weemen, B. K. and Schuurs, A. H. W. M. (1971) Immunoassay using antigen–enzyme conjugate. *FEBS Lett.*, **15**, 232–7.

Wilchek, M. and Bayer, E. A. (1984) The avidin–biotin complex in immunology. *Immunol. Today*, **5**, 39–43.

Wilson, A. T. (1964) Direct immunoelectrophoresis. *J. Immunol.*, **92**, 431–4.
Yalow, R. S. (1978) Radioimmunoassay: a probe for the fine structure of biologic systems. *Science*, **200**, 1236–45.
Yalow, R. S. and Berson, S. A. (1959) Assay of plasma insulin in human subjects by immunological methods. *Nature*, **184**, 1648–9.
Zuk, R. F., Ginsberg, U. K., Houts, T., Rabbie, J., Merrick, H., Ullman, E. F., Fischer, M. M., Sizto, C. C., Stiso, S. N. and Litman, D. J. (1985) Enzyme immunochromatography — a quantitative immunoassay requiring no instrumentation. *Clin. Chem.*, **31**, 1144–50.

13 Phagocytosis

Phagocytosis is a biological phenomenon that already occurs in unicellular organisms where it fulfils a basic nutritional function. Through this process unicellular organisms acquire nutrients that are necessary for metabolic processes. The term describes a general mechanism that enables transport of substances across the cytoplasmic membrane. During phylogenetic development, however, phagocytosis acquired other than simply *nutritional functions*. Even in simple multicellular animals, phagocytosis is responsible for clearance of the internal environment and also represents the main mechanism of nutrition and nutrient transport to other cells. In vertebrates, certain types of phagocytic cells are specialized for the execution of *defence functions*. These "professional" phagocytes can effectively take up and inactivate any foreign material present in the internal environment including pathogenic microorganisms, foreign (non-self) cells and autologous, but functionally and antigenically altered cells. Phagocytes became a key cell type of inflammation, and important *secretory* cells, since their products participate in the regulation of several physiological processes, maintaining homeostasis of the internal environment, and, finally, they participate in numerous pathological mechanisms.

The term *phagocytosis* was coined by Ilya Ilyich Metchnikoff in 1883. It is derived from the Greek "phagein" (eat) and "cytos" (the cell). During a visit to Messina in Sicily, Metchnikoff studied the life of sea stars which eventually led to his important discovery of the phagocyte defence function.

From the biological point of view, phagocytosis is a special type of endocytosis. The term **endocytosis** refers to the process of internalization of extracellular material into a space formed by invagination of the cytoplasmic membrane. Following invagination, a vesicle, containing the engulfed material and a small section of cytoplasma, is formed. It is generally called an endosome, or, more specifically, a phagosome, heterophagosome, phagocytic vacuole, digestive vacuole, pinosome or receptosome (de Duve and Wattiaux, 1966; Silverstein *et al.*, 1977).

Endocytosis is classically divided into **phagocytosis** (eating — ingestion of particles with a diameter more than 0.1 μm) and **pinocytosis** (drinking — ingestion of particles and molecules less than 0.1 μm). Pinocytosis is further divided according to whether the cell ingests substances that are in the fluid phase or bound to the pinosome membrane. The first type is called **fluid phase pinocytosis** (often also called *macropinocytosis*), the second type is known as adsorption pinocytosis. **Adsorption pinocytosis** is further divided into pinocytosis when the ingested substance is bound to the membrane non-specifically, and that mediated by specific receptors in the phagocyte cytoplasmic membrane. The latter type is known as *receptor-mediated pinocytosis* or coated pits-mediated endocytosis (*Fig. 13.1*).

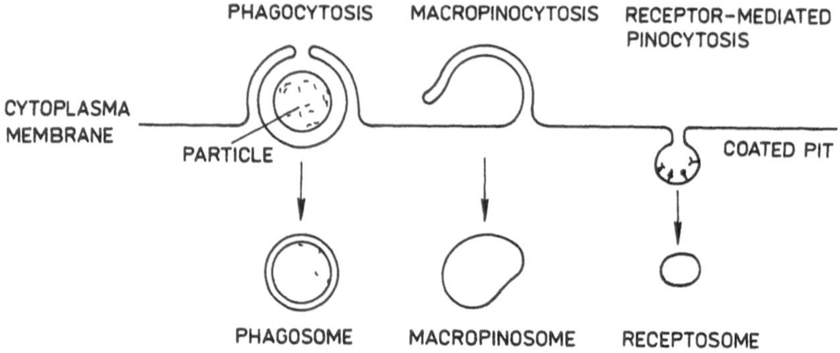

Fig 13 1 The three major types of endocytosis in eukaryotic cells

The term *"coated pits"* refers to depressions in the cell membrane containing clusters of various receptors. A typical eukaryotic cell has 500 –1 000 coated pits occupying about 1% of its surface (PASTAN and WILLINGHAM, 1985). Coated pits are the sites of ligand–receptor interactions as a first step of their transport into the cell. Within seconds of contact between ligands and receptors, the coated pit is closed and travels into the cytosol, which is then called a *receptosome*. Thus, receptosomes are transport organelles which transfer ligands and receptors from the surface to the inside of the cell. Their pH is low (about 4.5) but as they do not contain hydrolytic enzymes the ligand–receptor complex is degraded only slowly. During transport, the ligand passes information to all corresponding intracellular molecules. Eventually, the ligand may be incorporated into the Golgi apparatus. The undegraded ligand passes into the lumen and the membrane material is incorporated into the membrane system of the Golgi apparatus.

This mechanism conditions one of the basic properties of living cells — the ability to accept and respond to molecular signals from the external environment. A number of substances may act as a signal: hormones (insulin, growth hormone, calcitonin, thymosins, *etc.*), growth factors, serum transport proteins (for cholesterol transport — LDL; iron — transferrin; cobalt — transcobalamin, *etc.*), proteins altered by degradation (*e.g.* α_2-macroglobulin — proteases complexes), glycoproteins with special determinants (lysosomal enzymes, asialoglycoproteins) and certain antibodies (IgE, polymeric IgA, maternal IgG, other IgG — through Fc-receptors). In addition to these physiologically important molecules, three types of foreign substances also enter the cell by coated pits-mediated endocytosis: toxins (diphtheria, pseudomonas, cholera), lectins (ricin, concanavalin A) and viruses (Roux sarcoma, adenoviruses, *etc.*).

In multicellular organisms, pinocytosis mediated by receptors belongs to a system that is responsible for the biological turnover of secretory glycoproteins and participates significantly in the growth, nutrition and differentiation of cells. Thus, for example, mutation defects which disturb low-density lipoproteins (LDL) endocytosis, mediated by macrophage and neutrophil receptors, are a frequent cause of atherosclerosis and myocardial infarction.

Exocytosis is the opposite of endocytosis. Exocytosis is also a transport mechanism by which internal cell components pass into the extracellular environment. It enables the secretion of various substances which influence the activity of other cells, tissues and physiological systems, and secretion of substances generated by degradation of the engulfed material or degradation products of cell metabolism.

13.1 Professional phagocytes

The ability to phagocytose is a *general* biological phenomenon possessed even by cells of higher plants provided they are deprived of cell wall (protoplasts). All mammalian lysosome-containing cells, including muscle and myocardium cells, are capable of phagocytosis. However, the effectiveness of such phagocytosis is very low and usually occurs only under certain pathological conditions. **Professional phagocytes** play an important role in defence reactions. This term refers to phagocytes whose specific role is to ingest microorganisms or other foreign cells and particles. Professional phagocytes are equipped with special cytoplasmic membrane receptors for this purpose. As soon as the microbe is coated with C3b or specific antibody, this complex may be very effectively bound to complement receptors (CR) or receptors for immunoglobulin Fc-domains (FcR) which provide a signal for endocytosis. Such a mechanism is termed **immune phagocytosis** and it is much more effective than **non-immune phagocytosis** that does not utilize CR and FcR.

Substances that are able to coat microbes or other particles, and thus facilitate their attachment to the phagocyte surface, were termed **"opsonins"** — from the Greek *opsono* = to buy food, hence preparation for eating — (WRIGHT and DOUGLAS, 1903).

The function of professional phagocytes is ascribed to two cell types called microphages and macrophages by METCHNIKOFF in 1883. Nowadays, *"microphages"* refers to neutrophil granulocytes or neutrophil polymorphonuclear leukocytes (neutrophils); the latter type belong to the **mononuclear phagocyte system (MPS)** (VAN FURTH *et al.*, 1972) whose basic components are the blood monocytes and tissue macrophages. All the professional phagocytes differentiate from the multipotent haematopoietic stem cell in the bone marrow (*Fig. 13.2*). The blood elements that differentiate in multicellular organisms from the pluripotent haematopoietic stem cell must participate in three main functions: gas exchange (oxygen transport — erythrocytes), haemostasis (platelets), and defence mechanisms (leukocytes). Leukocytes arise either from the myeloid or lymphoid cell line. The myeloid line gives rise to granulocytes, monocytes and macrophages, the lymphoid line to lymphocytes. The **granulocytes** include neutrophils, eosinophils and basophils although ROITT (1981) also placed mastocytes tentatively into this cell line. Various *B*-lymphocyte and *T*-lymphocyte subpopulations are generated from the lymphoid line. In addition to these two leukocyte lines, there is also a "third" leukocyte population that shares many properties and similar markers with the lymphoid line. However, certain markers and properties suggest a possible relationship with cells of the myeloid line. To this third population belong various cells that are characterized on the basis of surface markers (null cells), function (*NK*-cells), or morphological characteristics (large granular lymphocytes — LGL).

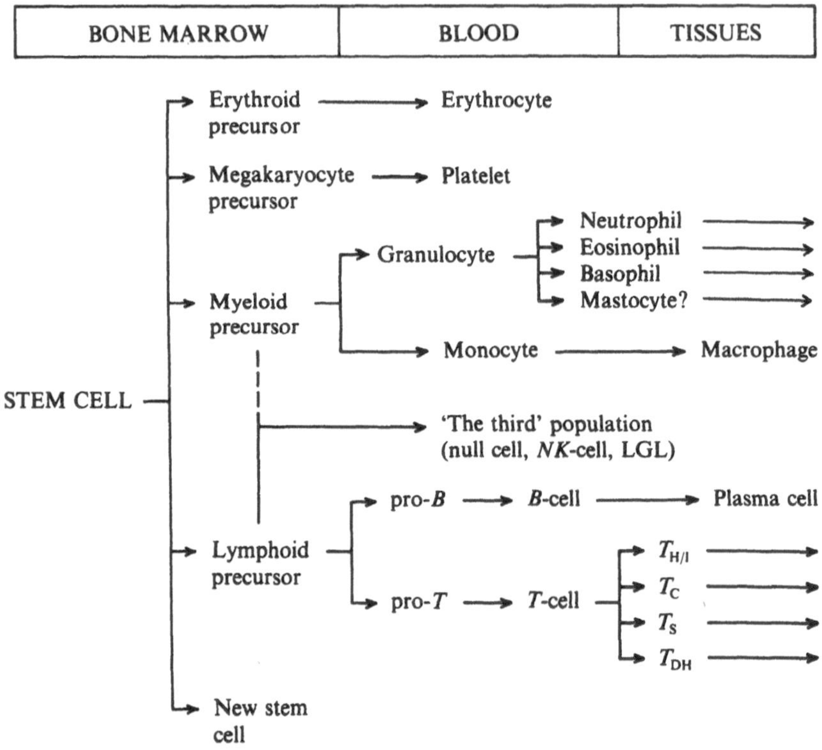

Fig. 13.2. Origin of blood elements from a multipotent haematopoietic stem cell.
pre-*B*, precursor *B*-cell; pre-*T*, precursor *T*-cell; $T_{H/I}$, helper and inducer *T*-lymphocyte; T_C, cytotoxic *T*-lymphocyte; T_S, suppressor *T*-lymphocyte; T_{DH}, effector *T*-lymphocyte engaged in delayed-type hypersensitivity reaction; LGL, large granular lymphocyte.

The endocytic activity is not equally expressed in all phagocytes. Neutrophils and macrophages highly effectively phagocytose various particles and cells and also possess significant secretory functions. In contrast, mastocytes and basophils have limited phagocytic activity but they readily release their content of secretory granules into the extracellular space. Eosinophils have significantly lower phagocytic capacity compared to neutrophils and macrophages, but their secretory products play an important role in regulatory and cytotoxic reactions. Professional phagocytes, *i.e.* neutrophils, eosinophils, monocytes and macrophages, possess both basic functions — endocytosis and exocytosis. Endocytosis is closely related to the ability of intracellular killing of microbes and other cells. The secretory products play an important role, particularly in cytotoxic reactions and regulation of inflammatory processes.

The main role of professional phagocytes is the defence of the host against pathogenic and opportunistic microorganisms. The defence function of pinocytosis has not yet been elucidated, but it is presumably not of great importance. Neutrophils possess only low pinocytic activity. On the other hand, macrophages are capable of producing a number of pinocytic vacuoles equivalent to 25% of their total cell volume every hour. Certain viruses may also enter the cell by this mechanism. Both eosinophils and basophils possess pinocytic activity.

Lymphocytes lack the ability to endocytose particles, but they can pinocytose through coated pits. In addition, they have a well-developed exocytosis system and secrete various substances (lymphokines, immunoglobulins *etc.*).

13.1.1 Neutrophils

Neutrophil polymorphonuclear leukocytes are considered to be the first line of defence against pathogenic microorganisms, particularly bacteria. Microorganisms invading the tissue are first recognized by neutrophils through chemotactic signals. They actively migrate towards the increasing signal intensity (increasing chemotaxin concentration) and start to engulf and kill the invading microbes. Later, after 5–6 hours, macrophages are attracted to the site of infection through different chemotactic signals.

Neutrophils are short-lived cells (Bainton, 1988). Their development in the bone marrow takes about two weeks; during this period, they undergo proliferation and differentiation. During maturation, they pass through six morphological stages: myeloblast, promyeloblast, myelocyte, metamyelocyte, non-segmented (band) neutrophil, segmented neutrophil. The *segmented neutrophil* is a fully functionally active cell. Only segmented neutrophils and a few non-segmented neutrophils are released into peripheral blood under physiological conditions. The functional differentiation of a neutrophil passes through the following steps: appearance of Fc-receptors (already present on myeloblasts and in higher numbers on promyelocytes), immune phagocytosis capacity (low in myeloblasts, fully active in metamyelocytes), CR3 (myelocytes), CR1 (non-segmented neutrophils), bacterial killing (present in myelocytes, fully active in non-segmented neutrophils), INT (iodonitrotetrazolium) or NBT (nitroblue tetrazolium) reduction and chemotaxis (present in metamyelocytes, fully active in segmented neutrophils).

The mature form, that is released into the blood circulation, has a multilobular nucleus and cytoplasm containing granules and glycogen particles (*Fig. 13.3*). It is, therefore, also called a *polymorphonuclear (PMN) leukocyte*. The term neutrophil and PMN leukocyte are often used synonymously, although PMN leukocyte is more accurately a synonym of granulocyte. Mature neutrophils circulate in the blood for approximately 10 h (half-life 6–7 h). Their viability in the tissues (after leaving the blood circulation) is unknown but is probably not longer than 1–2 days. The viability is significantly shorter in individuals suffering from infectious or acute inflammatory diseases when the tissue requirement for newly recruited neutrophils increases considerably.

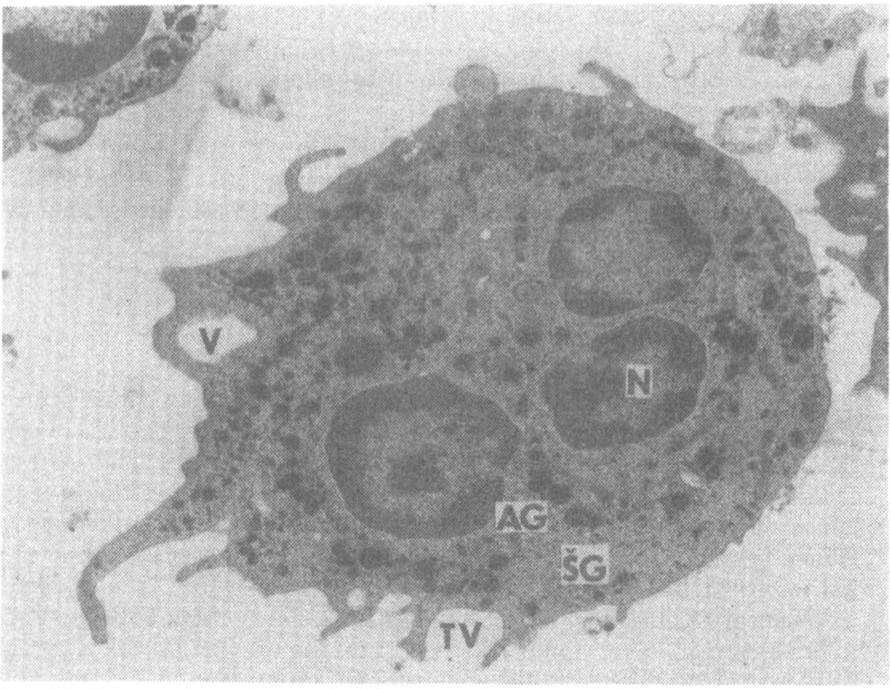

Fig. 13.3. Electron photomicrograph of human neutrophil, × 15 000 (according to P. Mraz). N, cell nucleus, V, phagocytic vacuole, TV, phagocytic vacuole being formed, AG, azurophil granules (larger and darker), SG, specific granules (smaller)

The neutrophil **granules** (*i.e. lysosomes*) are of major importance for their function (Ferenčík, 1986). When referring to phagocytes or leukocytes in general, the term granule is used more often than lysosome. The terms are not fully equivalent; the term granules was originally derived from morphological observations whereas the term lysosomes is based on functional and biochemical characteristics of these cell organelles. Not all organelles that look like granules are necessarily typical lysosomes.

The granules of professional phagocytes differ from lysosomes in other mammalian cells by their development, function and enzyme content. Granules are generated in polymorphonuclear leukocytes and mononuclear phagocytes during cell differentiation; they are produced for storage rather than continually. On the basis of function and enzyme content, human neutrophil granules can be divided into three main types — azurophil, specific and *small storage granules* (BAGGIOLINI, 1982; BAGGIOLINI and DEWALD, 1985). Their function is not just to provide enzymes for hydrolytic substrate degradation — as in classical lysosomes — but also to kill ingested bacteria and, finally, to secrete their contents to regulate various physiological and pathological processes. Their enzyme content correlates with these functions. Individual granule populations can be characterized morphologically (*e.g. azurophil granules* are larger and contain more electron-dense material than *specific granules*), or biochemically using enzyme markers or other substances (*Table 13.1*). The azurophil (primary) granules are formed during neutrophil development — and appear at the promyelocyte stage. Specific (secondary) granules are formed later — at the myelocyte stage.

Enzymes and other substances, present in segmented neutrophil granules, are synthesized during their maturation. After being synthesized, they remain in a latent state within the granules until the neutrophil receives an activation signal provided by the phagocytosable material, chemotactic factor or other soluble ligand. The signal induces release of the granule contents into the phagocytic vacuole (phagocytic release) or into the extracellular space (secretion) thereby activating the neutrophil. **Activated neutrophils** are rapidly mobilized against microbes that penetrate anatomical barriers of the host and migrate into the inflammatory reaction site. A number of enzymes and mediators are liberated during phagocytosis or after humoral stimulation (particularly by chemotaxins) that influence the course of the inflammatory reaction. In addition, the preformed substances that are present in granules and newly synthesized compounds with important regulatory functions are also released from activated neutrophils: superoxide and other reactive oxygen forms, arachidonic acid products (prostaglandins, thromboxanes, leukotrienes), platelet-activating factor (PAF), interleukin 1 (IL-1) and others.

Released neutrophil lysosomal enzymes, particularly neutral proteinases and reactive oxygen species, participate in the damage of cells and intercellular structures during immunopathological reactions. These enzymes are released not only during phagocytosis of foreign particles, but also when immune complexes attach firmly to large non-phagocytosable surfaces (*e.g.* glomerular basal membrane). This results in complement activation, generation of chemotaxins and accumulation of neutrophils. Neutrophils try to engulf the immune complexes, but fail because of its large surface. Thus open phagocytic vacuoles are generated, and lysosomal enzymes released into the extracellular space damage the host tissue. This mechanism was called **"frustrated phagocytosis"** by HENSON (1971b) and WEISSMANN *et al.* (1971).

Engulfment of phagocytosable material by neutrophils is mediated

Table 13.1 Enzymes and other constituents of human neutrophil granules

Constituents	*Granules*			
	Azurophil		*Specific*	*Small storage*
	A	*B*		
Antimicrobial	Lysozyme Defensins BPI	Myeloperoxidase	Lysozyme Lactoferrin	
Neutral proteinases	Elastase Cathepsin G	Proteinase 3 C5a-inactivator	Collagenase C5-activator	Gelatinase Plasminogen activator
Acid hydrolases	β-Glycerophosphatase *N*-acetyl-β-D-glucosaminidase α-Mannosidase α-Fucosidase Arylsulphatase Cathepsin B Cathepsin D Phospholipase A_2	β-Glucuronidase		β-glycerophosphatase β-D-glucuronidase *N*-acetyl-β-D-glucosaminidase α-Mannosidase Cathepsin B Cathepsin D
Cytoplasmic membrane receptors			Phospholipase A_2 CR3, CR4, FMLP receptors Laminin receptors	
Others	Chondroitin-4-sulphate		Cytochrome b_{558} FAD-flavoprotein Vitamin B_{12} binding protein Histaminase Monocyte-chemotactic factor Protein kinase C inhibitor Acid proteins	Cytochrome b_{558}

through opsonin-immunoadherence receptors (FcR, CR1 and CR3). Special receptors are required to receive the chemotactic signal. Until now, four main types of substances with chemotactic activity for neutrophils have been chemically defined: anaphylatoxin C5a and its desarginine derivative (generated during complement activation), *N*-formyl-methionyl-leucyl-phenylalanine (FMLP) and its analogues (produced by invading bacteria), leukotriene B_4 (LTB_4) and cytokine IL-8. These chemotaxins not only regulate the neutrophil movement towards their increasing concentration but they also induce neutrophil secretion. A special case is LTB_4 which is a product of activated neutrophils and, at the same time, acts as their specific chemotaxin. A similar autoregulatory function may also be exerted by PAF but this does not possess chemotactic activity and only stimulates neutrophil secretion.

Neutrophils participating in the inflammatory processes are thus considered to be activated (stimulated). Compared to resting neutrophils, they are more capable of adherence, spreading, aggregation, chemotaxis and chemokinesis, ingestion, lysosomal enzyme release and their metabolism, including production of oxygen-derived reactive products, is increased. Their intravascular behaviour is also changed. Activated neutrophils may aggregate, embolize microvascular regions, adhere to endothelium and produce endothelial damage by liberating proteinases and oxygen products. Because of these properties, activated neutrophils participate in the pathogenesis of many diseases (WRIGHT, 1982; BAGGIOLINI and DEWALD, 1985).

13.1.2 Eosinophils and other granulocytes

The development of eosinophils in the bone marrow takes place in three stages that are similar to neutrophil development. The mean generation time for eosinophils in the bone marrow is approximately 2–6 days. They mainly settle in the tissue where their number is about one hundred times higher than in the blood. Like other granulocytes, they possess a polymorphous nucleus, although with only two lobes and no nucleolus. The eosinophil cytoplasm contains large ellipsoid granules with an electron-dense crystalline nucleus and partially permeable matrix (*Fig. 13.4*). These arise at the beginning of eosinophil differentiation and their development is terminated after the myelocyte stage has been completed. In addition to these large primary crystalloid granules, another granule type that is smaller and lacks the crystalline nucleus starts to be formed in the metamyelocyte phase. Their number in the cytoplasm of tissue eosinophils may adaptively increase.

The biochemical composition of eosinophil granules is slightly different from that of the neutrophil granules. Eosinophils do not contain lysozyme, lactoferrin and cationic neutral proteinases. The large granules contain several cationic proteins (their enzyme activity has not yet been demonstrated), the most important being the **major basic protein (MBP** — the basic component of the crystalline nucleus of these granules) and a specific **eosinophil cationic protein** (ECP). In addition, **eosinophil peroxidase (EPO)**, histaminase and various hydrolytic lysosomal enzymes are also present in these granules.

Among the typical small granule enzymes are aryl sulphatase and acid phosphatase.

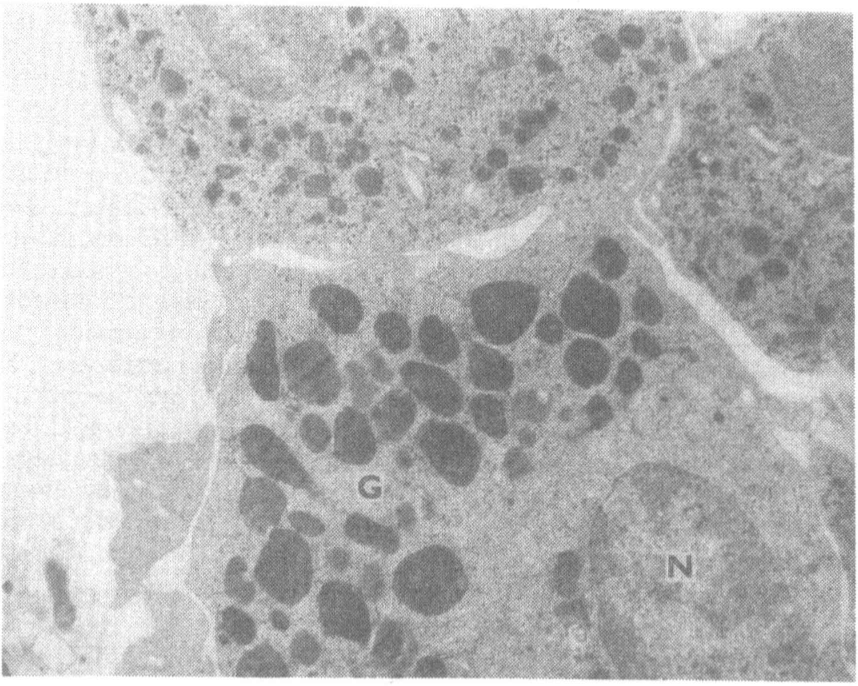

Fig. 13.4. Electron photomicrograph of human eosinophil, × 15 000 (according to P. Mráz). N, nucleus, G, granules

The eosinophil granule content is liberated following similar stimuli to neutrophil granules (*e.g.* during phagocytosis of opsonized particles and by chemotactic factors). However, whereas the neutrophil lysosomal enzymes act primarily on material engulfed in phagolysosomes, the eosinophil granule enzymes act mainly on extracellular target structures such as parasites and inflammatory mediators.

The eosinophil functional activity, like the immune response in general, may be beneficial or harmful for the organism. Compared to neutrophils, eosinophils have limited phagocytic activity which is mainly aimed at killing multicellular parasites (*e.g.* helminths) by release of granule content on to the parasite surface. Another beneficial activity is the inactivation of mediators of anaphylaxis. Thus, for example, arylsulphatase B may inactivate the **slow-reacting substance of anaphylaxis** (SRS-A — a mixture of leukotrienes LTC_4, LTD_4, LTE_4 and LTF_4), phospholipase D destroys the thrombocyte lytic factor, histaminase degrades histamine and lysophospholipase (phospholipase B) may inactivate the membrane-active lysophosphatides.

On the other hand, eosinophils may participate in host tissue damage during hypersensitivity reactions. Several factors are responsible for inducing damage: EPO and reactive oxygen products, generated during eosinophil metabolic activation, major basic protein (present in the crystalloid granules) and neurotoxin, which specifically damages the myelin coat of neurons. These cytotoxic substances are liberated from eosinophil granules mainly during the immediate-type hypersensitivity reactions, mediated by IgE homocytotropic antibodies. During these reactions eosinophils become activated.

The three cytokines that stimulate eosinophil production, IL-3, IL-5, and GM-CSF, can activate eosinophils *in vitro,* prolong their survival, reduce their density, enhance their ability to generate LTC_4, stimulate their helminthotoxic activities, and enhance the degranulation of eosinophils induced by immunoglobulin receptors. Thus, activation reflects not immaturity but stimulation, and the same cytokines that stimulate the intramedullary formation of eosinophils also stimulate the mature cells and activate their effector functions. Other cytokines, including a cytotoxicity-enhancing factor specific to eosinophils, and agents such as *platelet-activating factor (PAF)* can also activate eosinophils (WELLER, 1991).

Although eosinophils are capable of phagocytosing and killing bacteria and other small microbes *in vitro,* they do not have a major role in host defence against such microbial pathogens *in vivo* and cannot effectively defend against bacterial infections when neutrophil function is deficient (*e.g.* in drug-induced neutropenias or the leukocyte adhesion deficiency syndrome). According to ACKERMAN *et al.* (1982) eosinophils are part of a system that consists of homocytotropic antibodies (IgE in humans) and cells containing inflammatory mediators. The main function of this system appears to be defence against large, non-phagocytable organisms, most notably the multicellular helminthic parasites. In addition, eosinophils also participate in hypersensitivity reactions, especially through two lipid inflammatory mediators, LTC_4 and PAF. Both mediators contract airway smooth muscle, promote the secretion of mucus, alter vascular permeability and elicit eosinophil and neutrophil infiltration (HENDERSON, 1987). In addition to the direct activities of these eosinophil-derived mediators, major basic protein, by a non-cytotoxic mechanism, can stimulate the release of histamine from basophils and mast cells, and EPO from mast cells. Thus, once stimulated, eosinophils can serve as a local source of specific lipid mediators as well as induce the release of mediators from mast cells and basophils.

Basophils are the smallest circulating granulocytes with relatively the least known function. They arise in the bone marrow, and following maturation and differentiation, are released into the blood circulation. If they are adequately stimulated they may settle in the tissues. Of all granulocytes they have the lowest phagocytic activity, but they readily pinocytose solutions of certain substances.

The nucleus of basophils is not as well segmented as in the neutrophils and eosinophils. T ˇ cytoplasm contains a few mitochondria and other structures but is most noted for its numerous round or ovoid granules, which are basophilic in their staining properties. These granules are very important because their outer membrane is easily disrupted and the contents are released into the surrounding tissue and bloodstream. Among the important contents of the granules are heparin and histamine. **Heparin** is a powerful anticoagulant, and **histamine** is a vasoactive amine that contracts smooth muscle. These products are also found in tissue mast cells. The histamine of blood basophils and mast cells contributes to the severity of the IgE-dependent hypersensitivity reactions.

Mast cells settle in connective tissues and usually do not circulate in the bloodstream. Despite the fact that basophils and mast cells are different cell populations, they both possess high-affinity surface membrane receptors for the IgE Fc-domain. In addition, they contain special cytoplasmic granules which store various mediators of inflammation. As soon as IgE or other homocytotropic antibodies have bound to a mast cell or basophil, contact with the specific antigen (allergen) results in rapid release of histamine, serotonin and other granule-associated mediators into the external environment. Apart from these mediators, other substances which influence the course of inflammatory or anaphylactic reactions are also liberated during basophil and mastocyte degranulation, *e.g.* PAF (stimulates aggregation of platelets and liberates histamine from them), eosinophil chemotactic factor, plasminogen activator, leukotrienes (LTB_4, LTC_4) and others. In the cytoplasm of both mastocytes and macrophages are special organelles — **lipid bodies** — where metabolism of arachidonic acid occurs and where their products, including leukotrienes, are stored.

Purified basophil granules contain a mixture of neutral esterases and proteinases but no typical lysosomal hydrolytic enzymes. Peroxidase is present in the granules of human basophils, but is absent in guinea-pig basophils for example.

Biologically basophils and mastocytes play a key role in the pathogenesis of the *immediate type hypersensitivity reaction.* They also participate in various pathological processes in which homocytotropic antibodies probably do not have a role. Liberation of mediators during these processes is not dependent on the presence of specific antigen, but is caused by certain complement fragments, insect venom, basic peptides, and certain chemical substances and drugs.

Mastocytes are a highly heterogeneous cell population. Their morphological and biochemical properties are different in various animal species and even at various anatomical sites of an individual (GALLI, 1987; STEVENS and AUSTEN, 1989).

13.1.3 Monocytes and macrophages

Originally, monocytes and macrophages were classified as cells of the **reticulo-endothelial system (RES**; ASCHOFF, 1924). VAN FURTH *et al.* (1972) proposed the **mononuclear phagocyte system (MPS)**, and monocytes and macrophages became basic cell types of this system. Their development takes place in the bone marrow and passes through the following steps (VAN FURTH, 1982):

stem cell
↓
committed stem cell
↓
monoblast
↓
promonocyte
↓
monocyte (bone marrow)
↓
monocyte (peripheral blood)
↓
macrophage (tissues)

Monocyte differentiation in the bone marrow proceeds rapidly (1.5–3 days). Here monoblasts and promonocytes undergo one division. Before entering the bloodstream (where they circulate for about one day), monocytes do not divide again. During differentiation in the bone marrow, granules are formed in their cytoplasm; these can be divided as in granulocytes, into at least two types. However, they are fewer and smaller than their neutrophil counterparts (azurophil and specific granules). Their enzyme content is similar.

The **blood monocytes** (*Fig. 13.5*) are young cells that already possess migratory, chemotactic and phagocytic activity, as well as receptors for IgG Fc-domains and C3 complement component. One of the most critical stages of their life is transport into tissues where they settle, further differentiate (at least one day) and transform into multifunctional tissue macrophages. According to VAN FURTH (1982) they can be divided into normal and inflammatory macrophages. **Normal macrophages** include macrophages in connective tissue (histiocytes), liver (Kupffer's cells), lung (alveolar macrophages), lymph nodes (free and fixed macrophages), spleen (free and fixed macrophages), bone marrow (fixed macrophages), serous fluids (pleural and peritoneal macrophages), bone (osteoclasts), central nervous system (CSF macrophages, brain macrophages), skin (histiocytes, Langerhans' cell) and other tissues.

The macrophage population in a particular tissue is maintained in a steady-state by three mechanisms: influx of monocytes from the circulating blood, local proliferation and biological turnover. Originally, it was thought that tissue macrophages were long-living cells. More recently, however, it has been shown that depending on the type of tissue, their viability ranges between 6 and 16 days (VAN FURTH, 1985).

To date the exact mechanism of monocyte migration into the tissues remains unknown. Similarly, little is known about the ultimate fate of macrophages in the tissues. The life-history of mononuclear phagocytes can be divided into three distinct stages: (1) Division of monoblasts and promonocytes and production of monocytes in the bone marrow. (2) Monocyte transport into the blood and migration into the tissues and body cavities. (3) Differentiation into macrophages that are functional and terminal cells of the entire cell line.

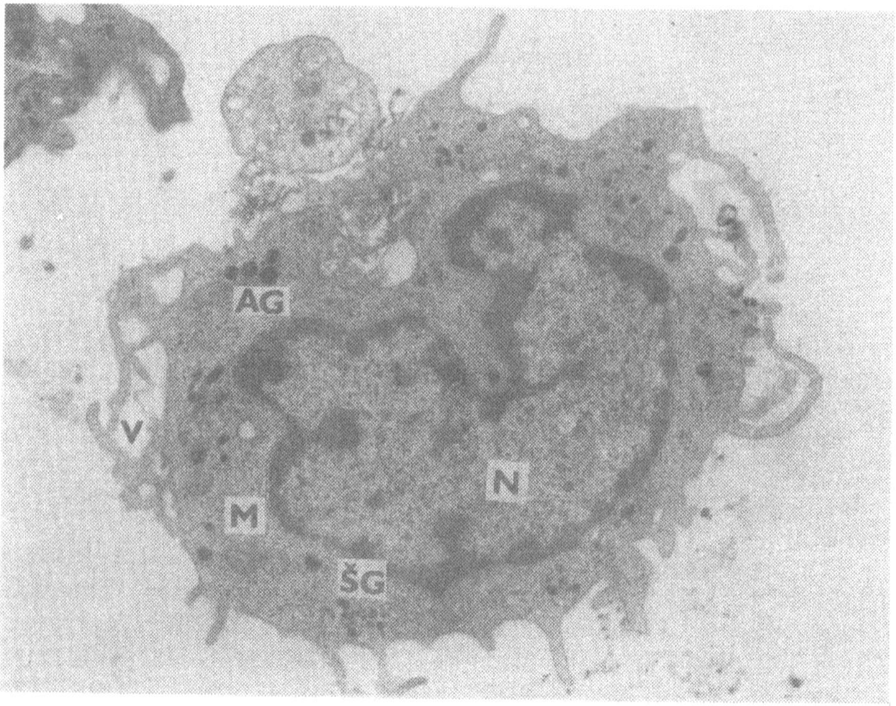

Fig. 13.5. Electron photomicrograph of a human monocyte, × 15000 (according to P. Mráz). M, mitochondria, other symbols as in Fig 13 3

During acute or chronic inflammation, the number of mononuclear phagocytes increases, both at the site of inflammation, and in the blood. The rate of this increase is dependent on the character of the stimulus and on the tissues where inflammation takes place. Circulating monocytes are mainly responsible for the increased numbers of macrophages in the inflammatory exudate. However, two other cell types also occur in the inflammatory exudate: epitheloid cells and multinuclear giant cells that arise from monocytes and are, therefore, classified as members of the MPS.

Macrophages are generally considered to be the basic effector cells of the host defence against intracellular bacteria, protozoa, viruses, multicellular parasites, tumour cells and clones of altered autologous cells. The **macrophage-mediated defence system** is not immunologically specific. Its effectiveness depends on the hereditary susceptibility of target cells to cytotoxic mechanisms of macrophages. The effector functions of MPS cells are modulated by functional changes of individual cells, by their predetermination to reach only a certain differentiation stage, and by local accumulation at sites of maximum need.

In addition, macrophages regulate various mechanisms of specific acquired immunity. They are responsible for antigen presentation to immunocompetent lymphocytes (UNANUE, 1989) and they may also activate *T*-lymphocytes, primarily through interleukin 1 (IL-1).

The term macrophage refers to a heterogeneous population of cells that differ in their origin, developmental stage (differentiation), local adaptation and thus also in their function and purpose. The nomenclature of individual macrophage types (particularly inflammatory) is rather confusing and terms such as stimulated, activated, induced, elicited *etc.* are often used interchangeably. Basically, two main macrophage groups can be distinguished: resident (normal) and inflammatory (exudate, activated, elicited) macrophages.

Resident macrophages, present at a given anatomical site, have not yet come into contact with any exogenous or endogenous inflammatory stimuli. These cells are also called normal macrophages. All other macrophages should be considered as inflammatory macrophages.

Inflammatory macrophages are present in various exudates. They may be characterized by various specific markers, *e.g.* peroxidase activity, and since they are derived exclusively from monocytes they share similar properties. The term *exudate macrophages* designates the developmental stage and not the functional state.

The term **activated macrophages** is reserved for macrophages possessing increased functional activity. This activation may be induced by various stimuli such as interaction with microbial products (*e.g.* LPS of Gram-negative bacteria), synthetic substances, immunoglobulins of various classes, and factors released from lymphocytes (lymphokines). IFN-γ is particularly important in immunologically-specific macrophage activation. Furthermore, CSF, IL-4, TGF-β and complement components (C5a), immune complexes, and prostaglandins can all activate macrophages and monocytes as well.

Activated macrophages possess increased antimicrobial and tumouricidal activity, as well as enhanced adherence activity, pinocytosis and phagocytosis. The higher tumouricidal activity of activated macrophages is based on the ability to recognize and destroy tumour cells both *in vitro* and *in vivo*. Enhanced adherence, phagocytosis and intracellular killing of engulfed bacteria is due to the modified membrane properties of activated macrophages. The increased tumouricidal and general cytotoxic activity results from stimulation of biochemical pathways leading to the elevated production of toxic oxygen forms, reactive nitrogen intermediates, cytolytic factors (tumour necrosis factor, neutral cytolytic proteinases) and release of lysosomal enzymes and various mediators. For intracellular killing, activation may be characterized as a process of stimulating those metabolic functions that are inevitable for intracellular killing, even when phagocytosis itself is not sufficient for such a stimulation. Both resident and exudate macrophages may be stimulated *in vitro* and *in vivo*. Activated macrophages play a vital role in host defence against intracellular parasitic bacteria, pathogenic protozoa, fungi and helminths as well as against secondary tumours (metastases).

Elicited macrophages are macrophages which have been attracted to a certain site, *e.g.* the peritoneal cavity, by irritating substances. This term refers only to a mononuclear phagocyte population accumulating at a certain site; it does not indicate their developmental or functional state. From the functional and developmental point of view, elicited macrophages are a heterogeneous cell population. Elicited macrophages may be designated in different ways, *e.g.* macrophages attracted into the peritoneal cavity by proteose peptone, thioglycolate, paraffin oil *etc.* The type of substance used for elicitation may influence specific functional and biochemical activities of these macrophages.

Besides the terms mentioned above, some authors also refer to "stimulated" or "induced" macrophages. However, the term *"stimulated macrophage"* is not unequivocal — stimulation may give rise to either activated or elicited macrophages. Similarly it should be understood that *"induced macrophages"*, have, compared to resident macrophages, altered (usually increased) functional or biochemical activity or both.

13.2 Morphological and biochemical events during phagocytosis

METCHNIKOFF recognized that phagocytosis was a complex event and that, for host defence, both a sufficient number and activity of mature phagocytes was important. He postulated that the phagocytic system comprised six components:

1. Generation of a sufficient number of phagocytic cells in the bone marrow.
2. Release of phagocytes from the bone marrow and passage into the blood circulation.
3. Spread and migration of blood phagocytes into the tissues.
4. The existence of a mechanism facilitating contact between phagocytic cells and invading microorganisms.
5. Phagocytosis itself.
6. Killing of ingested microorganisms.

Even today this classification remains basically correct. Progress in biochemistry and new approaches to ultrastructural studies on cells has made it possible further to differentiate the mechanisms of phagocytosis. It has been found that phagocytosis consists of several mutually biochemically and morphologically interconnected steps that also include contact between the phagocyte and microbe and killing of ingested microbial cells.

The individual steps of phagocytosis are schematically depicted in *Fig. 13.6*. Phagocytosis may be preceded by opsonization of the microorganism or another particle. Opsonization is required for immune phagocytosis. Chemotactic factors induce the phagocyte movement towards the object of phagocytosis. Following contact between the phagocyte and microorganism a series of morphological (adherence of microorganism to the phagocyte, ingestion, phagosome formation, lysosome degranulation and phagolysosome formation) and biochemical (respiratory burst and metabolic activation of phagocytes) events occur, resulting in killing and degradation of ingested microbes or other material (FERENČÍK and ŠTEFANOVIČ, 1977).

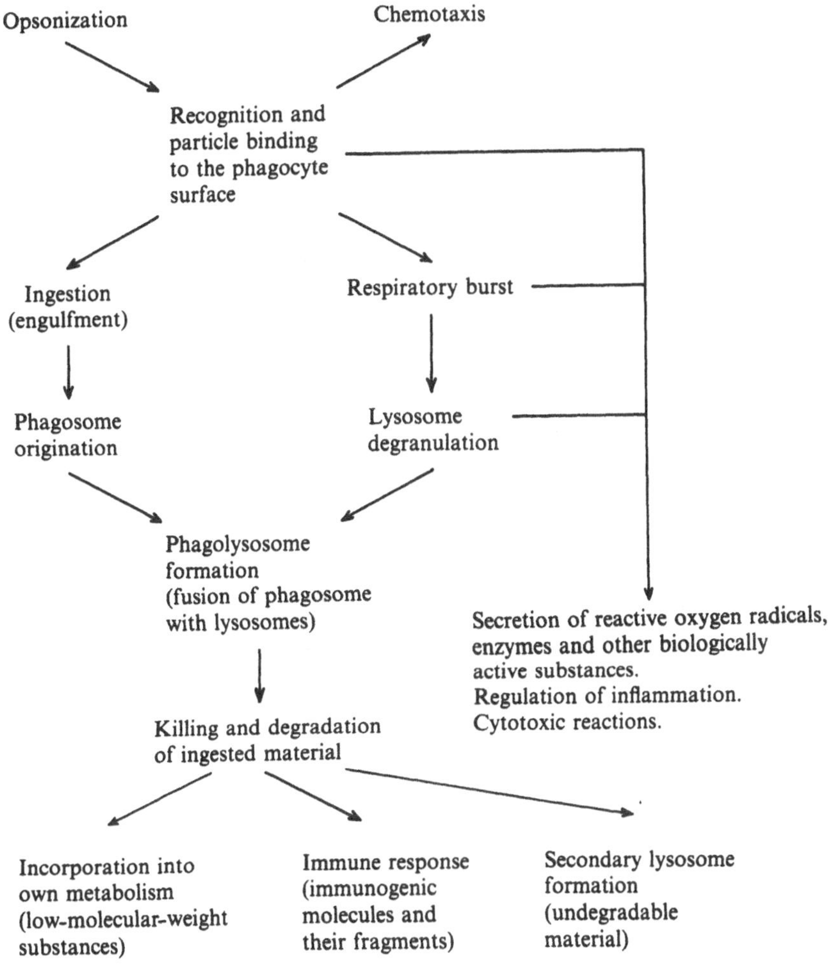

Fig. 13.6. Schematic representation of phagocytosis-associated events.

13.2.1 Opsonization and chemotaxis

Opsonization is defined as the process of adsorption of substances (**opsonins**) to the surface of bacteria or other particles. Binding of opsonins to the particle surface facilitates its adherence to the phagocyte cell membrane.

Originally it was thought that opsonins alter the particle surface charge, which in turn, facilitated adhesion. However, the current concept of the interaction between the opsonized particle and phagocyte is different and is based on the function of opsonin receptors in the phagocyte membrane. Specific binding between the particle and phagocyte which occurs during immune phagocytosis is mediated by these receptors. The serum-derived opsonins are either thermostable (IgG and IgM antibodies) or thermolabile (C3b, C4b). Specific IgG binds to antigen determinants on the microbe surface. Human phagocytes possess all types of FcR for the IgG (p. 91). surface. Human phagocytes possess all types of FcR for the IgG (p. 000). IgM-mediated opsonization proceeds according to the following scheme: first, IgM binds to the particle, forming an immune complex which in turn activates the classical complement pathway. The C3b and C4b fragments, generated during complement activation, bind to this immune complex and binding to the phagocyte is then mediated by CR1, CR3 or CR4 (p. 280). IgA and IgE antibodies may also possess opsonic activity, although this only works for some microorganisms or in special situations; it is, therefore, of rather less importance for antimicrobial host defence than IgG- or IgM-mediated opsonization. In addition to immunologically specific opsonins, non-specific opsonins (*e.g.* fibronectin, fibrin degradation products) also contribute to opsonization.

The term **chemotaxis** or **cytotaxis** refers to the movement of phagocytes (or cells in general), induced by a chemotactic stimulus. Besides chemotaxis (stimulated, directed migration), phagocytes also possess two other types of movement: **random migration** (undirected, spontaneous migration) and **chemokinesis** (stimulated, undirected migration). A chemotactic stimulus is provided by substances that can either attract or repulse the cells. Thus, cytotactic cell movement can be either positive or negative, *i.e.* the cells may move towards the source of chemotactic substances (towards an increasing concentration gradient) or in the opposite direction (BECKER *et al.*, 1982). Substances possessing cytotactic activity are called **cytotaxins** or **chemotactic factors** (chemotaxins).

In addition to phagocytes, other cell types, *e.g.* spermatozoa or microorganisms, also possess chemotactic activity. In prokaryotic cells, chemotaxis is used to search for food and migrate towards a nutrient source. Leukocyte chemotaxis (called also **leukotaxis**) is mainly responsible for their mobilization at the inflammatory site. Both exogenous and endogenous cytotaxins participate in phagocyte cytotaxis. *Exogenous chemotactic factors (CF)* include bacterial CF (oligopeptides of the FMLP type), synthetic *N*-formylmethionyl oligopeptides, lectins, denatured proteins, some lipids and lipopolysaccharides. *Endogenous chemotaxins* are produced by the host organism and are of humoral (*e.g.* complement fragments C5a, $C5a_{desArg}$ and Ba, fibrinopeptides, kallikrein and plasminogen activator) or cellular type (from neutrophils, eosinophils and monocytes — various arachidonic acid metabolites particularly LTB_4; from lymphocytes — chemotactic lymphokines for neutrophils, eosinophils, basophils, monocytes and macrophages; from alveolar macrophages — neutrophil CF; from activated macrophages — arachidonic acid metabolites; from mastocytes — neutrophil and eosinophil CF; from platelets — arachidonic acid metabolites). Endogenous chemotaxins can also be derived from connective tissue (peptides from collagen). Most chemotaxins are peptides or proteins and therefore limited proteolysis plays an important role in the regulation of their activity.

In addition to cytotaxins there are also cytotaxigens. **Cytotaxigens** differ from cytotaxins in influencing cell movement indirectly. The basis of cytotaxigen activity is the ability either to induce cytotaxin formation or unmask their activity. Cytotaxigens include immune complexes, bacteria, endotoxins from Gram-negative bacteria, heat-aggregated immunoglobulins, zymosan, lysosomal proteinases, trypsin, plasmin *etc.*

Interaction between the chemotaxin and its corresponding receptor triggers a series of coordinated biochemical events which include changes in the cell transmembrane potential, altered cyclic nucleotide levels and ion flow across the cytoplasmic membrane and increased glucose utilization and oxygen consumption. The composition of membrane phospholipids is altered and arachidonic acid, released by phospholipases, is metabolized into several biologically active intermediates and products, including prostaglandins and leukotrienes. Within a few minutes, the leukocyte changes from a round to a triangular shape (*Fig. 13.7*) that is oriented along the direction of chemotactic gradient. Reorganization of cytoskeletal contractile elements, particularly actin microfilaments and microtubular structures, contributes to this shape change. Activation of the contractile cell system not only results in migration but also in other forms of movement such as enhanced adherence, spreading, endocytosis and secretion of lysosomal enzymes.

The biological response of leukocytes to chemotactic factors may be generally divided into three areas: reactions related to cell motility, reactions involved in microbicidal and cytotoxic activities, and events resulting in stimulation of leukocyte secretory functions. All three types of response are regulated by different mechanisms and can be influenced by various pharmacological agents. Thus, low concentrations of chemotactic factor induce chemotaxis, whereas at least ten times higher concentrations are usually required to stimulate cytotoxic or secretory activity. For example, LTB_4, one of the most active chemotactic factors, only stimulates neutrophil aggregation at concentrations less than 10^{-10} mol/l, chemotaxis at concentrations less than 10^{-9} mol/l, and, at a concentration around 10^{-8} mol/l, stimulates both chemotaxis and secretion of lysosomal enzymes (GOETZL, 1983).

Such a differentiated response of a single receptor to different ligand concentrations opens the potential for effective regulation of various cell activities using a single substance. Another possibility for regulation is provided by receptor heterogeneity. All the receptors that are expressed at a given time on the leukocyte surface, do not possess the same affinity for the ligand. Thus, for example, the receptor for the chemotactic peptide FMLP, an analogue of chemotaxins produced by various bacteria, occurs in both low affinity and high affinity states. The dissociation constant K_D for the low affinity receptor is 1.1 and for the high affinity receptor 0.17 mol/l FMLP. The affinity state of the receptors is regulated by guanosine nucleotides. Stimulation of a high affinity receptor induces chemotaxis, whereas binding of FMLP to a low affinity receptor induces secretion of lysosomal enzymes and enhances superoxide production (SNYDERMAN and PIKE, 1984).

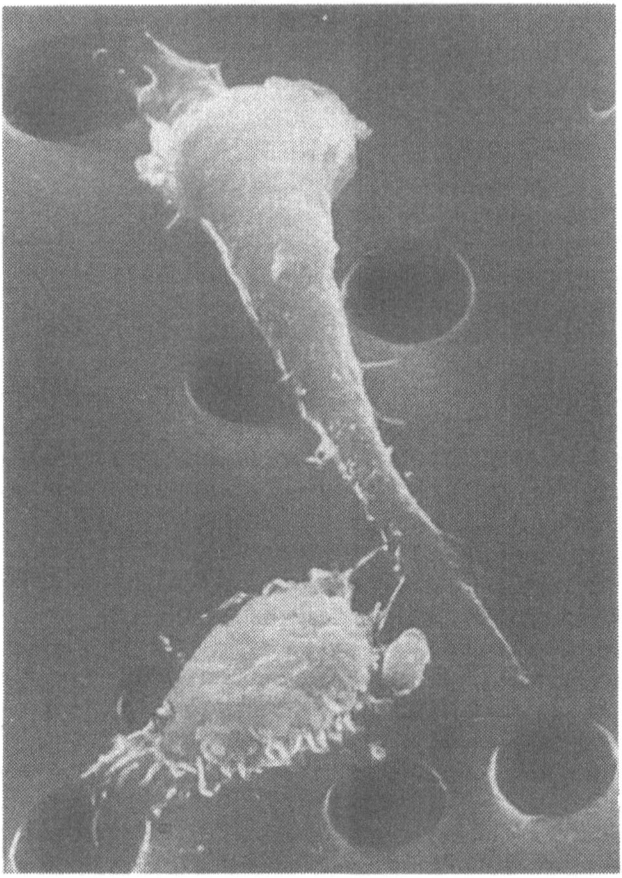

Fig. 13.7. Human blood monocytes migrating through pores of a polycarbonate filter towards the chemotactic signal (cells above the filter).
Scanning electron microscopy, ×4000 The upper monocyte has already passed through the pore, the lower monocyte is just emerging from the pore The upper cell has the typical shape of an elongated triangle

The role of cell migration is supposed to be the crucial process of inflammation. The locomotion of circulating cells (specifically leukocytes and monocytes) from the intravascular compartment to the extravascular space (*i.e.* tissue) involves a series of events, including ligand-induced cellular activation, deformity or polarization of the cells, adherence to the endothelium, and diapedesis or transport through the endothelium *via* a gradient-dependent migratory process. Together, these events result in a sequestration of the phagocytes in target tissue sites. The biological consequences of chemotaxis may be beneficial, as in traumatic injury and infection, or chronically detrimental, as in a variety of autoimmune and arthritic diseases (HUGLI, 1989).

13.2.2 The recognition systems of phagocytes

The recognition of material to be ingested is an initial and important step of phagocytosis; the recognition systems of phagocytic cells can be divided into non-specific and specific mechanisms. The basis of **non-specific recognition** are physical interactions of atoms present on the surfaces of phagocytes and particles. Electrostatic and hydrophobic forces in particular play an important role. Various particles may adhere to the phagocyte surface by non-specific interactions, *e.g.* inert particles (polystyrene latex beads) or even unopsonized bacteria. The surface structure of bacteria is important for the rate and stability of such bonds. For example, *E. coli* cells, possessing pili (fimbriae) on the cell surface, are phagocytosed faster than unopsonized bacteria lacking these organelles, although even specific binding, mediated by lectin receptors, cannot be excluded in this case.

Specific systems recognize the phagocytosable material using stereospecific complementarity. Four basic **stereospecific recognition systems** have been differentiated: the enzymatic, lectin–saccharide, morphogenetic and immune systems.

The **enzyme recognition systems** can be found in both the most primitive unicellular and the most developed multicellular organisms. This is illustrated by proteinases of microorganisms that degrade foreign and/or structurally altered own proteins faster than normal, fully functional own proteins. In phagocytes, the enzyme systems mainly participate in degradation of ingested material.

Receptors are the basic components of the other recognition systems. A receptor may be characterized as a cell surface structure (sometimes also present on other cell membrane structures or in the cytoplasm) that can receive an external signal (usually specific), process it, and using a second messenger transmit the signal inside the cell (or to some other site). A soluble *ligand* or an *effector* on the surface of another cell may act as an external signal. If the ligand, after being bound, induces an adequate receptor response, it is known as an *agonist*. If, in contrast, the ligand blocks the response (even in the presence of other ligands), it is referred to as an *antagonist*. From the chemical point of view, receptors are glycoproteins, glycolipids and proteins forming an integral component of the cell membrane. Their molecules can move laterally in the membrane which enables their aggregation into patches and caps of various size. The mobility and distribution of cell membrane receptors is controlled by microtubules and microfilaments that are basic components of the cell cytoskeleton apparatus.

The basis of a **morphogenetic system** is a positive recognition of "self" surface structures. The recognition unit consists of glycoprotein receptors. The role of recognized markers is played by oligosaccharides bound to a macromolecular carrier that may be part of the cell membrane, or free (soluble). Such a recognition system has been identified in primitive sea organisms such as sponges, Porifera or Coelenterata.

In fact, the recognition of foreign material in most invertebrates is a negative process, during which the defence mechanism is activated only if the individual does not recognize the self cellular marker on the particle surface. If such a marker is recognized, phagocytosis does not occur. It thus follows that phagocytosis may be considered a general cellular property. In primitive organisms, phagocytosis is initiated by contact of two cells. If autologous cells are involved, the process is terminated. Foreign cells, however, are engulfed and lysed. Presumably, both recognition units, *i.e.* the receptor and effector, are products of histocompatibility genes, and the vertebrate major histocompatibility complex, as well as the recognition systems of foreign surface markers, have developed from this primordial morphogenetic system (ROTHENBERG, 1978).

In vertebrates, the morphogenetic recognition system is supplemented by the **immune system** that can also actively recognize foreign surface markers. The recognition units of the immune system consist of antigen determinants that are recognized by *T*-cell membrane receptors for antigen or antibody combining sites. Unlike lymphocytes, professional phagocytes do not possess specific receptors for antigen. However, antigen can be recognized as an immune complex by receptors for immunoglobulin Fc-domains or for certain complement components or their fragments (immune phagocytosis). Recently, evidence has accumulated suggesting that the primitive immune system, capable of an adaptive response, is also present in some invertebrates.

The **lectin–saccharide recognition system** is mainly used by phagocytes during non-immune phagocytosis — *lectinophagocytosis* (WEIR, 1980; SHARON, 1984). In this system recognition is mediated by lectin — saccharide interaction. **Lectins** are proteins or glycoproteins that can specifically bind mono-, di-, tri- or tetrasaccharides. Such interaction may occur in three ways: (1) between saccharides on the phagocyte surface and lectins on the surface of their targets (other cells); (2) between lectins on the phagocyte surface and saccharides on other cells or particles; (3) lectins forming an extracellular bridge between saccharides on the surface of both cell types. Surface lectins are stereospecific receptors, each being specific for a certain saccharide only. Until now, eight types of lectins have been identified on the surface of eukaryotic cells: receptors binding D-glucose, D-galactose, D-mannose, L-fucose, *N*-acetylglucosamine, *N*-acetylgalactosamine, sialic acids (usually *N*-acetylneuraminic acid), or glucuronic acid. In macrophages the most important are receptors for mannose-terminated and galactose-terminated oligosaccharides.

For example, unopsonized *E. coli* or salmonella cells bind to macrophages through a mannose-specific receptor. However, this ability is only possessed by cells with surface fimbriae. *Fimbriae* are fine filamentous protrusions that are shorter and thinner than flagella. They are not an organ of motion but facilitate adherence of some Gram-negative bacteria to animal cells. The macrophage surface is not smooth but has an undulating shape. Lectin receptors are present in the membrane folds but become more accessible to the corresponding saccharide if the saccharide protrudes from the bacterial cell surface on the end of fimbria. Similarly, saccharide hidden in the phagocyte membrane folds is also more accessible to a fimbrial lectin receptor than a receptor actually on the bacterial cell membrane. This is why primarily fimbriated cells can adhere to macrophages through lectin receptors. Adherence of bacteria to the phagocyte surface through lectin receptors need not necessarily be followed by ingestion, which is another factor decreasing the effectiveness of non-immune phagocytosis. On the other hand, it is advantageous that the lectin–saccharide system acts immediately following invasion of an infectious agent into the body and that it is independent of specific immunity induction.

Receptors which participate in adsorption pinocytosis of proteins are also of lectin character. For this purpose macrophages have lectin receptors specific for D-mannose-6-phosphate (binding lysosomal enzymes and yeast mannans), D-glucose (binding zymosan and yeast glucans), D-galactose (binding several asialo-glycoproteins) and D-fucose (binding lactoferrin for example). Using specific receptors for D-galactose or mannose/*N*-acetyl-D-glucosamine, liver macrophages can bind and degrade effete erythrocytes.

Lectin–saccharide interactions also participate in macrophage and *T*-lymphocyte cytotoxic reactions called **LDCC (lectin–dependent cell cytotoxicity)**. LDCC is analogous to ADCC (p. 475) with the exception that the molecule which mediates the contact between the cytotoxic and target cell, is a lectin rather than an antibody.

Lectins are responsible for the following basic cell functions in animal cells. (1) Endocytosis and intracellular transport of glycoproteins. (2) Recognition and adherence of bacteria to phagocytes during lectinophagocytosis. (3) Binding of bacteria to epithelial cells (adherence of bacteria to host cells) as a prerequisite for their penetration into the organism. (4) Regulation of migration and mutual cell interactions (cooperation of *T*- and *B*-cells, macrophages and lymphocytes, effect of mitogenic lectins).

Stereospecific interaction between glycoproteins and proteins and, on the other hand, with oligosaccharides, participates in both the morphogenetic and lectin recognition mechanisms. However, there is a basic difference between the functions of these two systems. The morphogenetic system is highly specific for only "self" cell markers, whereas recognition by lectin systems is not — with respect to "self" — specific. For example, a saccharide that binds to a lectin receptor on the macrophage surface might be of either "self" or foreign origin, even from a highly unrelated organism.

13.2.3 Adhesion, ingestion and formation of the phagosome

The ability of the phagocyte to adhere is not only utilized in adherence of foreign particles to the phagocyte surface but also during adherence of PMN leukocytes to the surface endothelium of blood vessels and during migration out of the circulation towards the site of infection (*diapedesis*). Membrane adhesion glycoproteins play essential roles in the haemostatic and inflammatory responses to tissue injury. This group of **adhesion molecules** includes members of the *immunoglobulin superfamily* (Fc-receptors, p. 91), the *integrin family* (C3bi, fibronectin, laminin, vitronectin receptors, fibrinogen, platelet fibrin receptor *etc.*, RUOSLAHTI, 1991) and the recently described *selectin family* (MCEVER, 1991).

Three distinct receptors which mediate binding and phagocytosis of complement-coated particles by phagocytic leukocytes have been identified: CR1, a member of the regulators of complement activation, RCA (p. 266), CR3 and CR4, members of the integrin family. CR3 and CR4 together with **LFA-1** represent a group of adhesion molecules with well-characterized structure (GALLIN, 1985). Their molecules form heterodimers containing two chain types — α and β (*Table 13.2*). The β-chains (detectable by monoclonal

Table 13.2 Adhesive glycoproteins of the LFA-1 subfamily in the leukocyte and lymphocyte cytoplasmic membrane

Glycoprotein	*Chain molecular structure*	*Present in cells*	*Function*
LFA-1 (CD11a/18)	αL-β	T, B, NK, PMN, MO	Cytotoxicity Proliferation
Mo-1 (CD11b/18, Mac-1)	αM-β	PMN, MO, MA, NK	CR3
p150.85 (CD11c/18)	αX-β	PMN, MO, MA	CR4

LFA, leukocyte function-associated antigen, MO, monocytes; MA, macrophages. Relative molecular weight of chains αL (CD11a), 177000, αM (CD11b), 165000; αX (CD11c), 140000; β (CD18), 94000

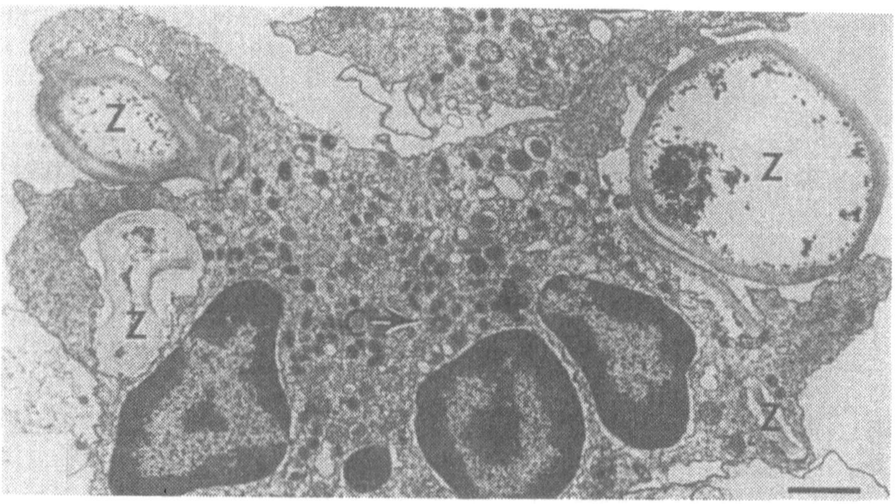

a

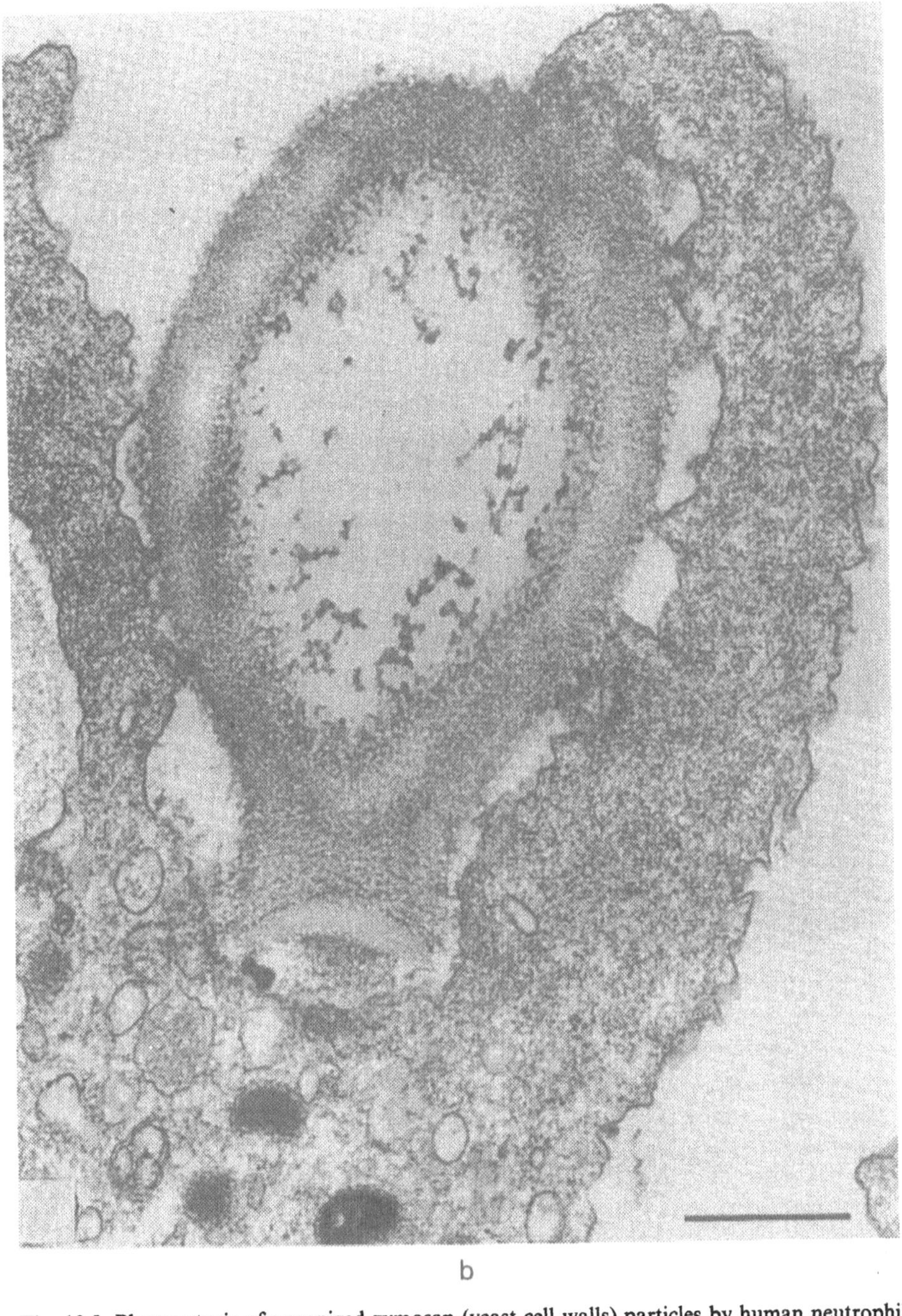

b

Fig. 13.8. Phagocytosis of opsonized zymosan (yeast cell walls) particles by human neutrophil viewed by transmission electron microscopy.

(a — precedent page) Two zymosan particles (Z, left and top right) start to be ingested A further two particles have already been engulfed in phagosomes. The bar represents 1 µm (× 11 000) (b) High power picture showing the phagocytic vacuole being formed around the zymosan particle shown in (a) (upper left) The bar represents 0 5 µm (× 48 500) antibody against the antigenic epitope CD18) are identical in all molecules of this subfamily, whereas the α-chains are distinct. So far, three types of α-chain have been discovered: αL (CD11a), αM (CD11b) and αX (CD11c). Their heterodimers form three glycoprotein types: LFA-1 (mainly on *T*-lymphocytes but also on *B*-cells, *NK*-cells, PMN leukocytes and monocytes), Mo-1 or Mac-1 (present mainly on PMN leukocytes, monocytes, macrophages and large granular lymphocytes) and Leu-M5 or p150,95 (on PMN leukocytes, monocytes and macrophages). It has been shown that Mo-1 is identical to CR3, and p150,95 is identical to CR4.

Similar heterodimeric structure with considerable sequential homology has been found in molecules of VLA proteins. There are at least six different **"very late antigens"** (VLA-1 to VLA-6) on leukocytes (especially on lymphocytes and monocytes). The LFA-1 family, VLA proteins (including fibronectin receptor) and the vitronectin receptor family are conserved subgroups in a superfamily of adhesive receptors, called **integrins**. An important feature of the integrin family is that many, if not all, are receptors for ligands containing the amino acid signal sequence Arg-Gly-Asp.

To date, three **selectins** have been identified: LAM-1 (mol. wt. 90–100 kD), ELAM-1 (115 kD), and GMP-140 (140 kD). All three molecules mediate interaction of leukocytes with blood vessel wall. LAM-1 is present on the surface of PMNs, monocytes and about 70% of lymphocytes; ELAM-1 on activated endothelial cells; GMP-140 on activated endothelial cells and platelets. The name "selectins" was proposed because they are hypothesized to mediate adhesion by lectin-type interaction with oligosaccharides on target cells: LAM-1 with activated endothelial cells, ELAM-1 and GMP-140 with PMNs and monocytes (McEver, 1991).

Receptors that recognize the Fc domains of IgG on professional phagocytes are involved not only in the ingestion of IgG-coated particles and immune complexes but also in the induction of superoxide anion generation, release of secretory products like cytokines and arachidonate metabolites, and cell-mediated cytotoxic processes. Human and mouse cells each have three distinct Fcγ receptors (p. 91). Ligation of the complement receptors of neutrophils and macrophages can induce phagocytosis but does not promote secretion of superoxide, H_2O_2 or arachidonate metabolites by these cells (Wright and Silverstein, 1983). Some microbial pathogens that are capable of growing inside mononuclear phagocytes, such as *Leishmania major*, *Mycobacterium tuberculosis*, *Mycobacterium leprae*, *Legionella pneumophila* and *Histoplasma capsulatum* bind to, and presumably are ingested *via*, CR3 on the surface of these cells. It is possible that inability of complement receptors to activate the respiratory burst allows these organisms to be engulfed without oxidant injury (Silverstein *et al.*, 1989).

Adherence of a microorganism or other particle to the surface of a phagocyte does not predestine these particles for ingestion (engulfment). When red blood cells or smooth encapsulated bacteria are attached to the surface of phagocytic cells with lectin, complement, or fibronectin, ingestion does not occur. Several lines of evidence indicate that particle engulfment requires engagement of specific receptors or generation of special signals. Ligation of specific plasma receptors, like those for Fc domains of IgG, causes a membrane response that is precisely localized in the segment of phagocyte plasma membrane in contact with the ligand-coated particle. Membrane pseudopods (lamellae) adhere closely to the surface of the particle as they advance around it, faithfully following every surface irregularity (*Figs. 13.8* and *13.9*). GRIFFIN *et al.* (1975, 1976) called the mechanism of particle enclosing the *"zipper mechanism"* because the movement of a phagocyte's plasma membrane over a ligand-coated particle is governed by the availability of unligated receptors on the surface of the phagocyte, and by the spatial distribution of ligands on the surface of the particle (*Fig. 13.10*).

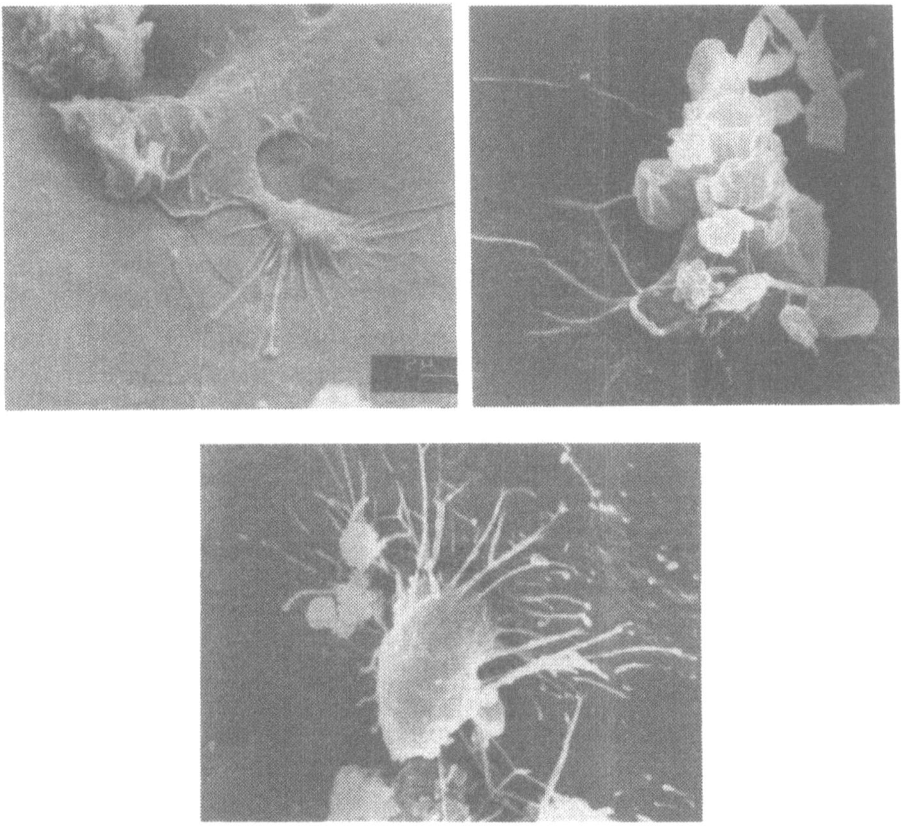

Fig. 13.9. A scanning electron photomicrograph of zymosan particle phagocytosis by a human neutrophil (according to S. Hoffstein).
(a) Non-phagocytosing neutrophil, (b) neutrophil 2 min after contact with zymosan particles, (c) neutrophil "loaded" with zymosan particles (30 min after contact).

In this way the **phagosome** (often also called a *phagocytic vacuole*) conforms precisely to the size and shape of the particle to be ingested. In order to kill the microbe and degrade the engulfed material, fusion of the phagosome with lysosomes must occur. Thus a **phagolysosome** is generated (*Fig. 13.11*). Phagosomes, lysosomes and phagolysosomes are components of the vacuolar-digestive system of phagocytes (*Fig. 13.12*).

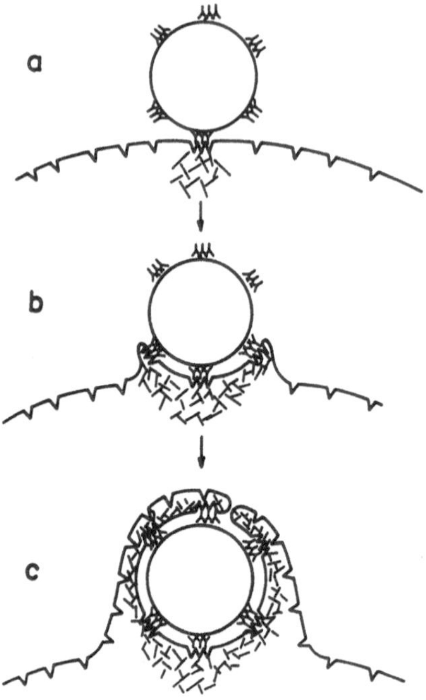

Fig. 13.10. Model depicting the zipper mechanism of Fcγ receptor-mediated phagocytosis (Silverstein *et al.*, 1989).

(a) A particle coated with aggregated immunoglobulin binds to and promotes clustering of Fcγ receptors on the surface of a macrophage Assembly and cross-linking of actin filaments occur in the cytoplasm subjacent to the ligated receptors, consequently, pseudopods begin to envelop the particle (b) Assembly of the underlying cytoskeleton drives additional plasma membrane into contact with the particle and promotes the interaction of additional receptors with ligands on the particle (c) After the particle is completely surrounded by macrophage membrane, a single membrane fusion event occurs to complete the phagosome

Phagosomes that are formed after engulfment of exogenous material are called *heterophagosomes* (or heteropinosomes). *Autophagosomes* (autopinosomes) are formed by internalization of endogenous material. Following fusion with lysosomes, damaged cell organelles or cytoplasmic fragments of autologous cells are degraded in autophagosomes. The mechanism of phagosome fusion with lysosomes is unknown. It appears to involve recognition of oligosaccharide components in the phagosome membrane by specific lectin receptors on lysosomes.

Lysosomes are intracellular organelles surrounded by a membrane, containing various hydrolytic enzymes usually requiring acid pH for optimal function. These hydrolases can degrade proteins, polynucleotides, polysaccharides and other biopolymers and complex substances into basic units.

They thus represent an effective digestive cell system. *In vitro* activity of hydrolases occurs only after disruption of the lysosomal membrane (latent activity).

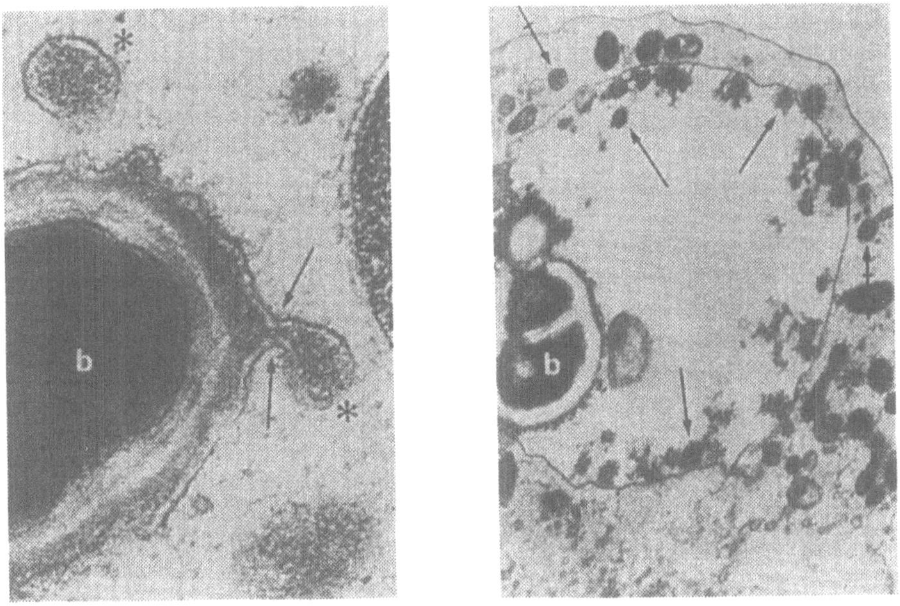

Fig. 13.11. (a) Fusion of bacterium (b)-containing phagosome with lysosome.
Fusion of phagosome and lysosome membrane is indicated by the arrows (× 54000).
(b) Phagolysosome containing ingested bacterium and content of lysosomes that had fused with the original phagosome.
Lysosomes, localized outside the phagolysosome, are sharply demarcated since they possess a membrane (two are marked by an arrow with a cross-line). The dense material, originating from lysosomes and present inside the phagolysosome, is not demarcated by a membrane which had fused with the phagolysosome membrane during fusion. Hydrolytic enzymes released from lysosomes coat the bacterium and start its digestion.

The existence of lysosomes was originally demonstrated using biochemical methods (De Duve *et al.*, 1955, Nobel prize in 1974) in the liver cells. Almost simultaneously these organelles were identified by electron microscopy (Novikoff *et al.*, 1956). The term lysosomes was coined by De Duve.

Lysosomes are present in all eukaryotic animal and plant cells. In mammals, lysosomes are present in all cells except mature erythrocytes. Despite the fact that a single cell may contain several hundreds of these organelles, it is convenient to consider lysosomes as a single membrane-bounded space which happens to be subdivided into numerous discrete compartments as part of its particular function. Lysosomes can fuse with each other to form larger lysosomes, or large lysosomes can divide into smaller ones. These processes of fusion and splitting proceed almost continually, thus enabling exchange of contents among individual lysosomes (De Duve, 1978).

The lysosomal cell apparatus consists of two lysosome types — primary and secondary. *Primary lysosomes* have not yet come into contact with the phagocytosable and pinocytosable material and function in the transport and storage of newly synthesized lysosomal enzymes. Their pH is neutral. *Secondary lysosomes* are actually phagolysosomes and pinolysosomes. They contain various amounts of ingested material at various stages of degradation. Their pH is acid. True cell digestion occurs in secondary lysosomes. From time to time, *tertiary lysosomes* are also mentioned. This term refers to post-lysosomes (residual bodies) that contain undegradable components of engulfed material.

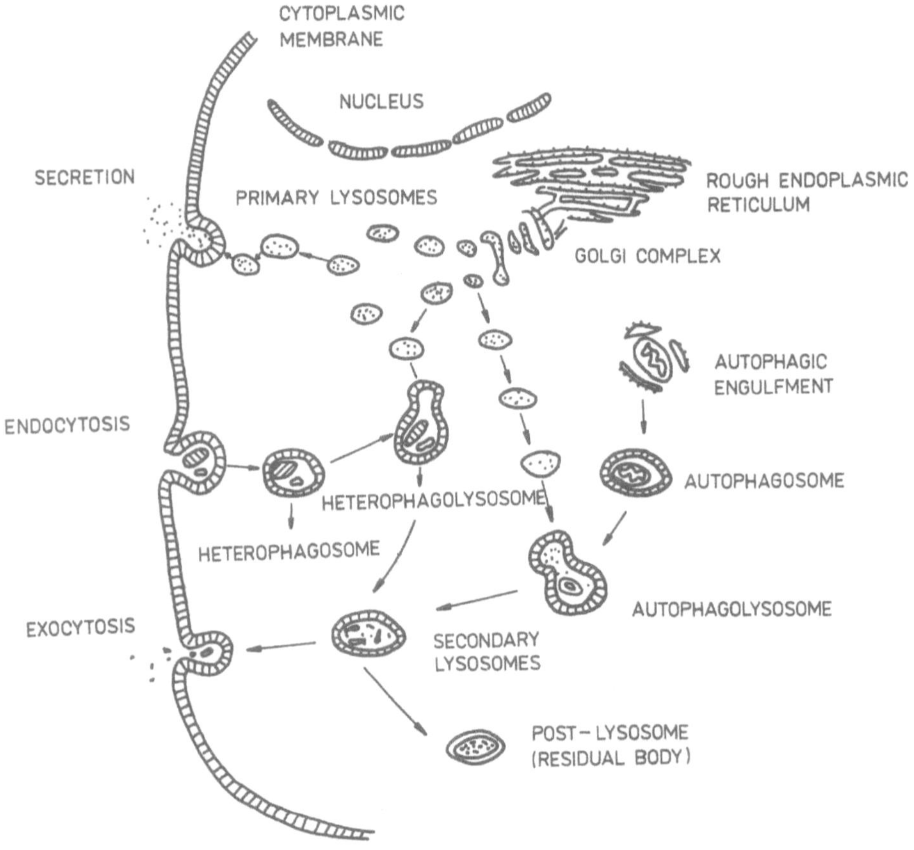

Fig. 13.12. The vacuolar-digestive apparatus of the phagocytic cells.

Substrates entering the lysosomes are broken down to small molecules which then diffuse or are transported across the lysosomal membrane into the cytoplasm. There they are metabolized or utilized for biosynthetic processes. This process requires perfect compatibility (both qualitative and quantitative) between substrates and enzymes. If there is some degree of incompatibility, lysosomal overloading occurs. This may occur mainly after ingestion of undegradable material or in defects of lysosomal enzyme activity. This may happen, for example, after ingestion of synthetic polymers against which lysosomal enzymes are lacking, or after engulfment of microorganisms that block fusion of the phagosome with lysosomes either because of their structure or secretory products.

Defects in lysosomal enzyme activity may be congenital or acquired. From the clinical point of view, the most important are diseases caused by genetically determined defects of one or more lysosomal enzymes — *lysosomal storage diseases*. These conditions are manifested by lysosomal overloading by some undegraded substrates, *e.g.* glucocerebroside (Gaucher's disease — a defect of glucocerebrosidase and β-D-glucuronidase), sphingomyelin (Nieman–Pick's disease — a defect of sphingomyelinase), heparan-*N*-sulphate (Sanfilippo A syndrome — a deficiency of heparan-*N*-sulphatase) and many others.

The time required for lysosome overloading to occur varies widely. It may take minutes up to years. It depends on the relative transport rate of substrates, their digestive products and enzymes, and on the degree of the enzyme defect. Lack of an enzyme always results in severe structural and functional disturbances of the lysosomal apparatus or even cell death.

Substances that are preferentially accumulated in lysosomes are termed *lysosomotropic substances*. These substances include some inorganic particles and organic low- and high-molecular-weight compounds. They may cause lysosome overloading or disturb their function in other ways. Conversely, some drugs may positively influence lysosome function.

A single neutrophil can engulf about 10–15 bacteria within a few minutes; the macrophage is quantitatively even more prodigious in its apetite. Within 15–30 min a macrophage can ingest enough particles to cause the internalization of up to 40% of its plasma membrane without evident change in its surface area. Additional plasma membrane components are derived from intracellular stores to maintain cell surface area. Neutrophils contain intracellular pools of membrane receptors, including CR1, CR3 and CR4. Macrophages contain sufficient internal stores of plasma membrane to increase their surface area by more than twofold in a few minutes.

13.2.4 The respiratory burst

The aerobic metabolism of resting (unstimulated) professional phagocytes is low. Although both polymorphonuclear leukocytes and mononuclear phagocytes contain mitochondria, both cell types derive the major portion of their metabolic energy from anaerobic glycolysis. However, following exposure to an activator, with or without involving specific membrane or cytoplasmic receptors, there is a dramatic increase in oxygen consumption, increased anaerobic glycolysis and superoxide anion radical formation. This event is known as the **respiratory burst**. Ligands which stimulate leukocytes through specific receptors may be soluble (chemotaxins, phorbol esters, lectins, soluble immune complexes) or insoluble (opsonized bacteria, yeasts, viruses, insoluble immune complexes, *Fig. 13.13*). In a normal healthy individual, the respiratory burst is always associated with phagocytosis, not only in neutrophils, but also in eosinophils, monocytes and macrophages. Enhanced leukocyte respiration during phagocytosis was described by BALDRIDGE and GERARD (1933). This phenomenon, however, was forgotten and subsequently rediscovered by SBARRA and KARNOVSKY (1959).

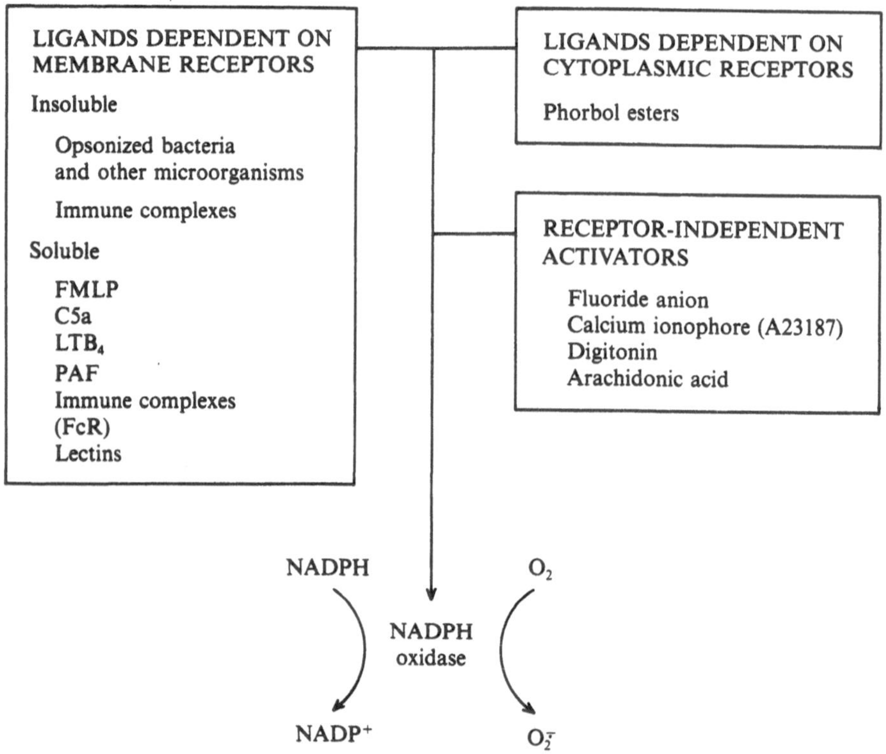

Fig. 13.13. Activators of the respiratory burst (adapted from Hurst, 1987).

The scheme of reactions which occur during the phagocyte respiratory burst is shown in *Fig. 13.14*. The important role is played by an enzyme system ("respiratory burst oxidase") that is present in resting phagocytes in an inactive form (dormant enzyme). After stimulation, this system becomes activated within 30–60 s, suggesting a multistep process. Cyanides do not inhibit the respiratory burst which makes it possible to differentiate this process from oxygen consumption in the mitochondrial respiratory chain.

The most widely held view is that the basic enzyme of respiratory burst is NADPH : O_2-oxidoreductase (commonly called *NADPH oxidase*). NADPH is generated in the hexose monophosphate shunt (HMPS), the activity of which increases about 10 times during the respiratory burst. A resting leukocyte only utilizes about 1% of glucose in the HMPS, whereas activated leukocytes use up to 10%. The NADPH level is regulated by the glutathione redox system and reduction of molecular oxygen. The basic function of

NADPH oxidase appears to be catalytic electron transport from NADPH to molecular oxygen which is thus reduced to the *superoxide anion radical* (often also known as the superoxide anion or simply superoxide). It results in a decreased NADPH pool ($NADP^+$ is generated), which in turn stimulates the HMPS to provide 12 moles of NADPH during oxidation of one mole of glucose.

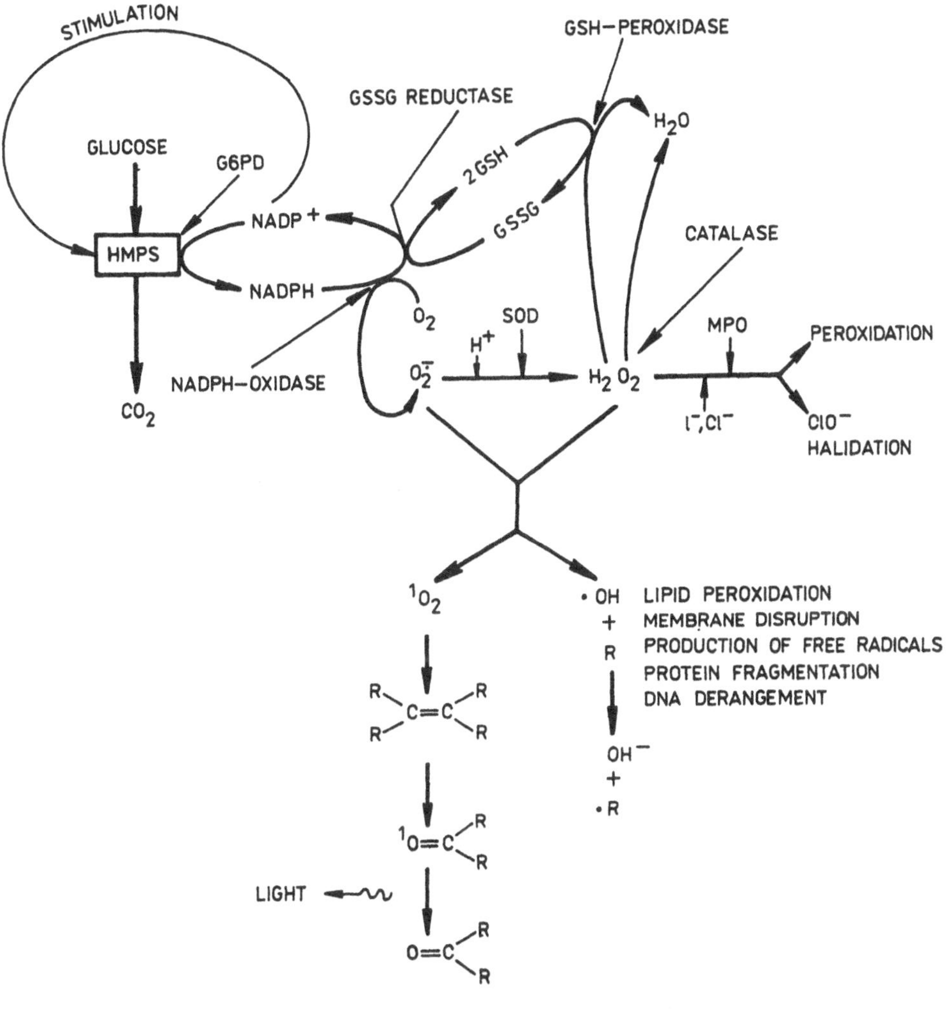

Fig. 13.14. Schematic representation of the phagocyte respiratory burst and formation of toxic forms of oxygen.

HMPS, hexose monophosphate shunt; G6PD, glucose-6-phosphate dehydrogenase; GSSG and GSH, reduced and oxidized form of glutathione respectively; NADPH and $NADP^+$, reduced and oxidized form of nicotinamide adenine dinucleotide phosphate respectively; SOD, superoxide dismutase; MPO, myeloperoxidase; $O_2^{\bar{\cdot}}$, superoxide anion radical; ·OH, hydroxyl radical; $^1O_2$, singlet oxygen.

Thus **NADPH oxidase** constitutes a transmembrane electron transport chain with cytosolic NADPH as the donor and molecular oxygen as the acceptor of a single electron (SEGAL, 1989; HURST and BARRETTE, 1989). The product, superoxide, is released at the outer surface of the plasma membrane, *i.e.* into the extracellular space or into phagocytic vacuoles. The active oxidase is associated with the plasma membrane, has much higher affinity for NADPH than for NADH, and contains a flavoprotein as the intermediate electron carrier and a *b*-type cytochrome (consisting of a 92 kD and a 22 kD subunit) with an unusually low negative redox potential of −245 mV, which is close to that of the O_2/O_2^- pair. In addition, this multicomponent oxidase includes at least two cytosolic factors, p47-*phox* and p67-*phox* (67-kD cytosolic *phagocyte oxidase* factor) and is schematically shown in *Fig. 13.15.*

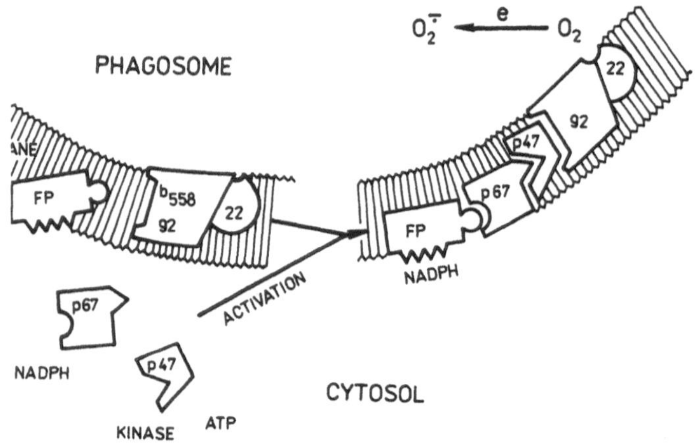

Fig. 13.15. A schematic representation of a model of the NADPH oxidase of professional phagocytes (adapted from Segal, 1989).
The oxidase represents an electron transport chain that in located between the substrate, NADPH, in the cytosol, and the lumen of the phagosome Electrons are pumped across this membrane to reduce oxygen to superoxide, thereby elevating the pH within the phagosome

It is highly likely that a *flavoprotein* is component linking the substrate, NADPH, and cytochrome b_{558} and that the cofactor is flavin adenine dinucleotide (FAD, ROSSI, 1986). Its molecular weight is 65–67 kD.

In humans, *cytochrome* b_{558} (originally named as cytochrome b_{-245}) was found in the professional phagocytic cells, neutrophils, macrophages, monocytes, and eosinophils, but not in a variety of other cell types (SEGAL *et al.*, 1981). In these cells it is located in the plasma membrane and in neutrophils an additional pool of the cytochrome is detected in the membrane of the specific granules. At −245 mV this cytochrome has the lowest midpoint potential of any mammalian cytochrome *b*, which provides it with the capability to directly reduce oxygen to superoxide (WOOD, 1987).

Some investigators have identified ubiquinone in extracts of neutrophils and claim it as a component of the electron transport chain, but its role in this oxidase is thermodynamically improbable because of its relatively high midpoint potential. BABIOR (1988) isolated several proteins visible in a "purified respiratory burst oxidase" from which three have been considered as likely subunits. Cytosolic factors p47-*phox* and p67-*phox* belong to them. *p47-phox* (47-kD cationic cytosolic phagocyte oxidase factor) is phosphorylated by a kinase using ATP as the substrate at multiple sites and moves into close association with the cytochrome b_{558} in the membrane during neutrophil activation. *p67-phox* protein also moves from cytosol into the membrane upon activation and is located very close to the NADPH-binding flavoprotein (SEGAL, 1989; NAUSEEF *et al.*, 1991).

Activation of NADPH oxidase. Under physiological conditions the activation process is initiated by the binding of an agonist to its receptor on the phagocyte surface. For a given agonist, the intensity of the response is related to the number of receptors occupied. Activation is likely to be a multi-step process including initiation, priming, assembly, and covalent modification. There are differences in signals generated by ligation of individual receptors and therefore several possible mechanisms for activation of the respiratory burst oxidase.

One mechanism postulates the involvement of protein kinase C and inositol phosphate metabolism. Furthermore, phosphorylation of component proteins of the NADPH oxidase have been shown to correlate with activation. In another mechanism activation is initiated by fusion of subcellular membrane components of the resting cells with plasma membrane to assemble an active NADPH oxidase (LAMBERTH, 1988; BAGGIOLINI and WYMAN, 1990).

The best characterized activators are *N*-formyl-methionyl peptides (*e.g.* FMLP), complement fragment C5a and two bioactive lipids, LTB_4 and PAF. Binding of these chemotaxins to corresponding membrane receptors stimulates *protein G* (GTP binding regulatory protein) and this in turn stimulates phosphoinositidase C (phosphatidylinositol-specific phospholipase C) that cleaves the membrane phosphatidylinositol-4,5-bisphosphate into *1,2-diacylglycerol* and *inositol-1,4,5-triphosphate* (IP_3). These two second messengers then trigger a series of events that dramatically increase superoxide formation by activated phagocytes (*Fig. 13.16*).

Guanine-nucleotide binding proteins constitute a family of receptor-associated proteins, each consisting of three polypeptide subunits, the α-, β-, and τ-subunits. The β- and τ-subunits are identical for all G-proteins, whereas GTP-binding α-subunits are distinct and determine the nature of the G-protein, whether stimulatory G_S or inhibitory G_I. When a receptor is complexed with ligand, the α-subunit of its associated G-protein complex will bind GTP and dissociate from the complex to perform its activity which may be activation or inactivation of adenylate cyclase or activation of phosphatidyl inositol-specific phospholipase C (SKLAR, 1986).

As soon as IP_3 has penetrated into cytosol, it induces a transient increase in Ca^{2+} ion concentration (Ca^{2+} is liberated from the endoplasmic reticulum). Ca^{2+} ions then trigger a number of processes, the most important being

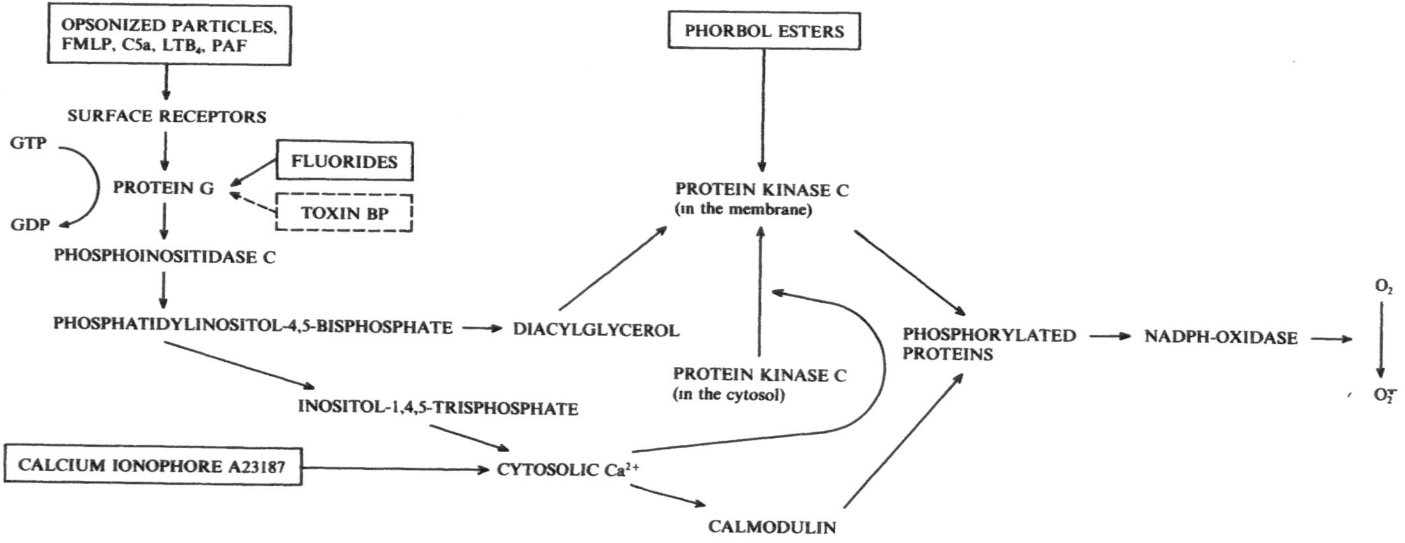

Fig. 13.16. Mechanisms of NADPH oxidase activation and transfer of intracellular signals for superoxide production in phagocytes.

Solid lines, stimulation, broken lines, inhibition Toxin BP, a toxin produced by *Bordetella pertussis* Solid boxes denote activators activation of phospholipase A_2 (which results in release of arachidonic acid from membrane phospholipids) and calmodulin-dependent protein phosphorylation. *Diacylglycerol* binds and activates Ca^{2+} and phospholipid-dependent *protein kinase C (PKC)*. In resting leukocytes, PKC is present mainly in the cytosol, whereas following activation, it binds firmly to the phagocyte cytoplasmic membrane and thus arrives in the proximity of its cofactor, phosphatidylserine. This transfer and PKC binding is catalysed by Ca^{2+} ions. Limited proteolysis of PKC by a Ca^{2+}-dependent proteinase (calpain) in stimulated neutrophils converts it to a form that is also fully active in the absence of Ca^{2+} and phospholipid. The Ca^{2+}/phospholipid-independent form of protein kinase C is released from the membrane into the cytosol where it can phosphorylate intracellular proteins.

It is known that **protein phosphorylation** plays an important role in the regulation of cell functions. Phosphorylation is performed by kinases and phosphatases that therefore act as target molecules, mediating the action of various growth factors, hormones and other external regulatory ligands which participate in the regulation of cell processes. One of the most important kinases, **protein kinase C**, plays a central role in regulating enzymatic events, dependent on cyclic nucleotides, Ca^{2+} and calmodulin. The second messenger, transferring a signal from the occupied receptor to the cell surface membrane, is diacylglycerol. The function of diacylglycerol in PKC activation may be replaced by *phorbol esters* (*e.g.* phorbol-12,13-myristate acetate, PMA) or other tumour promotors (substances strongly enhancing the effect of tumour-inducing carcinogens). Thus, PKC is considered to be a specific receptor for phorbol esters which are therefore very effective activators of NADPH oxidase. The calcium ionophore A23187 imitates the effect of IP_3 and increases the Ca^{2+} concentration in the cytosol. Fluorides are other respiratory burst activators that act directly on protein G and may thus induce hydrolysis of inositol lipids without any receptor involvement. Pertusis toxin (BP), which inactivates some G-proteins, inhibits chemotaxin-induced stimulation of NADPH oxidase (*Fig. 13.16*). Activated protein kinase C catalyses the cellular phosphorylation of many endogenous proteins including transferrin, IL-2 receptors and class I HLA-antigens with parallel NADPH oxidase activation.

A significant amount of evidence points to protein kinase C as the kinase responsible for the phosphorylation of NADPH oxidase components. This may explain their selective activation in the plasma membrane and phagosomal membrane, since PKC translocates only to the plasma membrane and thus cannot be expected to phosphorylate any proteins on the granule membrane before these have become incorporated into the plasma membrane.

In the resting neutrophils, 80% of the membrane-bound NADPH oxidase components are present in the membrane of the specific granules, 15% are present in the membrane of the secretory granules, and only 5% are in the plasma membrane. Activation of the neutrophil by secretagogues such as LTB_4 and FMLP causes exocytosis of secretory granules and incorporation of NADPH oxidase components into the plasma membrane, but mobilizes very few specific and azurophil granules. This probably explains the small and transient respiratory burst that these stimuli evoke. In contrast, more complete secretagogues like PMA, the ionophore A23187, or IgG-opsonized particles cause translocation of the NADPH oxidase components of the specific granules to the plasma membrane and both are capable of triggering a huge and long-lasting respiratory burst (BORREGAARD, 1988).

Besides PKC, **tyrosine protein kinase (TPK)** may also be involved in the activation of NADPH oxidase because treatment of human neutrophils with the chemotactic peptide FMLP or PMA induces a time- and concentration-dependent increase in the TPK activity found in the cell cytosol and cell particulate fraction (BERKOW and DODSON, 1991).

It is often assumed that each receptor transmits only one type of signal to the cytoplasm but this assumption is manifestly incorrect (SILVERSTEIN *et al.*, 1989). At a minimum, Fc receptors generate two different types of transmembrane signals. One of these mediates an increase in cytosolic free calcium and PKC activity, the second mediates phagocytosis. Not all phagocytosis-promoting receptors generate two or more signals during particle engulfment. The complement receptors appear to generate only one, the signal to ingest. There is no increase in cytosolic Ca^{2+} and ligation of complement receptors on neutrophils and macrophages does not activate superoxide production, prostaglandin formation, nor PKC activation associated with Fc-receptor phagocytosis (WRIGHT and SILVERSTEIN, 1983; BROZNA *et al.*, 1988). It follows that surface receptors other than the CR1 or CR3 themselves must be also ligated to enable the complement receptors to mediate full phagocytosis. This way of complement receptor regulation suggests that phagocytic cells, like other cells of the immune system (*T*- and *B*- cells), not only require the appropriate stimuli but must receive them in a specific order and in an appropriate context for them to be effective. For example, in neutrophils, stimulation of formylated chemotactic peptide receptors must precede ligation of fibronectin receptors for the latter to activate complement receptor-mediated phagocytosis (POMMIER *et al.*, 1984).

In the presence of protons, superoxide generated during the respiratory burst is reduced to hydrogen peroxide either spontaneously, or enzymatically by superoxide dismutase (EC 1.15.1.1, SOD). The H_2O_2 generated may become a component of the most effective microbicidal and cytotoxic system of leukocytes — the **myeloperoxidase system**. At the same time, however, it may also be toxic for the leukocyte, which therefore contains several protective mechanisms against H_2O_2. The most important role is played by *catalase* which is mainly present in peroxisomes. After fusion with the phagocytic vacuole (similarly as lysosomes), catalase can split H_2O_2 to O_2 and H_2O. Hydrogen peroxide that leaks into the cytosol is primarily reduced by the glutathione redox system which comprises glutathione reductase and glutathione peroxidase. In the leukocyte the H_2O_2 detoxifying effect is also possessed by *myeloperoxidase* (EC 3.11.1.7; MPO) which catalyses peroxidation and halidation reactions with molecules of engulfed microorganisms and thereby decreases free H_2O_2 concentration within the phagolysosome.

The **superoxide anion** ($O_2^{\cdot-}$) is a central metabolite which, as well as hydrogen peroxide, gives rise to further cytotoxic forms of oxygen such as the hydroxyl radical ($^{\cdot}OH$) and singlet oxygen ($^1O_2$).

As the superoxide anion is released from leukocytes 30–45 s after contact with the appropriate ligand (SMOLEN *et al.*, 1980), the antimicrobial mediators generated may also participate in killing of adherent bacteria or other cells, *i.e.* before enclosing the engulfed bacteria with the phagosome. Formation of $O_2^{\cdot-}$ and H_2O_2 continues even after closing the phagocytic vacuole that is formed from the cytoplasmic membrane; its outer surface forms the surface of the inner phagosome. Under appropriate experimental conditions stimulated neutrophils have the capacity to produce superoxide for several hours (BLACK *et al.*, 1991). During this reaction some generated $O_2^{\cdot-}$ may escape into the extracellular space before phagosome closure. A small amount of superoxide anion may pass into the extracellular space by diffusion across cell membranes. The activity of SOD, catalase and peroxidase in extracellular fluids is relatively low and it is therefore thought that the main scavenger of superoxide anion is *ceruloplasmin.*

After contact with the ligand (including, *e.g.* anti-IgE), superoxide formation is not only induced in professional phagocytes but also in basophils and mast cells, which suggests that the reactive oxygen intermediates might also participate in the immediate-type hypersensitivity reaction (HENDERSON and KALINER, 1978).

Superoxide dismutases (SOD) are the key enzymes that protect the cell against the oxidative stress provoked by superoxide and other reactive oxygen species. These are metalloenzymes which belong to the oxidoreductase group that catalyse dismutation of the superoxide anion:

$$2O_2^{\cdot-} + 2H^+ \rightarrow H_2O_2 + O_2$$

The result of their activity is not a synthetic degradation, or a transfer reaction as in other enzymes, but a detoxification reaction that removes toxic products generated in the organism. In 1939, MANN and KEILLIN discovered a specific copper-containing protein in mammalian liver and called it *erythrocuprein.* Its function, however, remained unknown. It was not until 1969 that MCCORD and FRIDOVICH showed that erythrocuprein was a superoxide-splitting enzyme and called it superoxide dismutase. Its molecule contains copper and zinc (Cu, Zn-SOD). Superoxide dismutases are composed of two or more subunits and according to the metal content, are divided into three classes:

1. SOD containing one atom of iron (Fe^{3+}) in each unit (*Fe-SOD*) occurs as a dimer particularly in prokaryonts and in some eukaryotic algae. The molecular weight of subunits is 20 000.
2. The primary structure of Fe-SOD possesses considerable homology with manganese-containing (probably trivalent) SOD (Mn-SOD). *Mn-SOD*

may occur in a dimeric or tetrameric forms. The dimer contains a single Mn atom while the tetramer contains two Mn atoms. Mn-SOD is present in cells of prokaryonts, mitochondria, chloroplasts and the cell nuclei of eukaryonts. In prokaryonts, Mn-SOD has two subunits, whereas mitochondrial Mn-SOD has four and has a molecular weight of about 80 000.

3. The third type is the *Cu,Zn-SOD* mentioned above, that is quite distinct from the two other forms. Cu,Zn-SOD forms a dimer with a molecular weight of 32 000. The enzyme is composed of two identical subunits joined by non-covalent bonds, and each contains one Cu^{2+} and one Zn^{2+} ion. It occurs particularly in the cytoplasm of eukaryotic cells. Its primary structure has only changed slightly during evolution. The amino acid sequence in Cu,Zn-SOD, isolated from human and horse spleen, for example, has approximately 80% homology, whereas homology between human and yeast Cu,Zn-SOD is about 50%. Human erythrocytes contain only Cu,Zn-SOD, whereas nucleated cells possess both Cu,Zn-SOD and Mn-SOD. The blood serum has a relatively weak superoxide dismutase activity. This not only includes Cu,Zn-SOD and Mn-SOD released from cells, but also ceruloplasmin, ferritin and transferrin, even though they have about 100 000 times less ability to degrade superoxide than Cu,Zn-SOD (Bergendi, 1988).

13.3 The antimicrobial and cytotoxic systems of phagocytes

The final task of phagocytosis is the killing and degradation of ingested microbes or other cells. This aim can only be accomplished if all stages of phagocytosis (opsonization, chemotaxis, recognition of the microorganism and its binding to the phagocyte cytoplasmic membrane, ingestion, respiratory burst, phagosome and phagolysosome formation, killing and degradation of engulfed microorganisms) are normal. Defects in any step, or in their mutual cooperation, result in delayed killing of the microbe or even its intracellular survival, or in overloading of secondary lysosomes by undegraded microbial components or other material. It is manifested by decreased defence and increased susceptibility to pathogenic and opportunistic microorganisms, or by other clinical symptoms.

Killing the ingested microorganism appears to be the key process in phagocytosis. This stage involves a complex of substances and factors, and requires the cooperation of several metabolic pathways and phagocytic cell structures. The antimicrobial and cytotoxic factors of professional phagocytes are schematically shown in *Table 13.3*.

Human neutrophils and most mammalian PMN leukocytes and macrophages contain three main antimicrobial and cytotoxic systems. The first is oxygen-dependent and may or may not cooperate with myeloperoxidase or other peroxidases. The second system is nitrogen-dependent and is represented by reactive nitrogen intermediates originated in L-arginine metabolism. The third system is oxygen- and nitrogen-independent and relies on the changes of pH in the phagolysosome and antimicrobial substances present in lysosomes or nuclei (KLEBANOFF, 1975; ELSBACH and WEISS, 1983; HIBBS *et al.*, 1988; MONCADA *et al.*, 1989; LEHRER and GANZ, 1990).

Table 13.3 Antimicrobial factors of neutrophil granulocytes and macrophages

I. Oxygen-dependent	
1. Cooperating with MPO	Metabolic readiness
MPO + H_2O_2 + Cl^-	of the phagocyte
MPO + H_2O_2 + I^-	
2. MPO-independent	
(a) Hydrogen peroxide (H_2O_2)	
(b) Superoxide anion ($O_2^{\bar{\cdot}}$)	
(c) Hydroxyl radical ($^{\bullet}OH$)	
(d) Singlet oxygen ($^1O_2$)	Inorganic oxidants
Peroxynitrite anion $ONOO^-$	with cytotoxic effects
II. Nitrogen-dependent	
(a) Nitric oxide ($NO^{\bullet}$)	
(b) Nitrogen dioxide (NO_2)	
(c) Nitrite (NO_2^-)	
(d) Nitrate (NO_3^-)	
III. Oxygen- and nitrogen-independent	
1. Alkaline and acid pH in the phagolysosome	
2. Granular antibiotic proteins (cationic proteins)	Lysosomal equipment
(a) Defensins	
(b) BPI	
(c) Neutral proteinases	
(d) Lysozyme	
(e) Lactoferrin	
3. Phospholipase A_2	
4. Histones	Nuclear equipment

BPI, bactericidal permeability-increasing protein

Thus, the cell mechanisms by which mammalian phagocytes inactivate and destroy invading microorganisms and other foreign cells consist of three basic components. The first component is the respiratory burst of effector cells coupled with the formation of oxygen-derived toxic products. The second component is mainly the nitric oxide produced by NO synthase and the third component is the lysis or degranulation of lysosomes (granules) with subsequent release of toxic substances onto the surface of target cells. These three components may mutually cooperate in various ways and they are responsible, not only for the intracellular killing of ingested microorganisms, but also for extracellular cytotoxic damage of target cells.

Microbicidal activity is mainly directed against mycoplasma, bacteria, fungi and protozoa. *Cytotoxic activity* is involved in removing virus-infected cells, resistance to tumours, elimination of infectious agents, rejection of allograft in the graft-versus-host reaction, and symptoms of delayed-type hypersensitivity. In addition to activated macrophages, mainly cytotoxic *T*-lymphocytes are involved in these reactions. Most effector cells possess both these activities, although they are not expressed equally in individual cell populations and one of them is usually prevalent.

Variability of microbicidal and cytotoxic mechanisms obviously reflects the diversity and variability of microorganisms that have come into contact with host phagocytes during evolution. All mechanisms are not equally effective against all pathogenic and opportunistic microorganisms. During the host–parasite interaction, two groups of microorganisms should be considered — *extracellular bacteria* and the *obligatory intracellular parasites*.

The first problem that faces the host infected with the first group of microorganisms, is rapidly to transfer the bacterial population from the extracellular environment into the intracellular space. Neutrophils and macrophages usually participate in this process. This process may be successful only if the rate of ingestion is faster than the rate of bacteria multiplication. The rate of ingestion is influenced by the frequency of contact between bacteria and phagocytes, the effectiveness of the association between phagocyte and bacterial cell, the presence of phagocytosis blocking factors on the surface of bacterial cells (capsules), the ability of bacteria to produce cyclic AMP or other substances that prevent the fusion of phagosomes with lysosomes, and/or other leukotoxic substances, the presence of opsonins and chemotaxins, and, finally, the activity of bactericidal and bacteriolytic systems in the phagocytes. With the second group of microorganisms, intracellular survival or even proliferation may be observed after ingestion. These microorganisms (intracellular parasites) are usually only killed by activated macrophages and the mechanism of killing is less clear than in the first group.

13.3.1 Oxygen-dependent microbicidal mechanisms

Oxygen-dependent microbicidal mechanisms are based on reactive oxygen forms (superoxide, hydroxyl radical, hydrogen peroxide and singlet oxygen) and the myeloperoxidase system (KLEBANOFF, 1975, 1982; BADWEY and KARNOVSKY, 1980; BABIOR, 1984; FERENČÍK and BERGENDI, 1984).

Understanding the origin and degree of reactivity of individual oxygen forms requires a detailed knowledge of electron localization and movement. The movement of an electron around the atom nucleus is geometrically described as *orbital*. Orbitals can be characterized by their four quantum numbers. *Figure 13.17* shows a simplified scheme describing the quantum numbers for atomic oxygen (O). The main quantum number n that characterizes the energy of an electron in the nucleus field may be 1 or 2 for oxygen. The azimuthal quantum number or angular momentum l determines the shape of the electron orbit and may reach the value of 0 up to n-1. For atomic oxygen the value of l may be 0 or 1, which is expressed by the notation s or p respectively. Thus when n equals 1, l equals 0, and the orbit is described as 1s; when n equals 2, l equals 0 and 1 and the orbitals are described as 2s and 2p. The magnetic quantum number m characterizes the behaviour of the electron in the magnetic field. It reflects the different vectorial orientation of the angular momentum, and this defines the orbit with reference to a given direction. For atomic oxygen, when $l = 1$, the m value is $+1$, 0, and -1, which is termed $2p_x$, $2p_y$ and $2p_z$. The spin quantum number S expresses the rotation movement of the electron around its own axis, both consistent $(+1/2)$ or inconsistent $(-1/2)$ with the direction of its orbit.

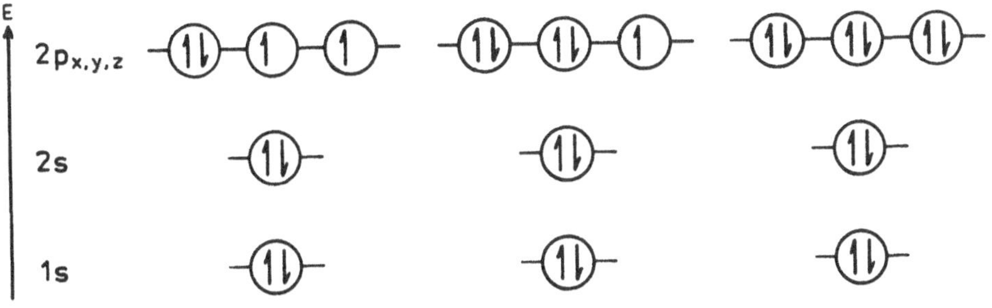

Fig. 13.17. Non-hybridized schemas of atomic oxygen orbits at various stages of reduction (Allen, 1979).
Orbits are illustrated as circles, electrons as arrows. Quantum numbers are on the left. The spin quantum number is characterized by the arrow direction. The large arrow (left) marks direction of electron energy increase on the orbit.

The filling of individual orbits by electrons is controlled by three rules: the Aufbau (building) principle, Pauli's exclusion principle and Hund's rule. According to the *Aufbau principle*, an electron will always occupy the lowest orbit available. *Pauli's exclusion principle* states that a maximum of two electrons can occupy any given orbit, and for two electrons to occupy the same orbit, their spins must be opposed, *i.e.* they must differ in spin number. The *Hund's rule* characterizes the behaviour of the electron during occupation of a degenerated orbit set. The term "degenerated" implies that each orbit of the set is of equal energy. For atomic oxygen, the degenerated set is represented by orbitals $2p_x$, $2p_y$ and $2p_z$. According to Hund's rule, one electron will occupy each orbit of a degenerate set before two electrons can occupy any given orbit of the set. Using this rule, it turns out that atomic oxygen contains one unpaired electron in orbitals $2p_y$ and $2p_z$ (*Fig. 13.17*). Similarly, molecular oxygen in the ground triplet state (designated $^3O_2$ or $^3\Sigma_g{}^-O_2$) possesses two unpaired electrons on two different orbitals π_g^x, $2p$ (*Fig. 13.18*).

During generation of molecular oxygen (O_2), each of two atoms supplies five orbits which results in the origin of 10 orbits. Only five of these orbits contribute to the interatomic bond. Orbits that do not participate in the bonding (antibonding) are marked in *Figure 13.18* by asterisks. *Figure 13.18* shows that both superoxide and peroxide dianion originate by gradual filling of antibonding π_g^x, $2p$ orbits by electrons.

The multiplicity of forms of a given atom or molecule is determined by its total spin quantum number J. It can be calculated from the equation $J = |2S| + 1$, where S is the sum of all electron spins. When $S = 0$, $J = 1$ and this state is called a singlet state. For practically all biogenic molecules, the singlet multiplicity appears to be a ground state. The exceptions are some free radicals and molecules containing transient metals, and even molecular oxygen. Free radicals are defined as atoms or molecules that possess at least one unpaired electron.

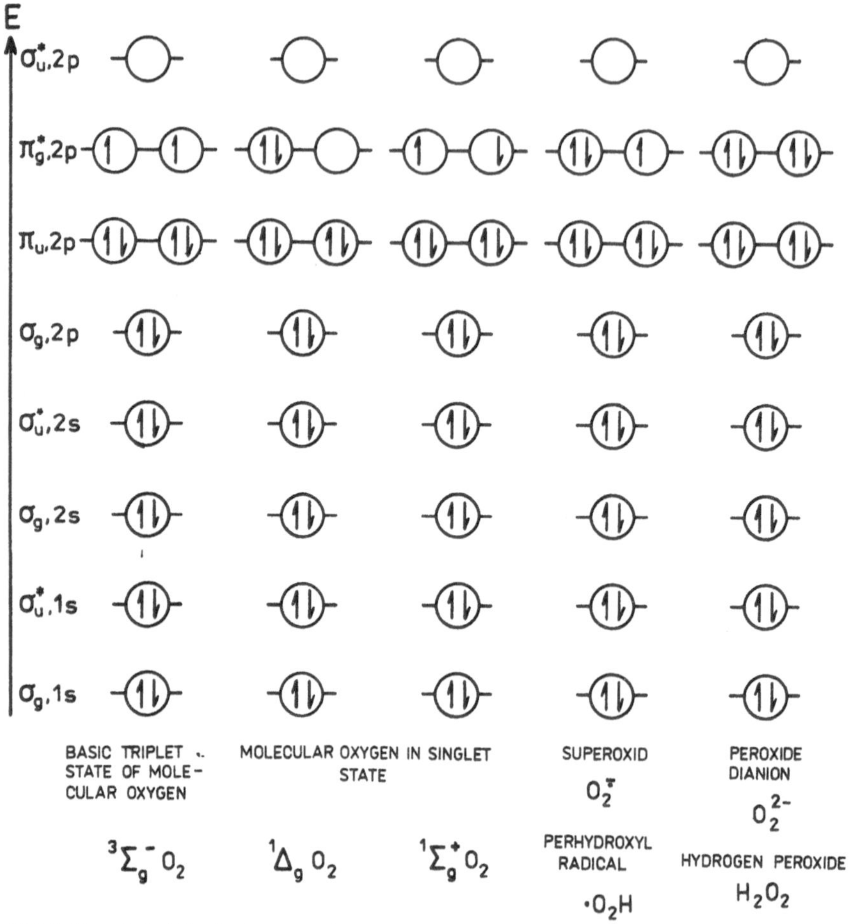

Fig. 13.18. Non-hybridized schemes of molecular oxygen orbits at various stages of electron excitation and reduction (adapted from Allen, 1979).
For explanations see *Fig. 13.17.*

Molecular oxygen in its ground electron state is actually a diradical as it possesses two unpaired electrons in degenerated orbits π_g^x, $2p$ (*Fig. 13.18*). In such an electron configuration, its molecule acquires the lowest energy. With respect to two unpaired electrons, the total spin number J is equal to 3, because in the above-mentioned equation, $|2S| = 2$. This state is therefore called a *triplet state,* designated by the index 3 ($^3O_2$). For simplicity, molecular oxygen is usually designated by the symbol O_2 only. The diradical character imparts a high reaction speed of molecular oxygen with other radicals.

Univalent reduction of molecular oxygen generates the *perhydroxyl radical* ($^{\bullet}O_2H$) or its conjugated base superoxide ($O_2^{\bullet-}$), which also has a radical character. The term **"superoxide"** does not mean that this species has an extraordinary high oxidizing activity. Depending on the particular reaction, $O_2^{\bullet-}$ may cause both oxidation (in this case a further electron is acquired and a peroxidase dianion is formed) and reduction (when one electron is lost and superoxide is oxidized to oxygen). Superoxide is in equilibrium with its protonated form, the perhydroxyl radical:

$$HO_2^{\bullet} \rightleftarrows O_2^{\bullet-} + H^+$$

The pK_a of the dissociation is 4.8 and the radical therefore exists almost entirely as $O_2^{\bullet-}$ at neutral or alkaline pH. When two molecules of superoxide interact, one is oxidized and the other reduced in a dismutation reaction with the formation of oxygen and H_2O_2

$$O_2^{\bullet-} + O_2^{\bullet-} + 2H^+ \rightarrow O_2 + H_2O_2$$

This reaction can occur spontaneously or can be catalysed by superoxide dismutase. Spontaneous dismutation is maximal at pH 4.8, where $O_2^{\bullet-}$ and $HO_2^{\bullet}$ are present in equal concentrations. The constant rate for spontaneous dismutation decreases when the pH is lowered and $HO_2^{\bullet}$ predominates, and it is particularly low at alkaline pH where $O_2^{\bullet-}$ predominates. Indeed, it is probable that spontaneous dismutation does not occur when $O_2^{\bullet-}$ is the only species. The rate constant for SOD-catalysed dismutation is hardly affected by pH over the range 5.0–10.0 and is therefore a particularly effective catalyst at neutral or alkaline pH where spontaneous dismutation is low.

The direct toxicity of $O_2^{\bullet-}$ is controversial. Superoxide reacts rather slowly with many important biological compounds (*e.g.* amino acids or carboxylic acids), leading to the suggestion that $O_2^{\bullet-}$ does not have the necessary reactivity to be directly cytotoxic. However, its chemical reactivity is considerably increased in a non-polar environment like that in the hydrophobic region of a biological membrane. Under these conditions $O_2^{\bullet-}$ is a powerful base with considerable nucleophilicity and reducing activity. Furthermore, the protonated form $HO_2^{\bullet}$ is a much stronger oxidant than $O_2^{\bullet-}$, raising the possibility that a local decrease in pH within a phagosome or at a membrane surface may cause a shift in the $HO_2^{\bullet} \rightleftarrows O_2^{\bullet-}$ equilibrium towards the more reactive protonated form with local damage of the target cell membrane.

Hydrogen peroxide is formed directly by dielectron reduction of molecular oxygen or by dismutation of superoxide anion. Some enzymes catalyse generation of H_2O_2 by both mechanisms (xanthine oxidase); others employ one mechanism only. Using leukocyte NADPH oxidase, H_2O_2 is formed only through the superoxide intermediate, although with glucose oxidase, H_2O_2 can be generated directly from oxygen. The degenerated orbitals π^x_g, $2p$ are occupied by electrons and therefore hydrogen peroxide does not have a radical character (*Fig. 13.18*).

Hydrogen peroxide is an oxidative reagent and at high concentration is germicidal. The germicidal H_2O_2 activity may be greatly enhanced in the presence of peroxidases and oxidizable cofactors. **Peroxidases** are generally haemoproteins that catalyse oxidation of many substances known as electron donors. This group also includes halides (iodide being the most effective on a molar basis) and anions such as thiocyanate. This system generates highly microbicidal and cytotoxic products. This effect was first discovered for milk peroxidase (*LPO — lactoperoxidase*) that may inhibit growth of various microorganisms in the presence of H_2O_2 and SCN^-. LPO is identical to the peroxidase of saliva. The most effective peroxidases, which participate in phagocyte microbicidal and cytotoxic reactions, are the neutrophil *myeloperoxidase (MPO)* and *eosinophil peroxidase (EPO)*. Effective peroxidases are also present in the granules of basophils and mast cells (KLEBANOFF, 1968, 1982).

The toxicity of H_2O_2 can also be considerably increased in the absence of peroxidases by several mechanisms, including reactions with ascorbic acid and a trace metal such as Cu^{2+}, synergism between H_2O_2 and proteinases, and reaction with ferrous iron (Fe^{2+}) leading to the formation of $^{\bullet}OH$ or other free radicals. Sublytic concentrations of H_2O_2 and neutral serine proteinases in neutrophil plasma membrane may act synergistically to lyse target cells. H_2O_2 also reacts synergistically with neutral proteinases released from activated macrophages to lyse tumour cells.

Free radicals, responsible for increase in H_2O_2 toxicity, namely $^{\bullet}OH$ and $I^{\bullet}$, may be generated during the following reactions:

$$\begin{array}{lcl} H_2O_2 + Fe^{2+} & \rightarrow & Fe^{3+} + OH^- + {}^{\bullet}OH \\ I^- \quad + Fe^{3+} & \rightarrow & Fe^{2+} + I^{\bullet} \\ \hline H_2O_2 + \ I^- & \rightarrow & OH^- + {}^{\bullet}OH \ + I^{\bullet} \end{array}$$

or:

$$\begin{array}{lcl} H_2O_2 + Fe^{2+} & \rightarrow & Fe^{3+} + OH^- + {}^{\bullet}OH \\ O_2^{\bar{\cdot}} \quad + Fe^{3+} & \rightarrow & Fe^{2+} + O_2 \\ \hline H_2O_2 + O_2^{\bar{\cdot}} & \rightarrow & O_2 \quad + OH^- + {}^{\bullet}OH \end{array}$$

The former reaction is called the Fenton's type of reaction and is a mechanism by which a significant amount of **hydroxyl radical ($^{\bullet}$OH)** could be generated in the cells. Actually, there is a disproportion between hydrogen peroxide and superoxide, described by HABER and WEISS (1934). The classical Haber–Weiss *reaction* ($H_2O_2 + O_2^{\bar{\cdot}} \rightarrow O_2 + OH^- + {}^{\bullet}OH$) proceeds under physiological conditions extremely slowly and cannot be measured. Only the presence of chelate-bound iron, acting as an oxidoreduction catalyst, permits a measurable increase in its rate. It is thought that the role of an oxidoreduction catalyst in neutrophils is played by lactoferrin.

Although $^{\bullet}OH$ is generally believed to be the product of the Haber–Weiss reaction, other radicals which are functionally similar to $^{\bullet}OH$ have been suggested. They include the "crypto-OH-radical" of unknown structure that mimics free $^{\bullet}OH$ but is more discriminating in its reactions, the ferryl radical ($FeOH^{3+}$ or FeO^{2+}), or the perferryl radical (FeO^{+} or $Fe^{3+}—O_2^{\bar{\cdot}}$). The *ferryl radicals* have an appreciably longer survival than $^{\bullet}OH$.

Fe^{3+} formed by Fenton's reaction is reduced by $O_2^{\bar{\cdot}}$ in the iron-catalysed Haber–Weiss reaction. However, under certain conditions, thiols such as glutathione or cysteine, the reduced pyridine nucleotides NADH and NADPH and ascorbic acid can replace superoxide as the reductant required for the formation of hydroxyl radical.

Iron has been most frequently implicated as the trace metal catalyst of the Haber–Weiss reaction. However, other metals may also be involved, particularly Co^{2+} or Cu^{2+}. Besides iron-saturated lactoferrin, transferrin may also catalyse $^{\bullet}OH$ formation when fully iron-loaded. When the Fe^{2+} concentration is high, recycling the iron by reducing the amount of Fe^{3+} formed is necessary. It has been suggested that haemoglobin (Hb)-Fe^{2+} can react with H_2O_2 to form $^{\bullet}OH$

$$\text{Hb-}Fe^{2+} + H_2O_2 \rightarrow \text{Hb-}Fe^{3+} + OH^- + {}^{\bullet}OH$$

The H_2O_2 needed for this reaction can be generated by interaction of ascorbic acid with oxyhaemoglobin. This mechanism may serve to generate $^{\bullet}OH$ in erythrocytes, but it cannot do so in leukocytes. An additional source of free or chelated iron is the microorganism. Bacteria grown in a high-iron medium increase their iron content and become more susceptible to destruction by H_2O_2.

The hydroxyl radical is an extremely powerful oxidant and, as such, is not discriminating in its action, reacting with the first molecule it meets. It can therefore be readily scavenged by compounds in the medium or by non-essential components of the target. However, there is no doubt that $^{\bullet}OH$ production by intact neutrophils contributes to their microbicidal and cytotoxic activities.

With respect to electron configuration, the hydroxyl radical may be generated either by univalent reduction of an oxygen atom, or by univalent oxidation of water molecules (*Fig. 13.17*).

Two different singlet states are generated during electronic excitation of molecular oxygen, simply called **singlet oxygen** ($^1O_2$). They have different electron arrangement on degenerated orbits $\pi_g^x,2p$ compared to the basic triplet state (*Fig. 13.18*). In addition, contrary to molecular oxygen, their spins are paired which results in the loss of the radical character of the singlet oxygen. Oxygen in the first (the lowest) electronically excited state has the symbol ${}^1\Delta_g O_2$ (*delta singlet oxygen*) and requires for its generation 94 kJ/mol more energy than oxygen in the ground triplet state. This energy is required to overcome limitation according to Hund's rule. The second electron-excited state is *sigma singlet oxygen* (${}^1\Sigma_g{}^+O_2$). Its energy is 157 kJ/mol higher than the energy of ${}^3\Sigma_g{}^-O_2$.

Both forms of singlet oxygen have relatively short survival after being generated. The survival of ${}^1\Sigma_g{}^+O_2$ in aqueous solution is about 10^{-11} s, whereas that of ${}^1\Delta_g O_2$ is longer and highly dependent on the environment (in water 2–4 μs). It is therefore probable that only ${}^1\Delta_g O_2$ is responsible for reactions in biological systems.

Electronically excited oxygen states are generated by transferring the electron-exciting energy from an appropriate sensitizer (polycondensed aromatic carbohydrates, some dyes, vitamins *etc.*) and during numerous chemical reactions.

Presumably, singlet oxygen may be generated during the modified Haber–Weiss reaction:

$$O_2^{\cdot -} + H_2O_2 \rightarrow {}^1O_2 + {}^{\cdot}OH + OH^-$$

A well-established mechanism for the formation of ${}^1O_2$ is the interaction of hypochlorite and H_2O_2:

$$OCl^- + H_2O_2 \rightarrow {}^1O_2 + Cl^- + H_2O$$

This is particularly important in leukocytes where hypochlorous acid is generated by the activity of the myeloperoxidase system, and the formation of ${}^1O_2$ by the interaction of HOCl and H_2O_2 or $O_2^{\cdot -}$ also has been proposed. This reaction emits a weak red chemiluminescence, and spectroscopic studies have established that the metabolizable product formed is ${}^1\Delta_g O_2$. Singlet oxygen may be generated during interaction of superoxide with diacyl peroxides or with hydroxyl radicals:

$$2\,O_2^{\cdot -} + R{-}\underset{\substack{\| \\ O}}{C}{-}O{-}O{-}\underset{\substack{\| \\ O}}{C}{-}R \longrightarrow 2\,{}^1O_2 + 2\,RCOO^-$$

$$O_2^{\cdot -} + {}^{\cdot}OH \longrightarrow {}^1O_2 + OH^-$$

The latter reaction, however, is improbable in biological systems because ${}^{\cdot}OH$ reacts immediately after generation with any adjacent molecule and therefore is not available for reaction with $O_2^{\cdot -}$. Singlet oxygen is also formed during PGG_2 reduction catalysed by prostaglandin hydroperoxidase:

$$PGG_2 + PGG_2 \longrightarrow 2\,PGH_2 + {}^1O_2$$

13.3.2 The myeloperoxidase system

Neutrophil myeloperoxidase (MPO) together with a halide or pseudohalide cofactor (I^-, Cl^-, Br^-, SCN^-) constitutes one of the most effective microbicidal and cytotoxic systems of mammalian leukocytes (including human). The MPO–H_2O_2–I^- or Cl^- complex originates from phagosome fusion with azurophil granules; only then are all components of this complex complete in the same compartment. The myeloperoxidase system may inactivate viruses and bacterial toxins, and kill bacteria, fungi, mycoplasma, chlamydia and multicellular parasites. The MPO–system is also toxic for some mamalian cells such as spermatozoa, erythrocytes, leukocytes, lymphocytes, platelets and tumour cells. In addition to these effector functions, it possesses important regulatory functions and it is involved in some pathological activities as well (*Table 13.4*). It may inactivate certain soluble mediators like chemotactic factors and α-1-proteinase inhibitor, decrease the number of FcR on the neutrophil surface, stimulate degranulation of platelets and mast cells, inactivate leukotrienes in neutrophils and activate latent collagenases and gelatinases. Many of these activities are involved in the regulation of inflammation.

Table 13.4 Function of the myeloperoxidase system (MPO + H_2O_2 + Cl^-)

Function	*Effect*
Effector	Kills mycoplasma, bacteria, fungi, protozoa, helminths.
	Inactivates viruses.
	Lyses tumour cells, spermatozoa, erythrocytes, leukocytes, lymphocytes and platelets.
	Detoxifies bacterial toxins (diphtheria toxin, pneumolysin, clostridial toxins).
Regulatory	Inactivates chemotactic factors (C5a, FMLP) resulting in decreased inflammation.
	Decreases the number of FcR and CR on neutrophil surfaces.
	Stimulates degranulation of platelets and mast cells.
	Inactivates leukotrienes generated by neutrophils.
	Binds to DNA resulting in protection of DNA from breakage.
Pathological	Activates latent collagenases and gelatinases resulting in increased inflammation.
	Inactivates α-1-proteinase inhibitor resulting in increased inflammation.
	May cause lung emphysema and glomerulonephritis.

Myeloperoxidase (MPO, EC 3.11.1.7) belongs to the class of oxyreductases and it is an enzyme that can bind two substrates — probably at various sites of its molecule. The first substrate is H_2O_2 and the second is halide. The enzyme–substrate complex MPO–H_2O_2 oxidizes halides into various toxic substances that may produce damage of the target cell by various halogenation and oxidation reactions. The exact cytotoxic mechanism remains unknown, although the MPO–H_2O_2–halide complex mainly produce hypohalous acids (HOCl, HOI, HOBr) which are able to oxidize haem proteins and

—SH groups in vitally important proteins of target cells and thus to change their molecular conformations. The other toxic products formed are halogens (Cl_2, Br_2, I_2), long-lived oxidants such as chloramines and aldehydes generated from amino acids, hydroxyl radicals and singlet oxygen formed from superoxide and H_2O_2. Halogenation reactions usually cause inactivation of many important compounds, *e.g.* chlorinated NADPH or NADH are no longer catalytically active. Chlorination of tryptophan residues of a protein by the MPO system results in chemiluminescence. Similarly, the MPO–H_2O_2–I^- complex can iodinate vitally important biopolymers in target cells (KLEBANOFF, 1967, 1968, 1982; ZGLICZYNSKI *et al.*, 1977).

The hydrogen peroxide which participates in these reactions arises either in phagocytes during the respiratory burst, or during microbial metabolism. Certain ingested bacteria (*e.g.* pneumococci, streptococci and lactobacilli) do not possess haem groups and therefore utilize flavoproteins that generally reduce oxygen to H_2O_2 for terminal oxidation. These microorganisms usually lack catalase and therefore H_2O_2 accumulates in their surrounding environment. Hydrogen peroxide of microbial origin is extremely important when the phagocyte metabolic pathway for producing H_2O_2 is defective, *e.g.* in *chronic granulomatous disease* (HOLMES *et al.*, 1967). The neutrophils and monocytes from patients suffering from this disease do not induce a respiratory burst after contact with microorganisms and therefore H_2O_2 is not formed, even though ingestion proceeds normally. If, however, the phagocytosed microorganisms form H_2O_2, it can be used with existing MPO and Cl^- or I^- for killing ("metabolic suicide").

The defective phagocytes of patients suffering from chronic granulomatous disease (CGD) are only able to kill catalase-negative microorganisms. Catalase-positive microorganisms (*Staphylococcus aureus, Eschericha coli, Salmonella, Aerobacter aerogenes, Hafnia, Serratia marcescens, Candida albicans* and others) each degrade hydrogen peroxide and are therefore not killed in the leukocytes of CGD patients. Such organisms therefore cause recurrent and usually therapy-resistant infections.

Neutrophils are the chief reservoir of MPO, representing 2–5% of cell weight. The MPO protein is localized entirely in the azurophilic granules and is also present in human monocytes but not in tissue macrophages. MPO has a different structure from that of milk and saliva (LPO) or eosinophil granulocytes (EPO; BAINTON *et al.*, 1971; LEHRER, 1975; BRETZ and BAGGIOLINI, 1974).

MPO is probably synthesized by a single protein from a single gene located on the long arm of chromosome 17 in the region of 17q12–23 (VAN TUINEN *et al.*, 1988). The human MPO gene contains 12 exons and 11 introns spanning 10 kb (MORISHITA *et al.*, 1987) and has an open reading frame of approximately 2 200 bp coding for the primary translation product of 745 amino acids with a calculated molecular weight of about 83 000. *In vitro* translation of mRNA selected by cDNA hybridization reveals the synthesis of a 74-kD protein. MPO molecule is composed of two chain types — heavy and light. The heavy chain consists of 467 amino acid residues located at the carboxy terminal of the protein; the light chain consists of 108 amino acids. For the formation of heavy and light chains, 164 amino acids of the pre-pro-sequence are removed; in addition, a small peptide of six amino acids between light and heavy chains and a single amino acid at the carboxy terminus of the heavy chain are removed (HASHINAKA *et al.*, 1988). Native MPO molecule is a tetramer with molecular weight approximately 140 000 consisting of two heavy and two light chains. An iron-chlorine haem prosthetic group is incorporated into each of the heavy subunits (TOBLER and KOEFFLER, 1991).

The optimum pH for MPO activity is 4.4–5.0 and it is an intense green. MPO may be replaced in antibacterial systems by LPO or EPO. MPO is found not only in human neutrophils but also in heterophilic leukocytes of all mammals studied so far. Avian heterophilic leukocytes do not contain MPO.

MPO can be separated by ion-exchange chromatography into three isoenzymes that vary in molecular weights, solubility, enzymatic activity, subunits structure, distribution in azurophil granule subpopulations, and release on neutrophil stimulation. This suggests that heterogeneity of MPO may exist in normal neutrophils. MPO itself has no direct effect on microorganisms, only in the complex with hydrogen peroxide and an oxidizable cofactor.

MPO forms three different complexes in reaction with H_2O_2 or certain other oxidants: compounds I, II, and III. H_2O_2 at relatively low (equimolar) concentrations reacts in a peroxidatic reaction with the iron of ferric MPO (MP^{3+}) to form compound I ($MP^{3+} + H_2O_2$). Compound I also appears to be formed during reduction of MPO by hypochlorous acid (HOCl). It is the primary catalytic peroxide compound of MPO and is highly unstable. Compound II is formed on addition of excess H_2O_2 to ferrous MPO (MP^{2+}). It is an inactive form of MPO in respect to oxidation of chloride. Compound III ($MP^{3+} + O_2^{-} = MP^{2+} + O_2$) is an oxyperoxidase, that, like oxyhaemoglobin, has oxygen attached to the same iron. It is formed by the reaction of compound II with H_2O_2, by aerobic oxidation of NADH, by reaction of ferrous MPO with oxygen, or by the reaction of ferric MPO with superoxide. Compound III is unstable, decaying to ferric peroxidase with a half-life of several minutes at room temperature. Compound III probably represents the catalytic form of MPO in neutrophils.

Several mechanisms are thought to be associated with the MPO-mediated killing of bacteria. Bacteria commonly lose their ability to divide within minutes of encountering phagolysosomes in leukocytes. Hypochloric acid itself is able to attack microorganisms at a variety of chemical sites. Potential targets are unsaturated carbon bonds, sulphydryl groups, amino groups, nucleic acids, and haem enzymes. Hypochlorous acid can further react with nitrogenous substances to form chloramines that may contribute to the MPO-mediated toxicity (ROOT and COHEN, 1981; KLEBANOFF, 1982). Another oxidative mechanism of cell damage is the alteration of cell membrane permeability.

The MPO system has not only intracellular but also extracellular effects on target molecules and cells. Cellular injury to normal and malignant cells can be mediated by neutrophils by two mechanisms: non-oxidatively through release of lysosomal antimicrobial proteins, such as defensins, cathepsin G or elastase and by reactive oxygen intermediates, including hydrogen peroxide (NATHAN *et al.*, 1979).

Eosinophil peroxidase (EPO) together with hydrogen peroxide and a halide cofactor constitute a cytotoxic system that is preferentially used in the extracellular killing of barely phagocytosable organisms such as helminths (KLEBANOFF, 1980). In addition, EPO is a relatively stable enzyme having — after extracellular release — a high affinity for negatively charged surfaces of some bacteria (*e.g.* staphylococci and *Legionella pneumophila*) that are subsequently killed by activated macrophages much faster than EPO-uncoated microbes.

When eosinophils adhere on the target, they discharge their granule contents, including EPO, as well as a number of other basic proteins. After stimulation, eosinophils generate more H_2O_2 than comparably stimulated neutrophils because the respiratory burst of eosinophils is greater than that of neutrophils. The interaction of EPO and H_2O_2 with a halide on the target cell surface might thus be performed.

Human EPO is a glycoprotein with a molecular weight of approximately 77 000 which can be separated into a large and small subunits under reducing conditions. The oligosaccharide moiety is associated with the large subunit and consists of mannose and *N*-acetylglucosamine residues. The haem prosthetic group of EPO is a protoporphyrin.

The blood monocyte most closely resembles the neutrophil in its antimicrobial systems. Monocytes respond to stimulation with a brisk respiratory burst, although of lesser magnitude than comparably stimulated neutrophils. The blood monocyte contains peroxidase in cytoplasmic granules that is identical to neutrophil MPO, although monocytes contain less MPO than neutrophils.

Blood monocyte-derived tissue macrophages usually lose lysosomal peroxidase during maturation and it therefore seems that this enzyme does not play a significant role in their microbicidal and cytotoxic activity.

Maturation of tissue monocytes into macrophages is not only associated with a loss of their granular peroxidase but also with a decrease in magnitude of the respiratory burst. This decrease results in a marked reduction in antimicrobial activity of resident macrophages against various pathogens. The respiratory burst of resident macrophages in response to stimulation is increased severalfold when the macrophage becomes activated. These changes in the respiratory burst, and thus also in the antimicrobial activity during macrophage differentiation and activation, appear to result from changes in the kinetics of NADPH oxidase.

Conversely, it has been shown that the tumouricidal activity of activated macrophages is directly dependent on the amount of liberated hydrogen peroxide and that their ability to kill ingested trypanosoma, toxoplasma and candida also correlates closely with H_2O_2 and superoxide formation. It therefore seems that activated macrophages may kill ingested microbes or other target cells even in the absence of peroxidase. The precursors of the cytotoxic substances involved in this mechanism appear to be superoxide and H_2O_2. In addition, macrophages may use receptor-mediated pinocytosis to take up peroxidases released from other cells and utilize them in cytotoxic mechanisms.

13.3.3 Nitrogen-dependent antimicrobial and cytotoxic mechanisms

Reactive nitrogen intermediates (RNI) synthesized from the semi-essential amino acid L-arginine include highly reactive oxides such as nitric oxide (NO•), nitrogen dioxide (NO_2) and nitrite (NO_2-). The key intermediate is NO• produced by NO synthase from the terminal guanidino nitrogen of L-arginine. This complex enzyme system is present in macrophages, neutrophils, endothelial cells, platelets and other cells. An L-arginine-dependent pathway, distinct from arginase, has been shown to be responsible for the cytotoxic activities of macrophages (HIBBS *et al.*, 1987, 1988). Later it was proposed (MONCADA *et al.*, 1989) that the synthesis of NO• from L-arginine might in fact be a widespread pathway for the regulation of cell function and intercellular communication.

RNI are of interest for at least three reasons: they are synthesized by enzymes novel in mammalian biochemistry, which are not yet well characterized; their production is under strict immunologic control (DING *et al.*, 1988, 1990; MILLS, 1991); and they appear to play an important role in some antitumour (HIBBS *et al.*, 1988; STUEHR and NATHAN, 1989), antimicrobial (GRANGER *et al.*, 1988; JAMES and GLAVEN, 1989; ADAMS *et al.*, 1990; LIEW *et al.*, 1990) and also carcinogenic (*N*-nitrosamines produced from amines) actions of the activated macrophages (IYENGAR *et al.*, 1987). RNI are also produced by neutrophils (MC CALL *et al.*, 1989; SCHMIDT *et al.*, 1989). Originally, macrophage RNI were thought to be generated *via* reactive oxygen intermediate-mediated non-enzymatic oxidation of a reduced nitrogen species such as ammonia. Subsequent work instead supported an enzymatic pathway involving NO synthase.

It is now generally accepted that RNI arise from oxidation of the terminal guanidino nitrogen of L-arginine by an enzymatic complex termed **NO synthase** producing NO and L-citrulline. A proposed biosynthetic pathway in macrophages is shown in *Fig. 13.19* (COLLIER and VALLANCE, 1989). Five steps of the pathway have been postulated. The first step in the mechanism of this L-arginine oxidation pathway involves *N*-hydroxylation by a specific monooxygenase-like enzyme. The next step is a two-electron oxidation to a nitroso amide-like intermediate and is followed by a one-electron oxidation resulting in fragmentation, liberating the ornithine diimide and nitric oxide. Finally, the ornithine diimide is hydrolysed to the end-product citrulline.

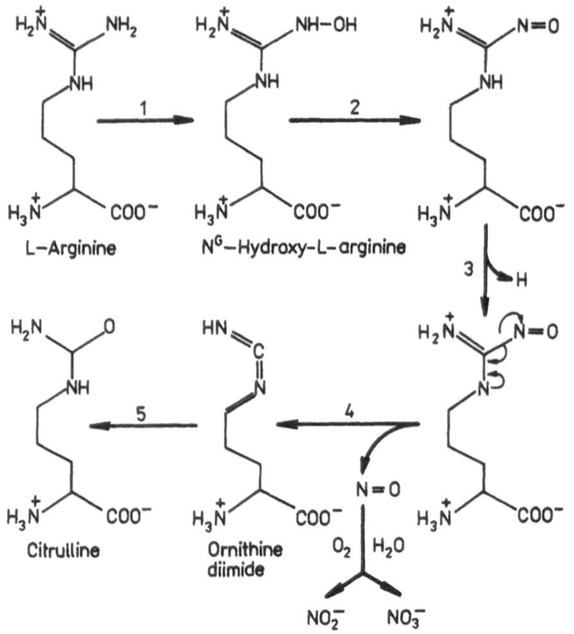

Fig. 13.19. Biosynthetic pathway of nitric oxide from L-arginine in macrophage (Collier and Vallance, 1989).

There is increasing evidence that there are two distinct NO synthases (*Table 13.5*). One is *constitutive*, Ca^{2+}-, calmodulin- and NADPH-dependent and synthesizes NO• as a transduction mechanism to regulate the activity of soluble guanylate cyclase (PALMER and MONCADA, 1989; MEYER *et al.*, 1989). It is present mainly in endothelial cells, neutrophils, hepatocytes, platelets *etc.* The other is *inducible* in macrophages and other cells (HIBBS *et al.*, 1988; BILLIAR *et al.*, 1990). Furthermore, this latter enzyme is Ca^{2+}-independent and requires NADPH, tetrahydrobiopterin, FAD and reduced glutathion as

Table 13.5 Two types of nitric oxide synthase and their occurrence in various cells

Constitutive	*Inducible*
Ca^{2+}-, calmodulin- and NADPH-dependent	Ca^{2+}-independent
Cofactors: NADPH, FAD, tetrahydrobiopterin	Cofactors: NADPH, FAD, GSH, tetrahydrobiopterin
Present in:	Present in:
Endothelial cells	Macrophages
Neutrophils	Hepatocytes
Hepatocytes	Endothelial cells
Kupffer cells	Platelets
Platelets	Neutrophils
Neurons (brain)	
Adrenal gland	
Tumour cells	

cofactors (TAYEH and MARLETTA, 1989; STUEHR *et al.*, 1990). Glucocorticoids inhibit the expression of an inducible but not the constitutive nitric oxide synthase (RADOMSKI *et al.*, 1990). Both types are cytosolic enzymes and their involvement in phagocytosis is unclear, although NO• produced by neutrophils may participate in cell chemotaxis (KAPLAN *et al.*, 1989).

L-Arginine is the natural substrate for NO synthase and, not surprisingly, arginine analogues can act as either inhibitors or alternative substrates for this enzyme. One such analogue, N^G-monomethyl-L-arginine (L-NMMA), which was initially shown to be an inhibitor of the formation of NO_2^- and NO_3^- from L-arginine in the macrophage (HIBBS *et al.*, 1987), has now been widely used as a tool to study this pathway in a variety of biological systems. Blockade of NO synthesis by L-NMMA is stereospecific, act at the level of the enzyme and is competitively reversed by L-arginine but not D-arginine (COLLIER and VALLANCE, 1989). Similar inhibitors are N^G-nitro-L-arginine methyl ester (L-NAME) and *N*-iminoethyl-L-ornithine (L-NIO).

Nitric oxide forms a paramagnetic relatively stable free radical (NO•) which can react rapidly with haemoglobin, myoglobin or oxygen. In the presence of H_2O and O_2, NO• is converted within seconds to the inactive end-products nitrite and nitrate:

$$2NO^\bullet + O_2 \rightarrow 2NO_2$$

$$NO + NO_2 \rightleftarrows N_2O_3 \xrightarrow{H_2O} 2NO_2^- + 2H^+$$

$$2NO_2 \rightleftarrows N_2O_4 \xrightarrow{H_2O} NO_2^- + NO_3^- + 2H^+$$

NO• also reacts with superoxide to form peroxynitrite anion in high yield:

$$NO^\bullet + O_2^- \rightarrow ONOO^-$$

In alkaline solutions, $ONOO^-$ is stable but has a pK_a of 7.5 at 37 °C and decays rapidly once protonated with a half-life of 1.9 s at pH 7.4 (BECKMAN *et al.*, 1990):

$$ONCO^- + H^+ \rightleftarrows ONOOH$$

$$ONOOH \rightarrow {}^\bullet OH + NO_2^\bullet \rightarrow NO_3^- + H^+$$

These reactions interconnect both types of inorganic oxidants with antimicrobial and cytotoxic effects — reactive oxygen intermediates and reactive nitrogen intermediates (*Table 13.3*).

The mechanism of the NO-mediated non-specific killing of tumour cells and pathogens has yet to be clarified. It has been postulated that NO• reacts with Fe-S groups resulting in the formation of iron–nitrosyl complexes that cause the inactivation and degradation of Fe-S prosthetic groups of aconitase of the citric acid cycle and complex I and complex II of the mitochondrial electron transport chain, as well as inhibition of thymidine incorporation into target cell DNA (KILBOURNE *et al.*, 1984; DRAPIER and HIBBS, 1986; HIBBS *et al.*, 1988; LIEW and COX, 1991).

The first biological activity demonstrated for NO• was its identity with endothelium-derived relaxing factor (EDRF; PALMER *et al.*, 1987). EDRF relaxes vascular smooth muscle and inhibits platelet aggregation and adhesion *via* stimulation of the soluble guanylate cyclase. The cytotoxic effect of macrophage-derived NO• on both microorganisms and tumour cells has been shown and now it is generally accepted that NO• also participates in the regulation of nervous, cardio-vascular and immune systems (MONCADA *et al.*, 1989).

The regulatory effects of the L-arginine / NO pathway are based on the existence of generator and target cells (*Fig. 13.20*). NO• synthesized in the generator cells acts as the endogenous and also exogenous stimulant of guanylate cyclase in the target cells. The effects of NO• depens on the site of release. If the generator cell is the vascular endothelium, NO• released by stimuli such as acetylcholine, bradykinin or pulsatile flow, activates guanylate cyclase in adjacent smooth muscle cells where it vasodilates and in platelets where it inhibits aggregation and adhesion. Neutrophils stimulated by FMLP or LTB_4 generate NO• which may activate guanylate cyclase in vascular smooth muscle and platelets. Other intercellular communication which may depend on NO• activation of guanylate cyclase includes that between Kupffer cells and hepatocytes, and also in excitatory neurotransmission (COLLIER and VALLANCE, 1989).

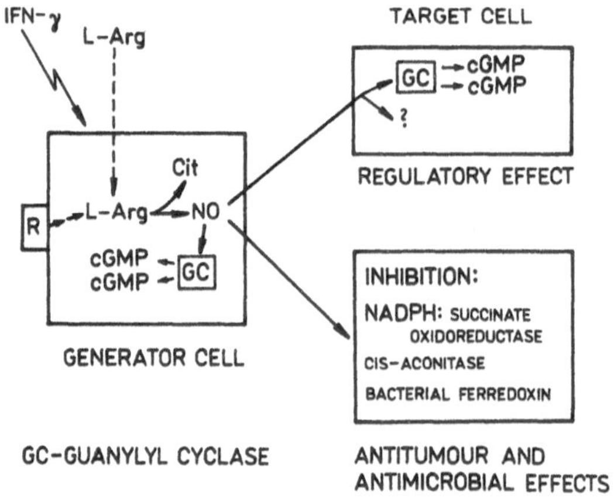

Fig. 13.20. The regulatory and effector effects of the L-arginine: NO pathway (adapted from Moncada *et al.*, 1989).

In macrophages the role of NO• seems to be different — cytotoxic on both microorganisms and tumour cells. The selectivity between cytotoxic and regulatory activities of NO• may lay in the different amounts of NO• produced. A high-output synthetic pathway that produces nanomolar concentrations of NO• could be cytotoxic, whereas a low-output pathway producing picomolar or femtomolar concentrations could act as a signal transduction system.

13.3.4 Oxygen- and nitrogen-independent antimicrobial mechanisms

Oxygen- and nitrogen-independent antimicrobial mechanisms are based on the changes of pH within the phagolysosome and numerous proteins of enzymatic and non-enzymatic character in phagocyte lysosomes (*Table 13.3*). These proteins may act on sensitive microorganisms in three ways: inhibition of multiplication, subtle increases in outer membrane permeability, and activation in the bacterial membrane of enzymes which degrade phospholipids and peptidoglycans (ELSBACH and WEISS, 1983; SPITZNAGEL, 1984).

It should be emphasized, however, that oxygen- and nitrogen-independent, oxygen-dependent and nitrogen-dependent mechanisms do not act separately, but are integral components of one system. However, they possess different specificity. The oxygen- and nitrogen-independent bactericidal proteins are highly active against certain species, while others are resistant. In contrast, oxygen-dependent microbicidal factors non-specifically damage all cells.

Mature PMN leukocytes lose most of their rough endoplasmic reticulum during their terminal differentiation and therefore possess only a limited ability to synthesize protein. Indeed, most antimicrobial proteins are synthesized during the development of these cells and are stored in lysosomes.

Phagocytin, the first described protein with bactericidal properties, was isolated by HIRSCH (1956) from rabbit PMN leukocytes. The phagocytin molecule has a positive charge and therefore such proteins were known as **granular cationic proteins** (present in granules). Subsequently, such proteins were also isolated from other sources, *e.g.* histone nuclear protein rich in arginine (*leukin*), which possessed bactericidal activity against several Gram-negative and Gram-positive bacteria. It was shown, however, that phagocytin from leukocyte granules comprised several molecular species, some being more active against Gram-negative, and others against Gram-positive bacteria (ZEYA and SPITZNAGEL, 1971). Originally, no enzyme activity could be demonstrated in cationic proteins. Using more precise separation techniques it was shown, however, that some of them possessed proteolytic activity. Neutral pH is optimal for the enzyme activity of these cationic proteins (therefore termed *neutral proteinases*). Neutrophil granules contain several neutral proteinases, the best defined so far being cathepsin G (EC 3.4.21.20), elastase (EC 3.4.21.37) and collagenase (EC 3.4.24.7). All these neutral proteinases possess antibacterial activity or amplify the myeloperoxidase system activity.

ILYA METCHNIKOV convinced the scientific world 110 years ago that phagocytosis has a major role in immunity to infection and that phagocytes destroy microbes with their enzymes, or cytases as he called them, but proof of this had to wait for new tools. NADPH oxidase, NO synthase, some serine proteinases, especially cathepsin G, and also lysozyme clearly belong to those "cytases". However, there are also proteins that share the capacity to kill bacteria independent of oxygen or nitrogen and evidently, contrary to METCHNIKOV's conjecture, to do so independent of enzyme action. At least three structurally different families associated with the azurophil granules of human neutrophils are currently in centre stage owing to their potent antimicrobial actions. Therefore SPITZNAGEL (1990) calls them *antibiotic proteins of human neutrophils*. They are: bactericidal permeability-increasing protein, azurocidin and defensins.

Bactericidal permeability-increasing protein (BPI) is effective against Gram-negative bacteria by enhancing the permeability of their cell wall (WEISS *et al.*, 1978). The BPI protein is rich in lysine and does not appear to possess an enzymatic activity but it can selectively activate enzymes that degrade phospholipids and peptidoglycans of bacterial outer membrane.

BPI was recently cloned (GRAY *et al.*, 1989) and its cDNA shows that the sequence predicts a 31-amino acid signal peptide, followed by a 456-residue mature protein. This encoded sequence predicts a protein of 50.6 kD but the estimated molecular weight of purified BPI is approximately 58 kD. This difference in apparent size may reflect the presence of two potential *N*-linked glycosylation sites at positions 122 and 249. BPI can bind to the lipid A region of LPS (p. 42) and the affinity of BPI for bacterial LPS most likely is responsible both for its antimicrobial specificity and for its ability to neutralize certain effects of endotoxin (LEHRER and GANZ, 1990). BPI manifests maximal bactericidal activity at neutral pH.

Eosinophils contain two *eosinophilic cationic proteins* (ECP) — one rich in arginine, the second in cysteine. Each mol of ECP contains approximately 2.5 mol Zn^{2+}. ECPs do not possess bactericidal, esterolytic and antihistamine activity, but they are cytotoxic for *S. mansoni* which is killed even at low concentrations.

Serine proteinases and their enzymatically inactive homologues. Human neutrophilic granules contain several proteinases that are active at neutral to slightly alkaline pH, including the chymotrypsin-like cathepsin G, collagenase, elastase, and proteinase 3. In addition, they contain azurocidin, a homologous but enzymatically inactive congener.

Cathepsin G is a 27 kD cationic glycoprotein that is found in the azurophil granules of human neutrophils and in much smaller amounts also in monocytes. Gene for cathepsin G has been localized to chromosome 14q11.2. It encodes mature protein with 236 amino acid residues. At concentrations of about 25 µg/ml, cathepsin G kills Gram-negative and Gram-positive bacteria as well as certain fungi, and is especially effective against *Neisseria gonorrheae*. Because enzymatically inactive cathepsin G retains bactericidal and fungicidal activity, the molecule's enzymatic activity is not necessarily required for its antimicrobial properties (ODEBERG and OLSSON, 1975; SHAFER *et al.*, 1986).

Human neutrophil *elastase* exists as a mixture of four isoenzymes that differ only in their carbohydrate content. It can degrade bacterial cell wall protein, potentiate the lytic activity of lysozyme and potentiate the microbicidal activity of myeloperoxidase and cathepsin G.

Azurocidin is a 29 kD glycoprotein that shares considerable *N*-terminal sequence homology with the human neutrophilic elastase and cathepsin G. Because it contains glycine instead of serine in the expected catalytic site, azurocidin lacks proteinase and peptidase activity. Purified azurocidin killed *E. coli, S. faecalis* and *C. albicans in vitro,* showing optimal activity under weakly acidic conditions (CAMPANELLI *et al.*, 1990).

Defensins are a family of microbicidal and cytotoxic peptides composed of 29–34 amino acid residues with molecular weight of approximately 3 500–4 000. They are variably arginine-rich, and contain six conserved cysteine residues that form three intracellular disulphide bonds. Because the cysteine residue closest to the amino terminus is disulphide-linked to the cysteine closest to the carboxy terminus of the peptide, defensins are cyclic molecules. More than 15 mammalian defensins derived from five species have been purified and sequenced to date. Human neutrophils contain several defensins that were also named human neutrophil peptide (HNP)-1, HNP-2 *etc.* Three of these (HNP-1, 2, and 3) differ from each other by a single amino-terminal amino acid. It appears that defensins are synthesized as 94–95 amino acid preprodefensins, with a characteristic hydrophobic signal sequence (GANZ *et al.*, 1985, 1990; LEHRER and GANZ, 1990).

Defensins constitute 5–7% of total cellular protein of human neutrophils, and 30–50% of total azurophil granule protein. Therefore, they are thought to constitute a key element in the non-oxidative killing of foreign organisms. Defensins are not known to be present in eosinophils, monocytes or lymphocytes. During phagocytosis of bacteria, defensins have been shown to enter the neutrophil's phagolysosomes. In addition, neutrophils may release defensins to the extracellular environment. Partially homologous proteins were purified from the haemolymph of infected or injured insects, suggesting that defensive-like molecules may be ancestral host defence components in the animal kingdom.

In vitro, defensins can kill a wide spectrum of Gram-positive and Gram-negative bacteria, many fungi, and inactivate certain enveloped viruses. HNP-1 and, to a lesser extent, HNP-2 are potently and selectively chemotactic for monocytes. Many defensins are cytotoxic to mammalian cells in culture and may play a role in inflammatory processes or antineoplastic defence.

Lysozyme (EC 3.2.1.17) participates both in antimicrobial and degradation processes. The lysozyme effect is based on cleavage of β-1,4-glycosidic bonds between *N*-acetylglucosamine and *N*-acetylmuramic acid residues which results in destruction of glycoprotein and polysaccharide components in microbial cell membranes.

The antibacterial effect of lysozyme was anticipated by FLEMING (1922) although it was subsequently shown that it was only directly effective against certain non-pathogenic Gram-positive bacteria, such as *Micrococcus lyso-*

deikticus, or *B. subtilis* and *B. megaterium.* Other bacteria are killed by lysozyme only in cooperation with additional factors such are antibodies, secretory IgA, complement, hydrogen peroxide, ascorbic acid *etc.* It seems that lysozyme is mainly of use in lysis of bacterial cells killed by other mechanisms, or in other than bactericidal processes. For example, lysozyme secreted by phagocytes may modify membrane glycoproteins and membrane receptors, modulate granulocyte function and interactions among lymphocytes, and potentiate the tumouricidal activity of monocytes. Because of its cationic character, lysozyme may form complexes with anionic proteoglycans, *e.g.* in the cartilage. The biological function of these complexes, however, is unknown.

Unlike the other major granule proteins, lysozyme is proportionally distributed between both azurophil and specific granules. It is not only a component of neutrophil, monocyte, and macrophage granules but it is also found in blood plasma, tears, saliva and airway secretions. The human lysozyme gene has been localized to chromosome 12 and it is composed from four exons and three introns.

Lactoferrin (LF) is a 78 kD slightly basic (pH = 8.7) glycoprotein that belongs to the transferrin family of iron binding proteins. It is a principal component of specific granules and is also found free in serum as well as in tears, semen, human milk and other secretions. The LF gene has been localized to chromosome 3q1–23, a region where the transferrin and transferrin receptor genes have also been localized. Current views on the physiological role of LF are dominated by its adequacy as an antimicrobial agent and immune modulator.

Its bacteriostatic and bactericidal effect against many bacteria is based on binding vitally important Fe^{3+} ions that are needed for their growth, or on an indirect effect based on the catalysis of hydroxyl radical generation from other oxygen-reactive intermediates. Lactoferrin also possesses immunoregulatory effects. It enhances the cytotoxic activity of *NK*-cells for tumour cells, increases the phagocytic ability of macrophages and inhibits the secretion of granulocyte-macrophage colony-stimulating factors.

When fully saturated, lactoferrin contains two Fe^{3+} atoms per molecule. Iron-saturated LF can act as a catalyst of the Haber–Weiss reaction with formation of •OH by intact neutrophils. However, LF saturated with an equivalent amount of iron has little or no stimulatory effect on •OH formation, and partially iron-saturated lactoferrin is inhibitory. Under physiological conditions, LF is largely unsaturated, containing 20% or less of its total iron capacity.

Phagolysosomal pH. Early studies seemed to suggest that the pH in phagolysosomes is relatively acid (3.0–5.5). Subsequently several authors found intralysosomal pH to be within the range 5.5–6.5. The differences, found by various authors, are given by distinct methods of measurement as well as by using different particles. CECH and LEHRER (1984) found that human neutrophils contained two types of phagolysosomes — sealed and unsealed. The *sealed* phagolysosomes do not communicate with the surrounding environment, whereas the *unsealed* types do. For candida killing, for example, the sealed phagolysosomes are important. Five minutes after initiation of phagocytosis with fluoresceinated zymosan particles (a pH probe), sealed vacuoles become alkaline with a pH of 7.80. Subsequently, their acidity increases; pH falling to 7.38 after 15 min, 6.35 after 30 min and 5.58 after 60 min.

These changes of pH in the phagolysosome do not have a direct antimicrobial effect, but create optimal conditions for the action of various antimicrobial factors such as the reactive oxygen intermediates, the myeloperoxidase system, defensins, BPI and lysosomal hydrolytic enzymes, which have different pH optimal activity. For example, the NADPH oxidase that generates the superoxide is maximally active at neutral pH, as are defensins and neutral proteinases. The production of H_2O_2 is maximal at pH values of 7.0–7.8. After one hour of phagocytosis, when the pH becomes acidic, the H_2O_2 production is reduced by 90% compared with the value at the beginning. Since high concentrations of H_2O_2 inhibit the chlorinating activity of the MPO system (Winterbourn *et al.*, 1985), the high concentrations observed during the initial phase of phagocytosis could therefore be restrictive for optimal chlorination activity. The weakly acid pH is optimal for the activity of lysozyme and many lysosomal hydrolases responsible for the degradation of engulfed material.

13.3.5 Cytotoxic factors of phagocytes and other cells

Effector cells can kill and lyse target cells by cytotoxic factors without phagocytosis. Activated neutrophils may possess cytotoxic activity, but it is expressed much more in activated macrophages and particularly in cells possessing *NK*-activity such as large granular lymphocytes (LGL), or other cells of spontaneous cytotoxicity, and cytotoxic *T*-lymphocytes (CTL) that are specialized for this function.

The target cells for the cytotoxic action of activated macrophages are mainly tumour cells and multicellular parasites. *NK*-cells are effective against tumour cells and virus-infected cells whereas CTL are responsible for the lysis of virus-infected cells, tumour cells and cells of allogeneic and xenogeneic grafts. Several factors participate in the cytotoxic activity of macrophages, *e.g.* tumour necrosis factors (TNF-α and TNF-β), reactive oxygen intermediates (H_2O_2, •OH), neutral proteinases, defensins, lysosomal enzymes, cytolytic fibronectin, thymidine, arginase and others. Until recently, relatively little was known about the CTL and LGL cytolytic molecules. However, the situation changed considerably when Podack and Konigsberg (1984) established that CTL and *NK*-cells contained cytoplasmic granules containing cytolytic proteins (*perforins*) capable of target cell lysis. Besides perforins, granules also contain further cytolytic substances such as neutral serine proteinases, esterases, monocyte chemotactic factor, and even typical lysosomal hydrolases, *e.g.* β-D-glucuronidase.

KOREN *et al.* (1987) recommended the term **"leukolysins"** for these leukocyte cytolytic molecules. The leukolysin molecules discovered so far may be divided into two main groups: molecules acting as direct mediators of cytolysis and molecules with an indirect effect on cytolysis (*Table 13.6*). According to the mode of action, two types of direct mediators of cytolysis can be distinguished: pore-forming molecules and molecules with another mechanism of action. The pore-forming leukolysins are known by some authors as "polyperforins" and by others as "cytolysins"; however, these substances are probably identical.

Table 13.6 Leukolysins — cytolytic molecules present in leukocytes (adapted from Koren *et al.*, 1987)

Group	*Leukolysin*
Direct mediators of cytolysis	
Molecules forming pores (holes)	Polyperforins
	Cytolysins
Molecules not forming pores	TNF-β (lymphotoxin)
	TNF-α (cachectin)
	Leukolexin
	NKCF
	Interferons
	IL-1
	Leukoregulin
	Reactive oxygen and nitrogen intermediates
Indirect mediators of cytolysis	Proteinases
	Esterases
	Proteoglycans
	Plasminogen activator

NKCF, natural killer cytotoxic factor.

Perforins are present in CTL and *NK*-cell granules and after receiving a corresponding signal through specific receptors, they are secreted on the cell surface. Afterwards, they polymerize to *polyperforins* on the target cell surface and punch round holes in the cytoplasmic membrane. These pores are similar to those formed by terminal complement components (p. 273). Granules of human CTL contain a perforin with the molecular weight of 65–70 kD. A cDNA clone of human perforin predicts a protein with 534 amino acids preceded by a leader peptide of 21 amino acids.

Activation, accompanied by generation of polyperforin, requires Ca^{2+} ions. The perforin that, following polymerization, forms tubular pores (with a height of 16 nm and diameter 5–20 nm) in the surface membrane of target cells, resembles the C9 complement component. It has a similar molecular weight, electric charge, cross-reacts with anti-C9 antibodies and its amino acid sequence has a 27% homology with C9. However, the size of the holes formed in the target membrane by both substances is different; perforin seems to be more effective than C9.

Other tubular polymers, called perforin 2, have been detected on membranes of cells lysed by an *NK*-cell line. These homogeneously sized tubules have a height of 12 nm and an inner diameter of 5 nm.

In the presence of one mmol Ca^{2+}, purified perforin exhibits potent lytic activity against cells as well as artificial lipid vesicles of different composition. Red blood cells are the most sensitive targets. Nucleated cells require between 6 and 100 times more perforin (PODACK, 1986; TSCHOPP and NABHOLZ, 1990).

Interaction between CTL and various target cells may be divided into three phases: recognition and adhesion, programming target cell lysis and the lysis itself that is independent of the presence of cytotoxic cells. The mechanism of perforin action clearly illustrates this action. During preparation for lysis, perforins are actually secreted from the effector cell and polymerize to polyperforins on the target cell membrane (lethal hit). Its lysis then proceeds automatically in the absence of CTL. There is a universal mechanism of lysis for all target cells; only the mechanism of recognition is different. Recognition of virus-infected cells or tumour cells requires that CTL simultaneously recognize surface virus-specific or tumour-specific antigens together with the class I histocompatibility antigens. Lysis of allogeneic and xenogeneic cells depends only on the recognition of foreign products of the major histocompatibility complex.

As well as perforins, CTL and LGL granules also contain a mixture of various cytotoxic molecules not forming pores (lymphotoxins) that are secreted after different stimuli, including the lectines. Among them, the best defined lymphotoxins so far include TNF-α, TNF-β (p. 232) and *NK*-cell cytotoxic factors (NKCF). The latter is a mixture of substances (mol. wt. 18 000 up to 60 000) secreted by *NK*-cells and *NC*-cells. Both populations of the natural killers may lyse certain normal, tumour or virus-infected cells, probably after recognition of corresponding receptors and without restriction by MHC products. Cytotoxic activity against target cells is also exhibited by interferons, IL-1, leukolexin, leukoregulin and reactive oxygen intermediates. Leukolexin forms a TNF-like molecule with molecular weight of 70 kD. It can induce calcium-independent slow lysis of some target cells.

CTL and *NK*-cells also contain further substances that do not induce direct lysis of target cells, although they can contribute to lysis, *e.g.* neutral serine proteinases, esterases, proteoglycans and plasminogen activator.

Granules of murine CTL lines contain seven serine proteinases called *granzymes A–G*. They represent approximately 90% of the total granule proteins. In humans only three homologous granzymes A, B and C have been identified so far. The biological functions of granzymes are unknown. Although lymphocyte-mediated cytolysis can be inhibited by protease inhibitors, it is not clear whether granzymes are the targets of the inhibitors. Therefore, it has been also postulated that granzymes may modulate perforin activity or DNA breakdown of target cells (TSCHOPP and NABHOLZ, 1990). Recently it has been shown that human granzyme B displays similarity to cathepsin G and possesses internal peptides analogous to His-Pro-Gln-Hyp-

Asn-Gln-Arg peptide with bactericidal action *in vitro* (SHAFER *et al.*, 1991). Thus the proteolytic activity of granzymes need not be important for their cytolytic effects.

Chondroitin sulphate proteoglycans may serve as carrier molecules for granule proteins and prevent autodegradation or proteolytic inactivation of perforin.

13.4 Secretory and regulatory functions of phagocytes

Until recently it was thought that phagocytes had at least two functions — defensive and nutritional. Later it was shown, however, that phagocytes can release numerous substances with significant bioregulatory functions during phagocytosis or after other, non-phagocytic stimuli. Thus phagocytes also possess important secretory functions.

Lysosomal enzymes are the most important leukocyte secretory products (SCHNYDER and BAGGIOLINI, 1978; WEISSMANN *et al.*, 1979; FERENČÍK and ŠTEFANOVIČ, 1979; WRIGHT, 1982). They are released during phagocytosis or after other types of stimulation, either into the phagolysosome, or into the extracellular space. Lysosomal enzymes released into phagolysosomes, participate in microbicidal and degradation reactions, while extracellularly released enzymes are components of cytotoxic mechanisms that operate without engulfment of target cells. Extracellularly released enzymes also play an important regulatory role in cell-to-cell communication and in inflammation, influence various physiological systems (complement, kinin, fibrinolysis, blood coagulation), and also participate in pathological damage to tissues. Release of lysosomal enzymes is not just a secondary phenomenon of phagocytosis, but is a component of its primary secretory function.

In general, the leukocyte granule content can be released by three mechanisms: after phagocyte death during autolytic processes, during phagocytosis and after stimulation with certain soluble mediators, even in the absence of phagocytosis. The latter two mechanisms require contact of the ligand with the corresponding receptor in the cytoplasmic membrane of neutrophils or macrophages, and fusion of granules with the cytoplasmic membrane. Neutrophil lysis with subsequent release of lysosomal enzymes is also caused by certain bacterial toxins such as streptolysins (from streptococci) or leukocidin (from staphylococci).

During phagocytosis, the neutrophil lysosomal content may be released into the extracellular space in three ways. First, some phagosomes, formed during ingestion of immune complexes, bacteria, zymosan or other particles, fuse with lysosomes before full closure. As a result, part of the lysosomal content may escape into the extracellular space. This mechanism was called *"regurgitation during feeding"* by WEISSMANN *et al.* (1971). Second, as soon as foreign particles (immune complexes, aggregated immunoglobulins) have attached firmly to non-phagocytosable surfaces that are too large to be engulfed by neutrophils, unclosed phagosomes are formed, into which the lysosomal content is poured. Lysosomal enzymes can then degrade not only the particles, but also the surface to which they were attached, *e.g.* the glomerular basal membrane and surrounding tissue. This so-called *"frustrated phagocytosis"* or *"reversed endocytosis"* operates particularly in immunopathological processes (HENSON, 1971a). Third, lysosomal enzymes are released during leukocyte migration. These enzymes, particularly neutral proteinases, participate in the chemotactic activation of neutrophils and during their response to appropriate chemotactic signal.

If phagocytes are allowed to react with cytochalasin B, which is an inhibitor of the microfilamentary system, the cells maintain their ability to recognize and adhere to the particles, and lysosomes can also fuse with the plasma membrane but no ingestion of the attached particle and no formation of the phagocytic vacuole takes place. As a result, the lysosomes release their content extracellularly (MALAWISTA *et al.*, 1971; WEISSMANN *et al.*, 1973).

All stimulators of extracellular lysosomal release studied so far also induce activation of HMPS and production of reactive oxygen intermediates. Certain secretory stimuli may induce release of both azurophil and specific granule contents. Such an ability is possessed by phagocytic stimuli, termed *"complete"* stimuli by SMOLEN and WEISSMANN (1981) — in contrast to *"incomplete"* stimuli that only induce release of specific granule content. The incomplete stimuli are due to various soluble substances. However, some incomplete stimuli may become complete stimuli in the presence of cytochalasin B (*Table 13.7*).

Table 13.7 Stimuli inducing extracellular secretion of neutrophil granule contents and formation of reactive oxygen intermediates (ROI)

Stimuli of secretion	*Secretion from granules*		*ROI production*
	Azurophil	*Specific*	
Phagocytic (complete)			
Opsonized particles	+	+	+
Immune complexes	+	+	+
Aggregated IgG	+	+	+
Soluble (incomplete)			
Chemotactic peptides (C5a, FMLP)	−	+	+
Chemotactic peptides + cytochalasin B	+	+	+
IL-8 or NAP-2 + cytochalasin B	+	+	+
A23187 (10 μmol/l) + Ca^{2+}	+	+	+
A23187 (1 μmol/l) + Ca^{2+}	−	+	+
Phorbol esters (PMA)	−	+	+
Lectins (Con A)	−	+	+
Arachidonic acid metabolites (5-HETE, 12-HETE, LTB_4)	±	+	+
PAF	−	+	+
C3b	−	−	+

FMLP, *N*-formyl-methionyl-leucyl-phenylalanine, A23187, calcium ionophore, PMA, phorbol myristate acetate, Con A, concanavalin A, 5-HETE and 12-HETE, 5-hydroxy- and 12-hydroxy-eicozatetraenic acids, PAF, platelet activating factor, NAP-2, neutrophil-activating peptide 2

It is thought that changes in intracellular cAMP, cGMP and free calcium concentrations play an important role in triggering secretion of lysosomal enzymes. All substances that increase the intracellular cGMP/cAMP ratio, stimulate this release, whereas substances that decrease this ratio, inhibit the release of lysosomal enzymes. To the first category belong adenylate cyclase inhibitors, as well as stimulators of guanylate cyclase and phosphodiesterase. In contrast, stimulators of adenylate cyclase and inhibitors of guanylate cyclase and phosphodiesterase inhibit lysosomal enzyme release. The second category includes isoprenalin, histamine and glucocorticoids that may influence phagocytes through specific receptors.

Studies of the signal transduction events associated with the activation of neutrophils for lysosomal enzyme secretion and superoxide production have implicated a role for increases in intracellular Ca^{2+}, diacylglycerol, and protein kinase C (KORCHAK *et al.*, 1984; ROSSI *et al.*, 1988).

Secretion of lysosomal enzymes from macrophages differs from lysosomal enzyme release from neutrophils in two respects: firstly, not all lysosomal enzymes are secreted by macrophages equally, secondly, there are also differences among individual types of macrophages. As far as the mechanism of secretion is concerned, macrophages contain three basic groups of lysosomal enzymes. 1. Enzymes that are continually secreted by macrophages in large amounts in the absence of phagocytic or soluble stimuli (constitutive enzymes). A typical example of this group is lysozyme. 2. Neutral proteinases (elastase, collagenase, plasminogen activator, angiotensinase) are continually secreted in small amounts, but their secretion may increase during phagocytosis or after other stimulation. 3. Acid hydrolases (β-D-glucuronidase, cathepsin D, *etc.*) are mainly released into the phagosome. Their extracellular secretion is lower than the first two groups and increases only slightly after stimulation.

The enzymes of the first group are secreted by resident, elicited and activated macrophages; neutral proteinases are secreted to a significantly higher degree by activated and inflammatory than resident macrophages; acid hydrolases are secreted most actively by inflammatory macrophages elicited by irritants, followed by activated macrophages and last by resting macrophages.

Besides lysosomal enzymes, macrophages and neutrophils also secrete several other substances (*Table 13.8*). These substances perform both effector and regulatory functions which cannot always be unequivocally differentiated. Thus, for example, the superoxide anion and other oxygen-derived free radicals are components of microbicidal, tumouricidal and generally cytotoxic mechanisms. At the same time, however, they significantly interact with the regulation of inflammatory and immune responses. They can release enzymes from lysosomes, including phospholipase A_2 that cleaves arachidonic acid from the cytoplasmic membrane. This acid may be metabolized *via* the cyclo-oxygenase or lipoxygenase pathways which result in the generation of endoperoxides (PGG_2, PGH_2), thromboxanes (TXA_2, TXB_2), prostacyclin (PGI_2), prostaglandins (PGE_2, $PGF_{2\alpha}$), leukotrienes (LTA_4, LTB_4, LTC_4,

LTD_4 and LTE_4 — the last three being identical in man with the slow-reacting substance of anaphylaxis — SRS-A) and hydroxy-eicosatetraenoic acids (12-HETE, 5-HETE). TXB_2, LTB_4, 5-HETE and 12-HETE have chemotactic activity, stimulate guanylate cyclase, HMPS, chemokinesis, CR1 expression and lysosomal enzyme liberation. PGE_2 and $PGF_{2\alpha}$ inhibit membrane NADPH oxidase by a feedback mechanism and thus inhibit superoxide production; they also inhibit lymphocyte transformation and release of MIF (migration inhibitory factor) and thus act as signals to stop the inflammatory response. PGE_2 and PGI_2 activate adenylate cyclase and thus inhibit lysosomal enzyme release. Various concentrations of PGG_2 (stimulatory) and PGE_2 (inhibitory) regulate macrophage accumulation in the inflammatory site.

Table 13.8 Secretory products of neutrophils and macrophages

Product	*Neutrophils*	*Macrophages*
Lysosomal enzymes (neutral proteinases, acid hydrolases)	+	+
Cytokines (IL-1, TNF-α, TGF-β, *etc.*)	+	+
Antimicrobial proteins (defensins)	+	±
Reactive oxygen intermediates	+	+
Reactive nitrogen intermediates	+	+
Chemotactic factors and their inhibitors	+	+
Bioactive lipids (PG, TX, LT, HETE, PAF)	+	+
Permeability factors (histamine)	+	+
Adhesive and binding proteins (fibronectin, transferrin, thrombospondin, avidin)	+	+
Factors influencing proliferation of leukocytes and other cells (G-CSF, G/M-CSF, TMF, erythropoietin)	+	+
Enzyme inhibitors (plasmin inhibitors, α-macroglobulin, α-1-proteinase inhibitor)	−	+
Complement components (C1, C4, C2, C3, C5, factors B, D, P, I, H)	−	+
Nucleosides (thymidine, uridine, neopterin)	−	+
Factors inhibiting replication of tumour and other cells and viruses (IFN, TNF)	+	+
Polypeptide hormones (ACTH, thymosin β_4, β-endorphin *etc.*)	+	+

PG, prostaglandins; TX, thromboxanes; LT, leukotrienes; HETE, hydroxyeicosatetraenoic acids; PAF, platelet activating factor; CSF, colony stimulating factor; IL-1, interleukin 1; TMF, thymocyte maturation factor; TNF, tumour necrosis factor; TGF-β, transforming growth factor-β.

Another source of complexity in the macrophage secretory repertoire is the propensity of the products to affect each other (NATHAN, 1987).

First, the products may affect each other's release. For example, TNF-α enhances macrophage capacity to release reactive oxygen intermediates, which share antitumour activity with TNF-α. Another cytokine whose antiproliferative action synergizes with that of TNF-α, interferon-γ, enhances macrophage release of TNF-α, and like TNF-α, enhances macrophage capacity to secrete reactive oxygen intermediates.

Second, macrophage products may affect each other's actions. Examples are synergy in killing bacteria by complement components and lysozyme, and in degrading matrix by elastase and plasminogen activator. On the negative side, macrophage (or neutrophil) elastase can cleave α-1-protease inhibitor and macrophage-derived oxidants can inactivate it.

Third, the macrophage products may affect the secretion by other cells of products macrophages also make. For example, TNF-α stimulates endothelial cells to release IL-1. IL-1 and TNF-α stimulate fibroblasts to release PGE_2, collagenase and GM-CSF.

Fourth, macrophage products like TNF-α may induce target cells to release cytokines that mediate or antagonize some of the actions attributed to the monokines.

These examples suggest that phagocyte secretory products may modulate the functional activity of cells of the immune system.

13.5 Defects of phagocytosis

Phagocytosis can only fulfil its biological role when all its phases are realized in the required intensity, adequately interconnected and mutually correlated. Disorders of any stage or in their cooperation results in delayed killing or in the survival of ingested bacteria giving rise to chronic, usually therapy-resistant infections and to defects of other effector or regulatory functions of phagocytes.

Defects of phagocytosis have been found in all types of phagocytes but most of the evidence concerns the neutrophil defects. The disorders of phagocyte function may affect all phagocytic stages, but the best-defined are disorders of phagocytic motility, respiratory burst, antimicrobial and degradation mechanisms.

The origin of phagocytic defects may be primary (congenital) or secondary (acquired) and they have a unimolecular or multimolecular cause. Unimolecular defects are rare compared to multimolecular defects. Primary defects may be caused by abnormal or missing genes, whereas secondary defects may be due to a natural deficiency in newborns and elderly individuals, inadequate nutrition, harmful environmental factors, various diseases and toxic effects of some drugs. Secondary defects can disappear after removal of unfavourable factors (Roberts and Gallin, 1983; White and Gallin, 1986; Ferenčík and Kotulová, 1988).

13.5.1 Primary disorders of phagocytosis

The most severe primary disorders of phagocytosis and intracellular killing are usually due to multimolecular defects and they also negatively influence other functions of neutrophils or other phagocytes (*Table 13.9*).

Chronic granulomatous disease (CGD) is a heterogeneous syndrome that may be caused by several molecular defects and their combinations (*Fig. 13.21*), which results in insufficient initiation of the respiratory burst (triggering defects, membrane potential abnormalities), disorders in the NADPH

Table 13.9 Primary disorders of phagocytosis

Disease	*Functional changes*	*Heredity*	*Note*
Chronic granulomatous disease (CGD)	↓ Microbicidal activity ↓ Metabolism of oxygen Cytochrome b_{558} or p47-phox and p67-phox missing	X AR AD	LTB_4, LTC_4 Inability to stimulate HMPS and H_2O_2 production during phagocytosis.
Leukocyte adhesion deficiency (LAD syndrome)	↓ Adherence, aggregation, chemotaxis, phagocytosis, microbicidal activity and oxygen metabolism	AR	*NK*-activity missing, *T*- and *B*-cell defects
Job (hyper-IgE) syndrome	↓ Chemotaxis ↓ Microbicidal activity ↓ Number of CR1	AR (X)	↑IgE, ↓SIgA, T_S-lymphocytes decreased
Chediak–Higashi syndrome	Neutropenia, giant azurophil granules, delayed lysosome degranulation ↓ Chemotaxis ↓ Microbicidal activity ↓ Elastase, cathepsin G, α-D-mannosidase	AR	*NK*-cells and platelets also affected. Beige mice — animal model Normal phagocytosis and HMPS activation (INT-test)
Specific granule deficiency	↓ Microbicidal activity ↓ Chemotaxis ↓ Oxygen metabolism after stimulation with FMLP only CR3, CR4, AP missing	AR	Lactoferrin and cobalophilin level in serum decreased, in saliva normal
MPO deficiency	↓ Candidacidal activity ↓ Immune phagocytosis ↑ Oxygen metabolism	AR	Significant disorders only when activity less than 1% of normal
Glucose-6-phosphate dehydrogenase deficiency	Delayed intracellular killing of ingested microorganisms	X	Associated with haemolytic anaemia
Pyruvate kinase deficiency	Inability to kill staphylococci		Normal *in vitro* phagocytosis and HMPS activation
Tuftsin deficiency	Decreased chemotaxis and phagocytosis		Normal INT- or NBT-test

AR, autosomal recessive heredity; AD, autosomal dominant heredity; X, X-linked heredity; HMPS, hexose monophosphate shunt; p47-phox and p67-phox, 47-kD or 67-kD cytosolic phagocyte oxidase factor; AP, alkaline phosphatase; ↓, decrease; ↑, increase.

oxidase activity required for the single-electron reduction of molecular oxygen into superoxide (deficiency of cytosolic proteins p47-phox and p67-phox, or absence of cytochrome b_{558}), and defects in oxygen metabolism (low affinity of NADPH oxidase for NADPH, disorders in NADPH supply, deficiency of glucose-6-phosphate dehydrogenase, glucose-6-phosphate translocase, glutathione reductase and glutathione peroxidase).

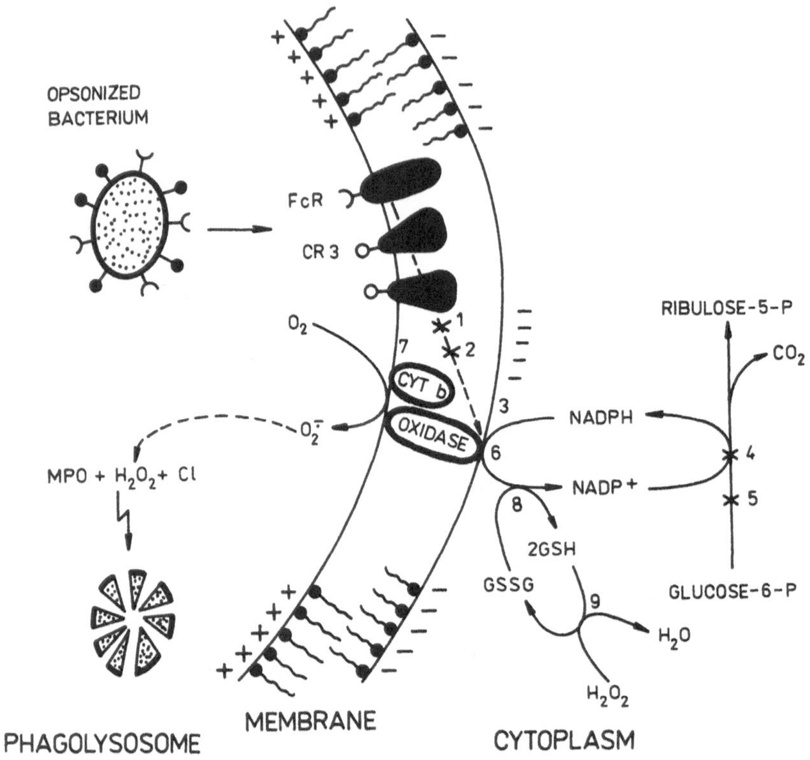

Fig. 13.21. Basic molecular defects in chronic granulomatous disease (adapted from Hitzing and Seger, 1983).

1, Triggering defect I, 2, triggering defect II, 3, defect in membrane potential, 4, glucose-6-phosphate dehydrogenase deficiency, 5, glucose-6-phosphate translocase deficiency, 6, disorder of NADPH oxidase activity caused by p47-phox and p67-phox deficiency, 7, cytochrome b_{558} deficiency, 8, glutathione reductase deficiency, 9, glutathione peroxidase deficiency

Generally, there are four main forms of CGD: 1. X-linked, cytochrome b_{558} negative form due to a mutation of the gene for the large β-subunit of the cytochrome. It results when both cytochrome subunits are missing. 2. X-linked, cytochrome positive form due to a different mutation in the same gene resulting in Pro/His exchange. 3. Autosomal recessive, cytochrome b_{558} negative form due to a mutation affecting the 22-kD α-subunit of the cytochrome; both cytochrome subunits are missing. 4. Autosomal recessive form due to mutations affecting the p47-phox and p67-phox cytosolic proteins. Cytochrome b_{558} is usually normal (MALECH and GALLIN, 1987; SEGAL, 1989; BAGGIOLINI and WYMANN, 1990).

The most frequent cause of CGD is *cytochrome b_{558} deficiency* that belongs to the most severe disorders of phagocytosis. Neutrophils, monocytes and macrophages of patients suffering from CGD are incapable of adequately stimulating the respiratory burst and HMPS during phagocytosis (and therefore do not produce hydrogen peroxide) and thus cannot kill catalasepositive bacteria like staphylococci, enterobacteria, pseudomonas, candida, aspergillus *etc.*

The reconstitution of the defective NADPH oxidase is the main task for the effective CGD treatment. A major advance in this was made in the use of cytokines, namely of IFN-γ which has been shown to have a corrective effect on phagocyte function in CGD. Treatment of autosomal recessive patients, whose phagocytes contained cytochrome b_{558}, with IFN-γ showed increased activity. The autosomal recessive patients without the cytochrome did not demonstrate activity before or after treatment. X-linked cytochrome b_{558}-deficient subjects fell into two groups: in the first, no response was visible; in the second, restitution of NADPH oxidase activity was partial. Thus, the knowledge of exact molecular defect is important for the effective treatment of CGD.

The changes in NADPH oxidase activity in neutrophils and monocytes that resulted from the administration of IFN-γ to patients with CGD were associated with marked improvements in bacterial killing which virtually returned at levels of NADPH oxidase activity at 10% of normal (SECHLER *et al.*, 1988; SEGAL, 1989). Interestingly, killing by neutrophil of some patients was enhanced in the absence of detectable induction of superoxide generation, suggesting some other influence on the biology of these cells. It may be caused by the fact that in CGD neutrophils the early elevation of pH in the phagolysosome is not observed and therefore the defensins and other cationic proteins cannot fully operate. The administration of IFN-γ may also improve the alkalization of phagolysosomes and in this way also kill engulfed microorganisms.

The overall defect in the ability to stimulate the HMPS is readily demonstrated by the significantly decreased reduction of INT or NBT (formazan dyes) by neutrophils, or the inability to increase chemiluminescence after both soluble (phorbol myristate acetate) or phagocytic (zymosan particles) stimuli. In most CGD patients, infections begin in early childhood and the prognosis is poor. However, some cases have been described when severe infections did not appear until the age of 20 years. This disease occurs mainly in boys (the male : female ratio is about 7 : 1).

Leukocyte adhesion deficiency (LAD) or adhesive membrane glycoproteins (LFA-1 family) deficiency occurs in two main forms; one with severe and the other with moderate clinical manifestations (ANDERSON *et al.*, 1985; TODD and FREYER, 1989). In the severe form, both α- and β-chains in the molecules of the LFA-1 family are completely lacking. The severe deficiency, affecting both boys and girls, is manifested by severe, life-threatening infections with high mortality (patients seldom survive beyond two years of age). Neutrophils and mononuclear phagocytes lack the CR3 receptor and cannot therefore respond with adequate mobility reactions and stimulation of oxygen metabolism to particles opsonized by this ligand. In contrast, migration, phagocytosis, respiratory burst and degranulation are normal if the phagocytes are stimulated with soluble ligands, *e.g.* chemotactic peptides or phorbol esters. The moderate deficiency is accompanied by partially preserved CR3 activity (patients express 2.5% to 6.0% of normal LFA-1, CR3 and CR4 levels) and usually have only recurrent skin infections. The CR3 molecules are stored within leukocytes (in specific granules) and can be mobilized and deposited in the cytoplasmic membrane during phagocytosis and degranulation. This CR3 supply is lacking in congenital deficiency of neutrophil-specific granules. Moderate forms of CR3 deficiency may be due to disorders of mobilization, function and biological turnover of the CR3 glycoprotein and not its absence.

Job's syndrome was originally recognized as a CGD variant. The syndrome was not named after the name of the discoverer but after the biblical Job who presumably suffered from this disease. The syndrome is mainly caused by severe disorders of chemotactic and microbicidal activity of neutrophils. Mononuclear phagocytes from these patients produce an inhibitor of neutrophil chemotaxis. The IgE serum level is usually elevated (more than 2 000 IU/ml) and, therefore, this disease is also called hyper-IgE syndrome (sometimes also the Job–Buckley syndrome). It is characterized by recurrent cutaneous or sinus–pulmonary infections caused by *S. aureus* or even *Haemophilus influenzae* and *Streptococcus pneumoniae*; 50% of patients suffer from mucocutaneous candidiasis. It is thought that specific IgE, directed against infectious microbes, can release histamine from mast cells (mostly in atopic patients). Histamine enhances the neutrophil cAMP level and thus inhibits chemotaxis.

Chédiak–Higashi syndrome (CHS) is a rare autosomal recessive disease characterized by a partial oculocutaneous albinism with photophobia, neutropenia, gingivitis, periodontal disease, recurrent pyogenic infections (*S. aureus*, streptococci, *H. influenzae*), neuropathy and early death in untreated patients. The disease often becomes malignant in childhood. The neutrophils of CHS patients possess giant azurophil granules, defective chemotaxis, delayed intracellular killing of ingested bacteria, and lower specific activity of certain lysosomal enzymes (elastase, cathepsin G, α-D-mannosidase). However, they have a normal or enhanced respiratory burst after phagocytic stimulus and exhibit hyperreactive phagocytosis of autologous blood cells after contact with bacteria. Monocytes also have impaired chemotaxis. These abnormalities not only affect the lysosomal apparatus of human cells, but a similar disease has been described in Aleutian mink, partial albino Hereford cattle, albino whales and beige mice. The primary defect responsible for the origin of CHS is unknown, but is thought to be due to disorders of the cell cytoskeletal apparatus, particularly a decrease in the number of centriole-associated microtubules and disorders in lysosome formation and function.

It appears that progressive aggregation and fusion of azurophil and specific granules occur in granulocytes during myelopoiesis. Therefore, the neutrophils of CHS patients have abnormal azurophil granules, whereas the specific granule content is normal, even though their number is lower (CHÉDIAK, 1952; BARAK and NIR, 1987).

Specific granule deficiency affects neutrophils and monocytes. Specific granules contain lactoferrin (a regulator of myelopoiesis and hydroxyl radical production), cobalofillin (a vitamin B_{12}-binding protein), proteinases which selectively degrade certain complement components into chemotactic fragments, CR3, CR4, receptors for the chemoattractant FMLP, laminin, cytochrome b_{558} and other important effector and regulatory substances. The presence of several receptors within the specific granule compartment led to the hypothesis that specific granules function as an important reservoir of certain plasma membrane receptors used during chemotaxis and phagocytosis. Deficiency of these factors results in decreased chemotaxis, aggregability and microbicidal activity which is manifested by severe pyogenic infections. A decrease in the number of specific granules and the lactoferrin level in neutrophils is also found in newborns, which may contribute to the decreased chemotaxis of these cells (GALLIN *et al.*, 1982).

Originally, it was assumed that **myeloperoxidase deficiency** was a rare autosomal-recessive disease; however, it has been recently shown that a partial or complete deficiency is relatively common, *e.g.* in the USA population, one case per 2 100 of the population (PARRY *et al.*, 1981). MPO-deficient leukocytes have decreased candidacidal activity, mainly in patients suffering from poorly controlled diabetes mellitus. The defect is usually manifested only when the MPO activity decreases below 1% of normal value. Compared to the previously described defects, the sequelae of MPO deficiency are much milder. This is because, besides the MPO system, leukocytes also possess other microbicidal mechanisms that can partially replace MPO activity. In addition, MPO-deficient leukocytes have enhanced phagocytic activity, an increased respiratory burst, enhanced superoxide production and increased lysosomal enzyme release (degranulation) after phagocytic stimuli. Increased respiratory burst and superoxide anion production is caused by elevated NADPH oxidase activity. This supports the hypothesis that MPO in normal leukocytes is supposed to terminate the respiratory burst through its ability to control NADPH oxidase activity. This mechanism simultaneously ensures the increased activity of oxygen-dependent (but MPO-independent) microbicidal mechanisms which compensates for the high degree of MPO deficiency.

Glucose-6-phosphate dehydrogenase deficiency occurs in neutrophils combined with its deficiency in erythrocytes. Together with other defects it may be one of the causes of CGD. In the case of a pure G6PD defect, clinical symptoms occur only when less than 1% of normal activity is present in leukocytes. Clinical symptoms are similar to CGD, but are milder. Deficient leukocytes do not reduce INT (NBT), and do not form hydrogen peroxide which results in shortage of the substrate (NADPH) for enzymatic reduction of oxygen.

Besides the above-mentioned defects there are deficiencies of other enzymes (alkaline phosphatase, glutathione reductase, pyruvate kinase *etc.*) and of phagocytosis-regulating substances (*tuftsin* — Thr-Lys-Pro-Arg, *rigin* — Gly-Gln-Pro-Arg). These defects, however, do not usually cause life-threatening infections or are quite rare.

13.5.2 Secondary disorders of phagocytosis

Secondary disorders may either be natural (during pregnancy, in newborns or elderly individuals), or caused by external agents or some diseases. Thus, defects of chemotaxis, ingestion and microbicidal mechanisms can be observed in protein-calorie malnutrition, deficiency of some metals (Zn^{2+}, Fe^{2+}, Cu^{2+}, Co^{2+}), following the ingestion of toxic chemicals in food or from the external environment (dibenzo-*p*-dioxins, dibenzofurans, polychlorinated and polybrominated biphenyls, polycyclic aromatic carbohydrates, organochlorinated insecticides, certain heavy metals such as lead, cadmium, mercury, organic tin compounds *etc.*), in alcohol intoxication, during immunosuppressive therapy and in some diseases like diabetes mellitus, Down's syndrome, rheumatoid arthritis, acute leukaemia *etc.*

Leukocytes of individuals with Down's syndrome have a 50% increase in superoxide dismutase (Cu,Zn-SOD) activity and decreased MPO activity. Increased Cu,Zn-SOD activity results in a decreased superoxide anion level which may cause increased susceptibility to bacterial, particularly staphylococcal infections. In patients with diabetes mellitus, insulin deficiency is the main cause of phagocytic disorders. Thus, for example, the leukocyte chemotactic response depends on the intracellular concentrations of potassium, glucose and cGMP; the higher the concentrations, the greater the response. Insulin stimulates potassium and glucose transport into the leukocyte and increases the intracellular level of cGMP. In insulin deficiency, there is a relative deficiency of these substances in leukocytes which in turn causes defective chemotaxis of both neutrophils and monocytes.

Deficient macrophage function has been observed, *e.g.* in virus infections, tuberculosis, leprosy, chronic candidiasis, sarcoidosis, systemic lupus erythematosus, liver cirrhosis and sickle-cell anaemia.

Secondary functional phagocyte defects may also be caused by some invading virulent microorganisms. This is due to their ability to resist killing and ingestion at various stages of the phagocytic process (*Table 13.10*).

Certain microbial surface structures may inhibit chemotaxis of phagocytes or block recognition and ingestion. Microorganisms possess several mechanisms that may prevent their killing by phagocytic cells. First, the fusion of phagosomes with lysosomes may be inhibited which prevents the necessary combination of microbicidal substances in the phagolysosome. Mycobacteria, for example, produce at least four different inhibitors of fusion: cAMP, poly-α-L-glutamic acid, 2,3,6,6′-tetra-acyltrehalose-2′-sulphate and ammonia. *Bordetella* form a soluble thermostable adenylate cyclase that may accumulate within phagocytes and catalyse uncontrolled cAMP formation.

Table 13. 10 Mechanisms by which certain virulent microorganisms are protected against killing by phagocytes

Phagocytosis step	*Antiphagocytic mechanism*	*Example*
Chemotaxis	Inhibitory effect of some surface lipids, glycoproteins, or exoproducts	*P. aeruginosa, Legionella pneumophila*
Recognition and ingestion	Coating of microbial surface with molecules of the host	*Schistosomula*
	FcR and CR blockade	Bacterial capsules Staphylococcal protein A (FcR only)
	C3b inactivation	M-protein of A-group streptococci
	Production of specific proteinases	*P. aeruginosa*
Killing and degradation	Inhibition of fusion of phagosomes with lysosomes (production of cAMP, polyglutamic acid, sulphatides, NH_4^+)	*Toxoplasma gondii*, enterotoxigenic *E. coli, Chlamydia* sp., *Nocardia*, mycobacteria
	Production of adenylate cyclase	*Bordetella* sp.
	Resistance to killing in phagolysosomes (membrane peptidoglycans, increase in osmotic pressure)	*Leishmania* sp., *M. lepraemurium, L. monocytogenes, S. typhimurium*
	Escape from phagolysosomes (damage of phagolysosomal membrane or its fusion with the pathogen)	*Trypanosoma crusi, M. leprae*, rickettsia, viruses
	Inability to induce respiratory burst and stimulate HMPS, or inhibition of NADPH oxidase	*Coxiella burnetii, Chlamydia trachomatis, L. pneumophila* (toxin), *Streptococcus pneumoniae* (pneumolysin), *M. leprae, T. gondii*
	Phagocyte damage, *e.g.* by release of lysosome content into the cytoplasm, by damage of cytoplasmic membrane, by mitochondria damage	Streptococcal streptolysin O and pneumolysin Staphylococcal leukocidin and α-haemolysin, cord-factor of virulent mycobacteria

Other virulent microorganisms (*Leishmania* sp., *Listeria monocytogenes, Mycobacterium lepraemurium, Salmonella typhimurium*) do not inhibit fusion of phagosomes with lysosomes, but are resistant to normal phagocyte microbicidal mechanisms. Protection against microbicidal mechanisms is provided by specific membrane peptidoglycans or by the ability to increase osmotic pressure within phagolysosomes which prevents the penetration of microbicidal substances into the phagolysosome.

As well as defects at the phagolysosomal level, certain virulent microorganisms (*Legionella pneumophila, Streptococcus pneumoniae*) can produce toxins that inhibit the respiratory burst, and thus also the production of oxygen microbicidal forms, or possess surface structures which do not induce a respiratory burst after coming into contact with the phagocyte (*e.g. Coxiella burnetii, Chlamydia trachomatis, Toxoplasma gondii*).

In addition, autoantibodies to leukocyte granule components have been detected in several diseases that are associated with an intense inflammatory component. Autoantibodies reactive with MPO or elastase have been described in various forms of glomerulonephritis. Their diagnostic, prognostic, or pathogenetic significance is not yet well known.

References

Ackerman, S. J., Durack, D. T. and Gleich, G. H. (1982) Eosinophil effector mechanisms in health and disease. In: Gallin, J. I. and Fauci, A. S. (eds) *Advances in Host Defense Mechanisms*. New York, Raven Press, vol. 1, pp. 269–93.

Adams, L. B., Hibbs, J. B. Jr., Taintor, R. R. and Krahenbuhl, J. L. (1990) Microbiostatic effect of murine-activated macrophages for *Toxoplasma gondii*. Role for synthesis of inorganic nitrogen oxides from L-arginine. *J. Immunol.*, **144**, 2725–9.

Allen, R. C. (1979) Reduced, radical, and excited state oxygen in leukocyte microbicidal activity. In: Dingle, T. T., Jacques, P. J. and Shaw, I. H. (eds) *Lysosomes in Applied Biology and Therapeutics*. Amsterdam, North Holland, vol. 6, pp. 197–234.

Anderson, D. C., Finegold, M. J., Hughes, B. J., Rothlein, R., Miller, A. S., Sheaver, W. T. and Springer, T. A. (1985) The severe and moderate phenotypes of heritable Mac-1, LFA-1, p150,95 deficiency: their quantitative definition and relation to leukocyte dysfunction and clinical features. *J. Infect. Dis.*, **152**, 668–89.

Aschoff, L. I. (1924) Das reticulo-endotheliale System. *Ergeb. Inn. Med. Kinderheilk.*, **26**, 1.

Babior, B. M. (1984) Oxidants from phagocytes: agents of defense and destruction. *Blood*, **64**, 959–71.

Babior, B. M. (1988) The respiratory burst. *Ann. Intern. Med.*, **109**, 127–42.

Babior, B. M., Kipnes, R. S. and Curnutte, J. T. (1973) Biological defense mechanisms. The production by leukocytes of superoxide, a potential bactericidal agent. *J. Clin. Invest.*, **52**, 741–4.

Badwey, J. A. and Karnovsky, M. (1980) Active oxygen species and the functions of phagocytic leukocytes. *Annu. Rev. Biochem.*, **49**, 695–737.

Baggiolini, M. (1982) Phagozyten und Phagozytose hundert Jahre nach Metschnikoff. *Schweiz. Med. Wschr.*, **112**, 1403–11.

Baggiolini, M. and Dewald, B. (1985) The neutrophil. *Int. Archs. Allergy Appl. Immun.*, **76**, suppl. 1, 13–20.

Baggiolini, M. and Wyman, M. P. (1990) Turning on the respiratory burst. *Trends Biochem. Sci.*, **15**, 69–72.

Bainton, D. F. (1988) Phagocytic cells: developmental biology of neutrophils and eosinophils. In: Gallin, J. I., Goldstein, I. M. and Snyderman, R. (eds) *Inflammation: Basic Principles and Clinical Correlates*. New York, Raven Press, pp. 265–80.

Bainton, D. F., Ullyot, J. L. and Farquar, M. G. (1971) The development of neutrophilic polymorphonuclear leukocytes in human bone marrow. *J. Exp. Med.*, **134**, 907–33.

Baldridge, C. W. and Gerard, R. W. (1933) The extra respiration of phagocytosis. *Amer. J. Physiol.*, **103**, 235–6.

Barak, Y. and Nir, E. (1987) Chédiak-Higashi syndrome. *Am. J. Pediat. Hematol. Oncol.*, **9**, 42–55.

Becker, E. L., Showell, H. J., Naccache, P. H., Freer, R. J., Walenga, R. W. and Sha'afi, R. I. (1982) Chemotactic factors: locomotory hormones. In: Karnovsky, M. L. and Bolis, L. (eds) *Phagocytosis — Past and Future*. London, Acad. Press, pp. 87–103.

Beckman, J. S., Beckman, T. W., Chen, J., Marshall, P. A. and Freeman, B. A. (1990) Apparent hydroxyl radical production by peroxynitrite: Implications for endothelial injury from nitric oxide and superoxide. *Proc. Natl. Acad. Sci. USA*, **87**, 1620–4.

Bergendi, Ľ. (1988) *Superoxide and other bioreactive forms of oxygen*. Bratislava, Veda, 198 pp. (*in Slovak*).

Berkow, R. L. and Dodson, R. W. (1991) Alteration in tyrosine protein kinase activities upon activation of human neutrophils. *J. Leukoc. Biol.*, **49** 599–604.

Billiar, T. R., Curran, R. D., Stuehr, D. J., Stadler, J., Simmons, R. L. and Murray, S. A. (1990) Inducible cytosolic enzyme activity for the production of nitrogen oxides from L-arginine in hepatocytes. *Biochem. Biophys. Res. Commun.*, **168**, 1034–40.

Black, C. D. V., Samuni, A., Cook, J. A., Krishna, C. M., Kaufman, D. C., Malech, H. L. and Russo, A. (1991) Kinetics of superoxide production by stimulated neutrophils. *Arch. Biochem. Biophys.*, **286**, 126–31.

Borregaard, N. (1988) The human neutrophil. Function and dysfunction. *Eur. J. Haematol.*, **41**, 401–13.

Bretz, U. and Baggiolini, M. (1974) Biochemical and morphological characterization of azurophil and specific granules of human neutrophilic polymorphonuclear leukocytes. *J. Cell. Biol.*, **63**, 251–69.

Brozna, J. P., Hauff, N. F., Phillips, W. A. and Johnston, R. B. Jr. (1988) Activation of the respiratory burst in macrophages. Phosphorylation specifically associated with Fc receptor-mediated stimulation. *J. Immunol.*, **141**, 1642–7.

Campanelli, D., Detmers, P. A., Nathan, C. F. and Gabay, J. E. (1990) Azurocidin and a homologous serine protease from neutrophils: Differential antimicrobial and proteolytic properties. *J. Clin. Invest.*, **85**, 904–12.

Cech, P. and Lehrer, R. I. (1984) Phagolysosomal pH of human neutrophils. *Blood*, **63**, 88–95.

Chédiak, M. M. (1952) Nouvelle anomalie leucocytaire de caractére constitutional et familial. *Revue Hématol.*, **7**, 362–7.

Collier, J. and Vallance, P. (1989) Second messenger role for NO widens to nervous and immune systems. *Trends Pharmacol. Sci.*, **10**, 427–31.

De Duve, C. (1978) An integrated view of lysosome function. In: *Molecular Basis of Biological Degradative Processes*. New York, Acad. Press, pp. 25–38.

De Duve, C., Pressman, B. C., Gianetto, R., Wattiaux, R. and Applemans, F. (1955) Tissue fractionation studies. Intracellular distribution patterns of enzymes in rat liver tissue. *Biochem. J.*, **60**, 604–17.

De Duve, C. and Wattiaux, R. (1966) Function of lysosomes. *Annu. Rev. Physiol.*, **28**, 435–92.

Ding, A., Nathan, C. F., Graycar, J., Derynck, R., Stuehr, D. J. and Srimal, S. (1990) Macrophage deactivating factor and transforming growth factors -β_1, -β_2, and -β_3 inhibit induction of macrophage nitrogen oxide synthesis by IFN-γ. *J. Immunol.*, **145**, 940–4.

Ding, A. H., Nathan, C. F. and Stuehr, D. J. (1988) Release of reactive nitrogen intermediates and reactive oxygen intermediates from mouse peritoneal macrophages. Comparison of activating cytokines and evidence for independent production. *J. Immunol.*, **141**, 2407–12.

Drapier, J. C. and Hibbs, J. B. (1986) Murine cytotoxic activated macrophages inhibit aconitase in human tumor cells. *J. Clin. Invest.*, **78**, 790–7.

Elsbach, P. and Weiss, J. (1983) A reevaluation of the roles of the O_2-dependent and O_2-independent microbicidal system of phagocytes. *Rev. Infect. Dis.*, **5**, 843–53.

Ferenčík, M. (1986) Lysosomal enzymes of professional phagocytes. *Folia Fac. Med. Univ. Comenianae Bratisl.*, **24**, 9–161.

Ferenčík, M. and Bergendi, Ľ. (1984) Biological importance of superoxide anion and other reactive forms of oxygen generated in the metabolism of aerobic cells. *Biol. Listy* (Prague), **49**, 1–25 (*in Slovak*).

Ferenčík, M. and Kotulová, D. (1988) Molecular reasons of phagocytosis disorders. *Prakt. Lék.* (Prague), **68**, 285–93 (*in Slovak*).

Ferenčík, M. and Štefanovič, J. (1977) Molecular bases of phagocytosis. *Biol. Listy* (Prague), **42**, 81–99 (*in Slovak*).

Ferenčík, M. and Štefanovič, J. (1979) Lysosomal enzymes of phagocytes and the mechanism of their release. *Folia Microbiol.*, **24**, 503–15.

Fleming, A. (1922) On a remarkable bacteriolytic element found in tissue and secretion. *Proc. Roy. Soc. Lond.* B, **93**, 306–17.

Galli, S. J. (1987) New approaches for the analysis of mast cell maturation, heterogeneity, and function. *Feder. Proc.*, **46**, 1906–14.

Gallin, J. I. (1985) Leukocyte adherence-related glycoprotein LFA-1, Mo-1 and p150,95: a new group of monoclonal antibodies, a new disease, and a possible opportunity to understand the molecular basis of leukocyte adherence. *J. Infect. Dis.*, **152**, 661–4.

Gallin, J. I., Fletcher, M. P., Seligmann, B. E., Hoffstein, S., Cehrs, K. and Mounessa, N. (1982) Human neutrophil-specific granule deficiency: a model to assess the role of neutrophil-specific granules in the evolution of the inflammatory response. *Blood*, **59**, 1317–29.

Ganz, T., Selsted, M. E., Szklarek, D., Harwig, S. S. L., Daher, K., Bainton, D. F. and Lehrer, R. I. (1985) Defensins. Natural peptide antibiotics of human neutrophils. *J. Clin. Invest.*, **76**, 1427–35.

Ganz, T., Selsted, M. E. and Lehrer, R. I. (1990) Defensins. *Eur. J. Haematol.*, **44**, 1–20.

Goetzl, E. J. (1983) Leukocyte recognition and metabolism of leukotrienes. *Feder. Proc.*, **42**, 3128–31.

Granger, D. L., Hibbs, J. B. Jr., Perfect, J. R. and Durack, D. T. (1988) Specific amino acid (L-arginine) requirement for the microbiostatic activity of murine macrophages. *J. Clin. Invest.*, **81**, 1129–37.

Gray, P. W., Flaggs, G., Leong, S. R., Gumina, R. J., Weiss, J., Ooi, C. E. and Elsbach, P. (1989) Cloning of the cDNA of a human neutrophil bactericidal protein. Structural and functional correlations. *J. Biol. Chem.*, **264**, 9505–9.

Griffin, F. M. Jr., Griffin, J. A., Leider, J. E. and Silverstein, S. C. (1975) Studies on the mechanism of phagocytosis. I. Requirements for circumferential attachment of particle-bound ligands to specific receptors on the macrophage plasma membrane. *J. Exp. Med.*, **142**, 1263–82.

Griffin, F. M. Jr., Griffin, J. A. and Silverstein, S. C. (1976) Studies on the mechanism of phagocytosis. II. The interaction of macrophages with anti-immunoglobulin IgG-coated bone marrow-derived lymphocytes. *J. Exp. Med.*, **144**, 788–809.

Haber, F. and Weiss, J. (1934) The catalytic decomposition of hydrogen peroxide by iron salts. *Proc. Roy. Soc.* A, **147**, 332–51.

Hashinaka, K., Nishio, C., Hur, S. J., Sakiyama, F., Tsunasawa, S. and Yamada, M. (1988) Multiple species of myeloperoxidase messenger RNAs produced by altered splicing and differential polyadenylation. *Biochemistry*, **27**, 5906–14.

Henderson, W. R. Jr., (1987) Lipid-derived and other chemical mediators of inflammation in the lung. *J. Allergy Clin. Immunol.*, **79**, 543–53.

Henderson, W. R. and Kaliner, M. (1978) Immunologic and nonimmunologic generation of superoxidase from mast cells and basophils. *J. Clin. Invest.*, **61**, 187–96.

Henson, P. M. (1971a) The immunologic release of constituents from neutrophil leukocytes. I. The role of antibody and complement on nonphagocytosable surfaces or phagocytosable particles. *J. Immunol.*, **107**, 1535–46.

Henson, P. M. (1971b) Interaction of cells with immune complexes: Adherence, release of constituents, and tissue injury. *J. Exp. Med.*, **134**, 114–35.

Hibbs, J. B., Taintor, R. R., Vavrin, Z. and Rachlin, E. M. (1988) Nitric oxide: a cytotoxic activated macrophage effector molecule. *Biochem. Biophys. Res. Commun.*, **157**, 87–94.

Hibbs, J. B. Jr., Vavrin, Z. and Taintor, R. R. (1987) L-Arginine is required for expression of the activated macrophage effector mechanism causing selective metabolic inhibition in target cells. *J. Immunol.*, **138**, 550–65.

Hirsch, J. G. (1956) Phagocytin: A bacterial substance from polymorphonuclear leukocytes. *J. Exp. Med.*, **103**, 589–611.

Hitzig, W. H. and Seger, R. A. (1983) Chronic granulomatous disease, a heterogeneous syndrome. *Hum. Genet.*, **64**, 207–15.

Holmes, B., Page, A. R. and Good, R. A. (1967) Studies of the metabolic activity of leukocytes from patients with a genetic abnormality of phagocytic function. *J. Clin. Invest.*, **46**, 1422–32.

Hugli, T. E. (1989) Chemotaxis. *Curr. Opin. Immunol.*, **2**, 19–27.

Hurst, N. P. (1987) Molecular basis of activation and regulation of the phagocyte respiratory burst. *Ann. Rheum. dis.*, **46**, 265–72.

Hurst, J. K. and Barrette, W. C. Jr. (1989) Leukocytic oxygen activation and microbicidal oxidative toxins. *Crit. Rev. Biochem. Molec. Biol.*, **24**, 271–327.

Iyengar, R. D., Stuehr, D. J. and Marletta, M. A. (1987) Macrophage synthesis of nitrite, nitrate, and *N*-nitrosamines: precursors and role of the respiratory burst. *Proc. Natl. Acad. Sci. USA*, **84**, 6369–73.

James, S. L. and Glaven, J. (1989) Macrophage cytotoxicity agains schistosomula of *Schistosoma mansoni* involved arginine-dependent production of reactive nitrogen intermediates. *J. Immunol.*, **143**, 4208–12.

Kaplan, S. S., Billiar, T., Curran, R. D., Zdziarski, U. E., Simmons, R. L. and Basford, R. E. (1989) Inhibition of chemotaxis with N^G-monomethyl-L-arginine: a role for cyclic GMP. *Blood*, **74**, 1885–93.

Kilbourne, R. G., Klostergaard, J. and Lopez-Berestein, G. (1984) Activated macrophages secrete a soluble factor that inhibits mitochondrial respiration of tumour cells. *J. Immunol.*, **133**, 2577–83.

Klebanoff, S. J. (1967) Iodination of bacteria: a bactericidal mechanism. *J. Exp. Med.*, **126**, 1063–78.

Klebanoff, S. J. (1968) Myeloperoxidase–halide–hydrogen peroxide antibacterial system. *J. Bacteriol.*, **95**, 2131–8.

Klebanoff, S. J. (1975) Antimicrobial mechanisms in neutrophilic polymorphonuclear leukocytes. *Semin. Hematol.*, **12**, 117–42.

Klebanoff, S. J. (1980) Oxygen intermediates and the microbicidal event. In: Van Furth, R. (ed) *Mononuclear Phagocytes: Functional Aspects.* Hague, Martinus Nijhoff Publ., pp. 1105–37.

Klebanoff, S. J. (1982) Oxygen-dependent cytotoxic mechanisms of phagocytes. In: Gallin, J. I. and Fauci, A. S. (eds) *Advances in Host Defense Mechanisms.* New York, Raven Press, vol. 1, pp. 111–62.

Korchak, H. M., Vienne, K., Rutherford, L. F., Wilkenford, C., Finkelstein, M. C. and Weissmann, G. (1984) Stimulus response coupling in the neutrophil: Temporal analysis of changes in cytosolic calcium and calcium effects. *J. Biol. Chem.*, **259**, 4076–81.

Koren, H. S. and others (1987) Proposed classification of leukocyte-associated cytolytic molecules. *Immunol. Today*, **8**, 69–71.

Lamberth, J. D. (1988) Activation of the respiratory burst oxidase in neutrophils: on the role of membrane-derived second messengers, Ca^{++}, and protein kinase C. *J. Bioenerg. Biomemb.*, **20**, 709–33.

Lehrer, R. I. (1975) The fungicidal mechanisms of human monocytes. I. Evidence for myeloperoxidase-linked and myeloperoxidase-independent candidacidal mechanisms. *J. Clin. Invest.*, **55**, 338–46.

Lehrer, R. I. and Ganz, T. (1990) Antimicrobial polypeptides of human neutrophils. *Blood*, **76**, 2169–81.

Liew, F. Y. and Cox, F. E. G. (1991) Nonspecific defense mechanism: the role of nitric oxide. *Immunoparazit. Today*, A17–A21.

Liew, F. Y., Millott, S., Parkinson, C., Palmer, R. M. J. and Moncada, S. (1990) Macrophage killing of *Leishmania* parasite *in vivo* is mediated by nitric oxide from L-arginine. *J. Immunol.*, **144**, 4794–7.

Malawista, S. E., Gee, J. B. L. and Bensch, K. G. (1971) Cytochalasin B reversibly inhibits phagocytosis: Functional, metabolic, and ultrastructural effects in human blood leukocytes and rabbit alveolar macrophages. *Yale J. Biol. Med.*, **44**, 286–92.

Malech, H. L. and Gallin, J. I. (1987) Neutrophils in human diseases. *New Engl. J. Med.*, **317**, 687–94.

Mann, T. and Keillin, D. (1939) Haemocuprein and hepatocuprein, copper-protein compounds of blood and liver in mammals. *Proc. Roy. Soc. Lond.* B, **126**, 303–15.

McCall, T. B., Boughton-Smith, N. K., Palmer, R. M. J., Whitte, B. J. R. and Moncada, S. (1989) Synthesis of nitric oxide from L-arginine by neutrophils. Release and interaction with superoxide anion. *Biohem. J.*, **261**, 293–6.

McCord, J. M. and Fridovich, I. (1969) Superoxide dismutase: an enzymic function for erythrocuprein (hemocuprein). *J. Biol. Chem.*, **244**, 6049–55.

McEver, R. P. (1991) Selectins: novel receptors that mediate leukocyte adhesion during inflammation. *Tromb. Haemostasis*, **65**, 223–8.

Metschnikoff, E. (1883a) Untersuchungen über die intrazellulare Verdaunung bei wirbellosen Tieren. *Arb. Zool. Inst. Univ. Wien*, **5**, 144–52.

Metchnikoff, E. (1883b) *Lectures on Comparative Pathology of Inflammation*. London, Paul, Kegan, Trench, Traubner and Co.

Meyer, B., Schmidt, K., Humbert, R. and Bohme, E. (1989) Biosynthesis of endothelium-derived relaxing factor, a cytosolic enzyme in porcine aortic endothelial cells Ca^{2+}-dependently converts L-arginine into an activator of guanylate cyclase. Biochem. *Biophys. Res. Commun.*, **164**, 678–85.

Mills, C. D. (1991) Molecular basis of "suppressor" macrophages. Arginine metabolism *via* the nitric oxide synthetase pathway. *J. Immunol.*, **146**, 2719–23.

Moncada, S., Palmer, R. M. J. and Higgs, E. A. (1989) Biosynthesis of nitric oxide from L-arginine. A pathway for the regulation of cell function and communication. *Biochem. Pharmacol.*, **38**, 1709–15.

Morishita, K., Tschiya, M., Asano, S., Kaziro, Y. and Nagata, S. (1987) Chromosomal gene structure of human myeloperoxidase and regulation of its expression by granulocyte colony-forming factor. *J. Biol. Chem.*, **262**, 15208–13.

Nathan, C. F. (1987) Secretory products of macrophages. *J. Clin. Invest.*, **79**, 319–26.

Nathan, C. F., Brukner, L., Silverstein, S. and Cohn, Z. A. (1979) Extracellular cytolysis by activated macrophages and granulocytes. II. Hydrogen peroxide as a mediator of cytotoxicity. *J. Exp. Med.*, **149**, 84–99.

Nauseef, W. M., Volpp, B. D., McCormick, S., Leidal, K. G. and Clark, R. A. (1991) Assembly of the neutrophil respiratory burst oxidase. *J. Biol. Chem.*, **266**, 5911–7.

Novikoff, A. B., Beaufay, H. and De Duve, C. (1956) Electron microscopy of lysosome-rich fractions from rat liver. *J. Biophys. Biochem. Cytol.*, Suppl. 2, 179–84.

Odeberg, H. and Olsson, I. (1975) Antibacterial activity of cationic proteins from human granulocytes. *J. Clin. Invest.*, **56**, 1118–25.

Palmer, R. M. J., Ferridge, A. G. and Moncada, S. (1987) Nitric oxide release accounts for the biological activity of endothelium-derived relaxing factor. *Nature*, **327**, 524–6.

Palmer, R. M. J. and Moncada, S. (1989) A novel citrulline-forming enzyme implicated in the formation of nitric oxide by vascular endothelial cells. *Biochem. Biophys. Res. commun.*, **158**, 348–52.

Parry, M. F., Root, R. K., Metcals, J. A., Celaney, K. K., Kaplow, L. S. and Richar, W. J. (1981) Myeloperoxidase deficiency. Prevalence and clinical significance. *Ann. Intern. Med.*, **95**, 293–301.

Pastan, I. and Willingham, M. C. (1985) The pathway of endocytosis. In: Pastan, I. and Willingham, M. C. (eds) *Endocytosis*. New York, Plenum Publ. Corp., pp. 1–44.

Podack, E. R. (1986) Molecular mechanisms of cytolysis by complement and by cytolytic lymphocytes. *J. Cell Biochem.*, **30**, 133–70.

Podack, E. R. and Konigsberg, P. J. (1984) Cytolytic *T*-cell granules. Isolation, structural, biochemical, and functional characterization. *J. Exp. Med.*, **160**, 695–710.

Pommier, C. G., O'Shea, J., Chused, T., Yancey, K., Frank, M. M., Takahashi, T. and Brown, E. J. (1984) Studies on the fibronectin receptors of human peripheral blood leukocytes. *J. Exp. Med.*, **159**, 137–51.

Radomski, M. W., Palmer, R. M. J. and Moncada, S. (1990) Glucocorticoids inhibit the expression of an inducible, but not the constitutive, nitric oxide synthase in vascular endothelial cells. *Proc. Natl. Acad. Sci. USA*, **87** 10043–7.

Roberts, R. and Gallin, J. I. (1983) The phagocytic cell and its disorders. *Ann. Allergy*, **50**, 330–45.
Roitt, I. M. (1981) *Essential Immunology*. Martin, Osveta, 320 pp. (*in Slovak*).
Root, R. K. and Cohen, M. S. (1981) The microbicidal mechanism of human neutrophils and eosinophils. *Rev. Infect. Dis.*, **3**, 565–98.
Rossi, F. (1986) The O_2^--forming NADPH oxidase of the phagocytes: nature, mechanisms of activation and function. *Biochim. Biophys. Acta*, **853**, 65–89.
Rossi, A. G., McMilan, R. M. and MacIntyre, D. E. (1988) Agonist-induced calcium flux, phosphoinositide metabolism, aggregation and enzyme secretion in human neutrophils. *Agents Actions*, **24**, 272–82.
Rothenberg, B. E. (1978) The self recognition concept: an active function for the molecules of the major histocompatibility complex based on the complementary interaction of protein and carbohydrate. *Develop. Comp. Immunol.*, **2**, 23–37.
Ruoslahti, E. (1991) Integrins. *J. Clin. Invest.*, **87**, 1–5.
Sbarra, A. J. and Karnovsky, M. L. (1959) The biochemical basis of phagocytosis. I. Metabolic changes during the ingestion of particles by polymorphonuclear leukocytes. *J. Biol. Chem.*, **234**, 1355–62.
Schmidt, H. H. H. W., Seifert, R. and Bohme, E. (1989) Formation and release of nitric oxide from human neutrophils and HL-60 induced by a chemotactic peptide, platelet activation factor, and leukotriene B_4. *FEBS Lett.*, **244**, 357–62.
Schnyder, J. and Baggiolini, M. (1978) Secretion of lysosomal hydrolases by stimulated and nonstimulated macrophages. *J. Exp. Med.*, **148**, 435–50.
Sechler, J. M. G., Malech, H. L., White, C. J. and Gallin, J. I. (1988) Recombinant human interferon-γ reconstitutes defective phagocyte function in patients with chronic granulomatous disease of childhood. *Proc. Natl. Acad. Sci. USA*, **85**, 4874–8.
Segal, A. W. (1989) The electron transport chain of the microbicidal oxidase of phagocytic cells and its involvement in the molecular pathology of chronic granulomatous disease. *J. Clin. Invest.*, **83**, 1785–93.
Segal, A. W., Garcia, R., Goldstone, H., Cross, A. R. and Jones, O. T. (1981) Cytochrome b-245 of neutrophils is also present in human monocytes, macrophages and eosinophils. *Biochem. J.*, **196**, 363–7.
Shafer, W. M., Onunka, V. C. and Martin, L. E. (1986) Antigonococcal activity of human neutrophil cathepsin G. *Infect. Immunol.*, **54**, 184–92.
Shafer, W. M., Pohl, J., Onunka, V. C., Bangalore, N. and Travis, J. (1991) Human lysosomal cathepsin G and granzyme B share a functionally conserved broad spectrum antibacterial peptide. *J. Biol. Chem.*, **266**, 112–6.
Sharon, N. (1984) Surface carbohydrates and surface lectins are recognition determinants in phagocytosis. *Immunol. Today*, **5**, 143–7.
Silverstein, S. C., Greenberg, A., DiVirgilio, F. and Steinberg, T. H. (1989) Phagocytosis. In: Paul, W. E. (ed) *Fundamental Immunology*. 2nd ed. New York, Raven Press, pp. 703–20.
Silverstein, S. C., Steinman, R. M. and Cohn, Z. A. (1977) Endocytosis. *Annu. Rev. Biochem.*, **46**, 669–722.
Sklar, L. A. (1986) Ligand-receptor dynamics and signal amplification in the neutrophil. *Adv. Immunol.*, **39**, 95–143.
Smolen, J. E., Korchak, H. M. and Weissmann, G. (1980) Initial kinetics of lysosomal enzyme secretion and superoxide anion generation by human polymorphonuclear leukocytes. *Inflammation*, **4**, 145–63.
Smolen, J. E. and Weissmann, G. (1981) Stimuli provoke secretion of azurophil enzymes from human neutrophils induce increments in adenosine cyclic 3′,5′-monophosphate. *Biochim. Biophys. Acta*, **672**, 197–206.
Snyderman, R. and Pike, M. C. (1984) Chemoattractant receptors on phagocytic cells. *Ann. Rev. Immunol.*, **2**, 257–81.
Spitznagel, J. K. (1984) Nonoxidative antimicrobial reactions of leukocytes. *Contemp. Top. Immunobiol.*, **14**, 283–343.
Spitznagel, J. K. (1990) Antibiotic proteins of human neutrophils. *J. Clin. Invest.*, **86**, 1381–6.

Stevens, R. L. and Austen, K. F. (1989) Recent advances in the cellular and molecular biology of mast cells. *Immunol. Today*, **10**, 381–6.

Stuehr, D. J. and Nathan, C. F. (1989) Nitric oxide. A macrophage product responsible for cytostasis and respiratory inhibition in tumour target cells. *J. Exp. Med.*, **169**, 1534–55.

Stuehr, D. J., Kwon, N. S., Cho, H. J. and Nathan, C. F. (1990) FAD and GSH participate in macrophage synthesis of nitric oxide. *Biochem. Biophys. Res. Commun.*, **168**, 558–65.

Tayeh, M. A. and Marletta, M. A. (1989) Macrophage oxidation of L-arginine to nitric oxide, nitrite, and nitrate: tetrahydrobiopterin is required as a cofactor. *J. Biol. Chem.*, **264**, 19654–8.

Tobler, A. and Koeffler, P. (1991) Myeloperoxidase: localization, structure, and function. In: Harris, J. R. (ed). *Blood Cell Biochemistry*. New York, Plenum Publ. Corp., vol. 3, pp. 255–88.

Todd, R. F. and Freyer, D. R. (1989) The CD11/CD18 leukocyte glycoprotein deficiency. *Hematol./Oncol. Clin. N. Amer.*, **2**, 13–31.

Tschopp, J. and Nabholz, M. (1990) Perforin-mediated target cell lysis by cytolytic *T* lymphocytes. *Annu. Rev. Immunol.*, **8**, 279–302.

Unanue, E. R. (1989) Macrophages, antigen-presenting cells, and the phenomena of antigen handling and presentation. In: Paul, W. E. (ed). *Fundamental Immunology*. 2nd ed. New York, Raven Press, pp. 95–115.

Van Furth, R. (1982) Current view on the mononuclear phagocyte system. *Immunobiology*, **161** 178–85.

Van Furth, R. (1985) Cellular biology of pulmonary macrophages. *Int. Archs. Allergy Appl. Immunol.*, **76**, Suppl. 1, 21–7.

Van Furth, R., Cohn, Z. A., Hirsch, J. G., Humphrey, J. H., Spector, W. G. and Langevoort, H. L. (1972) The mononuclear phagocyte system. A. new classification of macrophages, monocytes and their precursor cells. *Bull. WHO*, **46**, 845–53.

Van Furth, R., Gond, T. J. L. M., Van der Meer, J. W. T., Blussé van Oud Alblas, A., Diesselhoff-den Dulk, M. M. C. and Schadewijk-Nieuwstad, M. (1982) Comparison of the *in vivo* and *in vitro* proliferation of monoblast, promonocytes, and the macrophage cell line J774. In: Norman, D. J. and Sorkin, E. (eds). *Macrophages and Natural Killer Cells*. New York, Plenum Press, pp. 175–87.

Van Tuinen, P., Johnson, K. R., Ledbetter, S. A., Nussbaum, R. L., Rovera, G. and Ledbetter, D. H. (1988) Localization of myeloperoxidase to the long arm of human chromosome 17: Relationship to the 15; 17 translocation of acute promyelocytic leukemia. *Oncogene*, **1**, 319–22.

Weir, D. M. (1980) Surface carbohydrates and lectins in cellular recognition. *Immunol. Today*, **1**, 45–51.

Weiss, J., Elsbach, P., Olsson, I., Odelberg, H. (1978) Purification and characterization of a potent microbicidal and membrane active protein from the granules of human polymorphonuclear leukocytes. *J. Biol. Chem.*, **253**, 2664–72.

Weissmann, G., Korchak, H. M., Perez, H. D., Smolen, J. E., Goldstein, I. M. and Hoffstein, S. T. (1979) The secretory code of neutrophil. *J. Reticuloendothel. Soc.*, **26**, Suppl., 687–700.

Weissmann, G., Zurier, R. B. and Hoffstein, S. (1973) Leukocytes as secretory organs of inflammation. *Agents Actions*, **3**, 370–9.

Weissmann, G., Zurier, R. B., Spieler, P. J. and Goldstein, I. M. (1971) Mechanisms of lysosomal enzyme release from leukocytes exposed to immune complexes and other particles. *J. Exp. Med.*, **134**, 149–65.

Weller, P. F. (1991) The immunobiology of eosinophils. *N. Engl. J. Med.*, **324**, 1110–8.

White, C. J. and Gallin, J. I. (1986) Phagocyte defects. *Clin. Immunol. Immunopathol.*, **40**, 50–61.

Winterbourn, C. C., Garcia, R. C. and Segal, A. W. (1985) Production of the superoxide adduct of myeloperoxidase (compound III) by stimulated human neutrophils and its activity with hydrogen peroxide and chloride. *Biochem. J.*, **228**, 583–92.

Wood, P. M. (1987) The two redox potentials for oxygen reduction to superoxide. *Trends Biochem. Sci.*, **12**, 250–1.

Wright, A. E. and Douglas, S. R. (1903) An experimental investigation of the role of the body fluids in connection with phagocytosis. *Proc. Roy. Soc. Lond.*, **72**, 357–62.

Wright, D. G. (1982) The neutrophil as a secretory organ of host defense. In: Gallin, J. I. and Fauci, A. S. (eds). *Advances in Host Defense Mechanisms*. New York, Raven Press, vol. 1, pp. 75–110.

Wright, S. D. and Silverstein, S. C. (1983) Receptors for C3b and C3bi promote phagocytosis but not the release of toxic oxygen human phagocytes. *J. Exp. Med.*, **158**, 2016–23.

Zeya, H. I. and Spitznagel, J. K. (1971) Characterization of cationic protein-bearing granules of polymorphonuclear leukocytes. *Lab. Invest.*, **24**, 229–38.

Zgliczynski, J. M., Selvaraj, R., Paul, B. B., Stelmaszynska, T., Poskitt, K. and Sbarra, A. J. (1977) Chlorination by myeloperoxidase-H_2O_2-Cl-antimicrobial system at acid and neutral pH. *Proc. Soc. Exp. Biol. Med.*, **154**, 418–22.

Abbreviations

A	accessory cells
Ab	antibody
AB0	system of blood groups A, B, AB and 0
ACTH	adrenocorticotropic hormone
ADA	adenosine deaminase
ADCC	antibody-dependent cell-mediated cytotoxicity
Ag	antigen
AIDS	acquired immunodeficiency syndrome
Am	allotype marker of human heavy IgA chains
ARIS	apoenzyme reactivation immunoassay system
B	factor B of the alternative pathway of complement activation
B-cell	*B* lymphocyte
BCDF	*B*-cell differentiating factor
BCG	Calmett-Guérin bacillus
BCGF	*B*-cell growing factor
BPI	bactericidal permeability-increasing protein
C	complement
cAMP	cyclic adenosine-3′,5′-monophosphate
CD	cluster of differentiation
cDNA	complementary DNA
CDR	complementarity determining region
CEA	carcinoembryonic antigen
CEDIA	combined enzyme donor immunoassay
CF	chemotactic factor
CGD	chronic granulomatous disease
cGMP	cyclic guanosine-3′,5′-monophosphate
CH_{50}	unit of haemolytic complement activity
C_H	constant part of an immunoglobulin heavy chain
CHS	Chédiak–Higashi syndrome
CIA	chemiluminescence immunoassay
C_L	constant part of an immunoglobulin light chain
CML	cell-mediated cytolysis
ConA	concanavalin A
CR	complement receptor

CRP	*C*-reactive protein
CSF	colony-stimulating factor
CTL	cytotoxic *T*-lymphocyte
D	gene segment for heavy immunoglobulin chain
D	factor of the alternative pathway of complement activation
DAF	decay-accelerating factor
DELFIA	dissociation-enhanced lanthanide fluoroimmunoassay
DIBA	dot-immunobinding assay
DNA	deoxyribonucleic acid
DTH	delayed-type hypersensitivity
E	erythrocyte or eosinophil
ECF	eosinophil chemotactic factor
ECP	eosinophil cationic protein
EDRF	endothelium-derived relaxing factor
EDTA	ethylene diamine tetra-acetic acid
EIA	enzyme immunoassay
ELISA	enzyme-linked immunosorbent assay
ELISPOT	enzyme-linked immunospot
EMIT	enzyme multiplied immunoassay technique
Eo	eosinophil
EPO	eosinophil peroxidase
EPRIA	enzyme potentiated radioimmunoassay
FA	Freund's adjuvant
Fab	antigen-binding fragment of an immunoglobulin molecule
Fc	crystallizable fragment of an immunoglobulin molecule
FcR	receptor of immunoglobulin Fc-domains
FETIA	fluorescence excitation transfer immunoassay
FIA	fluorescence immunoassay
FITC	fluoresceine isothiocyanate
FMLP	*N*-formylmethionyl-leucyl-phenylalanine
FPIA	fluorescence polarization immunoassay
FRAT	free radical assay technique
FTS	thymulin
G6PDH	glucose-6-phosphate dehydrogenase
Gal	galactose
GalN	galactosamine
G-CSF	granulocyte colony-stimulating factor
Glc	glucose
GlcN	glucosamine
Gm	allotype marker of human heavy IgG chains
GM-CSF	granulocyte-macrophage colony-stimulating factor
GTP	guanosine triphosphate
GVH	graft versus host (reaction)
H	factor of the complement system
H	hapten
H	histamine
H	histocompatibility

H-antigens	histocompatibility antigens
H-chain	immunoglobulin heavy chain
H-2	main mouse histocompatibility system
HAT-medium	selective culture medium containing hypoxanthin, aminopterin and thymidine
HBsAg	surface antigen of hepatitis B virus
HETE	hydroxyeicosatetraenic acid
HGPRT	hypoxanthine(guanosine)-phosphoribosyl transferase
HLA	human leukocyte antigens
HMPS	hexose monophosphate shunt
hnRNA	heterogeneous nuclear ribonucleic acid
HR	factor H receptor
HRF	homologous restriction factor
HTC	homozygote typization cells
HVG	host versus graft (reaction)
I	factor I (C3b-inactivator)
Ia	Ia-antigen
ICLMA	immunochemiluminometric assay
IEF	immunoelectrophoresis
IFMA	immunofluorometric assay
IFN	interferon
Ig	immunoglobulin
IgA, IgD, IgE, IgG, IgM	immunoglobulin classes
IL-1–IL-11	interleukins 1–11
IP_3	inositol-1,4,5-triphosphate
Ir	Ir-gene
IRMA	immunoradiometric assay
IU	international unit
IUIS	International Union of Immunological Societies
IVIgG	therapeutic intravenous preparations of IgG
J	junction part of immunoglobulin chain
J	joining chain in IgA and IgM molecules
K	*K*-cell
kbp	kilobase pairs — a unit indicating thousands of nucleotide pairs (in DNA or RNA molecules)
K-cell	killer cell
Km	allotype marker of human immunoglobulin $\varkappa$ light chains
LAD	leukocyte adhesion deficiency
LAK	lymphokine-activated *K*-cells
LAT	latex agglutination test
LF	lactoferrin
LGL	large granular lymphocytes
LIF	leukocyte migration inhibition factor
LT	leukotrien or lymphotoxin
Ma or M	macrophage

MAC	membrane attack complex (C5b6789)
MAF	macrophage activating factor
MCF	macrophage chemotactic factor
MCP	membrane cofactor protein
MCP	monocyte chemoattractant protein
M-CSF	macrophage colony-stimulating factor
MDH	malate dehydrogenase
MHC	major histocompatibility complex
MHS	major histocompatibility system
MIA	metalloimmunoassay
MIF	macrophage migration inhibitory factor
MIP	macrophage inflammatory protein
MIRF	macrophage Ia-positive recruiting factor
MLC	mixed lymphocyte culture
MPO	myeloperoxidase
MPS	mononuclear phagocytic system
M_r	relative molecular weight
mRNA	messenger ribonucleic acid
NAD	nicotinamide adenine dinucleotide (oxidized form)
NADH	nicotinamide adenine dinucleotide (reduced form)
NADP	nicotinamide adenine dinucleotide phosphate (oxidized form)
NADPH	nicotinamide adenine dinucleotide phosphate (reduced form)
NAP	neutrophil-activating protein
NC	*NC*-cells — natural cytotoxic cells
NK	*NK*-cells — natural killer cells
NKCF	*NK*-cell cytotoxic factor
O	O-antigens
P	properdin
PACIA	particle counting immunoassay
PAF	platelet activating factor
PAGE	polyacrylamide gel electrophoresis
PBP	platelet basic protein
PBS	phosphate buffered saline
PCA	passive cutaneous anaphylaxis
PEG	polyethyleneglycol
PETIA	particle enhanced turbidimetric immunoassay
PETINIA	particle enhanced turbidimetric inhibition immunoassay
PFC	plaque-forming cells
PG	prostaglandin
PGLIA	prosthetic-group-label immunoassay
PHA	phytohaemagglutinin
PHI	passive haemolysis inhibition
PK	Prausnitz–Küstner test

PKC	protein kinase C
PMA	phorbol-12-myristate-13-acetate
PMN	polymorphonuclear
PNP	purine nucleotide phosphorylase
R	receptor
RAST	radioallergosorbent test
RES	reticuloendothelial system
Rh	blood group system Rh (rhesus)
RIA	radioimmunoassay
RID	radial immunodiffusion
RNA	ribonucleic acid
RNI	reactive nitrogen intermediates
ROI	reactive oxygen intermediates
S	sedimentation constant
SC	secretory component
SCID	severe combined immunodeficiency
SCPN	serum carboxypeptidase *N*
SD	serologically detectable (markers, antigens)
SDS	sodium dodecyl-sulphate
SIA	spin immunoassay
SIgA	secretory immunoglobulin A
SLE	systemic lupus erythematosus
SLFIA	substrate-labelled fluoroimmunoassay
SLO	streptolysin O
SOD	superoxide dismutase
SPALT	solid-phase antigen luminescence technique
SPIA	sol particle immunoassay
SRS-A	slow-reacting substance of anaphylaxis
T	*T*-cell
T-cell	*T*-lymphocyte
T_C	cytotoxic *T*-lymphocyte
TF	transfer factor
TGF	transforming growth factor
THF	thymus humoral factor
$T_{H/I}$	helper and inducer *T*-lymphocytes
TK	thymidine kinase
TNF	tumour necrosis factor
TPK	tyrosine protein kinase
T_S	suppressor *T*-lymphocytes
TS	thymostimulin
TX	thromboxane
V	variable region of immunoglobulin chains
VIA	virus immunoassay

Glossary of terms commonly used in immunochemistry

AB0 — a system of blood groups based on alloantigens present on the surface of human erythrocytes.

Accessory cells — cells belonging to the mononuclear phagocytic system (macrophages), dendritic and other cells presenting antigen and cooperating in antibody formation or in other immune reactions with *T*- and *B*-lymphocytes.

Acquired immunodeficiency syndrome (AIDS) — a viral disease in which the human immunodeficiency virus (HIV-1 or HIV-2) attacks lymphocytes and some other cells (*e.g.* brain cells, epithelial cells of the digestive tract). The virus selectively destroys the helper *T*-cell subgroup in particular leading to a serious impairment in the regulation of immune processes. The affected individual loses the ability to defend himself against microorganisms that do not usually cause disease in healthy subjects as well as against oncogenic viruses.

Activated lymphocytes — lymphocytes stimulated towards growth, cell division and differentiation by a specific antigen or non-specific mitogen.

Activated macrophages — macrophages exhibiting an increased antimicrobial, tumouricidal, adherent, phagocytic and pinocytic activity induced by different stimuli, such as interaction with microbial products, lymphokines released from lymphocytes, immunoglobulins and other humoral factors.

Acute phase proteins — proteins whose level in blood serum sharply increases in inflammation. They include mainly *C*-reactive protein (CRP), α_2-macroglobulin, α_1-antitrypsin, orosomucoid, ceruloplasmin and haptoglobin.

ADCC — antibody-dependent cell-mediated cytotoxicity. A process performed by *K*-cells and some T_C-lymphocytes that kills target cells coated by specific antibodies.

Adhesins — membrane glycoproteins which play an essential role in cell adherent activities. On the surface of human leukocytes the LFA-1 family of adhesive glycoproteins has been best characterized.

Adjuvants — compounds increasing the immune response to a simultaneously administered antigen.

Affinity chromatography — a separation technique using specific binding between a pair of substances, *e.g.* antigen–antibody or enzyme–substrate. One of the pair is covalently bound to an insoluble carrier. This is known as a ligand or affinant. When a mixture of compounds is passed through such a carrier, only the substance with a specific activity towards the ligand (the second compound in the pair) is adsorbed.

Agammaglobulinaemia — a condition in which the blood serum total immunoglobulin concentration is less than 1 g/l.

Agar — a polysaccharide isolated from seaweed. It is dissolved in boiling water and then cooled to room temperature to form semisolid gels.

Agarose — a component of agar.

Agglutination — the reaction between an insoluble (cellular) antigen and a soluble antibody. The antigens are usually cells (bacteria, erythrocytes) which are agglutinated by antibodies and sediment in a characteristic way. The sediment is called an agglutinate. Antigens reacting during agglutination are called agglutinogens and corresponding antibodies are known as agglutinins.

Allele — a form of gene. Every gene has at least two alleles differing in their phenotypic effects. A diploid individual has two or more alleles of every gene that can be functionally identical (homozygote) or different (heterozygote).

Allergen — an antigen inducing allergic hypersensitivity and usually mediated by IgE antibodies.

Allergic reactions — hypersensitivity reactions in which immune mechanisms damage the host cells and tissues. According to Coombs and Gell four basic types of allergic reactions exist. 1. Reactions which are in man mediated mainly by IgE antibodies. They are also known as reactions of immediate or anaphylactic type of hypersensitivity. 2. Reactions in which IgG or IgM antibodies reacting with antigens of target cells are involved. Thus, complement is immediately activated, damaging the target cells and the surrounding tissues. This is the cytotoxic type of hypersensitivity. 3. Complement can also be activated by immune complexes. During complement activation, chemotactic factors are formed, which attract granulocytes to the site where the immune complexes are deposited. Lysosomal enzymes (particularly proteinases) are released damaging the host tissue. Reactive oxygen products released particularly from neutrophils are also involved in this tissue damage. This type of reaction is known as hypersensitivity to immune complexes. 4. Cell-mediated hypersensitivity. The antigen induces a sequence of reactions, during which various cell types interact. Specific *T*-lymphocytes are also involved through production of lymphokines and other bioregulators. The tissue is then infiltrated by lymphocytes, monocytes and macrophages often resulting in an inflammatory reaction and necrosis of the tissue. There are usually 1–2 days between administration of the antigen to a sensitized individual and the resultant reaction. This type is therefore known as delayed hypersensitivity.

Allergy — altered immunological reactivity of the organism, usually manifested as hypersensitivity with pathological consequences.

Alloantigens — antigens distinguishing two genetically non-identical individuals of the same biological species. These antigens were previously termed isoantigens or homologous antigens. The term "isoantigen" is still used in haematology.

Allograft — a tissue or organ transplant (graft) between two genetically different individuals of the same biological species.

Allograft rejection — the immunological reaction of a recipient against a tissue transplant from a genetically different donor of the same biological species.

Allotypes — genetically determined markers of the antigenic structure of serum proteins, particularly immunoglobulins, that are present in some, but not all individuals of a given species. These markers are the products of individual alleles of an identical structural gene.

Amboceptor — a specific antibody against erythrocytes which induces their lysis in the presence of complement.

Amino acid sequence — the sequence (linear arrangement) of amino acid residues in the polypeptide chain.

Anaphylactic shock — a condition due to the presence of immune complexes resulting from the reaction of precipitating antibodies with the corresponding antigen. Such antibodies were formed after previous sensitization of an individual with the antigen and usually belong to the IgG class. The resultant immune complexes activate complement within a few seconds resulting in production of anaphylatoxins releasing histamine and other anaphylaxis mediators.

Anaphylatoxins — factors which are formed during complement activation (C5a, C3a, C4a) and which increase permeability of blood vessels, induce contraction of smooth muscles and stimulate release of histamine and other mediators from mast cells.

Anaphylaxis — an immediate hypersensitivity reaction to an antigen which occurs in almost all vertebrates. It is provoked by the binding of an antigen to cytotropic antibodies bound to the surface of mast cells followed by sensitization of these cells. This results in release of histamine and other mediators of anaphylaxis from the mast cells and basophils which, in turn, results in the contraction of smooth muscles, increased vascular permeability and other effects.

Antibody affinity — a measure of the strength of binding between the antigen determinant and the antibody binding site.

Antibody avidity — a measure of the strength of the bond between antigen and antibody molecules.

Antibody combining site — a space delineated by the hypervariable parts of one light and one heavy immunoglobulin chain to which the antigenic determinant is bound.

Antibody valency — the number of binding sites in the antibody molecule that can combine with the antigenic determinant.

Antigen presentation — a mechanism by which the antigen is presented to surface receptors of immunocompetent lymphocytes so as to stimulate them and induce an immune response.

Antigenic determinant — that part of an antigen molecule responsible for its ability to react specifically with an antibody combining site. There are two types of antigenic determinants — sequential and conformational.

Antigens — substances which are able to induce the specific immune response, *i.e.* antibody production, formation of effector cells of cell-mediated immunity, specific immunological tolerance, and to react specifically with the products of this response (antibodies and cells). Only complete antigens (immunogens) possess this ability. Incomplete antigens (haptens) cannot induce the immune response but can specifically react with its products. Thymus-dependent and thymus-independent antigens may be distinguished by the way they are recognized by *T*-lymphocytes.

Antigens of immune response (Ia antigens) — belong to the histocompatibility antigens (MHC class II products). They are responsible for antigen presentation and correct cellular cooperation during the immune response.

Antitoxins — antibodies neutralizing soluble bacterial toxins.

Atopic reaction — a reaction initiated by binding of the antigen (allergen) with IgE antibodies that have been bound to the surface of mast cells, followed by release of histamine and other mediators of anaphylaxis.

Atopy — a hereditary abnormal immediate-type hypersensitivity to a certain allergen or a group of allergens. It is due to an increased IgE antibody response. A particular individual (atopic) responds even to very low doses of the allergen, the resultant specific IgE persists for an abnormally long time without additional stimulation by the allergen and the level of total IgE also increases.

Autoantibodies — antibodies induced by autoantigens.

Autoantigens — antigens produced by the organism itself. Under certain conditions they can induce immune (usually immunopathological) reactions. In transplantation terminology such antigens are known as autochtonic.

Autoimmunity — an immunopathological state characterized by an immune response in which host tissues and cells are involved (response to autoantigens).

Autotolerance — the inability of the organism to stimulate immunocompetent cells by components of host tissues and cells.

Bacteriocidal — the ability to kill bacteria.

Bacteriolysis — damage of the bacterial cell envelope and release of the cytoplasmic contents into the surrounding environment.

***B*-cell** — *B*-lymphocyte, a type of lymphocyte derived from the Bursa of Fabricius in birds and from its equivalents in mammals. It carries antigen-specific immunoglobulin receptors on its surface. *B*-cells are precursors of plasma cells which produce antibodies.

Bursa of Fabricius — the central lymphatic organ of birds, located at the outer side of the intestine next to the cloaca, which regulates the development of *B*-lymphocytes.

Cachectin — one of the tumour necrosis factors.

Cell clone — a population of cells with identical genetic equipment (identical genotype) which originate from a single parent cell.

Cellular immunity — cell-mediated immunity; *T*-lymphocytes are the active cells of specific cellular immunity, whereas "professional" phagocytes are responsible for non-specific cell immunity.

Chemiluminescence — production of light in some exergonic (energy-releasing) chemical reactions. Products with excited electrons can be produced in such reactions. When the electrons regain their basic energy level, a light quantum (photon) is irradiated and can be measured in a luminometer.

Chemiluminescence immunoassay (CIA) — an immunochemical method to measure an antibody or antigen (hapten), one of which is labelled with a chemiluminophore (a compound capable of chemiluminescence).

Chemokinesis — stimulated, but undirected movement (migration) of cells.

Chemotaxis (cytotaxis) — the active migration of phagocytes (or of cells in general) in the direction (positive chemotaxis) or against the direction (negative chemotaxis) of the concentration gradient of chemotactic factors. It is a stimulated and controlled (directed) cell migration.

Chimeric antibodies — immunoglobulin molecules which are derived partly from one biological species, and partly from another.

Chronic granulomatous disease (CGD) — a serious, hereditary disease of children (boys in particular) based on the inability of neutrophilic granulocytes to induce the respiratory burst during phagocytosis. This results in insufficient production of hydrogen peroxide which is required for killing microorganisms. Children with this disease suffer repeated infections that are difficult to treat and which often lead to death.

Clonal selection theory of antibody formation — any circulating lymphocyte contains cytoplasmic membrane antigenic receptors with only a certain specificity. After stimulation it may produce antibodies with only one type of combining site or only one type of effector cell. Thus, each antigenic determinant stimulates only one type of *B*-lymphocyte giving rise to a cell clone producing antibodies with an identical specificity.

Complement — a system of serum glycoproteins representing the basic effector system of non-specific immunity. Individual complement components are inactive under physiological conditions but may be activated in the presence of immune complexes (classical activation pathway) or by other compounds (alternative activation pathway). Components of the complement system activated by the alternative mechanism are known as factors.

Complement fixation — a serological method used to detect an antigen–antibody reaction, during which available complement is activated and bound to the originating immune complexes. A known quantity of complement is added to the reaction mixture and its consumption (decrease) is therefore proportional to the antigen or antibody concentration.

Conformation determinant — an antigenic determinant whose specificity is determined by the spatial arrangement of the antigen molecule rather than by the sequence of its basic subunits (*e.g.* amino acids in protein antigens).

Conventional antibodies — antibodies produced by various animal species after accidental, intentional or experimental immunization. They are characterized by their heterogeneity.

Counter-immunoelectrophoresis — an immunoprecipitation technique which is a combination of one-dimensional double immunodiffusion with electrophoresis. It is suitable for the rapid demonstration of a particular antigen or antibody in the test sample.

Cross reaction — the reaction of an antibody against one antigen with another antigen which has identical or similar antigenic determinants.

Crossed immunoelectrophoresis (two-dimensional immunoelectrophoresis) — a combination of zonal electrophoresis with rocket immunoelectrophoresis.

Cytokines — a broad class of glycoprotein or protein cell regulators variously termed monokines, lymphokines, interleukins, and interferons, which control the amplitude and duration of immune responses. They form the cytokine network responsible for intercellular communication in the immune system.

Cytotaxigenes — precursors of cytotaxins.

Cytotaxins — chemotactic factors, compounds inducing chemotactic movement (regulated migration) of various cells.

Cytotoxic antibodies — antibodies which react with antigens on the surface of target cells leading to their damage by complement or cytotoxic cells cooperating with such antibodies.

Cytotoxic cells — various lymphocytes and macrophages which damage or kill target cells.

Cytotoxicity — the ability to damage a cell.

Cytotropic (cytophilic) antibodies — IgE or IgG antibodies which sensitize cells leading to a subsequent anaphylactic reaction.

Degradation of antigen — degradation of an antigen molecule to its basic components.

Degranulation — a process in which the cytoplasmic granules of phagocytes or mast cells release their contents.

Delayed hypersensitivity — an excessive reaction to the repeated administration of a particular antigen or group of antigens. The process involves *T*-lymphocytes, and the reaction is observed two, and occasionally three days, after exposure to the antigen.

Dendritic cells — a group of non-lymphoid mononuclear cells involved in the immune response in a similar way as macrophages. However, they differ from macrophages morphologically (dendritic projections), by their more efficient accessory function, substantially lower phagocytic capacity and by a very low number of lysosomes and low activity of lysosomal enzymes.

Determinant —*see* antigenic determinant.

Diapedesis — the passage of migrating cells (particularly granulocytes and monocytes) from the bloodstream across the walls of blood capillaries to the surrouding tissue.

Domains of immunoglobulins — parts of *H*- and *L*-chains with an identical tridimensional structure. Each domain consists of about 110 amino acid residues and is stabilized by a single disulphide bridge.

Effector cells — usually *T*-lymphocytes sensitized by the antigen and then acting cytotoxically on the target cells (which have this antigen on their surface) or exhibiting suppressive or stimulatory effects on the course of immune response.

Electrophoresis — a method to separate a complex mixture into its individual compounds by direct current. The compounds migrate in the electric field depending on the surface charge of their molecules.

Endocytosis — the process of engulfment of external material by the cell. It includes phagocytosis and pinocytosis.

Endotoxins — lipopolysaccharides contained in the cell walls of Gram-negative bacteria. After injection *in vivo* they induce toxic and pyrogenic effects affecting non-specific immunity and inducing formation of specific antibodies. They also exhibit mitogenic activity for *B*-lymphocytes.

Enzyme immunoassay (EIA) — an immunochemical method for determining an antigen (hapten) or antibody. The concentration of the test substance is determined by measuring the activity of an enzyme label with an appropriate substrate. When the activity of the enzyme labelling one of the reactants is changed after the antigen–antibody complex has been formed, the term homogeneous EIA is used. If the enzyme activity remains unchanged, the technique is known as heterogeneous EIA.

Enzyme-linked immunosorbent assay (ELISA) — a type of enzyme immunoassay in which one of the reactants (antigen or antibody) is immobilized by binding to a solid carrier (usually the wall of a test tube or well of a microtitre plate).

Epitope — synonym of antigenic determinant.

Equivalence — the antigen–antibody concentration ratio at which the highest quantity of an immunoprecipitate is formed.

Equivalent of Bursa of Fabricius — the unidentified anatomical site (probably bone marrow and some others areas of lymphoid tissue) which acts as the central lymphatic organ in mammals, and in which the population of *B*-lymphocytes develops.

Exon — a DNA region which is transcribed into mRNA and a specific protein.

Fab — antigen binding fragment. This fragment is formed during proteolytic degradation of the IgG molecule and contains a single combining site.

F(ab)$_2$ — a fragment of the IgG molecule containing both combining sites, in fact two Fab linked by disulphide bridges.

Fc-fragment — the *C*-terminal half of immunoglobulin heavy chains formed after proteolytic fragmentation of their molecules. It cannot bind antigen but carries other important biological functions.

Fc-receptor (FcR) — binding site for Fc-parts of immunoglobulin molecules. Such receptors are found on the surface of polymorphonuclear leukocytes, lymphocytes and other cells.

Fd-fragment — the *N*-terminal half of immunoglobulin heavy chains.

Fluorescence — the emission by a compound of light of a wavelength different from that used to irradiate the compound. It is a type of luminescence, during which light emission lasts for less than 10^{-8} s after excitation.

Fluorescence immunoassay (FIA) — an immunochemical method used for measuring antigen (hapten) or antibody concentration and utilizing principles similar to those of enzyme immunoassay. However, one of the reactants is labelled with a fluorophore instead of an enzyme. The concentration of the test substance is determined by measuring fluorescence.

Freund's adjuvant (FA) — an oil–water emulsion which increases the immune response to an antigen after mixing with the antigen solution and immunization of an animal. The incomplete FA consists only of oil, water, and emulsifier; complete FA also contains killed mycobacteria or other bacteria.

Fused rocket immunoelectrophoresis — a modification of rocket immunoelectrophoresis which is particularly useful to identify and measure antigens found in individual fractions after column chromatography.

Gammaglobulins — fractions of serum proteins having the lowest mobility on electrophoresis in alkaline pH. Most immunoglobulins are found in this fraction.

Gammopathy — the excessive production of immunoglobulins with pathological consequences for the patients. The conditions may be monoclonal or polyclonal. In a monoclonal gammopathy, a single cell clone (plasma cells or their precursors) multiplies in a non-coordinated manner. The clone then produces excessive amounts of a structurally and functionally homogeneous population of immunoglobulin molecules of a certain class, or parts of their molecules (light or heavy chains). In a polyclonal gammopathy, several clones of antibody-producing cells multiply in a non-coordinated manner.

Gel filtration chromatography — a method in which molecules may be separated on a gel column according to their relative molecular weights. Gels are made from insoluble polymers (*e.g.* Sephadex) which has the properties of a molecular sieve.

Gene — the basic hereditary unit localized in a certain chromosome site (locus). The gene can be characterized functionally (its product determines a certain ability or phenotypic marker) or structurally (a certain DNA region).

Genes of immune response (Ir-genes) — genes which determine the intensity of the immune response of an individual to specific antigens. They are components of the major histocompatibility complex having Ia antigens as their products.

Glycoproteins — proteins which contain saccharides in addition to amino acid residues.

Gnotobiont — an individual of a certain biological species living in an artificial germ-free environment.

Graft-versus-host reaction (GVHR) — the sequelae of immunopathological reactions induced by immunocompetent cells in a graft against the histoincompatible recipient (host) which is incapable of reacting against the graft by an immune response.

Gram staining (according to H. C. J. GRAM) — the staining of microorganisms to classify bacteria into two groups. Microorganisms stained positively with the Gram's stain are called Gram-positive, those that are not stained are termed Gram-negative.

Granules — membrane-bound organelles found in leukocytes, mast cells and some lymphocytes. They contain hydrolytic enzymes, antimicrobial compounds, different mediators and other substances. Neutrophils contain three main groups of granules — azurophilic, specific and small storage ones.

Granulocytes — (called also polymorphonuclear leukocytes) — white blood cells with a nucleus consisting of multiple lobes and containing characteristic cytoplasmic granules (lysosomes). They are divided into neutrophilic, eosinophilic and basophilic granulocytes according to their reaction to haematological stains.

Haemagglutination — a serological reaction resulting in agglutination of red blood cells by specific anti-erythrocyte antibodies.

Haemolysis — damage of the erythrocyte cell membrane and release of its contents into the external environment.

Haplotype — a cluster of genes closely linked within a particular region of one of two paired chromosomes and inherited complete from one of the parents.

Hapten — a low-molecular-weight compound which is not immunogenic (it does not induce antibody formation) but can specifically react with pre-existing antibodies. When a hapten is bound to a macromolecular carrier, an immunogen is usually formed. In this complex the hapten represents the determinant group.

Helper *T*-cells — a subgroup of regulatory *T*-lymphocytes which stimulate the immune response. They cooperate with *B*-cells and accessory cells in antibody formation. Some of them also control the development of T_C-lymphocytes.

Hepatitis — liver inflammation often accompanied by jaundice. Infectious hepatitis is caused by at least two viruses — virus A and virus B.

Heterobispecific antibodies — antibodies with two combining sites which have a different specificity.

Heterocytotropic antibodies — antibodies derived from one animal species which bind by their Fc-domains and passively sensitize cells and tissues of other animal species.

Heterogeneity of antibodies — differences in the primary structure of conventional antibodies produced in response to a multivalent antigen having several determinants. Individual populations of antibody molecules have a different affinity and avidity.

Heterophilic antigens — antigens of numerous phylogenetically unrelated biological species. They are responsible for the formation of antibodies yielding cross reactions with these antigens.

Hexosemonophosphate shunt (HMPS) — a metabolic pathway in which glucose is degraded *via* pentoses (thus it is also called the pentose cycle) to carbon dioxide and water. One mole of glucose yields 12 moles of $NADPH + H^+$.

Hinge immunoglobulin region — the region of heavy chains between the first and second constant domain. Its amino acid residues form a chain, usually with a linear structure, which is responsible for its flexibility. The hinge region contains disulphide bridges between the heavy chains of the immunoglobulin molecule and is the site at which its proteolytic cleavage to fragments Fab, $F(ab)_2$ and Fc occurs.

Histocompatibility — tissue compatibility, *i.e.* identity of transplantation antigens.

Histocompatibility antigens — antigens which occur on cell membrane surfaces and which are responsible for inducing transplantation reactions and rejecting allografts. However, their primary role is in cooperation interactions between cells. In this latter case, it is more appropriate to call them histocompatibility markers.

Histoincompatibility — tissue incompatibility (*e.g.* for transplant).

HLA complex — the major histocompatibility system in man which includes markers (antigens) of class I (HLA-A, HLA-B, HLA-C) and class II (HLA-DQ, HLA-DP, HLA-DR).

Homeostasis — the maintenance of biochemical and biophysical processes in a state of dynamic equilibrium with the outer environment. The nervous, humoral and immune systems maintain this equilibrium within specific physiological limits. Exceeding these limits results in pathological changes or even death of the organism.

Homocytotropic antibodies — antibodies which bind, *via* their Fc-domains, to cells of the species in which the antibodies were produced. In man they belong to the IgE class. When the organism re-encounters the antigen that induced their formation, an anaphylactic reaction occurs.

Host-versus-graft reaction (HVGR) — reaction of the host against the transplant of histoincompatible graft tissue or cells.

Humoral immunity — antibody-mediated immunity. In broad terms, those defence mechanisms constituting a part of non-specific (natural) immunity, and not associated with cell activity (*e.g.* complement), also belong to this group.

Hybridoma — a population of hybrid cells which produce a specific monoclonal antibody or other immunoregulatory factors. Hybridomas are formed by fusion of a functional cell with a myeloma cell followed by multiplication of this cell clone.

Hydrogen bonds — bonds formed when a hydrogen atom is localized between two electron-negative atoms and binds covalently to one of them.

Hydroxyl radical (•OH) — the product of monovalent reduction of atomic oxygen.

Hypergammaglobulinaemia — an increased concentration of circulating immunoglobulins.

Hypersensitivity (allergy) — abnormally increased sensitivity to an artificially administered antigen (drugs, diagnostic preparations) or one occurring naturally in the environment (dust, plant pollen, food components *etc.*). It is manifested by allergic reactions.

Hypervariable regions — regions of the variable domain of heavy and light immunoglobulin chains in which more than three different amino acids can be exchanged in any position (in molecules of antibodies with different specificities). The hypervariable regions are part of the combining sites of antibodies.

Hypogammaglobulinaemia — a deficiency (decreased level) of all immunoglobulin classes in the blood serum.

Ia antigens — immune response antigens encoded by Ir-genes.

Idiotypes — antigenically different immunoglobulin variants determined by various antigenic determinants in the hypervariable parts of light and heavy chains. Any antibody specificity has a different idiotype recognized by a set of corresponding anti-idiotype antibodies.

IgA — an immunoglobulin class occurring in the serum and mucosal secretions and which is involved in local immunity reactions. In mucosal secretions it occurs in the form of a dimer (secretory IgA).

IgD — the main immunoglobulin class on the surface of *B*-lymphocytes, where it forms antigenic receptors.

IgE — antibodies involved in protection against parasitic infections and reagins — the main immunoglobulin class responsible for atopic reactions in man.

IgG — the most common immunoglobulin class in extracellular fluids. IgG antibodies are produced in response to repeated administration of antigens, particularly of soluble ones.

IgM — immunoglobulin with the highest relative molecular weight. The IgM molecule consists of five identical subunits (forming a pentamer) and contains one *J*-chain. Antibodies of this class are produced particularly during the first contact of the organism with the antigen. They are also multivalent which facilitates the agglutination of bacteria and complement-dependent cytolysis.

Immediate hypersensitivity — a form of hypersensitivity mediated by homocytotropic antibodies. Such antibodies belong to the IgE class in man.

Immune complexes — complexes formed during the antigen–antibody reaction. Under *in vivo* conditions, or in the presence of serum, they may bind some complement components.

Immune response — the complex of reactions by which the organism responds to an antigen.

Immune serum — serum containing antibodies specific for the antigen that has induced their formation.

Immunity — the ability of an organism to react to an antigen in an appropriate and beneficial manner.

Immunity, anti-infectious — the resistance of the organism to pathogenic microorganisms and their toxins.

Immunity, antitumour — resistance to tumour cells and their antigens.

Immunity, non-specific (natural) — reactions at the tissue, cellular and molecular level, present in the organism since birth, and which are independent of previous contact with the antigen. Non-specific immunity mechanisms are involved in the removal of foreign compounds from tissues (inflammatory reactions), and in both anti-infection and antitumour protection.

Immunity, passive — resistance of the organism to certain infectious agents resulting from antibodies (or cells) passively transferred from another organism.

Immunity, specific (acquired) — the ability of the organism to respond to an antigen by a specific immune response using a specialized cell system (lymphatic tissue). The specific immune response results in the formation of specific antibodies, effector or regulatory cells able to react with the antigen that has induced their production, as well as in the occurrence of cells with immunological memory.

Immunity, transplantation — immunity (resistance) to cells and tissues transferred from genetically non-identical donors.

Immunization — a procedure used to induce a state of immunity. This may be active (where an antigen is introduced and the organism subsequently produces antibodies and effector cells) or passive (where antibodies produced in a different individual are administered).

Immunoadsorbent — an insoluble compound (carrier) to which an antigen or antibody are bound in the form of a ligand. This procedure can be used to separate a specific antigen or antibody (immunoadsorption chromatography).

Immunoblotting — a method to visualize compounds separated by electrophoresis (usually proteins) by their reaction with specific antibodies. Blotting is used to transfer molecules separated by electrophoresis or isoelectric focusing from the mobile phase of the gel (*e.g.* polyacrylamide or agarose) to a solid phase (*e.g.* a nitrocellulose membrane). The compounds can be transferred by simple diffusion or by direct current (electroblotting).

Immunocompetence — the genetically determined ability of some lymphocytes to react to an antigen with a specific immune response.

Immunodeficiency — immune insufficiency, impairment of immunity mechanisms.

Immunodiffusion — diffusion of antigen and antibody molecules, usually in agar or agarose gel. The migration velocity is dependent on the concentrations of both reactants and their diffusion constants. A precipitate is formed in the gel at the site where the antigen and antibody meet and its formation indicates the presence of the test antigen or antibody. The test component can be quantitated from the precipitate area.

Immunoelectrophoresis — the combination of electrophoresis and immunodiffusion in agarose or agar gel.

Immunofluorescence — a histochemical technique for the detection and localization of antigens using antibodies labelled with fluorescent compounds. Fluorescing immune complexes in tissues or cells can be observed with a fluorescence microscope.

Immunogen — a compound (complet antigen) which induces an immune response when administered to a suitable (immunocompetent) organism.

Immunogenetics — the interdiscipline between immunology and genetics which studies the gene systems responsible for tissue incompatibility (histoincompatibility), as well as for genetic control of the immune response.

Immunoglobulin classes — different types of immunoglobulins which can be classified according to the type of heavy chains present in their molecules. Five classes exist: IgA, IgD, IgE, IgG and IgM.

Immunoglobulin heavy (*H*) chains — one of two types of immunoglobulin peptide chains of higher relative molecular weight. There are five *H*-chains with different antigenic properties of their constant regions: α, δ, ε, γ and μ. The type of *H*-chain determines the immunoglobulin class.

Immunoglobulin types — the classification of immunoglobulins according to the primary structure of their constant light chain domains. There are two basic types, kappa ($\varkappa$) and lambda (λ).

Immunoglobulins — glycoproteins containing two identical light and two identical heavy polypeptide chains connected by disulphide bridges. All antibodies are immunoglobulins but not all immunoglobulins are antibodies. Immunoglobulins are divided into five classes: IgA, IgD, IgE, IgG and IgM. They all also contain a carbohydrate moiety.

Immunohormones — immunoregulatory agents of a hormonal nature. They are synthesized by one type of immunologically active cells and influence biological functions of other cell types through specific receptors.

Immunological memory — the ability of an organism to react to a repeated administration of an antigen by a more intensive and rapid immune response.

Immunological surveillance — an immune function which ensures the recognition and removal of impaired, effete or antigenically altered host cells and their components. This mechanism mainly serves to eliminate tumour cells that have been formed during the life of a given individual and involves *NK*-cells, *NC*-cells and activated macrophages.

Immunological tolerance — the specific unresponsiveness of the organism to an antigen, which would, under different conditions, induce an immune reaction in the same individual. The reactivity of the lymphocyte clone responsible for this antigen is decreased or totally suppressed, or the clone is absent from the organism.

Immunology — the scientific study of immune reactions in man and other animals.

Immunonephelometry — a method for quantifying immune complexes formed during the reaction of a soluble antigen with an antibody based on the intensity of scattered light passing through their suspension (nephelometrically). Nephelometry utilizes the precipitation reaction which enables the demonstration of the concentration of a soluble antigen with a known antibody and *vice versa.*

Immunopathology — the study of the pathological changes resulting from abnormal, insufficiently coordinated (hypersensitive) or defective (immunodeficiency) immune reactions, as well as the causes of these changes.

Immunoprecipitation — the formation of a precipitate during the reaction of a soluble antigen with an antibody.

Immunoradioassay — a technique similar to radioimmunoassay (RIA), in which the antibody, rather than the antigen, is labelled with a radioactive isotope.

Immunostimulation — a non-specific increase in natural or specific immunity manifested by an increased velocity and intensity of the immune response to different antigens, bacteria, viruses and tumour cells in particular.

Immunosuppression — the suppression of immunological reactions by external factors either intentionally (immunosuppressive therapy), or accidentally and undesirably (*e.g.* irradiation, treatments with certain drugs and bacterial toxins *etc.*). Immunosuppressive therapy is used after transplantation of organs and tissues and in the treatment of some immunopathological diseases.

Immunoturbidimetry — a precipitation method used to determine the amount of an antigen or antibody after their reaction in solution and which relies on changes in light transmittance due to turbidity of the solution caused by precipitated immune complexes.

Inbred line — a homogeneous animal line in which all individuals have identical genetic equipment (genotype). It is formed, for example, by repeated mating of siblings for many generations.

Inflammation — the protective reaction of an organism against physical, chemical or biological (infection) damage to tissues.

Ingestion — uptake of a particle by a phagocyte.

Interferon system — comprises three types of interferon: IFN-α (produced mainly by leukocytes following induction by extraneous cells, cells infected with viruses, tumour cells and bacteria), IFN-β (produced mainly by fibroblasts after induction with nucleic acids of viral or other origin) and IFN-γ (produced by activated lymphocytes during their response to specific antigens or mitogens). Interferons have specific immunoregulatory, antiviral and other biological effects.

Interleukins — immunoregulatory cytokines (glycoproteins or proteins) produced by macrophages (IL-1), lymphocytes (IL-2, IL-3 *etc.*) as well as other cells. Their main function is to transfer signals between different cells involved in the immune response.

Intron — a DNA region which is not transcribed in the protein product.

Ir-genes — genes controlling the immunological response to specific antigens.

Isoantigens — *see* alloantigens

Isoelectric focusing — a technique for separating proteins or other ampholytes based on differences in the isoelectric points of the individual compounds to be separated. The isoelectric point is the pH at which the total electric charge of the molecule is zero.

Isogeneic antigens — antigens of a genetically identical individual of the same biological species (monozygous twins).

Isotypes of immunoglobulins — antigenic variants of immunoglobulins identical in all individuals of a certain animal species. They characterize the individual classes and subclasses (determined by the C_H-domain structure), types and subtypes (determined by the primary C_L-domain structure).

***J*-chain** — polypeptide chain which is a component of polymeric immunoglobulins (IgM and secretory IgA).

***J*-region** — the part of heavy immunoglobulin chains which connects the variable domain and the first constant domain.

***K*-antigens** — surface (capsular) antigens of bacteria.

***K*-cells (killer cells)** — a heterogeneous cell population (particularly some macrophages and lymphocytes) which kills other (target) cells coated (labelled) by specific antibodies. They recognize them *via* Fc-receptors for these antibodies (usually of the IgG class). *K*-cells are responsible for antibody-dependent cell-mediated cytotoxicity (ADCC).

Kappa ($\varkappa$) chain — one of two types of light immunoglobulin chain.

Kinins — peptides produced by the activity of esterases on kallikreins and which have vasodilatory effects.

Kupffer cells — macrophages of the liver.

LAK-cells (lymphokine-activated killer cells) — a heterogeneous population of lymphocytes exhibiting cytotoxic effects on tumours including those resistant to *NK*-cells.

Lambda (λ) chain — one of two types of light immunoglobulin chain.

Langmuir adsorption isotherm — characterizes the strength of binding of a univalent hapten to the antibody combining site, or the binding of any ligand to the binding site.

Lectins — proteins and glycoproteins which specifically recognize and bind various mono-, di- and trisaccharides. They induce cell agglutination and are non-specific mitogens for lymphocytes. This is because most cells have different oligosaccharide chains on their surface. Lectins were originally isolated from higher plants, but later also discovered in animals and microorganisms where they are involved in various cellular interactions mediated by specific lectin receptors.

Leukocytes (white blood cells) — cells derived from the bone marrow, and which are involved in host defence. They are divided into lymphocytes, granulocytes (neutrophils, eosinophils, basophils) and mononuclear phagocytes (blood monocytes and tissue macrophages).

LGL **(large granular lymphocytes)** — characterized morphologically mainly by the granules and cytotoxic compounds in their cytoplasm. They are a population of non-adhering cells with receptors for IgG Fc-domains, but without surface immunoglobulin and complement receptors. Most LGL exhibit *NK*-activity.

LIF (leukocyte inhibitory factor) — a lymphokine which inhibits the migration of polymorphonuclear leukocytes from the site of the inflammatory reaction.

Ligand — a molecule which reacts specifically with a certain (binding) site on another molecule or with a receptor on the cell surface. From the chemical point of view, ligands are atoms, groups of atoms or molecules coordinately bound to a central atom or molecule of the complex. In some cases the central molecule acts as a carrier and the ligands have a specific functional site, *e.g.* immunoadsorbents.

Lipid A — the hydrophobic part of Gram-negative bacteria lipopolysaccharides.

Lipopolysaccharide — an endotoxin complex derived from the cell envelope of Gram-negative bacteria and consisting of lipid A and polysaccharide component.

Locus — a specific site of the chromosome carrying a gene. One locus contains only one allele of a given gene.

Lymphatic organs — organs composed of lymphatic tissue of mesenchymatic origin. They are termed central or primary (thymus, bursa of Fabricius and its equivalents) and peripheral or secondary (spleen and lymph nodes).

Lymphocytes — the basic cells of the lymphatic tissue, lymph and blood. They are circular (7–12 μm in diameter), with one large nucleus surrounded by a thin cytoplasm layer. The cytoplasm does not usually contain granules, although some lymphocyte populations (LGL) do. According to their functions and surface markers they are divided into three main populations: *B*-lymphocytes, *T*-lymphocytes and null lymphocytes.

Lymphokines — immunoregulatory cytokines secreted mainly by lymphocytes. They exhibit typical hormone activities: they are secreted from one cell type, react specifically with another cell type — target cells — and regulate a specific biological function by a feedback mechanism. They usually act *via* specific receptors on target cells.

Lymphotoxin — one of the tumour necrosis factors (TNF-β).

Lysosomes — organelles in the cytoplasm of all nucleated cells which contain hydrolytic enzymes. Lysosomes of phagocytes also contain various antimicrobial compounds.

Lysozyme (EC 3.2.1.17) — an enzyme which degrades β-1,4-glycosidic bonds in polysaccharides. It occurs particularly in saliva, tears, nasal secretions, serum and leukocyte granules. It can degrade cell wall glycopeptides of some sensitive (usually non-pathogenic) microorganisms and thus impair their activity. It damages other microorganisms only together with other antimicrobial factors (complement, secretory IgA, proteinases *etc.*).

Macroglobulin — a little used synonym for IgM.

Macrophages — large, usually mononuclear cells originating from monocytes. They have phagocytic and pinocytic activities and significant antimicrobial, regulatory and accessory functions. They are components of the mononuclear phagocytic system.

MAF (macrophage activating factor) — a mixture of cytokines (IFN-γ, IL-1, IL-2, IL-4, GM-CSF) which activates macrophages.

Major histocompatibility system — the set of genes which determine the histocompatibility antigens that are dominant for a given animal species.

Mast cells — tissue cells which resemble basophils and which contain granules from which histamine and other mediators are released during anaphylactic and atopic reactions.

Membrane attack complex (MAC) — a complex of the last five components of activated complement (C5b6789) which forms holes in the surface membrane of target cells thereby inducing their lysis. MAC is formed in both the classical and alternative complement activation pathways.

Memory cells — lymphocytes (*T* and *B*) proliferating after binding of an antigen to their receptors. However, they do not secrete antibodies and are not transformed to terminal effector cells of the cell-mediated immunity. They persist in the organism for a long period and "remember" the antigen that induced their formation. Therefore, on further contact with this antigen they react more readily and enable a faster and more intensive secondary immune response.

Microphages — Metchnikoff's original term for neutrophil polymorphonuclear leukocytes.

MIF (migration-inhibiting factor) — a cytokine which inhibits migration of macrophages. It is identical to MAF.

Mitogens — compounds which activate DNA synthesis and induce blastic transformation and division of lymphocytes or other cells.

Mixed lymphocytes cultures (MLC) — cultures of lymphocytes derived from two genetically non-identical donors.

Monoclonal antibodies (immunoglobulins) — products of a single clone of plasma cells or a clone of hybrid cells (hybridoma) produced by fusion of one plasma and one myeloma cell. They represent identical copies of molecules having the same primary structure, identical combining site specificity and identical function.

Monocyte — a large mononuclear cell found in the peripheral blood. It belongs to the mononuclear phagocytic system and is a precursor of the macrophage.

Monokines — immunoregulatory cytokines functioning as local hormones secreted by macrophages and monocytes.

Mononuclear cells — cells having one compact nucleus. Among leukocytes they include monocytes and macrophages, as well as lymphocytes and plasma cells.

Mononuclear phagocytic system (MPS) — a system formed by tissue macrophages, blood monocytes and their precursor cells in bone marrow. These cells are characterized by active phagocytosis, pinocytosis and the ability to adhere firmly to various surfaces. Monocytes and macrophages possess surface receptors for Fc-domains of immunoglobulins and for some complement fragments. They can therefore perform immune phagocytosis.

Multiple myeloma — a malignant disease of plasma cells or their precursors, manifested by the occurrence of undesirable monoclonal immunoglobulins or their polypeptide chains (paraproteins) in the blood.

Myeloma immunoglobulins — monoclonal immunoglobulins or their light or heavy chains in the blood serum of patients with multiple myeloma.

Myeloperoxidase — a peroxidase (EC 1.11.1.7) occurring in neutrophil-granulocytes. Together with hydrogen peroxide and an oxidizable co-factor(Cl^- or I^-) it forms the myeloperoxidase system, a highly effective microbicidal and cytotoxic mechanism in mammalian leukocytes.

***NC*-cells** — natural cytotoxic cells. Together with *NK*-cells these cells constitute the first protective barrier of the host against spontaneously arising tumour cells. However, *NC*-cells act on other target cells, have different surface characteristics and are influenced by control factors other than those affecting *NK*-cells.

Neutrophils — a type of granulocyte whose granules are readily stained with neutral stains. They belong to the group of "professional" phagocytes and play basic regulatory roles in inflammation. They may also be involved in immunopathological reactions. Neutrophils comprise some 60–70% of all leukocytes in human peripheral blood.

***NK*-cells (natural killer cells)** — cells which kill tumour and some other cells spontaneously, in the absence of specific antibodies to their surface structures, without earlier sensitization with the antigen and without restriction of their activity by antigens of the major histocompatibility complex. Carcinogenic compounds bring about a rapid decrease of *NK*-activity, which may be the basis of their effect. In addition to tumour cells, they also kill virus-infected cells, some foetal cells and certain subpopulations of normal lymphoid and haemopoetic stem cells.

Null cells — a subgroup of lymphocytes which does not carry surface markers (antigens) typical of *T*- and *B*-lymphocytes.

O-antigens — surface antigens of Gram-negative bacteria containing endotoxin.

Opsonization — a process, in which various compounds (opsonins) adsorb to the surface of bacteria or other particles, changing their surface properties, and facilitating adhesion of the particle to the phagocyte membrane and its subsequent engulfment.

Paraproteins — monoclonal immunoglobulins (complete moleculs of their chains) produced by myeloma cells.

Paratope — another term for antibody combining site.

Passive cutaneous anaphylaxis — a method used *in vivo* to demonstrate cytotropic antibodies inducing immediate-type hypersensitivity.

Passive haemagglutination — agglutination of red blood cells by antibodies against antigens that have been passively bound to the surface of erythrocytes (they are not their natural components).

Passive immunity — *see* immunity, passive.

Perforins — proteins which have cytolytic activity and which are present in granules of cytotoxic *T*-lymphocytes and *NK*-cells. After their release onto the surface of target cells they polymerize to polyperforins which are incorporated into the phospholipid bilayer of the cell membrane producing holes like those due to complement activation.

Peyer patches — clusters of lymphatic tissue in the small intestinal submucosa. They belong to the secondary (peripheral) lymphatic organs.

Phagocytes — cells able to engulf (phagocytose) and degrade extracellular material (particles) including pathogenic microorganisms. This ability is relatively characteristic for all nuclear cells. The cells specializing in engulfment, killing and degradation of pathogenic microorganisms are termed "professional" phagocytes.

Phagocytic index — a way of characterizing phagocytosis, *i.e.* the rate of uptake of foreign particles by phagocytes. It is expressed as the number of particles engulfed by a single phagocyte.

Phagocytic vacuole — another term for phagosome.

Phagocytosis — the mechanism by which phagocytes engulf microorganisms or other particles. It is a complex process including numerous, closely-associated steps.

Phagolysosome — the cell organelle formed after fusion of the phagosome with lysosomes. It is the site in the phagocyte at which microorganisms are killed and destroyed.

Phagosome — a cell organelle (vesicle) formed after engulfment of a particle by the inverted phagocyte plasma membrane.

Phytohaemagglutinin (PHA) — a lectin which is derived from the red kidney bean (*Phaseolus vulgaris*) and which agglutinates red blood cells and stimulated *T*-lymphocytes.

Pinocytosis — the engulfment of soluble material by cells. In macropinocytosis, drops of a liquid or particles of 0.1–1.0 μm in diameter are engulfed, whereas micropinocytosis is the absorption of molecules less than 0.1 μm diameter and is mediated by specific receptors (receptor-mediated pinocytosis).

Plaque technique — a method used to demonstrate individual cells producing antibodies against erythrocytes or bacteria. The cells are cultured in an agarose medium containing erythrocytes and complement. If the cell produces antierythrocyte antibodies a clear zone of haemolysis is formed around it (plaque). Production of antibodies against bacteria is assayed in a similar way, *i.e.* by bacteriolysis.

Plasma cells — cells producing and secreting antibodies and which are the final differentiation stage of *B*-cells.

Polymorphonuclear leukocytes (granulocytes) — cells which contain one nucleus divided into several lobes, and which have cytoplasm containing granules with hydrolytic enzymes and antibacterial compounds.

Precipitin — an antibody reacting during immunoprecipitation.

Precipitogen — an antigen reacting during immunoprecipitation.

Primary lymphatic organs — the central organs of the immune system (thymus, bursa of Fabricius or its equivalent).

Primary lysosomes — lysosomes that have not yet been involved in the digestive processes of a phagocyte.

Primary response — the specific immunological response of an organism which has encountered the antigen for the first time.

Primary structure of proteins — the sequence of amino acid residues in a polypeptide chain, type of bonds and number of polypeptide chains in one molecule of protein.

Productive phase of antibody formation — the period during which antibodies are actively synthesized in plasma cells and excreted into the extracellular space.

Professional phagocytes — leukocytes containing FcR and CR in their cytoplasmic membrane enabling them to perform immune phagocytosis. This group includes neutrophils, eosinophils, monocytes and macrophages.

Proliferation of cells — another term for division or formation of new cells.

Pyrogens — compounds which increase body temperature. They may be released endogenously from leukocytes (IL-1) or enter the host organism from the outside. Microbial products are usually involved in the latter case.

Quantitative precipitation — a method used to determine the relationship between the amount of the precipitate and the ratio of the antigen and antibody concentrations.

Radial single immunodiffusion (RID) — a quantitative immunoprecipitation technique in which a soluble antigen is pipetted into wells in an antibody-containing agarose gel. The antigen diffuses into the gel where it reacts with antibody and forms characteristic circular precipitates, whose diameter is proportional to the actual concentration or logarithm of antigen concentration (depending on whether the precipitation reaction has gone to completion).

Radioimmunoassay (RIA) — a technique to measure a soluble antigen or hapten by determining the radioactivity of immune complexes formed after reaction of an antigen with a specific antibody. Radiolabelled antigens (haptens) and their standard solutions with a known concentration are needed for the purpose.

Radioimmunoelectrophoresis — immunoelectrophoresis in which the immunoprecipitate is detected with an antigen or antibody labelled with a radioactive isotope.

Reagins — antibodies of the IgE class in man. They are of homocytotropic character and are responsible for allergic reactions of the immediate type.

Receptor — a complex of atoms or molecules with a stereochemical specific affinity for a certain compound which is known in pharmacology as a ligand or agonist. Receptors are usually localized in the cell surface but they may also occur on the internal side of the cytoplasmic membrane or in the cytoplasm. The occupation of the receptor by the ligand (agonist) is a signal for a specific cellular physiological response.

Rejection — another term for graft destruction by the host.

Respiratory burst — sudden metabolic changes (increased oxygen consumption, HMPS activation, increased superoxide and hydrogen peroxide production) in neutrophils and other professional phagocytes during phagocytosis or after stimulation by various soluble compounds (*e.g.* chemotactic factors).

Reticuloendothelial system (RES) — a cell system introduced by ASCHOFF in which the cells are classified according to their ability to phagocytose microorganisms and other particles penetrating the body. The term is no longer used.

Rheumatoid factor — an antibody against constant parts of immunoglobulin heavy chains and found in the serum of many individuals with rheumatoid arthritis.

Rocket immunoelectrophoresis — a quantitative immunoprecipitation technique, which represents a combination of single radial immunodiffusion and electrophoresis. The resultant immunoprecipitates are rocket-shaped and their height is proportional to the concentration of the test antigen.

Saturation (absorption) of polyspecific antisera — a simple procedure to prepare monospecific immune sera from polyspecific antisera.

Secondary immune response — the response of an organism following repeated contact with the antigens. Compared to the primary response, the secondary response is characterized by a faster onset and higher intensity, due to the involvement of memory cells produced during the primary response.

Secondary lysosomes — lysosomes that have been involved in digestive processes. In phagocytes, these are all phagolysosomes or pinolysosomes.

Secretory component — a protein (mol. wt. 70 000) synthesized by epithelial cells as part of their receptors for polymeric immunoglobulins. It is a part of the secretory IgA molecule.

Secretory IgA — an IgA dimer containing four heavy and four light chains, the secretory component and a *J* chain.

Selective theories of antibody formation — theories based on the assumption that the antigen, after entry into the organism, selectively stimulates only those cells that can produce specific antibodies. Therefore, the organism must contain lymphocytes which carry genes coding antibodies against all possible antigens.

Semihapten — a simple (inhibitory) hapten which does not visibly react with a specific antibody. It can, however, be demonstrated indirectly on the basis of inhibition of the reaction between the antibody and antigen or complete hapten.

Sephadex — dextran gels used in gel filtration chromatography.

Sepharose — gels derived from agarose and which are used particularly in the preparation of immunoadsorbents.

Sequential determinant — an antigenic determinant whose specificity is determined by the sequence of the basic units of the polymeric chain, *e.g.* of amino acids in protein antigens.

Serological methods — methods for the detection and measurement of an antigen or antibody performed in an aqueous environment.

Serum proteins — blood serum proteins.

Serum titre — the highest dilution of immune serum which still reacts with an antigen.

Singlet oxygen ($^1O_2$) — an oxygen form whose electrons are excited at a higher energy level compared to the normal (ground) triplet oxygen. When returning to the ground state they emit light (chemiluminescence) which may have antimicrobial and cytotoxic effects.

Slow reacting substance of anaphylaxis (SRS-A) — a mixture of leukotrienes (metabolic products of arachidonic acid and glutathione) LTC_4, LTD_4 and LTE_4 mainly involved in atopic reactions.

Spin immunoassay (SIA) — a technique for measuring antigen or hapten concentrations by their reaction with a specific antibody and which uses antigens or haptens labelled with a free radical whose electron has an unpaired spin.

Steric complementarity — complementary correspondence of the spatial arrangement of atom groups in two molecules, *e.g.* antigen and antibody. Thus, atoms forming convex shapes in the antigenic determinant have complementary atoms in the combining site of a specific antibody which form hollows of the same shape and *vice versa*.

Steric repulsive forces — forces which originate between two atoms which are not linked by chemical bonds, on the basis of mutual penetration of their electron clouds. These forces determine the strength of the bond (affinity) between the antibody combining site and the antigenic determinant. The lower the forces, the higher the steric complementarity of their electron clouds.

Superoxide — the univalent anion radical produced by the one-electron reduction of molecular oxygen. It is a precursor of other reactive oxygen species.

Superoxide dismutase (**SOD**, EC 1.15.1.1) — a group of metalloenzymes catalysing dismutation (conversion) of superoxide to hydrogen peroxide and molecular oxygen. They play a significant role in removing superoxide anion in aerobic cells. They may be classified according to the metals present in their molecules: Cu,Zn-SOD, Mn-SOD and Fe-SOD.

Suppressor *T*-cells — a subgroup of *T*-lymphocytes which can suppress the immunological response of *B*-cells (antibody formation) or stop the immune reactions of other effector *T*-cells.

Surveillance — *see* immunological surveillance.

Syngeneic — the genetic relationship among individuals of an identical inbred strain (identical genotype).

Synthetic antigens — synthetically prepared antigens. Such antigens have a known and relatively simple chemical structure and are therefore advantageous for use in experimental immunochemical studies.

Target cells — different types of cells which induce formation of specific antibodies or T_C-lymphocytes through their surface antigens, or cells which have otherwise become targets of the cellular and molecular mechanisms of specific and non-specific immunity. These mechanisms lead to damage or even death and lysis of target cells.

***T*-cells** — *T*-lymphocytes derived from the thymus and which are responsible for cell-mediated immunity. They may be divided into several *T*-cell subpopulations which differ in their development, function and properties. Helper ($T_{H/I}$) and suppressor (T_S) lymphocytes have a regulatory function, whereas T_C-lymphocytes are typical effector cells.

Thymectomy — the surgical removal of the thymus.

Thymopentin — the pentapeptide Arg-Lys-Asp-Val-Tyr; the active component of thymopoietin.

Thymopoietin — an immunohormone containing 49 amino acid residues and produced in the thymus. It induces differentiation of *T*-lymphocytes, exhibits immunonormalization effects and influences neuromuscular transfer (transfer of signals between nervous and muscle cells).

Thymosins — an excretable group of peptides and glycoproteins (mol. wt. 1 000–15 000) synthesized in the thymus. Some of them have an immunohormonal character.

Thymus — the primary endocrine lymphatic organ of the immune system found in the thorax. It mainly controls the ontogenetic development of *T*-lymphocytes.

Thymus-dependent antigens — antigens which require cooperation of *B*-lymphocytes with *T*-lymphocytes to induce antibody formation or other immune responses.

Thymus-independent antigens — antigens which can induce antibody formation without cooperation with *T*-lymphocytes.

Tolerance — *see* immunological tolerance.

Tolerogen — an antigen which can induce immunological tolerance.

Toxins — products of some microorganisms, plants and animals. Toxins are usually proteins which can damage human and animal organisms. They are antigenic and antibodies (termed antitoxins) can therefore be prepared against them.

Toxoids — non-toxic derivatives of bacterial toxins in which the original antigenic determinants have been preserved. They may be used to stimulate formation of antibodies (antitoxins) which also neutralize the original (harmful) toxin.

Transformation of cells — a change in properties and morphology of cells. The result of transformation is usually the conversion of normal cells into tumour cells.

Transplantation antigens — *see* histocompatibility antigens.

Transplantation immunity — *see* immunity, transplantation.

Triplet oxygen ($^3O_2$) — the ground electron state of molecular oxygen.

Tumour necrosis factors — TNF-α (cachectin) and TNF-β (lymphotoxin). These cytokines exert cytostatic and cytotoxic effects on tumour cells and other biological activities. Cachectin is an endogenous mediator of the toxic effects of Gram-negative bacterial endotoxins and can also induce cachexia.

Vaccination — immunization with antigens administered by injection, inhalation or orally with the aim of protection (prophylaxis) against infectious diseases caused by pathogenic microorganisms or viruses containing these antigens.

Van der Waals' forces — attractive forces between electron clouds of two atom proups with partial (not whole) electric charges of opposite sign.

Variable regions of immunoglobulins — the *N*-termini of heavy and light chains containing a variable sequence of amino acid residues. They contain parts of antibody combining sites (hypervariable regions).

Viroimmunoassay (VIA) — an immunochemical method used to determine an antigen (hapten) concentration using a specific antibody and a bacteriophage-labelled antigen (hapten).

Virulence — the quantitative expression of pathogenicity of a given microorganism for a certain animal species.

Waldenström's macroglobulinaemia — a disease characterized by the malignant growth of plasma cells producing one type of IgM. As a result, the blood serum of such patients contains high levels of this monoclonal IgM.

Xenogeneic — the genetic relationship between individuals of two phylogenetically different animal species.

Zonal electrophoresis — performed on various carriers (*e.g.* paper, cellulose acetate, agarose or polyacrylamide gel), in which proteins or other compounds are separated according to their size and molecular charge. After appropriate visualization the separated compounds form characteristic zones.

Author index

Subject index